Classical Field Theory

Classical field theory predicts how physical fields interact with matter and is a logical precursor to quantum field theory. This introduction focuses purely on modern classical field theory, helping graduate students and researchers build an understanding of classical field theory methods before embarking on future studies in quantum field theory. It describes various classical methods for fields with negligible quantum effects – for instance, electromagnetism and gravitational fields. It focuses on solutions that take advantage of classical field theory methods as opposed to applications of geometric properties. Other fields covered include fermionic fields, scalar fields, and Chern–Simons fields. Methods such as symmetries, global and local methods, the Noether theorem, the energy-momentum tensor, and important solutions of the classical equations – soliton solutions, in particular – are also discussed.

Horaţiu Năstase is a researcher at the Institute for Theoretical Physics at the State University of São Paulo, Brazil. To date, his career has spanned four continents. As an undergraduate, he studied at the University of Bucharest and Copenhagen University. He later completed his PhD at the State University of New York, Stony Brook, before moving to the Institute for Advanced Study, Princeton, New Jersey, where his collaboration with David Berenstein and Juan Maldacena defined the pp-wave correspondence. He has also held research and teaching positions at Brown University and the Tokyo Institute of Technology.

Classical Field Theory

HORAŢIU NĂSTASE

State University of São Paulo, Brazil

CAMBRIDGE
UNIVERSITY PRESS

Shaftesbury Road, Cambridge CB2 8EA, United Kingdom

One Liberty Plaza, 20th Floor, New York, NY 10006, USA

477 Williamstown Road, Port Melbourne, VIC 3207, Australia

314–321, 3rd Floor, Plot 3, Splendor Forum, Jasola District Centre, New Delhi – 110025, India

103 Penang Road, #05–06/07, Visioncrest Commercial, Singapore 238467

Cambridge University Press is part of Cambridge University Press & Assessment,
a department of the University of Cambridge.

We share the University's mission to contribute to society through the pursuit of
education, learning and research at the highest international levels of excellence.

www.cambridge.org
Information on this title: www.cambridge.org/9781108477017
DOI: 10.1017/9781108569392

First published 2019 (version 3, July 2023)

Printed in Great Britain by CPI Group (UK) Ltd, Croydon CR0 4YY, July 2023

A catalogue record for this publication is available from the British Library

Library of Congress Cataloging-in-Publication data
Names: Năstase, Horaţiu, 1972– author.
Title: Classical field theory / Horaţiu Năstase (Universidade Estadual Paulista, Sao Paulo).
Description: Cambridge ; New York, NY : Cambridge University Press, 2019. |
Includes bibliographical references and index.
Identifiers: LCCN 2018038600 | ISBN 9781108477017 (hardback)
Subjects: LCSH: Field theory (Physics)
Classification: LCC QC173.7 .N37 2010 | DDC 530.14–dc23
LC record available at https://lccn.loc.gov/2018038600

ISBN 978-1-108-47701-7 Hardback

To the memory of my mother,
who inspired me to become a physicist

Contents

Preface

The methods of classical field theory have been around for a long time, but it mostly meant electromagnetism and general relativity. The classic book that developed this point of view is Landau and Lifshitz's *The Theory of Classical Fields*, which, however, was first published almost 70 years ago. In the meantime, quantum field theory has developed into a very large subject. Electromagnetism, or more precisely classical electrodynamics, is a subject geared towards many physical applications, while general relativity has developed into a subject rich in Riemannian and differential geometry, as well as topology. Thus, in graduate school there usually is a specialized course that deals with electrodynamics – for instance, following the standard textbook of J. D. Jackson – as well as a specialized course in general relativity – perhaps following a textbook like Wald. On the other hand, in most physics departments in the United States and Europe, one usually jumps directly into a course in quantum field theory that offers a bit of classical methods interspersed here and there but mostly focuses in on the quantum aspects. It is my experience that this means students have a hard time adjusting to the large conceptual jump from classical mechanics to quantum field theory, and most modern methods of classical field theory don't make it into the standard curriculum.

At the Institute for Theoretical Physics of UNESP, we took a point of view that I find more sensible: we teach a semester of classical field theory, followed by two semesters of quantum field theory. This way, the transition is smoother, and one has time to learn some more modern methods of classical field theory. The course deals with electromagnetism and general relativity *only as examples of classical field theories*, so there are no applications of electromagnetism to a real medium and no geometry and topology for general relativity. Therefore, it is meant as a *complement* – and not a substitute – to electromagnetism and general relativity courses, one that focuses on the classical field theory aspects. The course I gave forms the basis for this book, to which I added more chapters, denoted here with an asterisk, to make it of use also to more advanced graduate students and researchers looking for a modern classical field theory reference. Thus, if one intends to use the book for teaching a one-semester course, one can safely drop the chapters with asterisk, which can be skipped until a second reading. There are other books on classical field theory out there, of which the most relevant I find Burgess's [5], which, however, spends little time on specific field theories and none on classical solutions; Rubakov's [58], which spends most of the time on gauge fields and on some quantum aspects; and Manton and Sutcliffe's [14], which deals only with solitons and is geared mostly towards researchers.

Acknowledgments

I want to thank all those who have guided me on my journey as a physicist, starting with my mother Ligia, the first physicist I knew who first introduced me to the wonderful world of physics. My high school physics teacher, Iosif Sever Georgescu, showed me that I can make physics into a career. My student exchange advisor at the Niels Bohr Institute, Poul Olesen, introduced me to string theory, my chosen field of study, and everybody at NBI showed me how interesting physics research can be. My PhD advisor Peter van Niewenhuizen showed me how to be a complete theoretical physicist; he introduced me to the value of rigor and the beauty in calculations; and he taught me that one should strive to understand field theory as a whole, no matter your specialization.

This book is an expanded version of a course I gave at the IFT in São Paulo, so I would like to thank all of the students in the class for their questions and input, which helped to shape this book.

I want to thank all of my collaborators on topics of research that I worked on and appear in this book. I want to thank my wife Antonia for her patience and encouragement while I wrote this book, mostly at home in the evenings. I also want to thank my students and postdocs for dealing with my reduced time for them during the period I wrote this book. A big thanks to my editor at Cambridge University Press, Simon Capelin, who has been always supportive of me and helped me get this book published. To all the staff at CUP, thanks for making sure this book, as well as my previous ones, is as good as it can be.

Introduction

As I said in the Preface, this book deals with classical aspects of field theory, including electromagnetism and general relativity, but it is not meant as a substitute for a course in either – rather a complement to them. It is meant as a bridge between the study of graduate classical mechanics and quantum field theory, albeit one that comes with many interesting problems of its own. As such, I assume a good knowledge of classical mechanics, in particular Lagrangean and Hamiltonian formulations. I mostly stay out of quantum topics, except when dealing with the spin statistics theorem and representations of the Lorentz group – since those are topics needed to describe the kinds of fields we will be using – and in the case of collective coordinate quantization – since this topic goes hand in hand with the topic of solitons. In these cases, we will deal with quantum *mechanics*, not quantum *field theory* issues. As such, I also assume a working knowledge of quantum mechanics at the advanced undergraduate (if not beginning graduate) level.

The representations of the Lorentz group and the spin statistics theorem are necessary to describe the types of fields we can have: scalars, gauge fields, symmetric and antisymmetric tensors, spinors, and so on. I will start in Part I by describing general properties of the fields and, after describing the classification, focusing on scalars and gauge fields; I will also describe the basic ideas of hydrodynamics, as viewed from the perspective of field theory. In Part II, I will continue with solitons and topological issues, which were previewed in Part I by the Hopfion solution of electromagnetism, as well as with non-Abelian gauge theory. In Part III, I will treat spinors, *p*-forms and anyons, and general relativity as a field theory only. That means that I will not describe standard geometrical and topological issues, only issues that can be easily described in terms of Lagrangeans and their equations of motion. With respect to spinors, I will only consider "classical" issues like solutions to the equations of motion – though, as I will explain, spinors are in some sense always quantum.

Central to classical field theory is the notion of *field*, which is a function of space and time that describes some physical interaction. The easiest example to understand is the electromagnetic field, which is responsible for electromagnetic (electric and magnetic) interactions. Classical field theory is the way to describe, using classical mechanics, these fields and their interactions via Lagrangeans and their equations of motion. That is why, besides some general formal aspects, a lot of the book will be about equations of motion for these Lagrangeans and their solutions. In the case of "soliton" solution, a solution can be on a similar footing as a "particle," or fundamental excitation, of the theory (in the case of electromagnetism, the photon). We will therefore study such solitons in all the theories that we will consider.

Another important notion is the notion of symmetry, encapsulated in the form of the Lagrangean of the system. That is why we will devote a lot of time to it, also, and we will introduce at the beginning of the book some notions of group theory and explain how we can write Lagrageans invariant under a symmetry. Throughout Part I we will deal with Abelian, or commuting, symmetries mostly and will analyze the more complicated non-Abelian symmetries in Part II. In Part III, besides the more standard fermions and general relativity, we will also deal with some more modern field theory issues: anyons and Chern–Simons fields have appeared a lot recently in condensed matter, while p-forms are mostly used in higher-dimensional theories (Kaluza Klein, supergravity, and string theory), though they also appear in some descriptions of QCD and other theories in 3+1 dimensions.

PART I

GENERAL PROPERTIES OF FIELDS;
SCALARS AND GAUGE FIELDS

1 Short Review of Classical Mechanics

In this first chapter, I will quickly review concepts of classical mechanics that we will need in order to understand classical field theory. Classical field theory is a generalization of classical mechanics, so by understanding well enough the concepts of classical mechanics, we will be ready for its generalization. We will also need to understand a bit about quantum mechanics, though we will deal (mostly) with classical concepts in this book, leaving the quantum concepts for quantum field theory.

1.1 A Note on Conventions

In most of this book, I will use field theorist's conventions, with $\hbar = c = 1$, unless needed to emphasize some quantum or (non)relativistic issues. We can always reintroduce $\hbar$ and c by dimensional analysis, if needed. In these conventions, there is only one dimensionful unit, namely $mass = 1/length = energy = 1/time = \dots$. When I speak of dimension of a quantity, I refer to mass dimension.

For the Minkowski metric $\eta^{\mu\nu}$, I use the mostly plus signature convention, so in the most relevant case of 3+1 dimensions, the signature is $(-++ +)$, for $\eta^{\mu\nu} = diag(-1, +1, +1, +1)$.

I also use the Einstein summation convention – i.e., repeated indices are summed over. The repeated indices will be one up and one down, unless we are in Euclidean space, when it doesn't matter, so we can put all indices down.

1.2 Lagrangean and Equations of Motion

In classical mechanics, one deals with a collection of particles i, $i = 1, \dots, N$, with positions $\vec{r}_i$ and velocities $\vec{v}_i = \frac{d\vec{r}_i}{dt}$. But in order to construct a useful classical mechanics description, we need to describe the system more abstractly, by constructing independent variables (generalized coordinates) $q_1, \dots, q_k, \dots, q_n$. Then we have

$$\vec{r}_i = \vec{r}_i(q_1, \dots, q_n, t); \quad i = 1, \dots, N, \tag{1.1}$$

and, of course, $3N \geq n$, since we have at most as many variables as positions in three-dimensional space. Then the velocities are, by the chain rule,

$$\vec{v}_i = \frac{d\vec{r}_i}{dt} = \frac{\partial \vec{r}_i}{\partial q_k} \frac{dq_k}{dt} + \frac{\partial \vec{r}_i}{\partial t} , \qquad (1.2)$$

and, therefore, the kinetic energy of the collection of particles is

$$T = \sum_{i=1}^{N} \frac{m_i \vec{v}_i^2}{2} = T(q_1, \ldots, q_n, \dot{q}_i, \ldots, \dot{q}_n; t). \qquad (1.3)$$

We consider the case when there are forces, and they are conservative, i.e., they can be obtained from a potential V,

$$\vec{F}_i = -\vec{\nabla}_i V. \qquad (1.4)$$

The potential is a function of the coordinates, thus of the generalized coordinates (and perhaps also explicitly of time)

$$V = V(\vec{r}_1, \ldots, \vec{r}_N; t) = V(q_1, \ldots, q_n; t). \qquad (1.5)$$

We could also define generalized forces

$$f_j = -\frac{\partial V}{\partial q_j}. \qquad (1.6)$$

To describe the classical mechanics system in a unified way, we must construct the fundamental quantity that describes the system, the *Lagrangean*,

$$L = T - V = L(\dot{q}_1, \ldots, \dot{q}_n; q_1, \ldots, q_n; t). \qquad (1.7)$$

From this function, we can calculate the *Euler–Lagrange equations of motion* describing the evolution of the system,

$$\frac{d}{dt}\left(\frac{\partial L}{\partial \dot{q}_j}\right) - \frac{\partial L}{\partial q_j} = 0. \qquad (1.8)$$

For each abstract coordinate (i.e., degree of freedom) q_k, we have an equation, so that the evolution of the system is completely determined.

Once we construct a Lagrangean, we have a complete description of a classical system, and we only need to find solutions to its equations of motion. Thus, classical mechanics is about the construction of Lagrangeans and finding solutions for them.

Classical field theory will be just a generalization of this formalism to a continuum of degrees of freedom, i.e., instead of having a discrete set of coordinates $\{q_k\}_{k=1,\ldots,n}$, we will have a continuum of them.

But there is an even more important concept, the one of *action*, and the associated Hamilton's principle of least action: we minimize the action, which is a functional of the path $\{q_k(t)\}$, with respect to the path – i.e., we look for the stationary paths, given fixed initial and final boundary conditions. The action is the integral of the Lagrangean,

$$S = \int_{t_1}^{t_2} L\,dt. \qquad (1.9)$$

The action is the fundamental starting point of the classical mechanics and classical field theory analysis and is more important than the Lagrangean, since the latter can be changed by a total derivative dF/dt without changing the equations of motion. The extra term in the action when doing this change is $\int_{t_1}^{t_2} dt\, dF/dt = F(t_2) - F(t_1)$, which is fixed, so doesn't contribute to the variation of the action.

To derive the equations of motion, we vary S through a variation $\delta q_k(t)$ of the path, given fixed endpoints $q_k(t_1)$ and $q_k(t_2)$, and put the result to zero. Then we obtain

$$\delta S = \sum_{k=1}^{n} \int_{t_1}^{t_2} dt \left(\frac{\partial L}{\partial \dot{q}_k} \delta \dot{q}_k + \frac{\partial L}{\partial q_k} \delta q_k \right) = 0. \tag{1.10}$$

We now do a partial integration, using the fact that δ and d/dt commute, so $\delta \dot{q}_k = d(\delta q_k)/dt$, and we obtain

$$\delta S = \sum_{k=1}^{n} \int_{t_1}^{t_2} dt \left[-\frac{d}{dt}\left(\frac{\partial L}{\partial \dot{q}_k} \right) + \frac{\partial L}{\partial q_k} \right] \delta q_k + \sum_{k=1}^{n} \int_{t_1}^{t_2} dt \frac{d}{dt}\left(\frac{\partial L}{\partial \dot{q}_k} \delta q_k \right), \tag{1.11}$$

and the last term vanishes, as we said, since by our assumption $\delta q_k(t_1) = \delta q_k(t_2) = 0$. Since the result is supposed to be zero for any variation $\delta q_k(t)$, this can only be true if the integrand multiplying it is zero at all times, i.e.,

$$\frac{d}{dt}\left(\frac{\partial L}{\partial \dot{q}_j} \right) - \frac{\partial L}{\partial q_j} = 0. \tag{1.12}$$

We have thus obtained the Euler–Lagrange equations from the principle of least action.

1.3 Systems with Constraints

Until now, we have assumed that the q_k were independent variables, so the variations with respect to them were considered independent as well. But this need not be the case. Perhaps we cannot solve for independent variables, but we are left with some constraints on the system,

$$\tilde{f}_a(\vec{r}_1, \ldots, \vec{r}_N; t) = 0, \quad \forall a = 1, \ldots, m, \tag{1.13}$$

or, in terms of our abstract variables,

$$f_a(q_1, \ldots, q_n) = 0, \quad \forall a = 1, \ldots, m. \tag{1.14}$$

The way to deal with them in the Lagrangean formalism is to introduce *Lagrange multipliers*, which are independent variables that multiply each of the constraints. Indeed, we can add the constraints times the Lagrange multipliers for free to the Lagrangean, since they vanish anyway, thus obtaining the modified action

$$S' = \int_{t_1}^{t_2} dt\, L' = \int_{t_1}^{t_2} dt \left(L + \sum_{a=1}^{m} \lambda_a f_a \right). \tag{1.15}$$

Since the λ_a are independent variables (independent on the $q_k(t)$), we just have an extra term in the Euler–Lagrange equations. Indeed, now we obtain the variation of the action as

$$\delta S = \sum_{k=1}^{n} \int_{t_1}^{t_2} dt \left[-\frac{d}{dt}\left(\frac{\partial L}{\partial \dot{q}_k}\right) + \frac{\partial L}{\partial q_k} + \sum_{a=1}^{m} \lambda_a \frac{\partial f}{\partial q_k} \right] \delta q_k = 0, \qquad (1.16)$$

from which we get the modified Euler–Lagrange equations,

$$-\frac{d}{dt}\left(\frac{\partial L}{\partial q_k}\right) + \frac{\partial L}{\partial q_k} + \sum_{a=1}^{m} \lambda_a \frac{\partial f}{\partial q_a} = 0. \qquad (1.17)$$

If before we had n equations (Euler–Lagrange) for n variables $q_k(t)$, now we have $n+m$ equations, n Euler–Lagrange and m constraints $f_a = 0$, for $n+m$ variables, n $q_k(t)$ variables and m Lagrange multipliers λ_a. So by solving them, we fix completely $q_k(t)$ and λ_a.

1.4 Canonically Conjugate Momentum and Conservation Laws

For our abstract variables, we can define a notion of momentum, called the canonical momentum conjugate to q_k,

$$p_k \equiv \frac{\partial L}{\partial \dot{q}_k}. \qquad (1.18)$$

If one of the variables is an actual particle position, $q_k = r_k$, then the Lagrangean is

$$L = m_k \frac{\dot{r}_k^2}{2} + \dots, \qquad (1.19)$$

so the momentum conjugate to q_k is

$$p_k = \frac{\partial L}{\partial \dot{r}_k} = m\dot{r}_k, \qquad (1.20)$$

which is indeed the usual particle momentum. That means that in general p_k is a generalization of the momentum, as we expected.

If the Lagrangean is independent on a variable q_k (though of course it can depend $\dot{q}_k$), we have $\partial L/\partial q_k = 0$, and then from the Euler–Lagrange equations we have

$$\frac{d}{dt}\frac{\partial L}{\partial \dot{q}_k} = 0 \Rightarrow \frac{d}{dt}p_k = 0, \qquad (1.21)$$

i.e., the canonically conjugate momentum is conserved (constant in time). In this case, we say that the variable q_k is *cyclic*.

Reversely, if we have a conserved quantity, we can look for a cyclic independent variable that has it as its canonically conjugate momentum.

The notion of symmetries – and, in particular, symmetries of the Lagrangean (or rather, of the action) – are really fundamental for modern theoretical physics. From the (sometimes abstract) symmetries of the action, we derive many interesting physical properties, so understanding the symmetries of the problem is very important.

We will describe symmetries more in the next chapter, but for the moment we will give a first taste through the simplest and most common case of rotational symmetry. A rotation is a transformation acting linearly on the Cartesian coordinates r_i^a, $a = 1, 2, 3$ (where $\vec{r}_i$ splits as (x_i, y_i, z_i), together r_i^a), as

$$r'^a_i = \Lambda^a{}_b r_i^b \,, \tag{1.22}$$

where $\Lambda^a{}_b$ is a matrix satisfying

$$\Lambda \cdot \Lambda^T = \Lambda^T \cdot \Lambda = \mathbb{1}. \tag{1.23}$$

Such a matrix is called *orthogonal* matrix. Note that here we have used *Einstein's summation convention*, where repeated indices are summed over.

Then we have

$$\vec{r}'^2_i = \sum_a (r'^a_i)^2 = \sum_{a,b,c} (\Lambda^a{}_b r_i^b)(\Lambda_{ac} r_i^c) = \sum_{b,c} r_i^b (\Lambda^T \Lambda)_{bc} r_i^c = \sum_{b,c} r_i^b \delta_{bc} r_i^c = \vec{r}^2_i \,, \tag{1.24}$$

so, indeed, $\vec{r}^2_i$ is invariant. In the same way, we show that

$$\left(\frac{d\vec{r}'_i}{dt} \right)^2 = \left(\frac{d\vec{r}_i}{dt} \right)^2. \tag{1.25}$$

That means that *if $V = V(\vec{r}^2_i)$ only*, i.e., if $V = V(|\vec{r}|)$ and is independent on the particle, a case known as a central potential (for instance, an electron around a nucleus, or a planet around the Sun), then the Lagrangean is rotationally invariant,

$$L(\vec{r}'_i, \dot{\vec{r}}'_i; t) = \sum_i \frac{m_i \vec{v}'^2_i}{2} - V(\vec{r}'^2_i) = \sum_i \frac{m_i \vec{v}^2_i}{2} - V(\vec{r}^2_i) = L(\vec{r}_i, \dot{\vec{r}}_i, t). \tag{1.26}$$

1.5 The Hamiltonian Formalism

The Lagrangean is the difference of the kinetic and potential energies, $L = T - V$, but we know that the total energy is the sum, $E = T + V$. We will, in fact, see that the energy is related to another important concept in classical mechanics, the Hamiltonian. To define it, we need first to understand the concept of Legendre transformation.

Generally, a function $f(x, y)$ has a differential

$$df = u \, dx + v \, dy \,, \tag{1.27}$$

where

$$u \equiv \frac{\partial f}{\partial x}; \quad v \equiv \frac{\partial f}{\partial y}. \tag{1.28}$$

But now we can forget where u, v come from, and consider them as independent variables. In that case, the Legendre transform of the function f is

$$g = f - ux. \tag{1.29}$$

Then its differential is

$$dg = df - u\,dx - x\,du = -x\,du + v\,dy. \tag{1.30}$$

From this differential, we can derive

$$x = -\frac{\partial g}{\partial u}; \quad v = \frac{\partial g}{\partial y}, \tag{1.31}$$

which means that the function g is a function of u and y instead,

$$g = g(u, y). \tag{1.32}$$

The Legendre transform was encountered, for instance, in thermodynamics, where the total energy U is a function of extensive quantities, $U = U(S, V, \ldots)$,

$$dU = TdS - PdV + \ldots, \tag{1.33}$$

but its Legendre transform in S, the free energy,

$$F = U - TS, \tag{1.34}$$

has the differential

$$dF = -SdT - PdV + \ldots, \tag{1.35}$$

so is a function of extensive quantities and one intensive one, T,

$$F = F(T, V, \ldots). \tag{1.36}$$

Similarly to the thermodynamics case, we now want to do a Legendre transform of the Lagrangean $L = L(q_k, \dot{q}_k, t)$, with differential

$$dL = \sum_i \frac{\partial L}{\partial q_i} dq_i + \sum_i \frac{\partial L}{\partial \dot{q}_i} d\dot{q}_i + \frac{\partial L}{\partial t} dt. \tag{1.37}$$

It is a bit unusual that the function depends both on coordinates and their derivatives, so it would be good to do a Legendre transform on the derivatives, to exchange them for independent variables. First, we note that in terms of the conjugate momenta $q_i \equiv \partial L / \partial \dot{q}_i$, the Lagrange equations of motion are

$$\dot{p}_i = \frac{\partial L}{\partial q_i}, \tag{1.38}$$

so using them, the Lagrangean differential can be rewritten as

$$dL = \sum_i \dot{p}_i dq_i + \sum_i p_i d\dot{q}_i + \frac{\partial L}{\partial t} dt. \tag{1.39}$$

We thus define the Hamiltonian as (minus) the Legendre transform of the Lagrangean,

$$H = \sum_i \dot{q}_i p_i - L(q_i, \dot{q}_i, t), \tag{1.40}$$

so its differential is

$$dH = -\sum_i \dot{p}_i dq_i + \sum_i \dot{q}_i dp_i - \frac{\partial L}{\partial t} dt. \tag{1.41}$$

The Hamiltonian is then a function of coordinates and their conjugate momenta, plus maybe an explicit time dependence,

$$H = H(q_i, p_i, t).\tag{1.42}$$

From the Hamiltonian differential, we derive the *Hamiltonian equations of motion*,

$$\dot{q}_i = \frac{\partial H}{\partial p_i}; \quad \dot{p}_i = -\frac{\partial H}{\partial q_i}; \quad \frac{\partial L}{\partial t} = -\frac{\partial H}{\partial t}.\tag{1.43}$$

The advantage of the Hamiltonian equations of motion is that the equations are first order in time derivatives, whereas in the Lagrangean case, the equations of motion were quadratic in time derivatives. The disadvantage is that we have doubled the number of variables, instead of q_i, we have now q_i and p_i, which together are called the *phase space*.

If the potential is conservative, independent on path or explicitly on time, $V = V(\{q_i\})$, and so the Lagrangean is independent explicitly on time, and if, moreover, there are no cubic or higher-order terms in the velocity in L, $\dot{q}^n$, $n \geq 3$, then $L = T - V$ implies

$$H = T + V = E,\tag{1.44}$$

so the Hamiltonian is the energy of the system.

Hamilton's principle of least action is also modified now, but only by replacing L with its form in terms of H – i.e.,

$$\delta S = \delta \int_{t_1}^{t_2} dt(p_i\dot{q}_i - H(q_i, p_i, t)) = 0.\tag{1.45}$$

1.6 Canonical Transformations

On the phase space, we can make coordinate transformations. If they preserve the Hamiltonian structure, they are called *canonical transformations*. That is, transformations

$$Q_i = Q_i(q_k, p_k, t); \quad P_i = P_i(q_k, p_k, t),\tag{1.46}$$

such that for the new phase space there is a Hamiltonian $\tilde{H}$, with Hamiltonian equations

$$\dot{Q}_i = \frac{\partial \tilde{H}}{\partial P_i}; \quad \dot{P}_i = -\frac{\partial \tilde{H}}{\partial Q_i},\tag{1.47}$$

and such that we have at the same time the principle of least action for both the original Hamiltonian,

$$\delta \int_{t_1}^{t_2} dt \left(\sum_k p_k\dot{q}_k - H(q_k, p_k, t) \right) = 0,\tag{1.48}$$

and the new one,

$$\delta \int_{t_1}^{t_2} dt \left(\sum_i P_i\dot{Q}_i - \tilde{H}(Q_i, P_i, t) \right) = 0,\tag{1.49}$$

for any boundary conditions (at t_1, t_2). That means that the integrands must be proportional up to a total derivative term (which doesn't change the equations of motion derived from the action), so we must have

$$\sum_k p_k \dot{q}_k - H(q_k, p_k, t) = \sum_i P_i \dot{Q}_i - \tilde{H} + \frac{dF}{dt}. \tag{1.50}$$

By comparing the coordinate dependence on both sides, we see that the function F is

$$F = F(q_k, Q_i, t). \tag{1.51}$$

We can invert the canonical transformation (it is assumed to be invertible, since we have the same number of coordinates on both sides), and find

$$q_k = q_k(Q_i, P_i); \quad p_k = p_k(Q_i, P_i). \tag{1.52}$$

With a bit more analysis, from the same equality (1.50), we can also find the consistency conditions

$$\frac{\partial Q_i}{\partial q_k} = \frac{\partial p_k}{\partial P_i} \frac{\partial Q_i}{\partial p_k} = -\frac{\partial q_k}{\partial P_i}$$

$$\frac{\partial P_i}{\partial q_k} = -\frac{\partial p_k}{\partial Q_i} \frac{\partial P_i}{\partial p_k} = \frac{\partial q_k}{\partial Q_i}. \tag{1.53}$$

1.7 Poisson Brackets

The concept of Poisson bracket is a very useful one in terms of the theory of the Hamiltonian equations, and the quantization of theories. For a couple of functions of phase space, $f(q_k, p_k)$ and $g(q_k, p_k)$, the Poisson bracket is defined as

$$\{f, g\}_{P.B.} = \sum_k \left(\frac{\partial f}{\partial q_k} \frac{\partial g}{\partial p_k} - \frac{\partial f}{\partial p_k} \frac{\partial g}{\partial q_k} \right). \tag{1.54}$$

As we can see, the Poisson bracket is antisymmetric,

$$\{f, g\}_{P.B.} = -\{g, f\}_{P.B.}, \tag{1.55}$$

just like the commutator.

Since, in particular, the q_k and p_k are functions of phase space, we can define their Poisson brackets, the *fundamental Poisson brackets* on the phase space. We easily find that

$$\{q_j, p_k\}_{P.B.} = \delta_{jk} \{p_k, q_j\}_{P.B.} = -\delta_{jk}$$
$$\{q_j, q_k\}_{P.B.} = 0 \{p_j, p_k\}_{P.B.} = 0. \tag{1.56}$$

Moreover, noticing that the objects appearing in the Hamiltonian equations of motion are of the same type as the ones appearing in the Poisson bracket, we find that the Hamiltonian equations of motion become

$$\dot{q}_i = \{q_i, H\}_{P.B.}; \quad \dot{p}_i = \{p_i, H\}_{P.B.}. \tag{1.57}$$

1.8 Hamilton–Jacobi Theory

We can obtain a powerful new formalism for solving classical mechanics systems by considering a canonical transformation after which $\tilde{H} = 0$. Then from the new Hamiltonian equations of motion, we obtain

$$\dot{Q}_i = \dot{P}_i = 0 \, , \tag{1.58}$$

i.e., the new coordinates and their conjugate momenta are conserved quantities. This previews the usefulness of the formalism.

From the relation (1.50), with the Legendre transform of the function $F(q_k, Q_i, t)$, written as

$$F(q_k, Q_i, t) = S(q_k, P_i, t) - \sum_i Q_i P_i \, , \tag{1.59}$$

we obtain (summing over repeated indices)

$$p_k \dot{q}_k - H = -Q_i \dot{P}_i - \tilde{H} + \frac{dS(q_k, P_k, t)}{dt} = -Q_i \dot{P}_i - \tilde{H} + \frac{\partial S}{\partial q_k} \dot{q}_k + \frac{\partial S}{\partial P_i} \dot{P}_i + \frac{\partial S}{\partial t}. \tag{1.60}$$

By comparing the dependence on $\dot{q}_k$ and $\dot{P}_i$ on both sides, we obtain

$$p_k = \frac{\partial S}{\partial q_k}; \quad Q_i = \frac{\partial S}{\partial P_i} \, , \tag{1.61}$$

and then the remainder of the equation gives

$$\tilde{H} = H + \frac{\partial S}{\partial t}. \tag{1.62}$$

Putting $\tilde{H} = 0$, and substituting p_k in the variables of H leads to the *Hamilton–Jacobi equation*,

$$H\left(q_1, \ldots, q_n, \frac{\partial S}{\partial q_1}, \ldots, \frac{\partial S}{\partial q_n}; t\right) + \frac{\partial S}{\partial t} = 0. \tag{1.63}$$

Here S is called *Hamilton's principal function*, and the solution of the equation is of the type (replacing the new momenta P_i by the corresponding constants)

$$S = S(q_1, \ldots, q_n; \alpha_1, \ldots, \alpha_n; t) \, , \tag{1.64}$$

where P_i and Q_i are *integrals of motion*, or constants (conserved quantities, independent on time),

$$P_i = \alpha_i; \quad Q_i = \beta_i = \frac{\partial S}{\partial \alpha_i}. \tag{1.65}$$

We can solve the Hamilton–Jacobi equation by writing an ansatz (usually of separation of variables type) for the principal function S. Once we have solved it in terms of integrals of motion α_i, β_i, we can invert the relations $P_i(q_k, p_k) = \alpha_i$ and $Q_i(q_k, p_k) = \beta_i$ to solve for the original coordinates in terms of the integrals of motion, i.e.,

$$q_k = q_k(\alpha_i, \beta_i; t); \quad p_k = p_k(\alpha_i, \beta_i; t). \tag{1.66}$$

Further Reading

Any good classical mechanics book, for instance Goldstein [1].

Exercises

(1)   Consider the Lagrangean

$$L[q] = \frac{\dot{q}^2}{2} - \alpha \frac{q^2}{2} - \lambda \frac{q^3}{3}. \tag{1.67}$$

(a) Calculate the Lagrange equations of motion.
(b) Calculate the Hamiltonian and the Hamiltonian equations of motion.
(c) If $Q = q + p$, what is $P(q,p)$ such that the transformation is canonical?

(2)   For

$$H = \frac{p^2}{2} + \frac{q^2}{2} \,, \tag{1.68}$$

write down the Hamilton–Jacobi equation, and solve it by an ansatz for the Hamilton principal function S.

(3)   Consider the Hamiltonian

$$H = \frac{p_1^2}{2} + \frac{p_2^2}{2} + \frac{(q_1 + q_2)^2}{2}. \tag{1.69}$$

(a) Calculate the Hamiltonian equations of motion, and solve them.
(b) Calculate the Lagrangean *for decoupled variables* Q_1, Q_2. Is the transformation between (q_1, q_2) and (Q_1, Q_2) canonical?

(4)   Consider the Lagrangean

$$L = \sum_{i=1}^{3} \left(\frac{\dot{q}_i^2}{2} - \alpha_i \frac{q_i^3}{3} \right) \,, \tag{1.70}$$

subject to the constraint

$$q_3 = q_1 + q_2. \tag{1.71}$$

Write the Lagrangean equations of motion *on the constraints*.

(5)   Consider the Lagrangean

$$L = \frac{\dot{q}_1^2}{2} + \frac{\dot{q}_2^2}{2} + \alpha q_1 \cdot q_2 + \lambda(q_1 + q_2) \,, \qquad (1.72)$$

for variables q_1, q_2, λ.

(a) Write the Lagrangean equations of motion.
(b) Is the system stable?
(c) Write a Hamiltonian for the system.
(d) Write a solution of the equations of motion.

2 Symmetries, Groups, and Lie algebras; Representations

In this chapter, we will learn about the mathematics of symmetry groups and Lie algebras, and how they relate to the description of physical systems via Lagrangeans.

As we said, the concept of symmetry is very important in theoretical physics. We look for symmetries that leave the action, describing the system under consideration, invariant. By finding symmetries, we can infer important physical phenomena. We will build examples of Lagrangeans invariant under various symmetries and belonging to some representation of a symmetry group.

2.1 Groups and Invariance in the Simplest Case

The simplest symmetry we can have is a discrete one, which means the associated group has a finite number of elements.

A **group** G is defined as a set of elements, together with a multiplication between them, $\cdot : G \times G \to G$, that has the following properties:

(a) It respects the group – i.e., $\forall f, g \in G$, $h = f \cdot g \in G$.
(b) It is associative – i.e., $(f \cdot g) \cdot h = f \cdot (g \cdot h)$.
(c) $\exists$ an element e called the identity, such that $e \cdot f = f \cdot e = f$, $\forall f \in G$.
(d) $\exists$ an inverse f^{-1}, $\forall f \in G$, such that $f \cdot f^{-1} = f^{-1} \cdot f = e$.

Example of group: The group $\mathbb{Z}_2$.

This is the simplest group, made up of only two elements, the identity $a = e$ and another one, b. These elements can be represented as real numbers by $a = +1$ and $b = -1$. That means that the *multiplication table for the group* (the set of all possible multiplications) is

$$a \cdot a = a, \quad b \cdot b = a, \quad a \cdot b = b \cdot a = b. \tag{2.1}$$

As we can see, the group is *Abelian* (named after the Norwegian mathematician Niels Henrik Abel, who also gave the name to the most important prize in mathematics, the Abel Prize), which means that the elements commute, $g_1 \cdot g_2 = g_2 \cdot g_1$, $\forall g_1, g_2 \in G$.

Invariance under the group means that if we multiply the variables of the theory by any element $g \in G$, the theory stays the same.

Example of Invariant System 1

For instance, consider a system defined by one real coordinate, $q \in \mathbb{R}$, but with potential depending only on the modulus, $V = V(|q|)$, so the Lagrangean is

$$L = m\frac{\dot{q}^2}{2} - V(|q|). \tag{2.2}$$

Then, replacing q with $q' = gq$, for $g = a$ or $g = b$, we must obtain the same Lagrangean. Since $g = a = +1$, invariance for it is trivial. For $g = b$, we get that $q' = -q$ should give the same result as q, so

$$L(q') = L(q), \tag{2.3}$$

which is indeed the case. Then we also have that the action is invariant, since

$$S = \int_{t_1}^{t_2} dt\, L = S'. \tag{2.4}$$

But the reverse need not be true. We can have a Lagrangean that is not invariant but an action that is. In fact, the only invariance that matters is invariance of the action, since it gives rise to the equations of motion through the principle of least action.

Example of Invariant System 2

For instance, consider a modification of the case above, as

$$L = m\frac{\dot{q}^2}{2} - V(|q|) + \alpha\frac{d}{dt}q, \tag{2.5}$$

together with the boundary condition $q(t_2) = q(t_1)$. Then the extra term vanishes in the action, since

$$\alpha\int_{t_1}^{t_2} dt\frac{d}{dt}q = \alpha(q(t_2) - q(t_1)) = 0. \tag{2.6}$$

However, the Lagrangean is not invariant, since $\dot{q}$ is not invariant under $b = -1$.

Note that in this case the Hamiltonian is the same, but that is not necessarily true in all cases (we could have chosen, for instance, $d(q^2\dot{q}^2)/dt$ as the extra term, and the Hamiltonian would be different). Also note that it is important that the action is invariant since, at the quantum level, it is the action that appears, as one will see in quantum field theory.

Otherwise we could have said that the symmetries of the equations of motion are the relevant ones. However, there are cases where the equations of motion have some symmetries, but the action has fewer, and then also the quantum theory has fewer.[*]

---

[*] A standard example is the case of a theory with a certain duality invariance called S-duality, relevant for Maxwell electromagnetism, which we will study later.

2.2 Generalizations: Cyclic Groups and Their Representations

The group $\mathbb{Z}_3 = \{e, a, b\}$ has three elements, which can be represented by the third roots of unity – i.e., the complex numbers x satisfying $x^3 = 1$. We can choose

$$e = 1 = e^0; \quad a = e^{\frac{2\pi i}{3}}; \quad b = e^{\frac{4\pi i}{3}}. \tag{2.7}$$

From this representation, we can derive the multiplication table for the group,

$$a^2 = b; \quad b^2 = a; \quad ab = ba = e \tag{2.8}$$

(multiplication by e is trivial), from which we also obtain $a^3 = b^3 = 1$, since they are third roots of unity.

For an example of a Lagrangean that is invariant under $\mathbb{Z}_3$, construct a complex variable $q = q_1 + iq_2$, $q_1, q_2 \in \mathbb{R}$, and then

$$|\dot{q}|^2 = \dot{q}_1^2 + \dot{q}_2^2. \tag{2.9}$$

Consider then the Lagrangean

$$L = m\frac{|\dot{q}|^2}{2} - V(q^3). \tag{2.10}$$

As an example, the potential could be

$$V = \frac{1}{q^3} + \frac{1}{\bar{q}^3}, \tag{2.11}$$

where the last term is needed since we need a real action. Then the Lagrangean is invariant under

$$q \to q' = gq, \quad g \in \{e, a, b\}. \tag{2.12}$$

This is so, since

$$|\dot{q}|^2 = |g\dot{q}|^2 = |\dot{q}|^2; \quad q'^3 = (gq)^3 = g^3 q^3 = q^3. \tag{2.13}$$

The *cyclic* group $\mathbb{Z}_N$ is a simple further generalization, with N elements $\{e, a_1, \ldots, a_{N-1}\}$, which can be represented by the Nth roots of unity, elements g such that $g^N = 1$ – i.e.,

$$e = 1, \quad a_1 = e^{\frac{2\pi i}{N}}, \quad a_{N-1} = e^{\frac{2\pi i(N-1)}{N}}. \tag{2.14}$$

This is still an Abelian group.

A Lagrangean invariant under it is defined is a similar way to the previous cases. For instance,

$$L(q, \dot{q}) = m\frac{\dot{q}^2}{2} - V(q^N) \tag{2.15}$$

is invariant, since

$$|\dot{q}'|^2 = |g\dot{q}|^2 = |\dot{q}|^2; \quad q'^N = (gq)^N = q^N \tag{2.16}$$

for all g.

Note that until now we have said that the various $\mathbb{Z}_N$ can be *represented* as the Nth roots of unity; that is, we can define the group elements as some complex numbers, which can multiply other complex numbers – for instance the q's – and give again complex numbers q'.

We therefore say that the representation of g – which, in general (for a representation R), we can write as $D_R(g)$ – acts on a one-complex dimensional vector space. We say that the dimension of this Nth root of unity representation is one (a one-complex dimensional vector space).

But for Z_N, $N \geq 3$, we also have other representations of the same group – that is, for the same multiplication table.

For instance, for $\mathbb{Z}_3$, we can define the *regular* representation, which is a three-dimensional representation in which the group elements are represented by the matrices

$$D(e) = \begin{pmatrix} 1 & 0 & 0 \\ 0 & 1 & 0 \\ 0 & 0 & 1 \end{pmatrix}, \quad D(a) = \begin{pmatrix} 0 & 0 & 1 \\ 1 & 0 & 0 \\ 0 & 1 & 0 \end{pmatrix}, \quad D(b) = \begin{pmatrix} 0 & 1 & 0 \\ 0 & 0 & 1 \\ 1 & 0 & 0 \end{pmatrix}. \tag{2.17}$$

We can check that the $D(g)$ matrices still satisfy the $\mathbb{Z}_3$ multiplication table. This is a three-dimensional representation, since it acts on column vectors $\begin{pmatrix} x \\ y \\ z \end{pmatrix}$. We can choose the obvious (Cartesian) basis on this space and denote its elements by relating them to the group elements, namely

$$|e\rangle = \begin{pmatrix} 1 \\ 0 \\ 0 \end{pmatrix} = |e_1\rangle, \quad |a\rangle = \begin{pmatrix} 0 \\ 1 \\ 0 \end{pmatrix} = |e_2\rangle, \quad |b\rangle = \begin{pmatrix} 0 \\ 0 \\ 1 \end{pmatrix} = |e_3\rangle. \tag{2.18}$$

The notation is not random; but, rather, it is used such that we have the property

$$D(g_1)|g_2\rangle = |g_1 g_2\rangle \equiv |h\rangle. \tag{2.19}$$

Indeed, by the group property, $h \in G$, so if g_1, g_2 are among $\{e, a, b\}$, so is h. Moreover, since $|e_i\rangle$ are orthogonal, we see that, in fact, we have

$$[D(g)]_{ij} = \langle e_i|D(g)|e_j\rangle = \langle e_i|D(g)e_j\rangle. \tag{2.20}$$

The roots of unity representation are obviously not equivalent with the regular representation, since they have different dimensionalities, yet they are representations for the same group.

Equivalent Representations

But if two representations have the same dimension, it is possible that the representations are equivalent if they can be related by a "similarity transformation," which is equivalent with a change of basis, namely

$$D(g) \to D(g)' = S^{-1}D(g)S. \tag{2.21}$$

Indeed, in that case,

$$\langle e_i'|D'(g)|e_j'\rangle = \langle e_i'|S^{-1}D(g)S|e_j'\rangle \equiv \langle e_i|D(g)|e_j\rangle \,, \tag{2.22}$$

where we have defined the change of basis

$$|e_j\rangle = S|e_j'\rangle \,, \tag{2.23}$$

together with the assumption of unitary matrices $S^\dagger = S^{-1}$ (or orthogonal, in the real case, $S^T = S^{-1}$). Unitarity of matrices is true in all the cases of the classical Lie groups, that we will be worried with here. The need for an unitary or othogonal matrix is related to the fact that we need to preserve orthonormality of the basis under it.

Reducible Representations

If there is an invariant subspace $\mathcal{H}$ of the representation space (space of vectors $|h\rangle$), associated with a subgroup $H \subset G$ – i.e., if $\forall h \subset H - D(g)|h\rangle \in \mathcal{H}$, we say that the representation is *reducible*. Otherwise, we say the representation is *irreducible*.

A reducible representation is always equivalent with one where the matrix elements are block diagonal – i.e.,

$$D(g) = \begin{pmatrix} D(h) & 0 & 0 & \cdots \\ 0 & D(\tilde{h}) & 0 \\ 0 & 0 & D(\tilde{\tilde{h}}) \\ & & & \cdots \end{pmatrix}. \tag{2.24}$$

As we can see, in this case, the representation vector space is split as $\begin{pmatrix} |h\rangle \\ |\tilde{h}\rangle \\ |\tilde{\tilde{h}}\rangle \end{pmatrix}$; and, indeed,

$D(g)|h\rangle \sim |h\rangle \in \mathcal{H}$.

2.3 Lie Groups and Abelian Lie Group Invariance

Until now, we have studied discrete groups, which have a discrete number of elements. But the case of most interest is of groups for which the elements $g \in G$ depend continuously on parameters α^a – i.e., $g = g(\alpha)$ – called *Lie groups*.

In this case, by convention, we will choose the parameters α^a such that

$$g(\alpha^a = 0) = e. \tag{2.25}$$

Then we can Taylor expand around the point $\alpha^a = 0$, and write

$$D(g(\alpha^a)) \simeq \mathbb{1} + id\alpha^a X_a + \cdots \tag{2.26}$$

That means that, by definition,

$$X_a \equiv -i\frac{\partial}{\partial\alpha^a}D(g(\alpha^a))\Big|_{\alpha^a=0} \tag{2.27}$$

and are called the *generators* of the Lie group. The i is introduced by convention, so that if $D(g)$ is unitary – i.e., $(D(g))^\dagger = (D(g))^{-1}$ – then X_a are Hermitian – i.e., $X_a^\dagger = X_a$.

This notion of Lie groups was defined by Sophus Lie, who showed that the generators can actually be defined independently of representation, though we will not show it here.

We can define the generators X_a such that finite elements with coefficients α^a, writing $d\alpha^a = \alpha^a/k$, for $k \to \infty$, are obtained by acting k times with the same infinitesimal factor – i.e., that

$$D(g(\alpha^a)) = \lim_{k\to\infty} \left(1 + \frac{i\alpha^a X_a}{k}\right)^k = e^{i\alpha^a X_a}. \tag{2.28}$$

In other words, the generators of the Lie group always appear in the exponent. The i is conventional, as we said. If we absorb it in the generators, the generators will be anti-Hermitian (that is sometimes called the mathematics convention).

The simplest example of a Lie group is the Abelian Lie group, $U(1)$. This is the case in which there exists a single generator, which can be taken to be the identity (or, rather, the number 1), $X_a = 1$. That means that the general group element in this representation is

$$D(g(\alpha)) = e^{i\alpha}. \tag{2.29}$$

We can easily construct a Lagrangean that is invariant under it. Since the group element is complex, we need to construct a complex coordinate variable, $q = q_1 + iq_2, q_1, q_2 \in \mathbb{R}$. Then the transformation is

$$q \to q' = D(g(\alpha))q = e^{i\alpha}q. \tag{2.30}$$

If, moreover, the potential depends only on the modulus of q – i.e.,

$$L = m\frac{|\dot{q}|^2}{2} - V(|q|), \tag{2.31}$$

then the Lagrangean is invariant. Indeed, we have

$$|\dot{q}'|^2 = |q^{i\alpha}\dot{q}|^2 = |\dot{q}|^2; \quad |q'|^2 = |e^{i\alpha}q|^2 = |q|^2 \Rightarrow L(q') = L(q). \tag{2.32}$$

2.4 Lie Algebra

The notion of Lie groups implies another important concept: the concept of the *Lie algebra*. This is the algebra satisfied by the generators X_a. We denote the Lie algebra by $\mathcal{L}(G)$ – i.e., $X_a \in \mathcal{L}(G)$. The Lie algebra is defined by the commutator, which acts on a product of Lie algebra and takes value again in the Lie algebra – i.e.,

$$[\,,\,] : \mathcal{L}(G) \times \mathcal{L}(G) \to \mathcal{L}(G). \tag{2.33}$$

We will now prove this statement.

By our definition of the Lie group, any group element takes the exponential form, so considering $g, h \in G$, we can write

$$g = e^{i\alpha^a X_a}; \quad h = e^{i\beta^a X_a}. \tag{2.34}$$

But by the group property, then also $g \cdot h \in G$, so it can also be written as an exponential, meaning we obtain

$$e^{i\alpha^a X_a} e^{i\beta^a X_a} = e^{i\gamma^a X_a}. \tag{2.35}$$

Taking the log, and adding and subtracting a 1 inside it, we obtain

$$i\gamma^a X_a = \ln\left[1 + e^{i\alpha^a X_a} e^{i\beta^b X_b} - 1\right] \equiv \ln[1 + Y]$$

$$Y \equiv e^{i\alpha^a X_a} e^{i\beta^b X_b} - 1 \simeq \left(1 + i\alpha^a X_a - \frac{(\alpha^a X_a)^2}{2}\right)\left(1 + i\beta^a X_a - \frac{(\beta^a X_a)^2}{2}\right) - 1$$

$$\simeq i\alpha^a X_a + i\beta^a X_a - \alpha^a X_a \beta^b X_b - \frac{1}{2}(\alpha^a X_a)^2 - \frac{1}{2}(\beta^b X_b)^2 + \ldots \tag{2.36}$$

Then, using that $\ln[1 + Y] \simeq Y - Y^2/2 + \ldots$, we get

$$i\gamma^a X_a = \simeq i\alpha^a X_a + i\beta^a X_a - \alpha^a X_a \beta^b X_b - \frac{1}{2}(\alpha^a X_a)^2 - \frac{1}{2}(\beta^b X_b)^2$$

$$+ \frac{1}{2}(\alpha^a X_a)^2 + \frac{1}{2}(\beta^b X_b)^2 + \frac{\alpha^a X_a \beta^b X_b + \beta^b X_b \alpha^a X_a}{2}$$

$$\simeq i\alpha^a X_a + i\beta^a X_a + \frac{\beta^b X_b \alpha^a X_a - \alpha^a X_a \beta^b X_b}{2}. \tag{2.37}$$

Putting the linear terms all on the left hand side, renaming the summed over indices for them as c, and multiplying the relation by -2, we obtain

$$-2i(\gamma^c - \alpha^c - \beta^c)X_c \simeq \alpha^a \beta^b [X_a, X_b]. \tag{2.38}$$

But this relation must be true for all α^a, β^a (in which case γ^a is determined by them), so it must be the case that $(\gamma^c - \alpha^c - \beta^c)$ is proportional to $(\alpha_a \beta_a)$, meaning we can write

$$-2(\gamma^c - \alpha^c - \beta^c) = \alpha^a \beta^b f_{ab}{}^c, \tag{2.39}$$

which defines $f_{ab}{}^c$. These are *fixed* coefficients (depending on the generators, not on the α^a, β^a, which are arbitrary). Substituting back in relation (2.39), we obtain

$$[X_a, X_b] = i f_{ab}{}^c X_c. \tag{2.40}$$

This is the Lie algebra that we promised, and $f_{ab}{}^c$ are called the *structure constants* and define the Lie algebra. Note from the definition (since $[X_a, X_b]$ is antisymmetric in a, b), that

$$f_{ab}{}^c = -f_{ba}{}^c. \tag{2.41}$$

Note that with the convention with Hermitian generators, we gct an i in the Lie algebra; whereas with the convention with anti-Hermitian generators (multiplying the generators by an i), we have no i in the Lie algebra ($[X_a, X_b] = f_{ab}{}^c X_c$).

Of course, in the above derivation, you might have asked why it is that we canceled the linear terms against the quadratic ones. But the answer is that if we do so, then all the higher-order terms cancel as well, such that the full relation (2.35) holds.

We can obtain a useful consistency condition on the structure constants as follows. There is a simple identity, called the *Jacobi identity*, which states that

$$[X_a, [X_b, X_c]] + \text{cyclic}(a, b, c) = 0, \tag{2.42}$$

which is left as an exercise to prove. It is an identity (i.e., of the type $0 = 0$), but if we substitute the Lie algebra twice inside it, we obtain an important consistency condition. Indeed,

$$[X_a, [X_b, X_c]] = [X_a, if_{bc}{}^d X_d] = -f_{bc}{}^d f_{ad}{}^e X_e \,, \tag{2.43}$$

which implies the consistency condition

$$f_{bc}{}^d f_{ad}{}^e + f_{ab}{}^d f_{cd}{}^e + f_{ca}{}^d f_{bd}{}^e = 0. \tag{2.44}$$

2.5 Representations for Lie Groups

Like in the case of finite groups, we can have different representations for Lie groups. Since the classical groups can all be represented as matrices – for instance, $SO(n)$ as the orthogonal special matrices, $SU(n)$ as the unitary special matrices, etc. – in these cases, there is a *fundamental* (or defining) representation in which the generators are these kinds of matrices.

In this and other matrix representations, the generators (from now on called T_a instead of X_a) in representation R would be

$$(T_a^{(R)})_{ij} \,, \tag{2.45}$$

where the indices i, j belong to the vector space of the representation.

But another important representation for Lie groups is called the *adjoint representation*, for which the indices i, j for the vector space of the representation are of the same type as the indices a labeling the generators of the Lie algebra, and moreover we can take

$$(T_a)_b{}^c = -if_{ab}{}^c. \tag{2.46}$$

This is indeed a representation, since we can check that then the consistency condition (2.44) becomes simply

$$([T_a, T_b])_c{}^e = if_{ab}{}^d (T_d)_c{}^e. \tag{2.47}$$

We can see that the dimension of the adjoint representation is the number of independent generators T_a of the Lie algebra.

Scalar Product in the Adjoint and Metric

We can define a scalar product on the space of generators in the the adjoint representation, $(T_a)_{bc}$, namely the trace in the adjoint representation,

$$\mathrm{Tr}_{\mathrm{adj}}(T_a T_b) \equiv (T_a)_c{}^d (T_b)_d{}^c = -f_{ac}{}^d f_{bd}{}^c. \tag{2.48}$$

One can then show that this gives

$$\mathrm{Tr}_{\mathrm{adj}}(T_a T_b) = k^a \delta_{ab} \,, \tag{2.49}$$

where no sum is taken over a. That means that after rescaling the generators T_a by numbers, and *if all $k_a > 0$*, a case that is known as a *compact* Lie algebra, then

$$\mathrm{Tr}_{\mathrm{adj}}(T_a T_b) = \lambda \delta_{ab}. \tag{2.50}$$

Here λ is a normalization. It is often the case that one takes $\lambda = 1$ (in which case, however, the trace in the fundamental representation will have a constant). Here $\lambda \delta_{ab} \equiv g_{ab}$ can be used as a metric on the Lie group, used to raise and lower indices.

The notion of a compact Lie algebra is related to the fact that Lie groups are also manifolds (parametrized by the coordinates α^a), and if $k^a > 0$ for all a, then the manifold is compact. If there are minuses among the k_a's, then the manifold is non-compact. In the compact case, there is an invariant metric on the manifold of the type δ_{ab} (Euclidean), whereas in the noncompact case, the invariant metric on the manifold is Lorentzian (with some minuses on the diagonal).

Another observation is that the Lie algebra used commutators, which was related to the implicit assumption that the coordinates α^a on the manifolds were complex and commuting numbers, and then so were the generators X_a. But there is an important generalization, in which the α^as are taken to be Grassmann (or anticommuting) numbers, and then so are the corresponding generators Q_a. In this case, the algebra of two Q's uses the anticommutator, like for any fermions, and of a Q with an X uses also the commutator, like for a boson with a fermion. This is an important case known as a *superalgebra*, and it is important for supersymmetry. This generalization was not realized in physics until the 1970s.

Important Concepts to Remember

- Invariance under a symmetry manifests itself as invariance of the Lagrangean – or, rather, of the action – under the action of the symmetry group on the variables of the theory, $q \rightarrow q' = gq$.
- $\mathbb{Z}_2$ is represented by $\{1, -1\}$, and in general $\mathbb{Z}_N$ can be represented by the Nth roots of unity, $x^N = 1$.
- The regular representation has the property that $D(g_1)|g_2\rangle = |g_1 g_2\rangle$.
- Equivalent representations are related by a similarity transformation or, equivalently, by a change of basis.
- Reducible representations have invariant subspaces, leading to block diagonal matrices. Irreducible representations do not have them.
- Lie groups depend on continuous parameters, and their elements can be written in terms of the generators in exponential form, $g = e^{i\alpha^a X_a}$. The simplest case is the Abelian Lie group, $U(1)$, for which $g = e^{i\alpha}$.
- For Lie groups, the generators obey a Lie algebra defined by the commutator, $[X_a, X_b] = if_{ab}{}^c X_c$.
- The adjoint representation of a Lie group is defined by $(T_a)_b{}^c = -if_{ab}{}^c$ and has the dimension equal to the number of independent generators.
- The trace in the adjoint representation gives a metric on the group, $\mathrm{Tr}[T_a T_b] = \lambda \delta_{ab}$.

Further Reading

For an easy to read introduction to group theory, see for instance the group theory book for particle physicists by Georgi [2].

Exercises

(1)   What are the symmetries of the Lagrangean ($q, \tilde{q} \in \mathbb{C}$)

$$L = \frac{m}{2}(|\dot{q}|^2 + |\dot{\tilde{q}}|^2) - V((q^2 + \tilde{q}^2)^2)?$$ (2.51)

What are the representations of the group in which $q, \tilde{q}$ belong?

(2)   Prove the Jacobi identity (2.42).

(3)   What are the symmetries of the Lagrangean (where q is complex)

$$L = \frac{m}{2}|\dot{q}|^2 - V(q) \,,$$ (2.52)

where the potential is

$$V = \frac{1}{q^3} + \frac{1}{\bar{q}^3} + \frac{1}{q^5} + \frac{1}{\bar{q}^5}?$$ (2.53)

What about if the potential is

$$V = \frac{1}{|q|^3} + \frac{1}{|q|^5}$$ (2.54)

instead? Answer the same question if

$$V = \frac{1}{q^3}.$$ (2.55)

(4)   Consider the matrices

$$A = \begin{pmatrix} 1 & 0 & 0 \\ 0 & 1 & 0 \\ 0 & 0 & 1 \end{pmatrix}, \quad B = \begin{pmatrix} 0 & 0 & 1 \\ 0 & 1 & 0 \\ 1 & 0 & 0 \end{pmatrix}.$$ (2.56)

(a) Do they form a representation of $\mathbb{Z}_2$? Why?
(b) If so, is the representation reducible?
(c) If so, is this a *regular* representation?
(d) If so, is this equivalent to the roots of unity representation?

Examples: The Rotation Group and $SU(2)$

After defining the general theory of groups and Lie algebras as related to symmetries and their representations, we now turn in this chapter to the most important examples, the rotation groups $SO(3)$ and $SU(2)$, as well as some simple generalizations.

3.1 Rotational Invariance

An important symmetry in physics is rotational invariance. It is easy to visualize, and we know simple examples. For instance, a *central* potential, $V = V(|\vec{r}|)$, is rotationally invariant – i.e., invariant under a rotation of coordinates. More generally, a potential that depends only on the distance between particles, $V = V(|\vec{r}_i - \vec{r}_j|)$ (for instance, the potential between electrons, and between them and nuclei, in a material) is also rotationally invariant.

We can say that a rotation is a linear transformation on the coordinates that leaves invariant the lengths $|\Delta \vec{r}_{ij}| = |\vec{r}_i - \vec{r}_j|$. Specifically, a rotation around a point O with position $\vec{r}_O$ is a linear transformation

$$(\vec{r} - \vec{r}_O)'_a = \Lambda_a{}^b (\vec{r} - \vec{r}_O)_b , \tag{3.1}$$

where $a, b = 1, 2, 3$ stand for the three spatial coordinates. It obviously implies also (by writing the same relation for i and for j and subtracting the two)

$$(\vec{r}'_i - \vec{r}'_j)_a = \Lambda_a{}^b (\vec{r}_i - \vec{r}_j)_b. \tag{3.2}$$

Having invariance of the lengths means that we need

$$|\vec{r}_i - \vec{r}_j|^2 = |\vec{r}'_i - \vec{r}'_j|^2 , \tag{3.3}$$

which gives the relation

$$\begin{aligned} |\vec{r}'_i - \vec{r}'_j|^2 &= (\Lambda_a{}^b (\vec{r}_i - \vec{r}_j)_b)(\Lambda^{ac} (\vec{r}_i - \vec{r}_j)_c) = (\vec{r}_i - \vec{r}_j)_b (\Lambda^T \Lambda)^{bc} (\vec{r}_i - \vec{r}_j)_c \\ &= (\vec{r}_i - \vec{r}_j)_b \delta^{bc} (\vec{r}_i - \vec{r}_j)_c , \end{aligned} \tag{3.4}$$

implying the necessary condition for the matrix Λ,

$$\Lambda^T \Lambda = \mathbb{1} \Rightarrow \Lambda^T = \Lambda^{-1}. \tag{3.5}$$

A matrix satisfying this condition is called an *orthogonal* matrix, and these matrices form a group, since if $\Lambda_1^{-1} = \Lambda_1^T$ and $\Lambda_2^{-1} = \Lambda_2^T$, then their product satisfies it as well,

$$(\Lambda_1 \Lambda_2)^{-1} = \Lambda_2^{-1} \Lambda_1^{-1} = \Lambda_2^T \Lambda_1^T = (\Lambda_1 \Lambda_2)^T. \tag{3.6}$$

The group of orthogonal matrices is called the orthogonal group, $O(3)$ in the case of 3×3 matrices acting on a three-dimensional space.

But, using that $\det(AB) = \det A \det B$ and $\det A^T = \det A$, we obtain that

$$\det(\Lambda^T \Lambda) = \det \Lambda \det \Lambda^T = (\det \Lambda)^2$$
$$= \det \mathbb{1} = 1, \tag{3.7}$$

so that

$$\det \Lambda = \pm 1. \tag{3.8}$$

That means that $O(3)$ contains also the element $-\mathbb{1}$, which is spatial parity – i.e., reflection in all directions $-x'_a = -x_a$, or

$$\vec{r}' = -\vec{r}. \tag{3.9}$$

But this element is not connected continuously with the identity, like it should be the case for a Lie group, where a group element should be written like $g = e^{i\alpha^a T_a}$, with $g(\alpha^a = 0) = \mathbb{1}$. To obtain such a group, we must eliminate the spatial parity. We can easily do that by imposing the *special* condition

$$\det \Lambda = +1 , \tag{3.10}$$

which also respects the group property (if $\det \Lambda_1 = \det \Lambda_2 = 1$, then $\det \Lambda_1 \Lambda_2 = 1$) and defines the *special orthogonal group*, $SO(3)$. Then $SO(3) = O(3)/\mathbb{Z}_2$, where the division by the group $\mathbb{Z}_2$ (composed of the identity and the parity – i.e., multiplication by -1) means that we identify elements related by parity. In other words, $O(3)$ contains two copies of $SO(3)$.

Note that everything that was said generalizes trivially to n dimensions, since nothing depended on the dimensionality. Thus, we define as $O(n)$ the group of orthogonal matrices in n dimensions, and as $SO(n)$ the (Lie) group of special orthogonal matrices in n dimensions – i.e., with $\det \Lambda = +1$.

We denote by $so(n)$ the Lie algebra of $SO(n)$. An element $g \in SO(n)$ can be represented by a orthogonal matrices acting on an n dimensional space $\vec{r}$.

3.2 Example: *SO*(2)

In our analysis, we will start with the simplest case, namely the Abelian group $SO(2)$. It can be represented as 2×2 matrices acting on a two-dimensional space. As is well known, the matrix representing a rotation by an angle θ is

$$M = \begin{pmatrix} \cos\theta & -\sin\theta \\ \sin\theta & \cos\theta \end{pmatrix} , \tag{3.11}$$

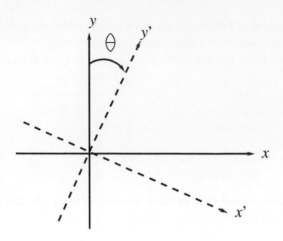

Figure 3.1 Rotation in a plane.

and it acts on a two-dimensional (column) vector space $\begin{pmatrix} x \\ y \end{pmatrix}$, corresponding to the components of $\vec{r} - \vec{r}_O$, (x, y), in a plane (in two dimensions), as in Figure 3.1, giving

$$\begin{aligned} x' &= x\cos\theta - y\sin\theta \\ y' &= x\sin\theta + y\cos\theta. \end{aligned} \tag{3.12}$$

As we can easily check, we indeed have invariance of the length – i.e.,

$$x'^2 + y'^2 = x^2 + y^2. \tag{3.13}$$

The group is in fact Abelian, as we can easily check, since $M(\theta)M(\phi) = M(\phi)M(\theta)$, $\forall \theta, \phi$. But rather, as we mentioned, there is a unique Abelian Lie group, so $SO(2)$ must be isomorphic to $U(1)$, as we will now show. As we said in the last chapter, we can define $M \in U(1)$ as a phase,

$$M = e^{i\theta} = \cos\theta + i\sin\theta \,, \tag{3.14}$$

acting on a complex space, $z = x + iy$, by simple multiplication, i.e.,

$$z \to Mz = x\cos\theta - y\sin\theta + i(x\sin\theta + y\cos\theta) \equiv z' = x' + iy'. \tag{3.15}$$

We see that, indeed, in this way we obtain the $SO(2)$ rotation (3.12) acting on the plane coordinates x, y.

We note that $U(1) = SO(2)$ is a subgroup of every compact Lie group. In particular, for $SO(n)$, $n \geq 2$, it is easy to see since we can pick a rotation of the n coordinates that only rotates two of them and leaves the other $n - 2$ ones invariant.

3.3 $SO(3)$ and Its General Parametrization

We now describe a general rotation in three-dimensional space – i.e., a parametrization of a general group element $g \in SO(3)$. It can be described by a usual planar, i.e., $SO(2)$, rotation of angle ϕ around a fixed axis z. In other words, in the coordinate system where we have rotated $\hat{z}$ to align it with the third direction,

$$\begin{pmatrix} 0 \\ 0 \\ 1 \end{pmatrix} = e_z,$$ (3.16)

the rotation, which leaves invariant z, is

$$R_z(\phi) = \begin{pmatrix} \cos\phi & -\sin\phi & 0 \\ \sin\phi & \cos\phi & 0 \\ 0 & 0 & 1 \end{pmatrix}.$$ (3.17)

But a general axis defined by a vector $\hat{z}$ in a fixed coordinate system with $\hat{1}, \hat{2}, \hat{3}$ orthonormal directions is defined by two angles, θ and ψ. Here θ is the angle made by the arbitrary $\hat{z}$ with the fixed $\hat{3}$. By projecting $\hat{z}$ into the plane of $\hat{1}$ and $\hat{2}$, $\hat{z}$ will make the angle $\pi/2 - \theta$ with it. And the projection will make the angle ψ with $\hat{2}$ (so $\pi/2 - \psi$ with $\hat{1}$); see Figure 3.2. Then ϕ, θ, ψ are called the *Euler angles*.

Then, if we first rotate by ψ around the $\hat{3}$ axis, to align the projection of $\hat{z}$ with $\hat{2}$, and then rotate around the direction $\hat{1}$ by θ, we align $\hat{z}$ with $\hat{3}$, so we can finally rotate by ϕ around it (see Figure 3.2). All in all, the parametrization of a general element $g \in SO(3)$ by the three Euler angles is

$$\begin{aligned} g(\phi, \theta, \psi) &= g(\phi)g(\theta)g(\psi) = R_3(\phi)R_1(\theta)R_3(\psi) \\ &= \begin{pmatrix} \cos\phi & -\sin\phi & 0 \\ \sin\phi & \cos\phi & 0 \\ 0 & 0 & 1 \end{pmatrix} \begin{pmatrix} 1 & 0 & 0 \\ 0 & \cos\theta & -\sin\theta \\ 0 & \sin\theta & \cos\theta \end{pmatrix} \begin{pmatrix} \cos\psi & -\sin\psi & 0 \\ \sin\psi & \cos\psi & 0 \\ 0 & 0 & 1 \end{pmatrix}. \end{aligned}$$ (3.18)

Note that, considering the most general axis $\hat{z}$ in three-dimensional space, we see that the range of ψ is $[0, 2\pi)$, and the range of θ is $[0, \pi)$. Then the range of the rotation in a plane is, as always, $\phi \in [0, 2\pi)$.

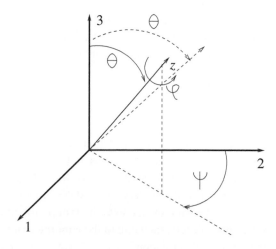

Euler angles parametrizing a rotation in three-dimensional space. Two angles, θ and ψ, relate to the Cartesian coordinate system, and one, ϕ, is a rotation around the axis itself.

From this parametrization, we see that there are three independent $U(1) = SO(2)$ subgroups for $SO(3)$ for the three Euler angles. We could have understood that there are three subgroups, since an $SO(2)$ subgroup of $SO(3)$ means that we pick a coordinate among the three and leave it invariant, rotating only the other two, and there are three ways to do that.

We can also easily write an $SO(3)$ invariant Lagrangean as

$$L = \sum_i m_i \frac{\dot{\vec{r}}_i^2}{2} - V(|\vec{r}_i - \vec{r}_j|) = \sum_i m_i \left[\frac{d}{dt}(\vec{r}_i - \vec{r}_O)\right]^2 - V((\vec{r}_i - \vec{r}_j)^2). \qquad (3.19)$$

This form makes manifest its invariance, since we have seen that $(\vec{r}_i - \vec{r}_j)^2$ is invariant, and so is its time derivative in the first term.

3.4 Isomorphism of $SO(3)$ with $SU(2)$ (Modulo $\mathbb{Z}_2$)

We consider the group of unitary matrices – i.e., matrices satisfying $U^\dagger = U^{-1}$. Indeed, as for $O(n)$, unitary matrices form a group, since if $U_1^\dagger = U_1^{-1}$ and $U_2^\dagger = U_2^{-1}$, then

$$(U_1 U_2)^\dagger = U_2^\dagger U_1^\dagger = U_2^{-1} U_1^{-1} = (U_1 U_2)^{-1}. \qquad (3.20)$$

This is the group $U(n)$ for $n \times n$ matrices acting on an n dimensional complex vector space. As before, we find that the determinant of unitary matrices satisfies a relation

$$\det(U^\dagger U) = \det \mathbb{1} = 1 \Rightarrow (\det U)^* \det U = 1 \Rightarrow \det U = e^{i\alpha}, \qquad (3.21)$$

so the determinant of a unitary matrix forms a $U(1)$ group.

We can then restrict to the group of matrices of $\det \Lambda = +1$, which again respects the group property (if $\det U_1 = \det U_2 = 1$, then $\det(U_1 U_2) = 1$), so considering unitary and special (of determinant 1) matrices, we obtain the group $SU(n)$. Unlike before, however, now, up to global issues, $U(n) \simeq SU(n) \times U(1)$, where $U(1)$ is the determinant of a unitary matrix. A more precise relation is $U(n) = (SU(n) \times U(1))/\mathbb{Z}_n$; however, we will not define what that means here. The relation is more precise in the Lie algebra, namely

$$u(n) = su(n) \times u(1). \qquad (3.22)$$

From now on, we will always use lowercase letters to refer to the Lie algebra and upper case to refer to the group.

Considering now the case $n = 2$, the group of $n \times n$ unitary special matrices, we will show by construction that it is the same as $SO(3)$, except for a global issue: $SU(2)$ contains two copies, or "winds twice" around $SO(3)$. Indeed, note that the element $g = -\mathbb{1}_{2\times 2}$ is an element of $SU(2)$, as it is unitary and of determinant 1. On the other hand, $-\mathbb{1}_{3\times 3}$ is not an element of $SO(3)$. Moreover, due to the group property, $-\mathbb{1}$ times any $SU(2)$ matrix is still an $SU(2)$ matrix, corresponding to the winding twice around $SO(3)$.

We can map an $SO(2)$ rotation into a $U(1)$ rotation, which is just a phase, as we saw. But we can embed the $SO(2)$ in the $SO(3)$, and the $U(1)$ in $SU(2)$, by considering a phase and its complex conjugate on the diagonal of the $SU(2)$ matrix. That means that we can map the rotation around the third axis (which is an $SO(2)$ rotation in a plane transverse to the axis) onto a "rotation," specifically a 2×2 dimensional complex matrix, acting on a two complex dimensional space $\begin{pmatrix} z_1 \\ z_2 \end{pmatrix}$ as

$$R_3(\phi) \to \begin{pmatrix} e^{i\frac{\phi}{2}} & 0 \\ 0 & e^{-i\frac{\phi}{2}} \end{pmatrix} \equiv \tilde{R}_3(\phi). \tag{3.23}$$

We can check that the matrix is unitary,

$$\tilde{R}_3^\dagger = \begin{pmatrix} e^{-i\frac{\phi}{2}} & 0 \\ 0 & e^{i\frac{\phi}{2}} \end{pmatrix} = \tilde{R}_3^{-1}. \tag{3.24}$$

We also see that it is special – $\det \tilde{R}_3(\phi) = e^{i\frac{\phi}{2}} e^{-i\frac{\phi}{2}} = 1$ – so, indeed, it belongs to $SU(2)$.

On the other hand, if we fixed this rotation, now we cannot represent a rotation around another axis as the same $U(1)$ rotation in the space of 2×2 complex matrices. Instead, the rotation around the first axis is now mapped to an abstract "rotation" in the two complex dimensional space by

$$R_1(\theta) \to \begin{pmatrix} \cos\frac{\theta}{2} & i\sin\frac{\theta}{2} \\ i\sin\frac{\theta}{2} & \cos\frac{\theta}{2} \end{pmatrix} \equiv \tilde{R}_1(\theta). \tag{3.25}$$

Then we can check that this matrix is also unitary,

$$\tilde{R}_1^\dagger = \begin{pmatrix} \cos\frac{\theta}{2} & -i\sin\frac{\theta}{2} \\ -i\sin\frac{\theta}{2} & \cos\frac{\theta}{2} \end{pmatrix} = \tilde{R}_1^{-1}, \tag{3.26}$$

and also that it is special, $\det \tilde{R}_1 = \cos^2\theta/2 + \sin^2\theta/2 = 1$.

Then we find that the general element of $SU(2)$, parametrized by the Euler angles, is

$$g(\phi, \theta, \psi) \to \tilde{R}_3(\phi)\tilde{R}_1(\theta)\tilde{R}_3(\psi)$$
$$= \begin{pmatrix} \cos\frac{\theta}{2}e^{i\frac{\phi+\psi}{2}} & i\sin\frac{\theta}{2}e^{i\frac{\phi-\psi}{2}} \\ i\sin\frac{\theta}{2}e^{-i\frac{\phi-\psi}{2}} & \cos\frac{\theta}{2}e^{-i\frac{\phi+\psi}{2}} \end{pmatrix}. \tag{3.27}$$

Like in the case of $SO(3)$, each of the three elements corresponds to three different $U(1)$ subgroups of $SU(2)$, though the way they embed is now less obvious.

However, as we said, the element $g = -\,\mathbb{1}_{2\times2}$ must belong to $SU(2)$, and it multiplies another copy of $SO(3)$. Therefore, the map from $SO(3)$ to $SU(2)$ is really

$$g(\phi, \theta, \psi) \to \pm \begin{pmatrix} \cos\frac{\theta}{2}e^{i\frac{\phi+\psi}{2}} & i\sin\frac{\theta}{2}e^{i\frac{\phi-\psi}{2}} \\ i\sin\frac{\theta}{2}e^{-i\frac{\phi-\psi}{2}} & \cos\frac{\theta}{2}e^{-i\frac{\phi+\psi}{2}} \end{pmatrix}, \tag{3.28}$$

if we consider the same ranges, $\theta \in [0, \pi), \phi \in [0, 2\pi), \psi \in [0, 2\pi)$. But since $\cos(\pi+\alpha) = -\cos\alpha$, $\sin(\pi+\alpha) = -\sin\alpha$, we can consider the fact that

$$\tilde{R}_3(\phi)\tilde{R}_1(\theta + 2\pi)\tilde{R}_3(\psi) = -\tilde{R}_3(\phi)\tilde{R}_1(\theta)\tilde{R}_3(\psi), \tag{3.29}$$

so just adding one more copy of the range in θ (thus winding twice) is enough to cover $SU(2)$.

There are three parameters α^a that can be mapped to (ϕ, θ, ψ), which means that there are three generators for the Lie algebra. Since we have a map between $SO(3)$ and $SU(2)$, the Lie algebras must be the same,

$$su(2) = so(3). \tag{3.30}$$

But three nontrivial elements can be mapped into a 3×3 antisymmetric matrix J_{ab} of generators (thus each element can be represented itself by a matrix), $a, b = 1, 2, 3$. But such a matrix can also be related to a vector J_a by

$$J_{ab} = \epsilon_{abc} J_c. \tag{3.31}$$

Such a set of generators J_c is the set of angular momenta used in quantum mechanics, which, as we know, satisfy an algebra, defined as the $su(2)$ algebra,

$$[J_a, J_b] = i\epsilon_{abc} J_c. \tag{3.32}$$

We note that this is of the form of a general Lie algebra, $[T_a, T_b] = if_{ab}{}^c T_c$, but with structure constants $f_{abc} = \epsilon_{abc}$ (only in this case, the structure constants, totally antisymmetric, form the Levi-Civitta tensor).

3.5 Construction of Representations of $SU(2)$ and Invariant Theories

The construction of representations of $SU(2)$, of spin j, is done as usual in quantum mechanics, which we will review here. By a similarity transformation, equivalent to a change of basis for the representation, $|i\rangle \to S|i\rangle$, we can diagonalize J_3. Note that we will use the same *bra* and *ket* notation for states in a representation of $SU(2)$ as used in quantum mechanics.

Consider then the highest value for the eigenvalue of J_3 in the representation, called j. Note that this only supposes that the representation is not infinite dimensional (in which case, it could be that there is no highest value for the eigenvalue of J_3), and we want only to define the finite dimensional representations. Then this maximal value j defines the representation, and the maximal J_3 eigenvalue state is defined by it and other labels, collectively called α, i.e.,

$$J_3 |j, \alpha\rangle = j|j, \alpha\rangle. \tag{3.33}$$

Moreover, define the states such that they are orthonormal, i.e.,

$$\langle j, \alpha | j, \beta \rangle = \delta_{\alpha\beta}. \tag{3.34}$$

Then, we denote

$$J^{\pm} = \frac{J_1 \pm iJ_2}{\sqrt{2}}, \tag{3.35}$$

called raising and lowering operators, respectively. The name is not incidental, since $J^{\pm}$ act similarly to $a^{\dagger}$ and a, respectively, on the states, and J_3 acts similarly to $N = a^{\dagger}a$. Indeed, the Lie algebra becomes

$$[J_3, J^\pm] = \pm J^\pm, \quad [J^+, J^-] = J_3 , \tag{3.36}$$

similar to the algebra of $a, a^\dagger, N$.

Define as m the eigenvalue of J_3, i.e.,

$$J_3|j, m\rangle = m|j, m\rangle. \tag{3.37}$$

Then, the commutation relations imply that

$$J_3(J^\pm|j, m\rangle) = J^\pm(J_3|j, m\rangle) \pm J^\pm|j, m\rangle = (m \pm 1)(J^\pm|j, m\rangle) , \tag{3.38}$$

so J^+ raises the eigenvalue of J_3 (from this point acting as $a^\dagger$ with respect to $a^\dagger a$), whereas J^- lowers the eigenvalue of J_3 (thus acting like a with respect to $a^\dagger a$).

On the other hand, if j is the maximum eigenvalue of J_3 and J^+ raises the eigenvalue, it means on this maximal state that it must give zero,

$$J^+|j, \alpha\rangle = 0. \tag{3.39}$$

This relation is then the equivalent of $a|0\rangle = 0$, (so from this point of view, J^+ is equivalent to a) and, as for the harmonic oscillator, we build the states by acting with $a^\dagger$. We now build the (states of the) representation by repeatedly acting with J^- on this $|j, \alpha\rangle$ state.

This simple way of representing groups can be generalized to any group. For any group, we can represent the generators in terms of some a and $a^\dagger$ values, and then find representations by defining a "vacuum" state $|0\rangle$ such that $a|0\rangle = 0$, and then define the representation by acting with $a^\dagger$ on it many times.

As usual, we find (from the fact that for every m we must also have $-m$) that j is semi-integer, $j = l/2, l \in \mathbb{Z}$. These are the *spin j representations of SU(2)*, the only finite dimensional representations. We, moreover, find the matrix elements of the generators as

$$\begin{aligned}
\langle j, m'|J_3|j, m\rangle &= m\delta_{mm'} \\
\langle j, m'|J^+|j, m\rangle &= \sqrt{(j+m+1)(j-m)/2}\delta_{m', m+1} \\
\langle j, m'|J^-|j, m\rangle &= \sqrt{(j+m)(j-m+1)/2}\delta_{m', m-1}.
\end{aligned} \tag{3.40}$$

We denote the generators in the spin j representation by J_a^j and their matrix elements by

$$[J_a^j]_{m'm} = \langle j, m'|J_a^j|j, m\rangle. \tag{3.41}$$

Specializing to the minimal spin $j = 1/2$ representation, we obtain

$$\begin{aligned}
J_1^{1/2} &= \frac{1}{2}\begin{pmatrix} 0 & 1 \\ 1 & 0 \end{pmatrix} = \frac{1}{2}\sigma_1 \\
J_2^{1/2} &= \frac{1}{2}\begin{pmatrix} 0 & -i \\ i & 0 \end{pmatrix} = \frac{1}{2}\sigma_2 \\
J_3^{1/2} &= \frac{1}{2}\begin{pmatrix} 1 & 0 \\ 0 & -1 \end{pmatrix} = \frac{1}{2}\sigma_3.
\end{aligned} \tag{3.42}$$

Here σ_i, $i = 1, 2, 3$ are the Pauli matrices, and $J_i^{1/2} = \frac{1}{2}\sigma_i$ are the generators of $SU(2)$ in this spin 1/2 representation, in terms of 2×2 matrices. However, this is nothing but the fundamental representation of $SU(2)$, since we have defined the group as 2×2 unitary and special matrices, with generators (in the exponent) J_i.

Indeed, the σ_i are Hermitean, $\sigma_i^\dagger = \sigma$, as we can easily check, and then we know that the group matrices $g = e^{i\vec{\alpha}\cdot\vec{J}}$ will be unitary, $g^\dagger = g^{-1}$. Moreover, an important matrix identity says that

$$\det M = e^{\mathrm{Tr}\ln M}, \tag{3.43}$$

and if $M = e^A$, we obtain

$$\det e^A = e^{\mathrm{Tr}A}. \tag{3.44}$$

Then if $g = e^{i\vec{\alpha}\cdot\vec{J}}$ has determinant 1, $\vec{J}$ must be traceless. Indeed, as we can check $\mathrm{Tr}\,\sigma_i = 0$.

Moreover, the Pauli matrices satisfy (as we can easily check)

$$\sigma_a\sigma_b = \delta_{ab} + i\epsilon_{abc}J_c, \tag{3.45}$$

which means that

$$[\sigma_a, \sigma_b] = 2i\epsilon_{abc}\sigma_c \Rightarrow \left[\frac{1}{2}\sigma_a, \frac{1}{2}\sigma_b\right] = i\epsilon_{abc}\frac{1}{2}J_c, \tag{3.46}$$

so indeed the $su(2)$ Lie algebra is satisfied.

Further, we then have

$$(\alpha^a\sigma_a)^2 = \alpha^a\alpha^b\sigma_a\sigma_b = \alpha^a\alpha^b\sigma_{(a}\sigma_{b)} = \alpha^a\alpha^b\delta_{ab} = \vec{\alpha}^2, \tag{3.47}$$

which means that

$$(\alpha^a\sigma_a)^{2n} = (\vec{\alpha}^2)^n; \quad (\alpha^a\sigma_a)^{2n+1} = \alpha^a\sigma_a(\vec{\alpha}^2)^n, \tag{3.48}$$

so finally

$$g = e^{i\frac{\vec{\alpha}\cdot\vec{\sigma}}{2}} = \sum_{n\geq 0}\frac{1}{n!}\left(\frac{i\alpha^a\sigma_a}{2}\right)^n = \cos\left|\frac{\vec{\alpha}}{2}\right| + i\frac{\alpha\cdot\sigma}{|\vec{\alpha}|}\sin\left|\frac{\alpha}{2}\right|. \tag{3.49}$$

We see that, while the generators of $SU(2)$ are σ_i only, the group elements are decomposed into σ_i and $\mathbb{1}$. This parametrization of $SU(2)$ by $\vec{\alpha}$ can be mapped into the one by Euler angles (ϕ, θ, ψ) defined before, but we leave it as an exercise.

After the fundamental representation, which was the spin 1/2 one, consider the adjoint representation, defined by

$$(T_a)_{bc} = -if_{abc} = -i\epsilon_{abc}, \tag{3.50}$$

which gives

$$T_1 = -i\begin{pmatrix} 0 & 0 & 0 \\ 0 & 0 & 1 \\ 0 & -1 & 0 \end{pmatrix}, etc. \tag{3.51}$$

One can find that this is in fact the same three-dimensional spin 1 representation (it is left as an exercise).

$SU(2)$ Invariant Theory

We can write an $SU(2)$ invariant Lagrangean as before, except, unlike the $SO(3)$ case, the invariance is an abstract one, not one of spacetime. Construct the two-dimensional complex vector

$$q = \begin{pmatrix} q_1 \\ q_2 \end{pmatrix}, \quad q_1 = x_1 + iy_1, \quad q_2 = x_2 + iy_2, \quad ; x_1, x_2, y_1, y_2 \in \mathbb{R}, \tag{3.52}$$

for which

$$\begin{aligned} q^\dagger q &= |q|^2 = |q_1|^2 + |q_2|^2 = x_1^2 + y_1^2 + x_2^2 + y_2^2 \\ \dot{q}^\dagger \dot{q} &= |\dot{q}|^2 = |\dot{q}_1|^2 + |\dot{q}_2|^2 = \dot{x}_1^2 + \dot{y}_1^2 + \dot{x}_2^2 + \dot{y}_2^2. \end{aligned} \tag{3.53}$$

Then the Lagrangean

$$L = m\frac{|\dot{q}|^2}{2} - V(|q|^2), \tag{3.54}$$

which, for instance, for $V(x) = kx^2/2$ is the Lagrangean of *two planar harmonic oscillators* (with plane coordinates x, y), is *SU*(2) invariant. Indeed,

$$\begin{aligned} q \to q' = gq &\Rightarrow |q|^2 \to |q'|^2 = |gq|^2 = q^\dagger g^\dagger gq = q^\dagger q = |q|^2; \\ &|\dot{q}|^2 \to |\dot{q}'|^2 = |g\dot{q}|^2 = \dot{q}^\dagger g^\dagger g\dot{q} = |\dot{q}|^2. \end{aligned} \tag{3.55}$$

This invariance is not like *SO*(3), an invariance of spacetime, but rather it is an abstract (internal) invariance, relating two different degrees of freedom, specifically the two harmonic oscillators.

Classical Lie Groups

The classical Lie groups can be divided in four series:

- the A_N series, namely $SU(N+1)$
- the B_N series, namely $SO(2N+1)$
- the C_N series, namely the group of *symplectic* matrices, $Sp(2N)$ (Their definition will be presented later.)
- the D_N series, namely $SO(2N)$

Besides them, we have the *exceptional groups*, the discrete E series, E_6, E_7, E_8, relevant for some grand unification models, in particular string theoretic ones, and the exceptional groups F_4 and G_2, with little applications (just some very abstract string theory ones). These exceptional groups will not be defined here.

Representations of the classical groups (or any groups, in fact), can be found by identifying $SU(2)$-type generators as part of the Lie algebra and acting with J^-'s on a "vacuum" type state (of maximal J_3).

Important Concepts to Remember

- The orthogonal group $O(N)$, made up of orthogonal (real) matrices, $\Lambda^T = \Lambda^{-1}$, leaves invariant the distances between points in an N dimensional (Euclidean) space. Restricting to the special matrices, of $\det \Lambda = 1$, we obtain the Lie group $SO(N)$.
- $SO(2)$, the group of rotations in a plane is isomorphic to the unique Abelian group $U(1)$. This group is a subgroup of any compact Lie group.

- $SO(3)$ can be parametrized by three Euler angles as $g = R_3(\phi)R_1(\theta)R_3(\psi)$.
- $U(N)$ is the group of unitary complex matrices and $SU(N)$ of special (determinant 1) unitary matrices. $u(N) = su(N) \times u(1)$.
- $SU(2)$ and $SO(3)$ are isomorphic, modulo global issues, so $su(2) = so(3)$, and the Lie algebra is $[J_a, J_b] = i\epsilon_{abc}J_c$.
- Representations of $SU(2)$ can be obtained from a state $|j, \alpha\rangle$ of maximal J_3 eigenvalue, j, defining the representation, and satisfying $J^+|j, \alpha\rangle = 0$, by repeatedly acting with J^-.
- The commutation relations of $J^\pm, J_3$, and analogous to $a^\dagger, a$, and $a^\dagger a$ for the harmonic oscillator, and $|j, \alpha\rangle$ to the vacuum $|0\rangle$.
- The fundamental representation of $SU(2)$ is the spin 1/2 representation, the adjoint one is the spin 1 representation, and the general representation is spin $j = l/2, l \in \mathbb{Z}$.
- The classical Lie groups are $A_N = SU(N+1), B_N = SO(2N+1), C_N = Sp(2N), D_N = SO(2N)$ and the exceptional E_6, E_7, E_8, F_4, G_2.

Further Reading

See, for instance, the group theory book for particle phyisicists by Georgi [2].

Exercises

(1) Write explicitly the matrices of the spin 1 and of the adjoint representations of $SU(2)$, and relate them.

(2) Write explicitly the 2×2 matrices $g = e^{i\frac{\vec{\alpha} \cdot \vec{\sigma}}{2}}$, and compare with $g(\phi, \theta, \psi)$ for $SO(3)$, to find an explicit map between α^i and the Euler angles (ϕ, θ, ψ).

(3) For a general matrix

$$A = \begin{pmatrix} a & b \\ c & d \end{pmatrix} \tag{3.56}$$

belonging to $SU(2)$ – i.e., $A^\dagger = A^{-1}$ and $\det A = 1$ – find the Euler angles in terms of a, b, c, d.

(4) Consider the complex fields q_1, q_2 forming a $q = \begin{pmatrix} q_1 \\ q_2 \end{pmatrix}$ and $A = \sum_{i=1}^{3} a_i \sigma_i, a_i \in \mathbb{C}$, with σ_i the Pauli matrices, and the Lagrangean for them

$$L = \dot{q}^\dagger \dot{q} + \det e^A - q^\dagger A q. \tag{3.57}$$

Prove that it is $SU(2)$ invariant.

(5)   Show that the spin j representation of $SU(2)$, with

$$\langle j, m'|J_3|j, m\rangle = m\delta_{mm'}$$
$$\langle j, m'|J^+|j, m\rangle = \sqrt{(j+m+1)(j-m)/2}\,\delta_{m',m+1}$$
$$\langle j, m'|J^-|j, m\rangle = \sqrt{(j+m)(j-m+1)/2}\,\delta_{m',m-1}\,, \qquad (3.58)$$

is indeed one of fixed eigenvalue for the total angular momentum squared $\vec{J}^2$.

4 Review of Special Relativity: Lorentz Tensors

In this chapter, we will review relevant concepts from special relativity, with a view of putting classical mechanics into a relativistic formulation. In particular, we will define Lorentz covariance and Lorentz tensors. Indeed, most of the classical field theory that we will describe in the following chapter is relativistic and will be described in terms of Lorentz covariant objects.

4.1 Special Relativity and the Lorentz Group

Special relativity starts with the experimental observation of the constancy of the speed of light c in all inertial (moving at constant speed relative to one another) reference frames. As a result, we can fix it to be $c = 1$, which amounts to a choice of dimensions, so that $[L] = [T]$.

Then the statement of special relativity is that the invariance of spacetime is not under independent changes of space and time but, rather, changes in *spacetime* $(\vec{x}, t) = (x^i, t) \equiv x^\mu$, $\mu = 0, 1, 2, 3$, which leave invariant the "length," or *spacetime interval*

$$\Delta s_{ij}^2 \equiv \Delta x_{ij}^2 - \Delta t_{ij}^2 = (\vec{r}_i - \vec{r}_j)^2 - (t_i - t_j)^2 \equiv (x_i^\mu - x_j^\mu)\eta_{\mu\nu}(x_i^\nu - x_j^\nu) \equiv (x_i^\mu - x_j^\mu)^2. \quad (4.1)$$

Here we have defined the *Minkowski metric*

$$\eta_{\mu\nu} = \begin{pmatrix} -1 & 0 & 0 & 0 \\ 0 & 1 & 0 & 0 \\ 0 & 0 & 1 & 0 \\ 0 & 0 & 0 & 1 \end{pmatrix} \equiv M, \quad (4.2)$$

which, as a matrix, is denoted by M.

Note that there is a convention involved here that the Minkowski metric is "mostly plus". The other possible convention is for $\Delta s^2 \to -\Delta s^2$, or $\eta_{\mu\nu} \to -\eta_{\mu\nu}$, the "mostly minus" convention. Different people use different conventions, so it is important to realize the convention one is working in, otherwise you get annoying minus signs wrong. I prefer the mostly plus convention, since then rotation to Euclidean space (a useful thing in quantum field theory) is simple, by just replacing $t \to it$.

Then the Lorentz transformations, the changes in spacetime that leave invariant Δs^2, are like generalizations of rotations – i.e., linear transformation of coordinate differences,

$$(x - x_O)'^\mu = \Lambda^\mu_{\ \nu}(x - x_O)^\nu \Rightarrow x_1'^\mu - x_2'^\mu = \Lambda^\mu_{\ \nu}(x_1^\nu - x_2^\nu). \quad (4.3)$$

Under these transformations of spacetime, we need invariance of the "lengths," or spacetime intervals, i.e.,

$$(x_1'^{\mu} - x_2'^{\mu})^2 = (x_1^{\mu} - x_2^{\mu})^2 \,, \tag{4.4}$$

which implies

$$\Lambda^{\mu}{}_{\nu}(x_1^{\nu} - x_2^{\nu})\eta_{\mu\rho}\Lambda^{\rho}{}_{\sigma}(x_1^{\sigma} - x_2^{\sigma}) = (x_1^{\nu} - x_2^{\nu})\eta_{\mu\sigma}(x_1^{\sigma} - x_2^{\sigma}) \,, \tag{4.5}$$

for any x_1, x_2, which means that the matrices Λ satisfy the condition

$$\Lambda^{\mu}{}_{\nu}\eta_{\mu\rho}\Lambda^{\rho}{}_{\sigma} = \eta_{\nu\sigma} \,, \tag{4.6}$$

or as a matrix relation

$$\Lambda^T M \Lambda = M. \tag{4.7}$$

These transformations form a group, like in the case of spatial rotations, since if Λ_1, Λ_2 satisfy the above relation, then so does $\Lambda_1 \Lambda_2$:

$$(\Lambda_1 \Lambda_2)^T M (\Lambda_1 \Lambda_2) = \Lambda_2^T \Lambda_1^T M \Lambda_1 \Lambda_2 = \Lambda_2^T M \Lambda_2 = M. \tag{4.8}$$

The group of Lorentz transformations is a non-compact version of $O(4)$ called $O(1,3)$, the *Lorentz group*.

The fact that the group is non-compact can be understood physically by saying that we can increase the momentum of a particle indefinitely by a Lorentz boost, so the group of boosts does not admit a compact parametrization.

We can also understand it intuitively in the following mathematical way. We can get a compact space, namely a sphere S^n, from the group manifold $SO(n+1)$, if we "divide" by $SO(n)$,

$$S^n = \frac{SO(n+1)}{SO(n)}. \tag{4.9}$$

Specifically, this means identifying elements of $SO(n+1)$ by elements of $SO(n)$ (formally, constructing "equivalence classes" under $SO(n)$, such that the ratio is a "coset space"). This is so, since in $(n+1)$−dimensional Euclidean space we can rotate a vector by a general, $SO(n+1)$ rotation, but we must consider the fact that the vector is left invariant by rotations around it – i.e., $SO(n)$ rotations. Thus, the sphere S^n is identified with the group of $SO(n+1)$ rotations, modulo $SO(n)$ rotations. In the case $n = 2$ from last chapter (Figure 3.2), the usual two-sphere equals $SO(3)$, defined by the Euler angles θ, ψ, ϕ, modulo the $SO(2)$ of rotations around the axis itself – i.e., ϕ. Indeed, the two-sphere is characterized by the angles θ and ψ.

Moreover, for the case of $SO(4)$, we can prove the identity $SO(4) \simeq SO(3) \times SO(3)$ (it is an isomorphism valid locally), so that

$$S^3 \simeq SO(3) \,, \tag{4.10}$$

and since S^3 is compact, it follows that $SO(3)$ is a compact space. Changing the locally Euclidean metric on the sphere S^3 into a three-dimensional locally Minkowski metric turns it into a hyperbolic space, which is non-compact, since we can have arbitrarily large null directions, with $\Delta s^2 = \Delta \vec{x}^2 - \Delta t^2 = 0$, and $\Delta \vec{x}$ as large as we want. That means that $SO(1,2)$ is non-compact, and similarly for $SO(1,3)$, the Lorentz group in 3+1 dimensions.

More informally, we can say that the Lorentz group $L = O(1,3)$ is the group of transformations that leave invariant the Minkowski metric $\eta_{\mu\nu}$ and is a modification of the group $O(4)$, satisfying

$$\Lambda^T \mathbb{1} \Lambda = \mathbb{1} , \tag{4.11}$$

i.e., leaving invariant the Euclidean metric $\mathbb{1}$.

As in the case of $O(n)$, we can take the determinant of the defining relation and obtain

$$\det(\Lambda^T M \Lambda) = \det \Lambda^T \det M \det \Lambda = -(\det \Lambda)^2$$
$$= \det M = -1 , \tag{4.12}$$

so again we find

$$(\det \Lambda)^2 = 1 \Rightarrow \det \Lambda = \pm 1. \tag{4.13}$$

In order to find the Lie group $SO(1,3)$, we must again consider the constraint

$$\det \Lambda = +1 ; \tag{4.14}$$

however, unlike the case of $SO(n)$, now this is not enough. This is so because if for $O(n)$ we only have overall parity $x^\mu \to -x^\mu$ among the group elements, for $O(1,3)$ we have separately spatial parity $\vec{x} \to -\vec{x}$ and time reversal, $t \to -t$ (both not continuously connected to the unity), as we can easily check. Imposing $\det \Lambda = +1$ still leaves the possibility of having both the spatial parity and the time reversal simultaneously. To eliminate that, we must also impose

$$\Lambda^0_{\ 0} > 0. \tag{4.15}$$

Then we find $SO(1,3)$ as the *proper* Lorentz group, sometimes denoted by L_{++} (for the two + sign conditions).

4.2 Characterizing the Lorentz Group

Among the Lorentz transformations, we certainly have the usual rotations $SO(3)$ (that leave time invariant), which have three generators, J_{ij} or J_i, and which we have described in the previous chapter.

Besides these, however, we have genuinely new transformations: the Lorentz *boosts*, or transformations by a velocity in a given direction. The transformations by $\beta = v/c$ and $\gamma = 1/\sqrt{1 - v^2/c^2}$ in the x direction are

$$t' = \gamma(t - \beta x)$$
$$x' = \gamma(x - \beta t) , \tag{4.16}$$

which means that the Lorentz matrix is

$$\Lambda^{\mu}{}_{\nu} = \begin{pmatrix} \gamma & -\gamma\beta & 0 & 0 \\ -\gamma\beta & \gamma & 0 & 0 \\ 0 & 0 & 1 & 0 \\ 0 & 0 & 0 & 1 \end{pmatrix},$$

(4.17)

when acting on $\begin{pmatrix} t \\ x \\ y \\ z \end{pmatrix}$. We can write similar ones for boosts on y or on z, which means there

are three new generators in the Lorentz group, which we can call K_i, besides the three for rotations J_i, for a total of six generators. Note that since $\gamma^2 - \beta^2\gamma^2 = 1$, we can represent

$$\gamma \equiv \cosh\theta, \quad \gamma\beta \equiv \sinh\theta,$$

(4.18)

and here we can have $\theta \to \infty$, which again proves the statement from before that the Lorentz group is non-compact, since we can increase the momentum indefinitely: we do so by taking $\theta \to \infty$, which directly shows the non-compactness of the manifold of $\Lambda^{\mu}{}_{\nu}$ values.

The six generators of the Lorentz group can be fit into an antisymmetric 4×4 matrix J_{ab}, with $a, b = 0, 1, 2, 3$, J_{ij} and $J_{0i} = K_i$, each element being also a matrix (when we represent the generators as matrices, acting on a vector space). Then we can write the group elements of $SO(1, 3)$ as

$$\Lambda^{\mu}{}_{\nu} = (e^{i\alpha^{ab} J_{ab}})^{\mu}{}_{\nu}.$$

(4.19)

We will see later that the full invariance of Minkowski spacetime includes translations (spatial and temporal), generated by generators $P_a = (P_0, P_i)$. Indeed, these translations $x^{\mu} \to x^{\mu} + a^{\mu}$, also leave invariant the intervals $(x_i^{\mu} - x_j^{\mu})^2$. The full invariance group, comprising of J_{ab} and P_a, is called the *Poincaré group*, and is usually denoted by $ISO(1, 3)$.

4.3 Kinematics of Special Relativity

We can define the *4-velocity* u^{μ} as a derivative with respect to τ, the proper time along the worldline of the particle,

$$u^{\mu} = \frac{dx^{\mu}}{d\tau}.$$

(4.20)

For a timelike interval, $ds^2 < 0$, as along the path of a massive particle, we write

$$ds^2 = -(c^2)d\tau^2.$$

(4.21)

In components, the four-velocity splits into

$$u^0 = \frac{(c)dt}{d\tau} = (c)\gamma; \quad u^i = \frac{dx^i}{d\tau} = \gamma v^i,$$

(4.22)

where $v^i = dx^i/dt$.

We can define also the *four-momentum* p^μ by multiplying with the rest mass m,

$$p^\mu = mu^\mu = (mc\gamma, m\gamma v^i) \equiv (E/c, p^i). \tag{4.23}$$

Then the energy is written in terms of the moving mass $(m\gamma)$ as

$$E = (m\gamma)c^2. \tag{4.24}$$

The invariant made from the four-velocity is

$$u^\mu u_\mu \equiv u^\mu u^\nu \eta_{\mu\nu} = \frac{dx^\mu dx^\nu \eta_{\mu\nu}}{d\tau^2} = \frac{ds^2}{d\tau^2} = -(c^2)1. \tag{4.25}$$

The invariant made from the four-momentum is then

$$\begin{aligned} p^\mu p_\mu &= m^2 u^\mu u_\mu = -m^2(c^2) \\ &= -\frac{E^2}{c^2} + \vec{p}^2\,, \end{aligned} \tag{4.26}$$

which means that

$$E^2 = m^2(c^4) + \vec{p}^2 c^2. \tag{4.27}$$

We can also construct good relativistic derivatives:

We have already used a relativistic *invariant* derivative, $d/d\tau$.

We can construct also a relativistically *covariant* derivative, which transforms under Lorentz transformations,

$$\frac{\partial}{\partial x^\mu} \equiv \partial_\mu. \tag{4.28}$$

A generalization of dx^μ and ∂_μ are relativistically "covariant" (here meaning that they transform according to the relativistic rules) objects called *tensors*.

Objects transforming as dx^μ,

$$dx'^\mu = \frac{\partial x'^\mu}{\partial x^\nu} dx'^\nu\,, \tag{4.29}$$

are called *contravariant tensors*, whereas objects transforming as ∂_μ,

$$\partial'_\mu = \frac{\partial x^\nu}{\partial x'^\mu} \partial_\nu\,, \tag{4.30}$$

are called *covariant tensors*.

For instance, a tensor $T^\mu{}_\nu$, with a covariant index ν and a contravariant index μ, transforms as

$$T'^\mu{}_\nu(x') = \frac{\partial x'^\mu}{\partial x^\rho} \frac{\partial x^\sigma}{\partial x'^\nu} T^\rho{}_\sigma(x). \tag{4.31}$$

Generalizing to a (p, q)-tensor $T^{\mu_1 \ldots \mu_p}_{\nu_1 \ldots \nu_q}$, the transformation is

$$T^{\mu_1 \ldots \mu_p}_{\nu_1 \ldots \nu_q}(x') = \frac{\partial x'^{\mu_1}}{\partial x^{\rho_1}} \cdots \frac{\partial x'^{\mu_p}}{\partial x^{\rho_p}} \frac{\partial x^{\sigma_1}}{\partial x'^{\nu_1}} \cdots \frac{\partial x^{\sigma_q}}{\partial x'^{\nu_q}} T^{\rho_1 \ldots \rho_p}_{\sigma_1 \ldots \sigma_q}. \tag{4.32}$$

The statement of special relativity is then that physical objects transform covariantly under Lorentz transformation – i.e., they form tensors.

Note that, so far, we excluded from the analysis fermions, since they classically have zero value. As we will see later, fermions transform in a different way than tensors (which are bosons) under Lorentz transformations, but since they do not take classical values, we do not care about them now.

4.4 Dynamics of Special Relativity

We need to write a four-vector generalization of Newton's force law. Given our previous kinematics analysis, it must be written as

$$\frac{dp^\mu}{d\tau} = F^\mu \, , \tag{4.33}$$

where we must define the four-force F^μ.

Since special relativity is based on electromagnetism, we will define it for the interaction of a particle with electromagnetism. The Lorentz force law is

$$\vec{F} = \frac{d\vec{p}}{dt} = q(\vec{E} + \vec{v} \times \vec{B}) \, , \tag{4.34}$$

or, in components,

$$\frac{dp_i}{dt} = q(E_i + \epsilon_{ijk} v_j B_k). \tag{4.35}$$

We need to generalize it to four-tensors. Since v_j appears, it is natural to define it as

$$\frac{dp^\mu}{d\tau} = q F^\mu{}_\nu u^\nu. \tag{4.36}$$

Note that when raising or lowering an index due to the Minkowski metric for a spatial index, it makes no difference, but a 0 index gets a minus sign. Here we define the antisymmetric tensor $F_{\mu\nu}$, called the electromagnetic *field strength* tensor as follows. First,

$$F^0{}_i = F^{0i} = +E_i = -F^{i0} = F^i{}_0 \, , \tag{4.37}$$

and then

$$F^i{}_j = \epsilon_{ijk} B_k. \tag{4.38}$$

Remember that the fields $\vec{E}$ and $\vec{B}$ are obtained from the electric potential ϕ and the vector potential $\vec{A}$ by

$$\vec{E} = -\partial_0 \vec{A} - \vec{\nabla}\phi; \quad \vec{B} = \vec{\nabla} \times \vec{A}. \tag{4.39}$$

Moreover, ϕ and $\vec{A}$ can be put together in the relativistically covariant *gauge field* $A_\mu = (-\phi, \vec{A})$.

Then, the spatial components of the four-force law are

$$\frac{dp^i}{d\tau} = q(F^i{}_0 u^0 + F^i{}_j u^j) = q\gamma(F^i{}_0 + F^i{}_j v^j) = q\frac{dt}{d\tau}(E_i + \epsilon_{ijk} v_j B_k) \, , \tag{4.40}$$

so we obtain the Lorentz force law. The time component of the four-force law is

$$\frac{dp^0}{d\tau} = \frac{dE}{d\tau} = qF^0{}_i u^i = qE_i\gamma v^i = \frac{dt}{d\tau}\vec{F}\cdot\vec{v}\,, \qquad (4.41)$$

which is the applied power law. In conclusion, the four-force law generalization is valid, at least for electromagnetism, as all its components are correct.

4.5 Relativistically Covariant Lagrangeans

In order to write relativistically invariant actions, we need to integrate a Lagrangean, as a function of some parameter λ on the world line (the path $x^\mu(\lambda)$ of the particle as it moves). Since we are talking about a moving particle, we can either have $ds^2 = -d\tau^2 < 0$ for massive particles, which move slower than the speed of light, or $ds^2 = 0$ for massless particles, which move at the speed of light. So if $ds^2 \neq 0$, we can choose λ to be the proper time τ; otherwise, we can choose some other parameter, generally called an affine parameter. In other words, the relativistically invariant action for a particle will be

$$S = \int d\tau\, L\left(x^\mu, \frac{dx^\mu}{d\tau}, \tau\right) \quad \text{or} \quad \int d\lambda\, L\left(x^\mu, \frac{dx^\mu}{d\lambda}, \lambda\right). \qquad (4.42)$$

In this case, the Euler–Lagrange equations, for the variables $x^\mu(\lambda)$ (parametrizing the path of the particle), will be

$$\frac{d}{d\lambda}\left(\frac{\partial L}{\partial \frac{dx^\mu}{d\lambda}}\right) - \frac{\partial L}{\partial x^\mu} = 0\,, \qquad (4.43)$$

in the case of a massless particle, or the same ones with λ replaced with τ for a massive particle.

In order to construct the Lagrangean of the relativistic free particle, we note that the free nonrelativistic particle has just a kinetic term,

$$dt\, T = dt\, m\frac{\dot{\vec{x}}^2}{2}\,, \qquad (4.44)$$

and we must obtain this in the nonrelativistic limit of our relativistic invariant Lagrangean.

We can try the simplest invariant we can think of along the world line, with dimensions of an action, $-mc^2 d\tau$. But, moreover, we must multiply it by some object that is nontrivial away from the real world line (when the equations of motion are not satisfied) but equals 1 on this world line. Therefore, we try

$$dS = -mc^2 d\tau = -mc\, d\tau\sqrt{-\frac{dx^\mu}{d\tau}\frac{dx^\nu}{d\tau}\eta_{\mu\nu}}\,, \qquad (4.45)$$

where the second equality is valid *only on the equations of motion*, i.e., on the world line, when $ds^2 = dx^\mu dx^\nu \eta_{\mu\nu} = -c^2 d\tau^2$. Away from the world line (that is, when we vary arbitrarily the action), the right hand side is a good action for $x^\mu(\tau)$.

Moreover, on the world line $x^\mu(\tau)$, we have

$$d\tau^2 = dt^2 - \frac{d\vec{x}^2}{c^2} = dt^2\left(1 - \frac{\vec{v}^2}{c^2}\right), \tag{4.46}$$

so the action becomes, in the nonrelativistic limit,

$$dS = -mc^2 d\tau = -mc^2\,dt\sqrt{1 - \frac{v^2}{c^2}} \simeq -mc^2 dt + m\frac{v^2}{2}dt + \cdots \tag{4.47}$$

Except for the first term, which is an irrelevant constant, the action is the one we expected, so our guess was correct, and the action of the free particle is

$$S = \int d\tau\, L_{\text{part.}} = -mc\int d\tau\sqrt{-\frac{dx^\mu}{d\tau}\frac{dx^\nu}{d\tau}\eta_{\mu\nu}}. \tag{4.48}$$

The equations of motion for the variable $x^\mu(\tau)$ are obtained by varying with respect to x^μ, obtaining

$$\begin{aligned}
\delta S &= -mc\int_{\tau_1}^{\tau_2} d\tau\,\delta(\sqrt{-\dot{x}^\mu\dot{x}_\mu})\\
&= -mc\int_{\tau_1}^{\tau_2} d\tau\left[\frac{-\eta_{\mu\nu}\dot{x}^\mu\delta\dot{x}^\nu}{\sqrt{-\dot{x}^\mu\dot{x}_\mu}}\right]\\
&= -mc\int_{\tau_1}^{\tau_2} d\tau\,\delta x^\mu\frac{d}{d\tau}\left[\frac{\eta_{\mu\nu}\dot{x}^\nu}{\sqrt{-\dot{x}^\mu\dot{x}_\mu}}\right] + \delta x^\mu mc\, u_\mu|_{\tau_1}^{\tau_2}.
\end{aligned} \tag{4.49}$$

In the last equality, we used a partial integration on the $d/d\tau$ derivative. We must put to zero the terms in δS. To put to zero the last term, we must choose a boundary condition, which can usually be taken to be

$$\delta x^\mu(\tau_1) = \delta x^\mu(\tau_2) = 0. \tag{4.50}$$

Then, in order for the first (bulk) term to vanish for any δx^μ, we must have the equation of motion

$$\frac{dp^\mu}{d\tau} = mc\frac{d^2 x^\mu}{d\tau^2} = 0. \tag{4.51}$$

Note that here we must use that $\dot{x}^\mu\dot{x}_\mu = -c^2$ on the world line (on the equations of motion) to get rid of the square root in the denominator.

We have thus obtained the correct equation of motion for a free relativistic particle. But it would be good to obtain also the interaction term with the electromagnetic field, in the Lorentz force law, $qF^\mu{}_\nu u^\nu$. We want to obtain it from an extra term in the action, which should be a coupling, along the world line, to the electromagnetic field, represented by the gauge field A_μ. Then, in order to obtain a relativistically invariant action, the simplest guess is the correct one,

$$S_1 = \int d\tau\, q\frac{dx^\mu}{d\tau}A_\mu(x^\rho(\tau)) = \int d\tau\, qu^\mu A_\mu(x^\rho(\tau)). \tag{4.52}$$

The action is a functional of $x^\mu(\tau)$, which appears explicitly and implicitly in $A_\mu(x^\rho(\tau))$. The variation of S_1 is

$$\delta S_1 = q \int d\tau \left[u^\mu \partial_\nu A_\mu \delta x^\nu - \delta x^\mu \partial_\nu A_\mu \frac{dx^\nu}{d\tau} \right]$$

$$= q \int d\tau (\partial_\mu A_\nu - \partial_\nu A_\mu) u^\nu \delta x^\mu. \tag{4.53}$$

In the first line, the first term is from the variation of the implicit $A_\mu(x^\nu(\tau))$ term, and the second from the variation of $dx^\mu/d\tau$, after the $d/d\tau$ is partially integrated (and assuming that the boundary term vanishes because of the same boundary condition noted earlier).

Moreover, since we have

$$F_{0i} = -F^0{}_i = -E_i = \partial_0 A_i + \partial_i \phi = \partial_0 A_i - \partial_i A_0 , \tag{4.54}$$

and

$$F_{ij} = \epsilon_{ijk} B_k = \epsilon_{ijk} \epsilon_{klm} \partial_l A_m = 2\delta^{lm}_{ij} \frac{1}{2} (\partial_l A_m - \partial_m A_l) = \partial_i A_j - \partial_j A_i , \tag{4.55}$$

where $\epsilon_{ijk} \epsilon^{klm} = 2\delta^{lm}_{ij} = \delta^l_i \delta^m_j - \delta^m_i \delta^l_j$, it means that

$$F_{\mu\nu} = \partial_\mu A_\nu - \partial_\nu A_\mu. \tag{4.56}$$

Then, finally, if the boundary condition is respected, the variation of the total, particle + interaction, action is (for $c = 1$)

$$\delta(S + S_1) = \int d\tau \delta x^\mu \left[-\frac{dp_\mu}{d\tau} + F_{\mu\nu} u^\nu \right] , \tag{4.57}$$

so, as equation of motion, we indeed obtain the four-vector Lorentz force law.

Finally, we note that we can rewrite the interaction action as

$$S_1 = q \int dx^\mu(\tau) A_\mu(x^\rho(\tau)) = q \int d^3x \, dt \, \delta^3(x^i - x^i_0(t)) A_\mu(x^\rho) \frac{dx^\mu}{dt}$$

$$\equiv \int d^4x A_\mu(x^\rho) j^\mu(x^\rho) , \tag{4.58}$$

where the four-current of the particle j^μ is defined as

$$j^\mu(x^\rho) = q \frac{dx^\mu}{dt} \delta^3(x^i - x^i_0(t)). \tag{4.59}$$

The temporal and spatial components of j^μ are the charge density j^0 and the current density j^i,

$$j^0 = q\delta^3(x^i - x^i_0(t))$$
$$j^i = qv^i \delta^3(x^i - x^i_0(t)) , \tag{4.60}$$

which integrate to the charge and current,

$$\int d^3x \, j^0 = q; \quad \int d^3x \, j^i = qv^i = \frac{dQ^i}{dt} = I^i. \tag{4.61}$$

- In special relativity, spacetime $x^\mu = (\vec{x}, t)$ transforms together under Lorentz transformations, which are defined as transformations that preserve the interval $\Delta s^2 = \Delta x^2 - \Delta t^2 = \Delta x^\mu \Delta x^\nu \eta_{\mu\nu}$.
- Lorentz transformations form the Lorentz group $L = O(1,3)$, which respects the Minkowski metric $\eta_{\mu\nu} = M$, so $\Lambda^T M \Lambda = M$, which is a non-compact group.
- To get to the *proper* Lorentz group $L_{++} = SO(1,3)$, we must impose $\det \Lambda = +1$ and $\Lambda^0{}_0 > 0$.
- There are three rotations and three boosts in $so(1,3)$, put together in J_{ab}.
- The statement of special relativity is that physical quantities transform covariantly – i.e., they form tensors (covariant, transforming like ∂_μ, or contravariant, transforming like dx^μ).
- The relativistic Lorentz force law is $dp_\mu/d\tau = qF_{\mu\nu}u^\nu$, where the antisymmetric electromagnetic field strength $F_{\mu\nu}$ unites $\vec{E}$ and $\vec{B}$ together, as $F^{0i} = E_i$, $F^{ij} = \epsilon^{ijk}B_k$.
- The gauge field A_μ is the fundamental field of electromagnetism, out of which $F_{\mu\nu} = \partial_\mu A_\nu - \partial_\nu A_\mu$.
- The relativistic action for the particle is $S = -mc \int d\tau \sqrt{-\frac{dx^\mu}{d\tau}\frac{dx^\nu}{d\tau}}$, and the coupling to electromagnetism is through $S_1 = q \int d\tau \frac{dx^\mu}{d\tau} A_\mu(x^\rho(\tau))$.
- The interaction term can be rewritten as $S_1 = \int d^x A_\mu(x^\rho)j^\mu(x^\rho)$, where the four-current is $j^\mu = q\frac{dx^\mu}{dt}\delta^3(x^i - x_0^i(t))$.

Further Reading

Any book on special relativity. See, for instance, *Classical Theory of Fields* of Landau and Lifshitz [3] chapters 1, 2, and 3.

Exercises

(1)   Derive the transformation law for $\vec{E}, \vec{B}$, from the transformation law of the tensor $F_{\mu\nu}$ and the Lorentz force law $\vec{F} = q(\vec{E} + \vec{v} \times \vec{B})$ in a different Lorentz frame.

(2)   Write the nonrelativistic limit of the coupling $S_1 = \int d\tau q u^\mu A_\mu$ in terms of $\phi, \vec{A}.\vec{v}$, and re-derive the Lorentz force law in this case.

(3)   Consider a relativistic symmetric tensor field, $B_{\mu\nu} = B_{\nu\mu}$. Write down its transformation law under a Lorentz transformation, and then, breaking it up under spatial and temporal (nonrelativistic) components (like $F_{\mu\nu}$ into $\vec{E}$ and $\vec{B}$), write the transformation law for these nonrelativistic components.

(4)   Consider a symmetric tensor field $B_{\mu\nu}$, and an action for the interaction of it with a particle

$$S_1 = \int d\tau \sqrt{\det B_{\mu\nu}(x^\rho(\tau))}. \qquad (4.62)$$

Calculate the equation of motion for the particle.

5 Lagrangeans and the Notion of Field; Electromagnetism as a Field Theory

In this chapter, we will finally take our first step toward defining classical field theory as a generalization of classical mechanics. We will define fields and Lagrangeans for them, and, in particular, we will consider electromagnetism as our quintessential physical example.

5.1 Fields and Lagrangean Densities

Classical mechanics was defined for point particles, defined on world lines parametrized by $x_i^\mu(\tau)$, and the variables were these $x_i^\mu(\tau)$. The extension to field theory is an extension to considering instead of (functions of) particle world lines, functions of spacetime $(\vec{x}, t) = x^\mu$ as variables. These functions are called *fields*.

We saw an example in the last chapter. We finished the chapter by defining the coupling of a particle with an external electromagnetic field, described by the gauge field (or gauge potentials) $A_\mu(x^\rho) \equiv A_\mu(\vec{x}, t)$. This electromagnetic field is defined throughout spacetime, and the coupling on the particle world line was written formally as an integral over the whole spacetime,

$$q \int d\tau\, u^\mu A_\mu(x^\rho(\tau)) = \int d^4x A_\mu(x^\rho) j^\mu(x^\rho). \tag{5.1}$$

So, even though for a single particle the term was of the type

$$\int d\tau\, L\left(x^\mu, \frac{dx^\mu}{d\tau}, \tau\right), \tag{5.2}$$

we rewrote it in a way that can be generalized to many particles, as an integral over spacetime. When we have several particles, we just have a more complicated $j^\mu(x^\rho)$; namely, it becomes a sum of particle terms. Moreover, the result was written in a Lorentz covariant way, with the resulting action $S = \int d\tau L$ being Lorentz invariant.

In the above, $A_\mu(x^\rho)$ was a fixed (external) field, but now we want to generalize to the case with a dynamics for A_μ. That means that we want to have variables $A_\mu(x^\rho)$ instead of $x^\mu(\tau)$. Since we have functions of spacetime, the derivatives of the variables, inside the Lagrangean, are also $\partial_\nu A_\mu(x^\rho)$ instead of just $\frac{d}{d\tau}$. It also follows that, because we want a Lorentz invariant action, we want an action that is an integral of spacetime, not just of time (note that for the particle, the action was an integral over *proper* time), of the type

$$S = \int d^4x \mathcal{L}(A_\mu(x^\rho), \partial_\nu A_\mu(x^\rho), t) = \int dt\, d^3x \mathcal{L}\left(A_\mu(x^\rho), \frac{\partial}{\partial t} A_\mu(x^\rho), \frac{\partial}{\partial x^i} A_\mu(x^\rho), t\right). \tag{5.3}$$

Here we have put a possible explicit time dependence, though that is unlikely (and would also break relativistic invariance), so we will ignore it in the following. To make contact with the Lagrangean formalism from before, we write

$$S = S[A_\mu(x^\rho), \partial_\mu A_\nu(x^\rho)] = \int dt \, L\left[A_{\mu,\vec{x}}(t), \frac{\partial}{\partial t}A_{\mu,\vec{x}}(t)\right]. \tag{5.4}$$

Here $S[f(x)]$ means a *functional* – i.e., unlike a function ($\mathcal{L}$ is a function of $A_\mu(x^\rho)$, for instance) – it depends on the functional form of a function $f(x)$ but is integrated over the function's variables. Moreover, we put $\vec{x}$ as an "index" on A_μ, to stress the similarity with the classical mechanics formalism, though now L is a functional of $A_{\mu,\vec{x}}(t)$, since it is integrated over $\vec{x}$.

So both the action S and the Lagrangean L are now functionals of the fields, and $\mathcal{L}$, called the *Lagrangean density*, is a function. It is a density, since the Lagrangean is its integral over space. Usually, however, $\mathcal{L}$ is named the Lagrangean, too, though it is obviously a misnomer.

The Lagrangean density will be written like a functional of the electromagnetic field (gauge potential)

$$A_\mu = (-\phi, \vec{A}) \Rightarrow A^\mu = \eta^{\mu\nu} A_\nu = (\phi, \vec{A}). \tag{5.5}$$

Note that when raising or lowering indices, we will do it with the Minkowski metric. Since $\eta_{\mu\nu}$ has a minus only in the time direction, only when raising/loweing a 0 index, we get a minus; otherwise, we get nothing. Also note that we work with $c = 1$; otherwise, we would have a $1/c$ in front of $-\phi$ for A_0.

5.2 Maxwell's Equations in Covariant Formalism

The Maxwell equations in vacuum are written as

$$\begin{aligned}
\vec{\nabla} \times \vec{E} &= -\frac{\partial}{\partial t}\vec{B} \\
\vec{\nabla} \cdot \vec{E} &= 0 \\
\vec{\nabla} \times \vec{B} &= +\frac{1}{c^2}\frac{\partial}{\partial t}\vec{E} \\
\vec{\nabla} \cdot \vec{B} &= 0.
\end{aligned} \tag{5.6}$$

Using indices and putting $c = 1$, they become

$$\begin{aligned}
\epsilon_{ijk}\partial_j E_k + \partial_0 B_i &= 0 \\
\partial_i E_i &= 0 \\
\epsilon_{ijk}\partial_j B_k - \partial_0 E_i &= 0 \\
\partial_i B_i &= 0.
\end{aligned} \tag{5.7}$$

We can define the relativistically covariant field strength $F_{\mu\nu}$ as we saw in the last chapter,

$$E_i = E^i \equiv F^{0i} = -F_{0i}$$

$$B_i = B^i \equiv \frac{1}{2}\epsilon^{ijk}F_{jk}. \tag{5.8}$$

Note that, since we raise indices with $\eta_{\mu\nu}$,

$$F_{\mu\nu} = \eta_{\mu\rho}\eta_{\nu\sigma}F^{\nu\sigma}. \tag{5.9}$$

Also note that the above formula matches the definition from the last chapter,

$$F_{ij} = \epsilon_{ijk}B_k\,, \tag{5.10}$$

since the totally antisymmetric symbol, the Levi-Civitta tensor ϵ_{ijk}, defined by

$$\epsilon^{123} = +1\,, \tag{5.11}$$

satisfies

$$\epsilon_{ijk}\epsilon^{klm} = \epsilon_{ijk}\epsilon^{lmk} = \delta_i^l\delta_j^m - \delta_i^m\delta_j^l = 2\delta_{[ij]}^{[lm]}. \tag{5.12}$$

Here we used the fact that an index hopping over another in ϵ^{ijk} gives a minus sign, as does interchanging two indices. We have also used the fact that, since k is the same up and down, $[ij]$ must be the same as $[lm]$, but not necessarily in the same order. We have defined the object $\delta_{[ij]}^{[lm]}$, with *strength 1*, as the object that multiplied with an antisymmetric object just changes the indices, for instance

$$\delta_{[ij]}^{[lm]}F^{ij} = F^{lm}. \tag{5.13}$$

With the previously noted definitions, and using $F^{0k} = -F_{0k} = +F_{k0}$, the first Maxwell equation (5.7) becomes

$$\epsilon_{ijk}\partial_j F^{0k} + \frac{1}{2}\epsilon_{ijk}\partial_0 F_{jk} = 0 \Rightarrow$$
$$\frac{1}{2}\epsilon^{ijk}[2\partial_j F_{k0} + \partial_0 F_{jk}] = 0. \tag{5.14}$$

But, defining the four-dimensional (Minkowski) totally antisymmetric Levi-Civitta tensor by

$$\epsilon^{0123} = +1\,, \tag{5.15}$$

which means that we can extend the three-index tensor to a 4-index one by

$$\epsilon^{0ijk} = \epsilon^{ijk}\,, \tag{5.16}$$

the first Maxwell equation becomes

$$-\frac{1}{2}\epsilon^{ijk0}[2\partial_j F_{k0} + \partial_0 F_{jk}] = 0 \Rightarrow \epsilon^{\mu\nu\rho\sigma}\partial_\nu F_{\rho\sigma} = 0\,, \tag{5.17}$$

for $\mu = i$. Indeed, then 0 must be among $\nu\rho\sigma$, and the factor of 2 comes from the sum over F_{0k} and $F_{k0} = F_{0k}$.

The second Maxwell equation (5.7) becomes

$$\partial_i F^{0i} = 0 \Rightarrow \partial_\mu F^{\nu\mu} = 0\,, \tag{5.18}$$

for $\nu = 0$ (so that μ must be spatial, $\mu = i$).

The third Maxwell equation (5.7), using (5.12) becomes

$$\epsilon_{ijk}\partial_j\frac{1}{2}\epsilon^{klm}F_{lm} - \partial_0 F^{0i} = 0 \Rightarrow$$
$$\partial_j F^{ij} + \partial_0 F^{i0} = \partial_\mu F^{\nu\mu} = 0\,, \tag{5.19}$$

for $\nu = i$.

The fourth and final Maxwell equation (5.7) becomes

$$\epsilon^{ijk}\partial_i F_{jk} = \epsilon^{0ijk}\partial_i F_{jk} = 0 \Rightarrow$$
$$\epsilon^{\mu\nu\rho\sigma}\partial_\nu F_{\rho\sigma} = 0\,, \tag{5.20}$$

for $\mu = 0$.

Then the second and third Maxwell equation become

$$\epsilon^{\mu\nu\rho\sigma}\partial_\nu F_{\rho\sigma} = 0 \Rightarrow \partial_{[\mu}F_{\nu\rho]} = 0\,, \tag{5.21}$$

called the *Bianchi identity*, and the first and fourth become the *equations of motion*,

$$\partial_\mu F^{\mu\nu} = 0. \tag{5.22}$$

The Bianchi identity means that we can write $F_{\mu\nu}$ as the antisymmetric derivative of another field, i.e.,

$$F_{\mu\nu} = \partial_\mu A_\nu - \partial_\nu A_\mu \equiv 2\partial_{[\mu}A_{\nu]}\,, \tag{5.23}$$

since $\partial_{[\mu}\partial_{\nu]} = \partial_\mu\partial_\nu - \partial_\nu\partial_\mu = 0$ (derivatives commute).

The field A_μ is the gauge field from before, $A_\mu = (-\phi, \vec{A})$, so

$$E_i = -F_{0i} = -\partial_0 A_i + \partial_i A_0 = -\partial_0 A_i - \partial_i\phi \Rightarrow$$
$$\vec{E} = -\vec{\nabla}\phi - \partial_t\vec{A}$$
$$B^i = \frac{1}{2}\epsilon^{ijk}F_{jk} = \epsilon^{ijk}\partial_j A_k \Rightarrow$$
$$\vec{B} = \vec{\nabla}\times\vec{A}. \tag{5.24}$$

So we solved the Bianchi identity by writing $F_{\mu\nu}$ in terms of A_μ, and we are left with the the equations of motion (5.22), to be obtained from a Lagrangean density written in terms of the variables A_μ.

5.3 Euler–Lagrange Equations in Field Theory

We now take a step back and find the equations of motion for a general field theory. For a general field theory, the variables will be the set of fields – i.e., functions of spacetime, $\{\phi^i(x^\rho)\}_i$ – so the action principle is written as

$$S[\{\phi^i(x^\rho)\}, \{\partial_\mu\phi^i\}] = \int d^4x \mathcal{L}(\phi^i(x^\rho), \partial_\mu\phi^i(x^\rho)). \tag{5.25}$$

Varying the action with respect to $\phi^i(x^\rho)$, we obtain

$$
\begin{aligned}
\delta S &= \delta \int d^4x \mathcal{L}(\phi^i(x^\rho), \partial_\mu \phi^i(x^\rho)) \\
&= \int d^4x \left[\frac{\partial \mathcal{L}}{\partial \phi^i(x^\rho)} \delta \phi^i(x^\rho) + \frac{\partial \mathcal{L}}{\partial(\partial_\mu \phi^i)} \delta \partial_\mu \phi^i \right].
\end{aligned}
\tag{5.26}
$$

Using the fact that δ and ∂_μ commute, $\delta \partial_\mu \phi^i = \partial_\mu \delta \phi^i$, and partially integrating ∂_μ, we obtain

$$
\delta S = \int d^4x \left[\frac{\partial \mathcal{L}}{\partial \phi^i} - \partial_\mu \left(\frac{\partial \mathcal{L}}{\partial(\partial_\mu \phi^i)} \right) \right] \delta \phi^i + \int_M d^4x \partial_\mu \left[\frac{\partial \mathcal{L}}{\partial(\partial_\mu \phi^i)} \delta \phi^i \right].
\tag{5.27}
$$

The last term is a boundary term, and in order to vanish, we must impose a boundary condition.

If as is common, the space M stands for $\mathbb{R}^{3,1}$, Minkowski space, then we can impose $\delta \phi^i|_{\partial M} = 0$, since fields (which represent some physical system) must vanish at infinity – both space-like, $|\vec{x}| \to \infty$, and time-like, $t \to \pm\infty$ – and then so do their variations. But in a more general situation, it might be that the fields simply are defined over only a compact domain (a finite region), like, for instance, the "Cooper pair" field in a superconductor (existing only inside the superconductor and describing the superconducting state), or it might be that at a certain surface we have some physical condition, like in the case that we put a conductor material in a wall, implying the electric potential ϕ is constant throughout the conductor. In that case, we must impose some physical boundary condition at the surface, and imposing the boundary condition must make the boundary term vanish.

The vanishing of the bulk term in δS, for any $\delta \phi^i(x^\rho)$, means that the integrand must be zero, i.e.,

$$
\frac{\partial \mathcal{L}}{\partial \phi^i(x^\rho)} - \partial_\mu \left(\frac{\partial \mathcal{L}}{\partial(\partial_\mu \phi^i(x^\rho))} \right) = 0.
\tag{5.28}
$$

These are the *Euler–Lagrange equations* for the fields ϕ^i, which are functions of spacetime, so the equations are defined at each point x^ρ. Note that, strictly speaking, the Euler–Lagrange equations are the equations associated with the Lagrangean L, so integrated over d^3x, but by extension, we call the equations of motion of the Lagrangean density $\mathcal{L}$ in the same way.

5.4 Lagrangean for the Gauge Field A_μ

We now apply for A_μ, with $\mathcal{L} = \mathcal{L}(A_\mu, \partial_\nu A_\mu)$, the formalism above (above, ϕ^i was a generic collection of fields, and now $\phi^i \to A_\mu$), obtaining the equations of motion

$$
\frac{\partial \mathcal{L}}{\partial A_\mu} - \partial_\nu \left(\frac{\partial \mathcal{L}}{\partial(\partial_\nu A_\mu)} \right) = 0.
\tag{5.29}
$$

We want to find $\mathcal{L}$ such that these equations to coincide with (5.22). Since the ∂_μ can be partially integrated from acting on A_ν, we expect that

$$\mathcal{L} \propto F_{\mu\nu}F^{\mu\nu} \equiv F_{\mu\nu}F_{\rho\sigma}\eta^{\mu\rho}\eta^{\nu\sigma} \; ; \tag{5.30}$$

more precisely,

$$\mathcal{L} = cF_{\mu\nu}F^{\mu\nu}. \tag{5.31}$$

But in this case, the Lagrangean doesn't depend explicitly on A_μ, so the equations of motion are

$$\frac{\partial \mathcal{L}}{\partial A_\mu} = 0 \Rightarrow -\partial_\nu \left(\frac{\partial \mathcal{L}}{\partial(\partial_\nu A_\mu)} \right) = 0. \tag{5.32}$$

But the Lagrangean can be rewritten as

$$\mathcal{L} = c(\partial_\mu A_\nu - \partial_\nu A_\mu)^2 = 2c[(\partial_\mu A_\nu)^2 - (\partial_\mu A_\nu)(\partial^\nu A^\mu)] \,, \tag{5.33}$$

which means that

$$\frac{\partial \mathcal{L}}{\partial(\partial_\nu A_\mu)} = 4c[\partial^\nu A^\mu - \partial^\mu A^\nu] = 4cF^{\nu\mu} \,, \tag{5.34}$$

thus the equations of motion are

$$- 4c\partial_\nu F^{\nu\mu} = 0. \tag{5.35}$$

This suggests (though it is only a suggestion) that $c = -1/4$, which is, in fact, true. To show that, consider that

$$\int d^3x \mathcal{L} = L = T - V, \tag{5.36}$$

and, since $F_{0i} = \dot{A}_i - \partial_i A_0$, for $A_0 = 0$, $F_{0i}F_{0i}/2 = \dot{A}_i^2/2$ is a kinetic term,

$$T = \int d^3x \frac{\vec{E}^2}{2} = \int d^3x \frac{(F^{0i}F^{0i})}{2} = -\frac{1}{4}\int d^3(F_{0i}F^{0i} + F_{i0}F^{i0}) \,, \tag{5.37}$$

and the potential is

$$V = \int d^3x \frac{\vec{B}^2}{2} = \int d^3x \frac{1}{2}\left(\frac{1}{4}\epsilon^{ijk}\epsilon_{ilm}F_{jk}F^{lm}\right) = \frac{1}{4}\int d^3x F_{jk}F^{jk}. \tag{5.38}$$

Moreover, the normalization is correct, since then the total energy is

$$E = T + V = \frac{1}{2}\int d^3x(\vec{E}^2 + \vec{B}^2). \tag{5.39}$$

The Lagrangean is thus

$$L = \int d^3x \mathcal{L} = -\frac{1}{4}\int d^3x[F_{0i}F^{0i} + F_{i0}F^{i0} + F_{jk}F^{jk}] = -\frac{1}{4}\int d^3x F_{\mu\nu}F^{\mu\nu} \,, \tag{5.40}$$

as we expected.

5.5 Adding Sources to Maxwell's Equations and Their Lagrangean

The Maxwell's equations that get modified by the addition of sources are the equations of motion,

$$\vec{\nabla} \cdot \vec{E} = \frac{\rho}{\epsilon_0} \rightarrow \partial_\mu F^{0\mu} = \frac{\rho}{\epsilon_0} \,, \tag{5.41}$$

as well as

$$\vec{\nabla} \times \vec{B} - \frac{1}{c^2} \frac{\partial \vec{E}}{\partial t} = \mu_0 \vec{j}. \tag{5.42}$$

We work with $\epsilon_0 = 1$ and, since also $c = 1$ (and $c^2 \epsilon_0 \mu_0 = 1$), also $\mu_0 = 1$, so the previous equation becomes

$$\partial_j F^{ij} + \partial_0 F^{i0} = \partial_\mu F^{i\mu} = \vec{j}. \tag{5.43}$$

Together, the equations become

$$\partial_\mu F^{\nu\mu} = j^\nu \Rightarrow \partial_\mu F^{\mu\nu} + j^\nu = 0. \tag{5.44}$$

The first term was $-\partial_\mu(\partial\mathcal{L}/\partial(\partial_\mu A_\nu))$, and the second term can be obtained as $\partial\mathcal{L}/\partial A_\nu$, if one adds a term to the Lagrangean,

$$\mathcal{L} \rightarrow \mathcal{L} + A_\mu j^\mu \,, \tag{5.45}$$

which is exactly like the coupling to particle current from the last chapter that we started this chapter with.

In conclusion, the Lagrangean density for electromagnetism coupled to a particle current is

$$\mathcal{L} = -\frac{1}{4} F_{\mu\nu} F^{\mu\nu} + A_\mu j^\mu. \tag{5.46}$$

Important Concepts to Remember

- A field is a function of spacetime, $\phi(\vec{x}, t)$, like the electromagnetic (gauge field) potential $A_\mu(\vec{x}, t) = A_\mu(x^\rho) = (-\phi, \vec{A})$.
- The Maxwell equations in vacuum in relativistic formulation are the Bianchi identity $\partial_{[\mu} F_{\nu\rho]} = 0$, and the equations of motion $\partial_\mu F^{\mu\nu} = 0$.
- The Bianchi identity is solved by $F_{\mu\nu} = \partial_\mu A_\nu - \partial_\nu A_\mu$, so one needs a Lagrangean for A_μ that will give the equations of motion.
- The Lagrangean for electromagnetism, like for any field, is of the type $L = \int d^3x \mathcal{L}[A_\mu(x^\rho), \partial_\nu A_\mu(x^\rho)]$, the integral of a Lagrange density, and a functional of the fields.
- For an action $S = \int d^4x \mathcal{L}(\phi^i(x^\rho), \partial_\mu \phi^i(x^\rho))$, the equations of motion (Euler–Lagrange equations) are $\partial\mathcal{L}/\partial\phi^i - \partial_\mu(\partial\mathcal{L}/\partial(\partial_\mu\phi^i)) = 0$.
- The Lagrangean for pure electromagnetism is $L = -\frac{1}{4} \int d^3x F_{\mu\nu} F^{\mu\nu} = \int d^3x [\vec{E}^2 - \vec{B}^2]/2$.
- The Maxwell equations of motion with sources are $\partial_\mu F^{\nu\mu} = j^\nu$, so the Lagrangean is $\mathcal{L} = -\frac{1}{4} F_{\mu\nu} F^{\mu\nu} + A_\mu j^\mu$.

Further Reading

See, for instance, the *Classical Theory of Fields* of Landau and Lifshitz [3], chapters 4 and 4.

Exercises

(1)   Write the equivalent of the Maxwell's equations (equations of motion and Bianchi identities) coming from the Lagrangean

$$\mathcal{L} = -\frac{1}{4}f(\phi^i)F_{\mu\nu}F^{\mu\nu} \,, \tag{5.47}$$

where $f(\phi^i)$ is a function of the other fields.

(2)   Putting $c = 1$, write the Maxwell's equations in vacuum in terms of a single complex field

$$\vec{F} = \vec{E} + i\vec{B}. \tag{5.48}$$

If one looks for solutions of the Maxwell's equations of the type

$$\vec{F} = \vec{\nabla}\alpha \times \vec{\nabla}\beta \,, \tag{5.49}$$

$\alpha, \beta \in \mathbb{C}$, show that half the Maxwell's equations are satisfied.

(3)   Consider the Lagrangian term

$$\mathcal{L}_1 = \theta(x)\epsilon^{\mu\nu\rho\sigma}F_{\mu\nu}F_{\rho\sigma} \,, \tag{5.50}$$

where $\epsilon^{\mu\nu\rho\sigma}$ is totally antisymmetric and $\epsilon^{0123} = +1$. Calculate the resulting action in terms of $\vec{E}$ and $\vec{B}$. If we add this to the Maxwell action, calculate the equations of motion for A_μ.

(4)   Consider the action (the *Born–Infeld action*, written in 1932)

$$S = -b^{-4} \int d^4x \left[\sqrt{-\det\left(\eta_{\mu\nu} + b^2 F_{\mu\nu}\right)} - 1 \right]. \tag{5.51}$$

Write it *explicitly* in terms of $\vec{E}$ and $\vec{B}$.

6 Scalar Field Theory, Origins, and Applications

In this chapter, we will define the simplest possible field theory, scalar field theory, from its origins to its applications. If electromagnetism is something that we can relate to easily, scalar fields are a bit more abstract but nevertheless have many applications.

6.1 Scalar Fields and Their Equations of Motion

We saw that, in general, we describe a field theory by a Lagrangean density, a function of "fields," which are functions of spacetime, $\phi^i(x^\rho)$, where i is any type of index. Moreover, we have

$$\mathcal{L} = \mathcal{L}(\phi^i(x^\rho), \partial_\mu \phi^i(x^\rho)). \tag{6.1}$$

We looked at the case of A_μ, when μ is a Lorentz index – i.e., it transforms under Lorentz transformations like boosts.

However, the simplest case is if there is no index, at least not a Lorentz index, and the field doesn't transform. In this case, we say that we have a *scalar field*. More precisely, under Lorentz transformations, we have

$$\phi'(x'^\rho) = \phi(x^\rho) , \tag{6.2}$$

where x'^ρ is the Lorentz transformed coordinate $\Lambda^\rho{}_\nu x^\nu$, and ϕ' the Lorentz transformed field. Thus, the relation means that the value of the field at the point is numerically the same, whether we use the original or the transformed coordinate system.

The reason we haven't started our field theory description with scalar fields but, rather, with A_μ (gauge field) in the last chapter is that A_μ is a field associated with electromagnetism, which is more common. At the quantum level, it is associated with the electromagnetic force that is realized through the quantum of A_μ, a photon.

But there are other forces in nature, and they are all related to some fields in spacetime: the strong, weak, gravitational, and Higgs forces (or *interactions*). Only the Higgs force, the latest to be found (in 2012), is described by a scalar field *at a fundamental level*.

The strong force has short range of the order of the nuclear distance (the characteristic size of a nucleon) of about 1 fermi. It is associated *at a fundamental level* with a non-Abelian gauge field $A_\mu^a(x^\rho)$, which will be studied later in the book. Its quantization gives *gluons*, the non-Abelian equivalent of photons.

However, on distances that are not too short, but a bit larger (so intermediate, of the order of 1 fermi), we can use an effective description, in terms of a scalar field, instead of the fundamental description given earlier. More precisely, it is a *pseudoscalar field* (which means that it changes sign under parity, the inversion of spatial coordinates), though the difference is not important for us now. Even more precisely, there are three fields, $\pi^i(x^\rho)$, transforming under a global $SU(2)$ symmetry acting on the index i.

The associated quantum particles are the pions π^+, π^-, π^0, where the index $(+, -, 0)$ now refers to electric charge. However, we will neglect all of that and use a single scalar field $\phi(\vec{x}, t) = \phi(x^\rho)$ for the pion field. This is a standard example of a scalar used in particle physics. It is an effective, composite field, which means it is composed of the more fundamental particles of the theory and breaks apart at sufficiently large energies.

Now that we have discovered the Higgs boson, there is also the Higgs field associated with it (which was postulated much earlier, in the 1970s, which is why people were looking for the Higgs particle until it was discovered in 2012), $H(x^\mu)$. It is the only fundamental scalar in particle physics observed so far, though there are indications that it might actually also be composite, not fundamental, meaning that there might be a more fundamental description without this scalar field.

In condensed matter physics, we have another important composite scalar example, though a nonrelativistic one – the BCS scalar, condensate or "Cooper pair" field $\phi = \psi_\uparrow \psi_\downarrow$, understood as the pairing of two fermionic fields, for a spin-up and a spin-down. The resulting particle is a boson, since two fermions have Bose-Einstein statistics, and has no spin, since the total spin is zero. There are also more nonrelativistic examples, since condensed matter physics has almost exclusively nonrelativistic models (ones that change under the action of a Lorentz transformation).

In cosmology, there are other examples of scalar fields. Any good cosmological model can be usually described in terms of a scalar field. Examples include inflation, ekpyrotic and cyclic models, holographic cosmology, etc. In all of these cases, the use of scalar fields is understood nowadays as just an effective way of dealing with a complicated situation, the simplest model that we can write and describe. The restriction to scalars is thus not for some fundamental reason but for reasons of simplicity of description. The scalar can be, besides a fundamental scalar, a simple mode of a more complicated field, a combination of several fields, or an effective description (like for the composite field noted earlier). The most popular incarnation of the most common cosmological model, inflation, is defined by a scalar field, the inflaton. There are other incarnations – like, for instance, models using gauge fields – but they are difficult and not very good.

We have considered a Lagrangean that depends only on ϕ and $\partial_\mu \phi$, and that is correct for fundamental fields. But for composite fields, in effective descriptions, we can have even higher derivatives, usually coming from quantum corrections, i.e.,

$$\mathcal{L} = \mathcal{L}(\phi(x^\rho), \partial_\mu \phi(x^\rho), \partial_\mu \partial_\nu \phi(x^\rho), \ldots). \tag{6.3}$$

We will see how this can happen in quantum field theory, but for now we just want to see how to deal with this situation in the classical theory.

To obtain the equations of motion in this case, we proceed like before and vary $\mathcal{L}$ with respect to arbitrary variations $\delta\phi$. Then,

$$\delta S = \delta \int d^4 x \mathcal{L}(\phi, \partial_\mu \phi, \partial_\mu \partial_\nu \phi, \dots)$$
$$= \int d^4 x \left[\frac{\partial \mathcal{L}}{\partial \phi} \delta \phi + \frac{\partial \mathcal{L}}{\partial(\partial_\mu \phi)} \delta \partial_\mu \phi + \frac{\partial \mathcal{L}}{\partial(\partial_\mu \partial_\nu \phi)} \delta \partial_\mu \partial_\nu \phi + \cdots \right]. \tag{6.4}$$

Using that $\delta \partial_\mu = \partial_\mu \delta$ and partially integrating, we obtain

$$\delta S = \int d^4 x \delta \phi \left[\frac{\partial \mathcal{L}}{\partial \phi} - \partial_\mu \left(\frac{\partial \mathcal{L}}{\partial \partial_\mu \phi} \right) + \partial_\nu \partial_\mu \left(\frac{\partial \mathcal{L}}{\partial(\partial_\mu \partial_\nu \phi)} \right) + \cdots \right] + S_{\text{boundary}}. \tag{6.5}$$

Considering that the boundary term is set to zero by the boundary conditions, then for arbitrary $\delta \phi$, we obtain the Euler–Lagrange equations

$$\frac{\partial \mathcal{L}}{\partial \phi} - \partial_\mu \left(\frac{\partial \mathcal{L}}{\partial \partial_\mu \phi} \right) + \partial_\nu \partial_\mu \left(\frac{\partial \mathcal{L}}{\partial(\partial_\mu \partial_\nu \phi)} \right) + \cdots = 0. \tag{6.6}$$

The terms with more than one derivative are quantum corrections, as we mentioned.

6.2 Constructing Lagrangeans

We know that the Lagrangean is written as $L = T - V$, and in the standard classical mechanics case, $T = m\dot{q}^2/2$. In order to generalize to classical field theory, we replace $q(t)$ with the fields $\phi(t, \vec{x})$.

In the vast majority of cases, we are concerned with *local* field theories, which means that the Lagrangean is the integral of a Lagrangean density,

$$L(t) = \int d^3 x \mathcal{L}(\vec{x}, t). \tag{6.7}$$

Note that this is not the only case possible. Sometimes one considers, for instance, *bi-local* theories, for instance something like

$$L(t) = \int d^3 x d^3 y F(\vec{x}, \vec{y}, t). \tag{6.8}$$

Or we can consider a *nonlocal* theory, which involves operators that don't act locally on a field. For example, by "integrating out" in the quantum theory a massless fermion interacting with a gauge field (taking into account the fermion's effects so that we don't need to write it in the action anymore; we will see how in quantum field theory), we obtain a term

$$\int d^4 x F_{\rho\sigma} \frac{1}{\partial_\mu \partial_\nu \eta^{\mu\nu}} F^{\rho\sigma}. \tag{6.9}$$

In here, the inverse of the operator $\Box \equiv \partial_\mu \partial_\nu \eta^{\mu\nu}$ doesn't act locally on $F^{\rho\sigma}$ (it is some sort of "double integral"). To see why something that is not a finite positive power law of $\Box$ can act nonlocally, consider as an example the operator $e^{a\Box} = e^{a\partial^\mu \partial_\mu}$, or, even simpler, consider $e^{a^\mu \partial_\mu}$ acting on a function $f(x^\rho)$. Then the result is the Taylor expansion,

$$\sum_{n \geq 0} \frac{(a^\mu \partial_\mu)^n}{n!} f(x^\rho) = f(x^\rho + a^\rho), \tag{6.10}$$

so, in effect, the operator involves a different position than the apparent one – i.e., it is nonlocal.

We will also consider mostly *relativistic* field theories in this book, since the intended applications are for particle physics. In condensed matter physics, most models are nonrelatistic, so we describe them by nonrelativistic field theories, but here we will not consider them; we will only mention them in later chapters.

For a relativistic theory, the Lagrangean density is a function of Lorentz covariant objects, i.e.,

$$\mathcal{L}(\vec{x}, t) = \mathcal{L}(\phi(\vec{x}, t), \partial_\mu \phi(\vec{x}, t), \dots) \,, \tag{6.11}$$

but not separately of $\dot{\phi}$ and $\partial_i \phi$, for instance.

The action is then

$$S = \int dt \, L(t) = \int dt \, d^3x \, \mathcal{L}(\phi(\vec{x}, t), \partial_\mu \phi(\vec{x}, t), \dots) = \int d^4x \mathcal{L}(\phi(x^\rho), \partial_\mu \phi(x^\rho), \dots). \tag{6.12}$$

In order to understand better how to proceed, we make more precise the classical mechanics analogy, like we suggested in the last chapter, by discretizing space – i.e., by replacing

$$\phi(\vec{x}, t) \to \phi_{\vec{x}_a}(t) \equiv q_a(t). \tag{6.13}$$

Then the Lorentz covariant derivative of the scalar field $\partial_\mu \phi(\vec{x}, t)$ splits into time derivative,

$$\partial_0 \phi(\vec{x}, t) = \dot{\phi}_{\vec{x}_a}(t) \equiv \dot{q}_a(t) \,, \tag{6.14}$$

which has the same form as the time derivative in classical mechanics, and spatial derivative,

$$\partial_i \phi(\vec{x}, t) \to \frac{1}{\Delta \vec{x}_a} (\phi_{\vec{x}_{a+1}} - \phi_{\vec{x}_a}) = \frac{q_{a+1} - q_a}{\Delta \vec{x}_a}. \tag{6.15}$$

Then the Lagrangean, written as an integral over space of a Lagrangean density, becomes a sum over Lagrangeans for different modes,

$$L(t) \to \sum_a \mathcal{L}_a(q_a(t), \dot{q}_a(t)). \tag{6.16}$$

The relation is somewhat fuzzy, since dimensions don't quite work out (a Lagrangean density has different dimensions from a Lagrangean), so let's ignore masses. Then a natural Lagrangean would be a standard classical mechanics kinetic term at $m = 1$, plus potential terms,

$$L = \sum_a \frac{\dot{q}_a^2}{2} + \cdots \tag{6.17}$$

This means that the Lagrangean density should be

$$\mathcal{L}_a = \frac{\dot{q}_a^2(t)}{2} + \cdots \to \mathcal{L} = \frac{(\partial_0 \phi)^2}{2} + \cdots \tag{6.18}$$

But the theory is relativistic, so we must have the relativistically invariant term that contains the above, i.e.,

$$\mathcal{L} = \frac{1}{2}[(\partial_0\phi)^2 - (\partial_i\phi)^2] = -\frac{(\partial_\mu\phi)^2}{2} + \cdots \equiv -\frac{1}{2}\partial_\mu\phi\partial_\nu\phi\eta^{\mu\nu} + \cdots \tag{6.19}$$

The new term, which as we see is part of the potential (since $L = T - V$), discretizes to

$$\frac{(\partial_i\phi)^2}{2} \propto \frac{1}{2}(q_{a+1} - q_a)^2. \tag{6.20}$$

This is then like a system of *coupled* harmonic oscillators,

$$L = \sum_a m\frac{\dot{q}_a^2}{2} - \sum_a k\frac{(q_{a+1} - q_a)^2}{2}. \tag{6.21}$$

Thus, in a certain sense, a theory for an usual scalar field is an infinite number of (perturbed) harmonic oscillators. Moreover, in the discretization, it seems like it is a discrete infinite, but, in fact, we have a continuous infinite number of harmonic oscillators.

But moreover, we can add a term corresponding to the harmonic oscillators at each site, suggested by the kinetic term $\sum_a mq_a^2/2$. The dimensions don't quite work, since we are replacing L with $\mathcal{L}$, as we said, but this suggests the term

$$\sum_a m^2\frac{q_a^2}{2} \to \int d^3x\, \frac{m^2\phi^2}{2} \tag{6.22}$$

to be added to the potential V.

We can now define the *potential for a field* ϕ, $V(\phi)$, which is a *potential density*, excluding the $(\partial_i\phi)^2/2$ term. Note that we used the same V to describe the potential, and the potential for a field (which is a density), since these are the usual notations for them. We hope there is no confusion, since we use $V(\phi)$ to distinguish from V. Then we have

$$V(\phi) = \frac{m^2\phi^2}{2} + \Delta V(\phi). \tag{6.23}$$

Here $\Delta V(\phi)$ can contain for instance powers of ϕ,

$$\Delta V(\phi) = \sum_{n\geq 3}\frac{\lambda_n\phi^n}{n!} = \lambda_3\frac{\phi^3}{3!} + \lambda_4\frac{\phi^4}{4!} + \cdots \tag{6.24}$$

In the context of quantum theory, there are consistent models called renormalizable (we will see in quantum field theory what that means), and they correspond to a maximum power in the potential depending on dimension; in the case of $d = 4$, we have a maximum ϕ^4 term. If the model is just effective, however, we can have an arbitrary power.

Then the potential is

$$V = \int d^3x\left[\frac{1}{2}(\vec{\nabla}\phi)^2 + V(\phi)\right], \tag{6.25}$$

and the Lagrangean is

$$L(t) = T - V = \int d^3x\frac{\dot{\phi}^2}{2} - V. \tag{6.26}$$

Finally, the action is

$$S = \int dt\, L(t) = \int d^4x\, \mathcal{L} = \int d^4x \left[-\frac{1}{2}(\partial_\mu\phi)(\partial_\nu\phi)\eta^{\mu\nu} - V(\phi) \right]$$
$$\equiv \int d^4x \left[-\frac{1}{2}(\partial_\mu\phi)^2 - V(\phi) \right]. \tag{6.27}$$

Its equations of motion are found in the usual way, by varying the action as

$$\delta S = \int d^4x [-(\partial_\mu\phi)\delta\partial_\nu\phi\eta^{\mu\nu} - V'(\phi)\delta\phi]$$
$$= \int d^4x [(\eta^{\mu\nu}\partial_\mu\partial_\nu\phi) - V'(\phi)]\delta\phi(x^\rho) + S_{\text{boundary}}. \tag{6.28}$$

Aa usual, we have used $\delta\partial_\nu = \partial_\nu\delta$, and partially integrated. We assume the vanishing of the boundary term because of boundary conditions, resulting in the equation of motion

$$\eta^{\mu\nu}\partial_\mu\partial_\nu\phi = V'(\phi). \tag{6.29}$$

The operator

$$\Box \equiv \eta^{\mu\nu}\partial_\mu\partial_\nu \tag{6.30}$$

is called in this context the *Klein-Gordon operator*. In a more general mathematics context, it is the d'Alembertian operator. In the case of $\Delta V(\phi) = 0$, i.e., just a mass term,

$$V(\phi) = \frac{m^2\phi^2}{2}, \tag{6.31}$$

the resulting equation,

$$(\Box - m^2)\phi = 0 \tag{6.32}$$

is called the *Klein-Gordon (KG) equation*. This is the case of a *free massive scalar* (free means no interactions higher than quadratic).

6.3 Specific Models (Applications)

We showed that the potential can be a power law. But it is not necessary; if we admit an arbitrary power law for an effective model, we can consider more general potentials – for instance, exponential or cosine.

Sine-Gordon Model

In the cosine case, we obtain an important model, the *sine-Gordon model*, with

$$V(\phi) = \frac{\alpha}{a^2}[1 - \cos(a\phi)] \geq 0. \tag{6.33}$$

The constant (α/a^2) was added to the cosine potential in order for the potential to be positive definite, as in Figure 6.1.

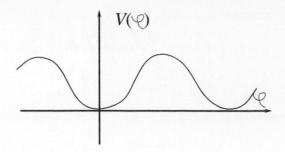

Figure 6.1   Potential for the sine-Gordon model.

Indeed, for consistency of the physics, we need the potential energy, and thus the potential $V(\phi)$, to be positive definite. In that case, the equation of motion is

$$\Box\phi = V'(\phi) = \frac{\alpha}{a}\sin(a\phi)\,,\tag{6.34}$$

hence the name sine-Gordon model (which was actually derived from a bad joke: the Klein-Gordon model was modified by having a sine term instead of a mass term, hence "sine-Gordon," even though Gordon had nothing to do with it). Expanding for small ϕ, we obtain

$$\phi = \alpha\phi - \alpha a^2\frac{\phi^3}{3!}+\cdots\,,\tag{6.35}$$

where the right hand side must be compared with the result of a general power law potential, $V'(\phi) = m^2\phi + \lambda_4\phi^3/3! + \cdots$, which means that

$$\alpha = m^2,\quad \lambda_4 = -\alpha a^2.\tag{6.36}$$

Note that $\lambda_4 < 0$, though the full potential is positive definite.

Higgs Model

Another important model, perhaps the most important, has the Higgs potential, with a minimum at a $\phi_0 \neq 0$,

$$V(\phi) = \frac{\lambda}{4}\left(\phi^2 - \frac{\mu^2}{\lambda}\right)^2 = \frac{\lambda}{4}\phi^4 - \frac{\mu^2\phi^2}{2} + \frac{\mu^4}{4\lambda} \geq 0.\tag{6.37}$$

Again, we see that the potential is positive definite (see Figure 6.2) so it is consistent. But in this case, we have

$$m^2 = -\mu^2 < 0\,,\tag{6.38}$$

so already the simplest term is negative. It is related to *symmetry breaking*, which will be considered later on in the book. Indeed, the vacua of the theory are

$$|\phi| = |\phi_0| = \frac{\mu}{\sqrt{\lambda}} \Rightarrow \phi_0 = \pm\frac{\mu}{\sqrt{\lambda}}\,,\tag{6.39}$$

so choosing one of them means breaking the symmetry $\phi \to -\phi$ of the model.

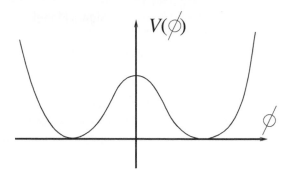

Figure 6.2 Potential for the Higgs model with a real scalar field.

In case the potential is for a complex field, with $\phi^2 \to |\phi|^2$, the Higgs potential is also called a "Mexican hat (sombrero) potential" or "wine bottle potential" due to its double well shape, rotated around the vertical axis.

DBI Model

Until now, we have considered generalizations with different $V(\phi)$. But also the kinetic terms can be modified. We don't need to have only $-(\partial_\mu\phi)^2/2$; we can have higher order terms in $\partial_\mu\phi$ as well.

One such example is the *Dirac-Born-Infeld (DBI)* Lagrangean,

$$\mathcal{L} = -\frac{1}{L^4}\sqrt{1 + L^4[(\partial_\mu\phi)^2 + m^2\phi^2]}. \tag{6.40}$$

The model is a scalar field variant of the Born–Infeld Lagrangean, which was written in 1932 as a nonlinear version of Maxwell's electrodynamics (see Exercise 4 of Chapter 5) that admits a nonsingular electron electromagnetic field. Dirac studied the model further and in particular the scalar version in Eq. (6.40). In string theory, the model was rediscovered as an important effective model for the motion of some objects called D-branes.

But the first physical application of the model was in Heisenberg's 1952 model for high-energy scattering of nucleons as a scalar model for the pions (which describe interactions of nucleons – i.e., for intermediate nuclear distances, as we said). This model will be described later on in the book.

Expanding the Lagrangean, we obtain

$$\mathcal{L} \simeq -\frac{1}{L^4}\left\{1 + \frac{L^4}{2}[(\partial_\mu\phi)^2 + m^2\phi^2] - \frac{L^8}{8}[(\partial_\mu\phi)^2 + m^2\phi^2]^2 + \cdots\right\}$$
$$= -\frac{1}{L^4} - \frac{1}{2}[(\partial_\mu\phi)^2 + m^2\phi^2] + \frac{L^4}{8}[(\partial_\mu\phi)^2 + m^2\phi^2]^2 + \cdots \tag{6.41}$$

As we see, the first term is an irrelevant constant, the second is the usual kinetic term for the scalar, but the next has a piece with four derivatives, $[(\partial_\mu\phi)^2]^2$. However, in this case, there are still only terms with $\partial_\mu\phi$, though with higher powers of this object.

Sigma Model

But in principle, we can also consider Lagrangeans with higher derivatives of the field, i.e.,

$$\mathcal{L}[\phi, \partial_\mu \phi, \partial_\mu \partial_\nu \phi, \partial_\mu \partial_\nu \partial_\rho \phi, \ldots]. \tag{6.42}$$

Note that in the order L^4 term in the DBI Lagrangean, we have the term $\phi^2(\partial_\mu \phi)^2$ term. That suggests that there is another possible generalization: we can consider the usual kinetic term, but multiplied by an arbitrary function of the fields (prefactor), i.e.,

$$\mathcal{L} = -\frac{1}{2}g(\phi)(\partial_\mu \phi)(\partial_\nu \phi)\eta^{\mu\nu}. \tag{6.43}$$

This is called the *sigma model* and is a generalization of the classical mechanics case of $g(q)\dot{q}^2/2$.

O(N) Model

We can also consider more than one scalar fields, $\phi^I(x^\rho)$.

The simplest case is the $O(N)$ *model*, with scalar fields $\phi^I, I = 1, \ldots, N$. The Lagrangean is simply the sum of Lagrangeans of the free massive scalars,

$$\mathcal{L} = \sum_{I=1}^{N} \mathcal{L}_I = -\frac{1}{2}\sum_{I=1}^{N} \partial_\mu \phi^I \partial_\nu \phi^I \eta^{\mu\nu} - \frac{1}{2}\sum_{I=1}^{N} m^2 \phi^I \phi^I. \tag{6.44}$$

The model is invariant under *abstract*, or internal, $O(N)$ rotations of the fields (unlike the $SO(3)$ example we have started the description of group theory with, which was a spatial rotation – i.e., a symmetry of spacetime), acting on the index I. The $O(N)$ rotation acts as

$$\phi^I \to \phi^{\prime I} = \Lambda^I{}_J \phi^J, \tag{6.45}$$

such that

$$\phi^{\prime I} \phi^{\prime I} = \phi^J \Lambda^I{}_J \Lambda^I{}_K \phi^K = \phi^J \phi^J, \tag{6.46}$$

which implies the $O(N)$ condition

$$\Lambda^I{}_J \delta_{II'} \Lambda^{I'}{}_K = (\Lambda^T \Lambda)_{JK} = \delta_{JK} \Rightarrow \Lambda \in O(N). \tag{6.47}$$

The kinetic term is similarly invariant under the same $O(N)$ rotation.

Nonlinear Sigma Model

With more than one scalar field, we can construct a more general (nonlinear) sigma model

$$\mathcal{L} = -\frac{1}{2}g_{IJ}(\phi^K)\partial_\mu \phi^I \partial_\nu \phi^J \eta^{\mu\nu}, \tag{6.48}$$

where $g_{IJ}(\phi^K)$ is a "metric in field space"; we will see why towards the end of the course.

The original (and simplest) nonlinear sigma model (NLSM) is obtained by imposing an $O(N)$-invariant constraint on the $O(N)$ model,

$$\phi^I \phi^I = A^2 = \text{fixed}. \tag{6.49}$$

When solving the constraint as $\phi^N = \phi^N(\phi^1, \ldots, \phi^{N-1})$ and substituting back in the Lagrangean, we get a NLSM for the independent fields $(\phi^1, \ldots, \phi^{N-1})$.

Important Concepts to Remember

- Scalar fields have numerical values invariant under Lorentz transformations, i.e., $\phi'(x'^\rho) = \phi(x^\rho)$.
- Scalar fields have Lagrangeans that depend on ϕ and (in principle) all of their derivatives.
- Local Lagrangeans are $L(t) = \int d^3x \mathcal{L}(\vec{x}, t)$, and relativistic ones are functions of $\phi, \partial_\mu \phi$, etc.
- The kinetic term for a scalar, generalizing $\dot{q}^2/2$, is $-\frac{1}{2}(\partial_\mu \phi)^2$, and contains T and a gradient potential energy term.
- In the potential energy, we add a term that is integral of a potential $V(\phi) = m^2 \phi^2/2 + \Delta V(\phi)$.
- For a free massive scalar $m \neq 0$, $\Delta V(\phi) = 0$, the equation of motion is the Klein-Gordon equation $(\Box - m^2)\phi = 0$.
- Besides polynomial potentials, the sine-Gordon model has a $(1 - \cos(a\phi))$ potential, and the Higgs model has a $(\phi^2 - \mu^2/\lambda)^2$ potential, with $m^2 = -\mu^2 < 0$, leading to symmetry breaking.
- The DBI Lagrangean has kinetic terms with more derivatives, though not more on a single scalar.
- The $O(N)$ model is the sum of N free massive scalar and has an abstract $O(N)$ symmetry that rotates them.
- The (nonlinear) sigma model has a metric in field space, so $\mathcal{L} = -\frac{1}{2}g_{IJ}(\phi^K)\partial_\mu \phi^I \partial^\mu \phi^J$.
- Imposing an $O(N)$ invariant constraint on the $O(N)$ model, we get a nonlinear sigma model.

Further Reading

In Peskin and Schroeder's book [4], there is not much about scalar fields, just a few examples in chapters 11.1, 13.3, and 20.1. In Burgess's book [5], we have a bit in chapter 18.

Exercises

(1)   Expand the DBI Lagrangean up to order L^8, and calculate the equations of motion up to that order.

(2)   (a) What are the vacua of the sine-Gordon model?
       (b) Consider a solution $\phi(x)$ (independent of time t) of the equations of motion of the sine-Gordon model that interpolates between two different vacua, ϕ_1 and ϕ_2, at $x = +\infty$ and $x = -\infty$, respectively. Can it have an arbitrarily small energy? Why?

(3)   Write the original nonlinear sigma model, obtained from the $O(N)$ model, in terms of just the first $N-1$ fields, solving for ϕ^N.

(4)   Calculate the equations of motion for the modified DBI model (with a function $g(\phi)$)

$$\mathcal{L} = -\frac{1}{L^4}\sqrt{1 + L^4[g(\phi)(\partial_\mu \phi)^2 + m^2 \phi^2]}. \tag{6.50}$$

(5)   Is the particular nonlinear sigma model Lagrangean

$$\mathcal{L} = -\frac{1}{2}g_{IJ}(\Phi^K \Phi^K)\partial_\mu \Phi^I \partial^\mu \Phi^J \tag{6.51}$$

invariant under $O(N)$?

7 Nonrelativistic Examples: Water Waves and Surface Growth

In the last chapter, and in general until now, we have considered relativistic real scalar fields. In this chapter, we will consider nonrelativistic examples of scalar fields that are easier to relate to, and we can easily see their physical interpretation.

We first take a nonrelativistic limit of the relativistic scalar action to find a nonrelativistic scalar one. We then move to hydrodynamics and construct a field theory for water waves. An example is the KdV equation, which admits a solitonic water wave, describing the real one. We finally describe the growth of surfaces in a classical field theory way that also usually includes random driving forces.

7.1 Nonrelativistic Scalar Action

We start off by taking a nonrelativistic limit of the relativistic action from the last chapter. However, it turns out that for this purpose, we must consider complex versions of the scalar fields rather than the real scalars. Consider then the Lagrangean for the scalar fields $\tilde{\phi} = \phi_1 + i\phi_2$,

$$\mathcal{L} = -|\partial_\mu \tilde{\phi}|^2 - m^2 |\tilde{\phi}|^2 = -(\partial_\mu \phi_1)^2 - (\partial_\mu \phi_2)^2 - m^2 \phi_1^2 - m^2 \phi_2^2 , \qquad (7.1)$$

which, therefore, is of the same form considered before in the real case. In order to take the nonrelativistic limit, we first reintroduce dimensions by replacing $t \to ct$, so $\partial_t \to \partial_t/c$, as well as $m \to mc/\hbar$. Then the Lagrangean is

$$\mathcal{L} = \frac{1}{c^2} |\partial_t \tilde{\phi}|^2 - |\vec{\nabla} \tilde{\phi}|^2 - \frac{m^2 c^2}{\hbar^2} |\tilde{\phi}|^2. \qquad (7.2)$$

The nonrelativistic limit is formally $c \to \infty$, but we must be careful how we take the limit. In particular, we must write an ansatz for the relativistic field $\tilde{\phi}$ in terms of the nonrelativistic field ϕ, which factors out the rest mass of the particle. Indeed, a wave function in quantum mechanics has a phase $e^{-i\frac{E}{\hbar}t}$, but in a nonrelativistic theory, we need to drop the rest energy piece. We also add a useful normalization factor; we then write

$$\tilde{\phi} = \frac{\hbar}{\sqrt{2m}} e^{-i\frac{mc^2}{\hbar}t} \phi. \qquad (7.3)$$

Under this rescaling, the kinetic term becomes

$$\frac{1}{c^2} |\partial_t \tilde{\phi}|^2 = \frac{m^2 c^2}{\hbar^2} \frac{\hbar^2}{2m} |\phi|^2 + i\hbar \phi^* \partial_t \phi + \mathcal{O}\left(\frac{1}{c^2}\right), \qquad (7.4)$$

where we have partially integrated the $\partial_t \phi^*$ term. Adding the kinetic and mass terms, the $\mathcal{O}(c^2)$ term above cancels, and taking the nonrelativistic limit $c \to \infty$, the action becomes, after a partial integration of $\vec{\nabla}$,

$$S_{\text{nonrel.}} = \int d^4x \left[i\hbar \phi^* \partial_t \phi + \frac{\hbar^2}{2m} \phi^* \vec{\nabla}^2 \phi \right] . \tag{7.5}$$

We see that its equation of motion is

$$\frac{\delta S}{\delta \phi^*} = 0 \Rightarrow i\hbar \partial_t \phi = -\frac{\hbar^2}{2m} \vec{\nabla}^2 \phi , \tag{7.6}$$

which is nothing but the Schrödinger equation for a free massive particle. That means that the field ϕ could, in this case, be interpreted as a the nonrelativistic wave function of the system. In fact, one can show that even for fermions, the same result is obtained: taking the nonrelativistic limit on the action for fermions, we obtain the same nonrelativistic action above, so, again, we can interpret the field as a wave function.

Then the relativistic field theory we studied before is, in some sense, a relativistic version of quantum mechanics, though this interpretation is not quite right. But since it is a *classical* field theory, the field must be made up of a large number of particles, such that can be in a classical situation, and quantum fluctuations to be small. For instance, an electromagnetic field made up of many photons is approximately classical and has small fluctuations.

The aforementioned observations are the start of the setup for quantum field theory, which is also called second quantization just because of the following observation: it amounts to a quantization of the field, interpreted as a wave function (therefore giving a *first* quantization). This reasoning will be developed further in quantum field theory.

7.2 Hydrodynamics

But, since this is a course in *classical* field theory, we must have nonrelativistic examples that are independent of quantum mechanics. This is indeed the case. Being relieved of the constraint of having Lorentz invariance, we don't need to consider the transformation properties of the fields (other than the usual transformations under the Galilean group of Newtonian mechanics), and we can consider any fields as simply functions of space and time (transforming correctly under rotations, if need be).

The examples we will start with come from fluid dynamics; namely, we consider *water waves*. Indeed, this is a good starting point, since the idea of soliton, to be described in detail in Part II of the book, arose from the observation of solitonic water waves in a canal in Scotland. For the current purposes, it will suffice to describe the fluid by the continuity and dynamics equations.

The *fields* in the case of general hydrodynamics are the density $\rho(\vec{x}, t)$, pressure $p(\vec{x}, t)$ and the velocity of the particles in the fluid, $\vec{v}(\vec{x}, t)$. They obey the continuity equation,

$$\frac{\partial \rho(\vec{x}, t)}{\partial t} + \vec{\nabla} \cdot (\rho(\vec{x}, t) \vec{v}(\vec{x}, t)) = 0 , \tag{7.7}$$

which describes just the conservation of energy, since if we integrate it over a volume V (as we will see in more detail later on in the book, when we will describe hydrodynamics more in depth), we obtain that the loss of energy in the volume equals the flux of energy through its surface, and we obtain the dynamics equation.

For an *ideal* fluid, the relevant dynamics equation is the Euler equation, obtained just as the force law per unit volume of the fluid,

$$\frac{d\vec{v}}{dt} \equiv \frac{\partial \vec{v}}{\partial t} + (\vec{v} \cdot \vec{\nabla})\vec{v} = \vec{f} - \frac{1}{\rho}\vec{\nabla}p \,, \tag{7.8}$$

where $\vec{f}$ is the external force per unit mass and p is the pressure.

Note that, since $(\vec{v} \cdot \vec{\nabla})\vec{v} = \vec{\nabla}(v^2/2) + (\vec{\nabla} \times \vec{v}) \times \vec{v}$, we obtain also the Lamb form of the dynamics (Euler) equation,

$$\frac{\partial \vec{v}}{\partial t} + \vec{\nabla}\left(\frac{v^2}{2}\right) + (\vec{\nabla} \times \vec{v}) \times \vec{v} = \vec{f} - \frac{1}{\rho}\vec{\nabla}p. \tag{7.9}$$

For a viscous fluid, the dynamics equation takes on an additional term, second order in derivatives on the velocity fields, obtaining the *Navier–Stokes equation*

$$\frac{d\vec{v}}{dt} \equiv \frac{\partial \vec{v}}{\partial t} + (\vec{v} \cdot \vec{\nabla})\vec{v} = \vec{f} - \frac{1}{\rho}\vec{\nabla}p + \frac{\eta}{\rho}\Delta\vec{v} \,, \tag{7.10}$$

where η is called shear viscosity. We also can define the coefficient $\nu \equiv \eta/\rho$.

We will not explain further the fundamentals of fluid dynamics here, since we will discuss it in more generality later on in the book. For now, we will just examine the consequences of these equations.

For an irrotational flow – i.e., with zero *vorticity*, $\vec{w} \equiv \vec{\nabla} \times \vec{v} = 0$ – we can define a potential function ϕ, such that

$$\vec{v} = -\vec{\nabla}\phi(\vec{x}, t). \tag{7.11}$$

The potential function is then a nonrelativistic scalar field $\phi(\vec{x}, t)$. Note that the vorticity is really a measure of the presence of vortices in the fluid, as then the fluid velocity lines form circles, and $\vec{w}$ is nonzero.

If the flow is also incompressible, i.e., – with constant density ρ = constant, the continuity equation then gives also $\vec{\nabla} \cdot \vec{v} = 0$, so the irrotational, incompressible flow obeys just the Laplace equation,

$$\vec{\nabla}^2\phi = 0 \,, \tag{7.12}$$

just like the electrostatic potential in vacuum in electromagnetism.

For an irrotational flow, (7.9) becomes

$$-\partial_t\vec{\nabla}\phi + \vec{\nabla}\left(\frac{v^2}{2}\right) - \vec{f} + \frac{1}{\rho}\vec{\nabla}p = 0. \tag{7.13}$$

If the flow is also incompressible (ρ = constant), integrating Eq. (7.13) gives the *Bernoulli law*,

$$-\dot{\phi} + \frac{v^2}{2} + \frac{p}{\rho} - \int \vec{f} \cdot d\vec{x} = C(t). \tag{7.14}$$

For only a gravitational force, $\int \vec{f} \cdot d\vec{x} = -gz$.

If we are in two spatial dimensions, the vorticity is a scalar, and $\vec{w} = \vec{\nabla} \times \vec{v}$ becomes

$$w = \epsilon^{ij}\partial_i v_j \,, \tag{7.15}$$

where ϵ^{ij} is the antisymmetric Levi-Civitta symbol in two dimensions, with $\epsilon^{12} = +1$.

Then, moreover, for incompressible flows (ρ = constant, so $\vec{\nabla} \cdot \vec{v} = 0$, i.e., $\partial^i v_i = 0$), we can write the velocity field in terms of a *stream function* ψ by

$$v^i = \epsilon^{ij}\partial_j \psi. \tag{7.16}$$

Indeed, then $\partial^i v_i = 0$. But, combining with the definition of vorticity, we find

$$w(\vec{x}, t) = \epsilon^{ij}\epsilon_{jk}\partial_i\partial_k\psi(\vec{x}, t) = \Delta\psi(\vec{x}, t). \tag{7.17}$$

So, unlike the potential ϕ, the more general stream function ψ obeys the Poisson equation.

As we know from electrostatics, in order to solve either the Laplace or the Poisson equation, we need to consider boundary conditions for the equations. However, we will not show here how to do this explicitly, since the methods are very similar.

7.3 Water Waves

For a system of water waves, namely waves for a fluid inside some space, we can consider the fluid (water) with some boundary conditions, relevant for the space.

Boundary Conditions

The boundary conditions must be given at a surface, like the surface of a lake or the surface of a moving ship. In general then, we consider a time-dependent surface given by an equation

$$F(x, y, z, t) = 0. \tag{7.18}$$

If the surface is time independent, the boundary condition must be that the velocity normal to the surface, $\vec{v} \cdot \vec{n} = 0$ (where $\vec{n}$ is the normal) at the surface. For a time-dependent surface (like the hull of a moving ship inside the water), the condition must be that the normal velocity equals the normal velocity of the moving surface, i.e.,

$$\left(\vec{v} \cdot \vec{n} - \frac{\partial_t F}{|\vec{\nabla}F|}\right)_{F(x,y,z,t)=0} = 0. \tag{7.19}$$

If the surface $F = 0$ is parametrized as

$$z = \eta(x, y, t) \,, \tag{7.20}$$

the boundary condition is written as

$$\frac{dz}{dt} \equiv v_z = \left.\frac{d\eta}{dt}\right|_{z=\eta(x,y,t)} \equiv (\partial_t\eta + v_x\partial_x\eta + v_y\partial_y\eta)_{z=\eta(x,y,t)} \,. \tag{7.21}$$

In this form, the condition is called the *kinematic free surface boundary condition (KFSBC)*. Note that for a lake, we have also a bottom, $z = -h$, where h is the height, and the vertical velocity must vanish, $v_z = 0$ at $z = -h$. Then in KFSBC we can also neglect the nonlinear term and write $v_z \simeq \partial_t \eta$ at $z = \eta$.

If we consider the fluid (water) in contact with the air, there is a surface tension σ at their interface, which means that for a curved surface with radii of curvature R_1 and R_2, the difference in pressures is

$$p_{\text{fluid}} = p_{\text{air}} + \sigma \left(\frac{1}{R_1} + \frac{1}{R_2} \right). \tag{7.22}$$

If, moreover, we consider that $1/R = 1/R_1 + 1/R_2 \simeq \partial_x^2 \eta$ on the surface (of a lake), we obtain from (7.14)

$$-\dot\phi + \frac{v^2}{2} + \frac{p_{\text{air}}}{\rho} + g\eta - \frac{\sigma}{\rho}\partial_x^2 \eta = C(t). \tag{7.23}$$

Taking the time derivative ∂_t, neglecting $\vec{v} \cdot \partial_t \vec{v}$, considering that $\dot{p}_{\text{air}} = 0$, and using the KFSBC $v_z \simeq \partial_t \eta$, we obtain at the surface $z = \eta$ the boundary condition

$$\ddot\phi = -g\partial_z \phi + \frac{\sigma}{\rho}\partial_x^2 \partial_z \phi. \tag{7.24}$$

Waves

Considering waves only in x and z (planar waves, y is an isometry), the solution of the Laplace equation, using separation of variables, is written as

$$\phi = e^{i\omega t} F(x) G(z). \tag{7.25}$$

For a wave solution of momentum k, the separated solution of the Laplace equation $(\partial_x^2 + \partial_z^2)\phi = 0$ is

$$\phi = e^{i\omega t}(A \cos kx + B \sin kx)(Ce^{kz} + De^{-kz}). \tag{7.26}$$

Imposing the bottom boundary condition $\partial_z \phi = 0$ at $z = -h$, we obtain

$$\phi = e^{i\omega t}(A \cos kx + B \sin kx)C \cosh k(h + z). \tag{7.27}$$

Then, imposing the surface boundary condition (7.24), we obtain the *dispersion relation*

$$\omega^2 = gk \left(1 + \frac{\sigma}{g\rho}k^2 \right) \tanh kh \equiv gk \left(1 + \frac{(ak)^2}{2} \right) \tanh kh \equiv c^2(k)k^2 , \tag{7.28}$$

where

$$a = \sqrt{\frac{2\sigma}{g\rho}} \tag{7.29}$$

is called the capillary length (and is about 4 mm for water at normal temperature), and $c(k)$ is the speed of propagation of the wave.

The two extreme approximations that we can have are:

• deep water, $kh \gg 1$, and
• shallow water, $kh \ll 1$.

In both cases, it is most likely to have $ka \ll 1$, since it means $\lambda \gg 2\pi a$, which, for water, is about 2.5 cm.

In the case of shallow water, with $ka \ll 1$, we obtain

$$c(k) \simeq \sqrt{gh}. \tag{7.30}$$

7.4 Korteweg–de Vries (KdV) Equation and Solitonic Water Wave

But we can consider higher-order terms in the expansion of the dispersion relation, and write

$$\omega^2 = c^2(k)k^2 \simeq gk\tanh kh \simeq gk^2h - \frac{1}{3}gk^4h^3 \Rightarrow \omega \simeq c_0 k - \gamma k^3. \tag{7.31}$$

We can then derive an equation for η as follows. The continuity equation for an incompressible fluid (independent on y) is $\vec{\nabla} \cdot \vec{v} = \partial_x v_x + \partial_z v_z = 0$. But the KFSBC is $v_z \simeq \partial_t \eta$, so, replacing it in the continuity equation and integrating over z, we obtain

$$\partial_x \int_0^{\eta(x)} v_x dz = -\partial_t \eta. \tag{7.32}$$

If v_x is independent of z (true, if true at $t = t_0$), then we obtain

$$\partial_x(v_x \eta) + \partial_t \eta = 0. \tag{7.33}$$

Moreover, for a wave generated by an initial height disturbance, one can find that the horizontal fluid velocity v_x satisfies $v_x - 2c = \text{constant} = -2c_0$ (we will not explain this further here, as it is a bit complicated to prove), so ($c_0 = \sqrt{gh_0}$)

$$v_x = 2\sqrt{g\eta} - 2\sqrt{gh_0}. \tag{7.34}$$

Substituting it in (7.33), we obtain

$$\partial_t \eta + (3\sqrt{g\eta} - 2c_0)\partial_x \eta = 0. \tag{7.35}$$

Expanding $\eta = h_0 + \tilde{\eta}$, so that

$$\sqrt{g\eta} = c_0 \left(1 + \frac{1}{2}\frac{\tilde{\eta}}{h_0}\right), \tag{7.36}$$

and substituting in (7.36), we find

$$\partial_t \tilde{\eta} + c_0 \left(1 + \frac{3}{2}\frac{\tilde{\eta}}{h_0}\right)\partial_x \tilde{\eta} = 0. \tag{7.37}$$

But this, in Fourier modes, is a dispersion relation, so we replace $c_0 k \rightarrow c_0 k - \gamma k^3$, as in (7.31), and move back to position space to finally obtain

$$\partial_t \tilde{\eta} + c_0 \left(1 + \frac{3}{2}\frac{\tilde{\eta}}{h_0}\right)\partial_x \tilde{\eta} + \gamma \partial_x^3 \tilde{\eta} = 0. \tag{7.38}$$

This is called the *Korteweg–de Vries (KdV) equation.*

It, in fact, has "solitonic" solutions that maintain their shape as they travel at a constant speed. The solutions are

$$\tilde{\eta} = \tilde{\eta}_0 \cosh^{-2}\left[\left(\frac{3\tilde{\eta}_0}{4h_0^3}\right)^{1/2} (x - Ut)\right],$$ (7.39)

where

$$\frac{\tilde{\eta}_0}{h_0} = 2\left(\frac{U}{c_0} - 1\right)$$ (7.40)

defines the velocity U of the traveling soliton. It is left as an exercise to prove that these are solutions. These solitonic water waves have been experimentally observed. In fact, the first observation of a soliton, and the name of soliton, was due to John Scott Russel, who followed for a long time a traveling soliton wave in a canal in Scotland.

We note a few properties of these solitons:

- The height $\tilde{\eta}_0$ increases with the velocity U.
- The width $\propto \tilde{\eta}_0^{-1/2}$ decreases with $\tilde{\eta}_0$ and U.
- In fact, solitons act as particles, scattering and reemerging from it having the same shape. This is due to a remarkable property of the KdV equation, *integrability*, which will be discussed in the next chapter.

The standard (abstract) form of the KdV equation is

$$\partial_t \phi + 6\phi \partial_x \phi + \partial_x^3 \phi = 0.$$ (7.41)

7.5 The Kuramoto–Sivashinsky (KS) Equation

Consider a linear flow of a viscous fluid down an inclined plane of angle θ and a reference system relative to the inclined plane, so x is parallel to the plane and z is perpendicular to it. Then one can show that the perturbation theory for the flow for the height $\eta = 1 + ab\tilde{\eta}$, where a, b are constants depending on the parameter of the system, reduces to the equation

$$\partial_t \tilde{\eta} + \tilde{\eta}\partial_x\tilde{\eta} + \partial_x^2\tilde{\eta} + \partial_x^4\tilde{\eta} = 0.$$ (7.42)

We will not prove this here. This equation is called the Kuramoto–Sivashinsky (KS) equation.

Another form for it is obtained by writing $\tilde{\eta} = -2\partial_x\phi$ and then integrating the equation over x, to obtain

$$\partial_t \phi = (\partial_x\phi)^2 - \partial_x^2\phi - \partial_x^4\phi.$$ (7.43)

This is another important example of a nonrelativistic field theory, and much studied one, though we only mention it here.

7.6 Growth of Surfaces

The KS equation is relevant for another system of physical importance, the growth of surfaces – i.e., interfaces of two media. This can be, for instance, a sand surface being deposited, a colony of bacteria growing, etc. Either way, it describes the one-dimensional spread of a surface.

If the system would be at equilibrium, we would expect to be able to use the one-dimensional diffusion equation for the height $h(t, x)$ of the one-dimensional interface in two dimensions,

$$\frac{dh}{dt} = \nu \frac{d^2 h}{dx^2}. \tag{7.44}$$

A way to model it would be to add a random term $\eta(x, t)$, for a random uncorrelated noise, i.e., with statistical average

$$\langle \eta(x, t) \eta(x', t') \rangle \simeq \Gamma \delta(x - x') \delta(t - t'), \tag{7.45}$$

on the right hand side of the diffusion equation. But to emphasize that the growth is a nonlinear process, not one of equilibrium, we can add yet another term on the right hand side, obtaining

$$\frac{dh}{dt} = \nu \frac{\partial^2 h}{dx^2} + \frac{\lambda}{2} (\partial_x h)^2 + \eta. \tag{7.46}$$

This is called the *Karder-Parisi-Zhang (KPZ) equation*.

But this is a somewhat unsatisfactory equation, since it contains a random component. In order to have a deterministic equation, we can use the same KS equation for fluid flow on inclined planes in the form (7.43) for $h = \phi$. We see that we need to change the sign of the diffusing term, which now exponentially increases instead of exponentially decreasing, but to compensate, we added a fourth-order derivative to cut off large momenta and the nonlinear term to saturate the growth. This equation, though deterministic, is highly chaotic, so it shares some common features with the solutions of the KPZ equation.

In conclusion, in this chapter, we have seen that we can have nonrelativistic field theories with nontrivial equations of motion. They describe a variety of physical situations, from wave functions to hydrodynamics and water waves, growth of surfaces, and others. In the following chapters, however, we will mostly consider relativistic systems and field theories.

Important Concepts to Remember

- The nonrelativistic limit $c \to \infty$ of a massive relativistic scalar, rescaled by $e^{-i\frac{mc^2}{\hbar}t}$, is the Lagrangean for a free Schrödinger wavefunction, $\int [\phi^* (i\hbar \partial_t \phi + \hbar^2/(2m) \vec{\nabla}^2) \phi$.
- Hydrodynamics is defined by the continuity equation, and the dynamics equation, Euler for ideal fluids, and Navier–Stokes for viscous fluids.
- Irrotational flow is potential, $\vec{v} = -\vec{\nabla}\phi$, and if it's also incompressible, it satisfies the Laplace equation $\Delta \phi = 0$.

- In two spatial dimensions, incompressible flow has $v^i = \epsilon^{ij}\partial_j\psi$, and then $\Delta\psi = w$ (vorticity).
- For water waves, we solve the Laplace equation with some boundary conditions. For a lake, we use the bottom boundary condition $v_z = 0$ at $z = -h$, and a condition at a moving surface, $v_z = d\eta/dt$ at $z = \eta$.
- From the boundary condition, one finds the dispersion relation of the water waves, $\omega^2 = k^2 c^2(k) = gk(1 + a^2k^2/2)\tanh kh$.
- For nonlinear effects in water waves, we obtain the KdV equation, $\partial_t\eta + c_0(1 + 3\eta/2h_0)\partial_x\eta + \gamma\partial_x^3\eta = 0$.
- The KdV equation has solitonic traveling waves.
- For viscous flow down a plane, one finds the KS equation, $\partial_t u + u\partial_x u + \partial_x^2 u + \partial_x^4 u = 0$, or $\partial_t\phi + \partial_x^2\phi + \partial_x^4\phi - (\partial_x\phi)^2 = 0$.
- Deterministic (yet chaotic) growth of surfaces can also be defined by the KS equation. Otherwise, it can be described by the KPZ equation, which has a random component $\eta(x, t)$.

Further Reading

For hydrodynamics, see Landau and Lifshitz's *Fluid Mechanics* volume [6]. For a detailed description of the dynamics of interfaces, the KPZ and KS equations (and for turbulent behavior in water waves) see [7].

Exercises

(1)   Show that, for a shallow water wave, one has the equation

$$\partial_t v_x + v_x\partial_x v_x = -g\partial_x\eta. \tag{7.47}$$

(2)   Show that the KdV soliton (7.39) solves the KdV equation.

(3)   Try writing a Lagrangian for the KS equation (such that the KS equation is its Euler–Lagrange equation). Then write one in the case we drop *one* of the terms in the equation.

(4)   Write a spherically symmetric stream function ψ in the case of a delta-function vorticity $w(\vec{x}, t) = q\delta^3(\vec{x})$.

Classical Integrability: Continuum Limit of Discrete, Lattice, and Spin Systems

To understand better the notion of field, besides having concrete examples like in the last chapter, it will also be useful to have a discretization of the field. In the case of hydrodynamics from the last chapter, it was implicit that there is a discretization. In terms of the particles composing the fluid, we just looked at larger distances than those of the particles. In this chapter, we will instead consider discrete systems and see how the fields emerge in the limit of large number of points. We will also examine the notion of (classical) integrability, which was touched on in the last chapter for discrete systems, for which it is easier to describe.

8.1 Classical Integrability

Consider a system with a finite number n of degrees of freedom, and with Hamiltonian $H(x_i, p_j)$, where x_i are positions and p_j are momenta. Its Poisson brackets, defined in Chapter 1, will be

$$\{x_i, p_j\}_{P.B.} = \delta_{ij}. \tag{8.1}$$

Its Hamiltonian equations of motion are

$$\dot{x}_i = \{x_i, H\}_{P.B.} = \frac{\partial H}{\partial p_i}; \quad \dot{p}_i = \{p_i, H\}_{P.B.} = -\frac{\partial H}{\partial x_i}. \tag{8.2}$$

Then we say that this system is integrable if and only if *there are n integrals of motion* $I_i(x, p)$, $i = 1, \ldots, n$. That is, the integrals of motion are constant in time,

$$\dot{I}_i = \{I_i, H\}_{P.B.} = 0, \tag{8.3}$$

and they are independent, i.e.,

$$\{I_i, I_j\}_{P.B.} = 0. \tag{8.4}$$

It is not always the case, but integrable systems sometimes have a *Lax pair*, namely a pair of $N \times N$ matrices $L(x, p)$ and $M(x, p)$, such that the equation

$$\dot{L} = [L, M], \tag{8.5}$$

called the Lax equation, is equivalent to the Hamiltonian equations of motion of the system. Here, N is unknown, and, moreover, there is no algorithm to tell us even if there is a Lax pair, much less to actually find it. The only thing we know is that $N \geq n$, since we need at least n independent integrals of motion. Indeed, knowing the Lax operator L, we can construct integrals of motion as

$$I_i = \text{Tr}[L^{n_i}],\tag{8.6}$$

where n_i is an integer. Indeed, then

$$\dot{I}_i = n_i \text{Tr}[\dot{L}L^{n_i-1}] = n_i \text{Tr}[[L,M]L^{n_i-1}] = 0,\tag{8.7}$$

where in the last equality we used cyclicity of the trace. For $n_i \geq N$, the I_i are functionally dependent on the lower n_i, since the trace can be expressed in terms of the lower power traces. That means that, indeed, we have $N \geq n_i$, so $N \geq n$.

The Lax pair is not unique. In fact, there is a "gauge invariance" changing the Lax pair but, leaving the Lax equation invariant,

$$L \to S^{-1}LS, \quad M \to S^{-1}MS - S^{-1}\dot{S}.\tag{8.8}$$

We can make the analogy to gauge invariance more precisely by introducing a spurious coordinate σ: L, M are independent of it. We then also define $\alpha = 0, 1$ standing for time t as $\alpha = 0$ and σ as $\alpha = 1$, and L_α is $L_0 = M, L_1 = L$. We can thus write (since $\partial_1 L_0 = \partial_\sigma M = 0$) the Lax equation as

$$\partial_0 L_1 - \partial_1 L_0 + [L_0, L_1] = 0 \Leftrightarrow \partial_\alpha L_\beta - \partial_\beta L_\alpha + [L_\alpha, L_\beta] = 0.\tag{8.9}$$

We will see later in the book that the left-hand side is the field strength of a Yang–Mills gauge field, which thus vanishes but must, moreover, be invariant. We can easily check explicitly (a fact that that will be important later on and is left as an exercise) that (8.9) is left invariant by the "gauge invariance"

$$L_\alpha \to hL_\alpha h^{-1} + (\partial_\alpha h)h^{-1}.\tag{8.10}$$

A stronger form of integrability appears if there is a complex parameter z, called a *spectral parameter*, such that $(L(z), M(z))$ are Lax pairs for all z. Then the integrals of motion can be recovered from the Taylor expansion of the traces of the Lax operators,

$$\text{Tr}\, L(z) = \sum_n L_n z^n.\tag{8.11}$$

8.2 Examples of Integrable Systems

1. **The nonperiodic Toda system**. It is a nonrelativistic system of $n + 1$ points on a *linear* "chain," with exponential nearest-neighbor interaction – i.e., with the Hamiltonian

$$H = \frac{1}{2}\sum_{i=1}^{n+1} p_i^2 - M^2 \sum_{i=1}^{n} e^{x_{i+1}-x_i}.\tag{8.12}$$

The center of mass coordinate $x_0 = \sum_i x_i$ decouples, leaving only n degrees of freedom. The system with $n = 1$ becomes

$$H = \left[\frac{(p_1 + p_2)^2}{4}\right] + \frac{(p_1 - p_2)^2}{4} - M^2 e^{x_2 - x_1}. \tag{8.13}$$

Dropping the center of mass motion, and defining $x = (x_1 - x_2)/\sqrt{2}$, we obtain

$$H = \frac{p^2}{2} - M^2 e^{-\sqrt{2}x}. \tag{8.14}$$

This system is called *Liouville theory*.

2. **The periodic Toda system**. It is a nonrelativistic system of $n + 1$ points on a *circular* "chain," with exponential nearest-neighbor interaction – i.e., with the Hamiltonian

$$H = \frac{1}{2} \sum_{i=1}^{n+1} p_i^2 - M^2 \sum_{i=1}^{n+1} e^{x_{i+1} - x_i}, \tag{8.15}$$

where $x_{n+2} = x_1$. As before, the center of mass decouples, leaving only n degrees of freedom. For $n = 1$, the Hamiltonian reduces to

$$H = \left[\frac{(p_1 + p_2)^2}{4}\right] + \frac{(p_1 - p_2)^2}{4} - M^2 (e^{x_2 - x_1} - e^{x_1 - x_2}). \tag{8.16}$$

Dropping the center of mass motion and defining, as before, $x = (x_1 - x_2)/\sqrt{2}$, we obtain

$$H = \frac{p^2}{2} - 2M^2 \cosh \sqrt{2}x, \tag{8.17}$$

which is just a *Sinh-Gordon system* (Sine-Gordon with Sine replaced with Sinh).

One can write down Lax pairs for these cases, but in order to do so, we would need to introduce some group theory and Lie algebra concepts, so we will not do it.

We can next consider nonrelativistic models of $n + 1$ particles on a complex line with general two-body interaction, generically called the *Calogero-Moser systems*.

3. **The rational Calogero-Moser system**, with Hamiltonian

$$H = \frac{1}{2} \sum_{i=1}^{n+1} p_i^2 - \frac{1}{2} m^2 \sum_{i \neq j}^{n+1} \frac{1}{(x_i - x_j)^2}. \tag{8.18}$$

4. **The trigonometric Calogero-Moser system**, with Hamiltonian

$$H = \frac{1}{2} \sum_{i=1}^{n+1} p_i^2 - \frac{1}{2} m^2 \sum_{i \neq j}^{n+1} \frac{1}{\sin^2(x_i - x_j)}. \tag{8.19}$$

5. **The ellyptic Calogero-Moser system**, with Hamiltonian

$$H = \frac{1}{2} \sum_{i=1}^{n+1} p_i^2 - \frac{1}{2} m^2 \sum_{i \neq j}^{n+1} \mathcal{P}(x_i - x_j; \omega_1, \omega_2), \tag{8.20}$$

where $\mathcal{P}$ is the Weierstrass elliptic (doubly periodic) function

$$\mathcal{P}(x; \omega_1, \omega_2) \equiv \frac{1}{x^2} + \sum_{(m,n) \neq (0,0)} \left[\frac{1}{(x + 2m\omega_1 + 2n\omega_2)^2} - \frac{1}{(2m\omega_1 + 2n\omega_2)^2} \right]. \quad (8.21)$$

Here $2\omega_1, 2\omega_2$ are periods, and for $\omega_2 \to \infty$, one gets the trigonometric case, and if also $\omega_1 \to \infty$, one gets the rational case.

The elliptic Calogero-Moser system admits a Lax pair with spectral parameter z,

$$L_{ij}(z) = p_i \delta_{ij} - m(1 - \delta_{ij}) \Phi(x_i - x_j; z)$$
$$M_{ij}(z) = d_i(x) \delta_{ij} + m(1 - \delta_{ij}) \Phi'(x_i - x_j; z) \quad (8.22)$$

where $\Phi'(x; z) \equiv \partial_x \Phi(x; z)$,

$$d_i(x) = m \sum_{k \neq i} \mathcal{P}(x_i - x_k), \quad (8.23)$$

$\Phi(x; z)$ is the Lamé function,

$$\Phi(x; z) = \frac{\sigma(z - x)}{\sigma(x)\sigma(z)} e^{x\zeta(z)}, \quad (8.24)$$

$\mathcal{P}(z) = -\zeta'(z)$, $\zeta(z) = \sigma'(z)/\sigma(z)$, and $\zeta(z)$ is the Weierstrass function.

Toda systems can be obtained as limits of elliptic Calogero-Moser systems, by

$$\omega_1 = -i\pi, \ Re(\omega_2) \to \infty, \ m = M e^{\delta\omega_2}; \ x_j = X_j + 2\omega_2 \delta j \quad (8.25)$$

with $M, X_j, 0 \le \delta \le 1/N$ fixed. In this way, we can write the Lax pairs for the Toda systems as well.

8.3 Classical Integrable Fields

We can now also construct integrable systems for *fields* in 1+1 dimensions by generalizing the case of the particles (0+1 dimensions). For instance,

1. *Toda field theory* for n scalars $\phi^I, I = 1, \ldots, n$ in 1+1 dimensions has the Lagrangean

$$\mathcal{L} = -\frac{1}{2} \sum_I \partial_\mu \phi^I \partial^\mu \phi^I - M^2 \sum_I e^{\phi^{I+1} - \phi^I}. \quad (8.26)$$

2. *Korteweg–de Vries (KdV) equation*, already studied. It is for a single scalar field in 1+1 dimensions, $u(t, x)$, with equation of motion

$$\partial_t u = 6u \partial_x u - \partial_x^3 u. \quad (8.27)$$

The equation has rational solutions, of the form

$$u(t, x) = 2 \sum_j \frac{1}{(x - x_j(t))^2}, \quad (8.28)$$

only if the $x_j(t)$ satisfy the equation

$$\partial_t x_j = \{x_j, \operatorname{Tr} L^3\}_{P.B.},\tag{8.29}$$

where L is the Lax operator of the rational Calogero-Moser system, as one can check (left as an exercise to prove).

We see then that, in some sense, we can construct (the solutions of) an integrable field theory based on a discrete integrable model.

8.4 Spin Systems and Discretization

We now consider a simple way to understand the 1+1 and higher-dimensional systems, namely a discretization. A discretized spatial dimension is simply a chain of sites, $j = 1, \ldots, N$. At each site, we have a variable ("field"), which we will call spin S^a even if it is a variable more general than the spin of a particle. Thus, we have the field $S_i^a(t)$, over discretized space i and time t.

If S_i^a is the actual spin 1/2 of a fermionic particle at site i, then the simplest model we can consider is the *Heisenberg XXX$_{1/2}$ model*, for interactions that are nearest-neighbor only, with Hamiltonian describing the interaction that is spin–spin (magnetic), i.e.,

$$H = J \sum_{j=1}^{L} \vec{S}_j \cdot \vec{S}_{j+1}.\tag{8.30}$$

This can be generalized, if the couplings for various directions of the spin are different, to the *XYZ$_{1/2}$ model*,

$$H = \sum_{j=1}^{L} \sum_{a=1}^{3} J_a S_j^a S_{j+1}^a.\tag{8.31}$$

The solution of the quantum model can be found by the ansatz proposed by Bethe in 1931 for the wave function, called *Bethe ansatz*, but we will not explain it here, as it is outside the scope of the book.

Here, we will note that we can consider the chain as a discretization, and then the derivative of a function becomes

$$f'(x) \to \frac{1}{a}(f_{j+1} - f_j),\tag{8.32}$$

where a is the step size in the chain. Moreover, the sum becomes an integral, since $dx = a$, so

$$\sum_{j=1}^{L} = \frac{1}{a} \int_0^{La} dx.\tag{8.33}$$

Then, in the continuum limit, $L \to \infty$, aL = fixed, the Hamiltonian

$$H = J|\vec{S}^2| \sum_{j=1}^{L} \cos\left(\theta_j - \theta_{j+1}\right) \simeq J|\vec{S}^2| \left[L - \frac{1}{2}\sum_{j=1}^{L}(\theta_j - \theta_{j+1})^2\right] \tag{8.34}$$

becomes

$$H = J|\vec{S}^2| \left[L - \frac{a}{2}\int_0^{La} dx(\partial_x\theta)^2\right]. \tag{8.35}$$

Here $|\vec{S}^2|$ is a constant, for spin 1/2 being (at the quantum level) $S(S+1) = 3/4$.

This doesn't have the right behavior as $a \to 0$ unless we redefine $Ja \equiv J' =$ finite in the limit.

However, if we consider a 2+1 dimensional system instead of the 1+1 dimensional one, it does have the right behavior anyway. Consider then a two-dimensional spatial *lattice*, and nearest neighbor interaction, where the nearest neighbors are denoted by $\langle ij \rangle$. Add also the constants $\sum \vec{S}^2$, to obtain a Hamiltonian

$$H = \frac{J}{2}\sum_{\langle ij \rangle}(\vec{S}_i - \vec{S}_j)^2 = J|\vec{S}^2|\sum_{\langle ij \rangle}[1 - \cos(\theta_i - \theta_j)] \simeq \frac{J}{2}|\vec{S}^2|\sum_{\langle ij \rangle}(\theta_i - \theta_j)^2. \tag{8.36}$$

Then, in the continuum limit, we obtain

$$H = \frac{J|\vec{S}^2|}{2}\int d^2x(\vec{\nabla}\theta)^2. \tag{8.37}$$

Adding also a kinetic term, this is just the Hamiltonian for the free massless scalar in 2+1 dimensions,

$$H \propto \frac{1}{2}\int d^2x[\dot{\theta}^2 + (\vec{\nabla}\theta)^2]. \tag{8.38}$$

We have described how to take a continuum limit of spin systems and obtain a field theory, but, as we said, "spin" here can be, in principle, anything. We can thus define any kind of degrees of freedom on the discrete "lattice" of sites, and in the continuum limit, we will find a corresponding field theory.

Important Concepts to Remember

- Integrable systems are systems with as many integrals of motion as degrees of freedom.
- Integrable systems sometimes have Lax pairs L and M, which have Hamiltonian equations of motion equivalent with the equations $\dot{L} = [L, M]$. The Lax pair is not unique; there is a gauge invariance acting on it.
- Integrable systems sometimes have a Lax pair, depending on a spectral parameter z (complex).
- Examples of integrable systems are Toda systems, with exponential type potentials, and Calogero-Moser, with inverse polynomial type potentials.
- In 1+1 dimensions, Toda field theories and the Korteweg–de Vries (KdV) equation are integrable systems related to the Lax pairs of 0+1 dimensional systems.

- The Heisenberg model is a model with spin interactions only in the nearest neighbors. The rotationally invariant fermion case is the $XXX_{1/2}$ model.
- The continuum limit of the Heisenberg model gives a free massless scalar.

Further Reading

See, for instance, the book on integrability [8] for a detailed description of the models presented here.

Exercises

(1)   Check that $L_\alpha \to hL_\alpha h^{-1} + (\partial_\alpha h)h^{-1}$ leaves invariant the equation

$$\partial_\alpha L_\beta - \partial_\beta L_\alpha + [L_\alpha, L_\beta] = 0. \tag{8.39}$$

(2)   Prove that the KdV equation admits rational solutions only if

$$\partial_t x_j = \{x_j, \mathrm{Tr}\, L^3\}_{P.B.}, \tag{8.40}$$

where L is the Lax operator of the ellyptic Calogero-Moser system.

(3)   Check that the elliptic Calogero-Moser system has a Lax pair given by

$$\begin{aligned}
L_{ij}(z) &= p_i \delta_{ij} - m(1 - \delta_{ij})\Phi(x_i - x_j; z) \\
M_{ij}(z) &= d_i(x)\delta_{ij} + m(1 - \delta_{ij})\Phi'(x_i - x_j; z),
\end{aligned} \tag{8.41}$$

as claimed in the text.

(4)   Find the field theory corresponding to a spin–spin interaction

$$\sum_{\langle ij \rangle, \langle jk \rangle} (\vec{S}_i - \vec{S}_j) \cdot (\vec{S}_j - \vec{S}_k), \tag{8.42}$$

in 1+1, and in 2+1 dimensions.

9 Poisson Brackets for Field Theory and Equations of Motion: Applications

In this chapter, we will expand our understanding of Poisson brackets, with which we are familiar from classical mechanics. We will first study them within the more formal description of the symplectic formulation; then we will learn how to generalize this description in the field theory case. Finally, we will study two concrete examples as applications of the formalism in order to see how we can make use of Poisson brackets.

As we saw in Chapter 1, for a classical mechanics system, we can define the Poisson brackets of two functions of phase space, $f(q,p), g(q,p)$, as

$$\{f,g\}_{P.B.} = \sum_k \left(\frac{\partial f}{\partial q_k} \frac{\partial g}{\partial p_k} - \frac{\partial f}{\partial p_k} \frac{\partial g}{\partial q_k} \right). \tag{9.1}$$

In particular, we can write fundamental Poisson brackets for the variables (q,p):

$$\begin{aligned} \{q_j, p_k\}_{P.B.} &= \delta_{jk} = -\{p_k, q_j\}_{P.B.} \\ \{q_j, q_k\}_{P.B.} &= 0 = \{p_j, p_k\}_{P.B.}. \end{aligned} \tag{9.2}$$

We have also defined canonical transformation, which are transformations on the phase space, $(q,p) \to (Q,P)$, that preserve the existence of a Hamiltonian. The fundamental Poisson brackets are invariant under canonical transformations.

9.1 Symplectic Formulation

We now develop a so-called symplectic formulation for classical mechanics in the Hamiltonian formalism. First, we define the *symplectic form*, a $2N \times 2N$ matrix,

$$\Omega = \begin{pmatrix} 0 & \mathbb{1} \\ -\mathbb{1} & 0 \end{pmatrix}. \tag{9.3}$$

Then, we can define the *symplectic group*, $Sp(2N, \mathbb{R})$, as the group of real matrices M satisfying the relation

$$M^T \Omega M = \Omega, \tag{9.4}$$

i.e., the analog of the orthogonal condition, except with a metric Ω instead of $\mathbb{1}$.

We can also define *unitary symplectic* matrices, $USp(2N)$, as the group of unitary (thus over $\mathbb{C}$) symplectic matrices.

Note that

$$\Omega^T = \begin{pmatrix} 0 & -\mathbb{1} \\ \mathbb{1} & 0 \end{pmatrix} \equiv -\Omega = \Omega^{-1} \Rightarrow \Omega^T\Omega = \Omega\Omega^T = \mathbb{1}. \tag{9.5}$$

That means also that

$$\Omega^2 = -\mathbb{1} \quad \text{but} \quad \det \Omega = +1. \tag{9.6}$$

In order to construct a symplectic formulation of the Hamiltonian formalism, we define phase space variables η by joining q's and p's together,

$$\eta_i = q_i; \quad \eta_{i+n} = p_i, \quad i \le n. \tag{9.7}$$

Then the Hamiltonian equations of motion can be written in the $2n \times 2n$ matrix form as

$$\dot{\eta} = \Omega \frac{\partial H}{\partial \eta}, \tag{9.8}$$

where we have defined in an obvious manner,

$$\left(\frac{\partial H}{\partial \eta}\right)_i = \frac{\partial H}{\partial q_i}; \quad \left(\frac{\partial H}{\partial \eta}\right)_{i+n} = \frac{\partial H}{\partial p_i}. \tag{9.9}$$

Moreover, then, the Poisson brackets can be written in symplectic form as

$$\{f,g\}_{P.B.} = \left(\frac{\partial f}{\partial \eta}\right)^T \Omega \frac{\partial g}{\partial \eta}. \tag{9.10}$$

With this notation, the equations of motion in Poisson bracket form,

$$\begin{aligned} \dot{q}_i &= \{q_i, H\}_{P.B.} \\ \dot{p}_i &= \{p_i, H\}_{P.B.}, \end{aligned} \tag{9.11}$$

can be written in symplectic form as

$$\dot{\eta} = \{\eta, H\}_{P.B.}. \tag{9.12}$$

Moreover, the fundamental Poisson brackets are

$$\{\eta, \eta\}_{P.B.} = \Omega. \tag{9.13}$$

A canonical transformation is then a change, $\eta \to \zeta(\eta)$, $\zeta = \begin{pmatrix} Q \\ P \end{pmatrix}$. If, moreover, the functional form of the Hamiltonian is unchanged, $\tilde{H} = H$ in the new variables, then it means that we also have

$$\dot{\zeta} = \Omega \frac{\partial H}{\partial \zeta}. \tag{9.14}$$

But, from the coordinate transformation $\zeta = \zeta(\eta)$,

$$\dot{\zeta}_i = \frac{\partial \zeta_i}{\partial \eta_j} \dot{\eta}_j \equiv M_{ij} \dot{\eta}_j, \tag{9.15}$$

which also implies

$$\frac{\partial H}{\partial \eta} = M^T \frac{\partial H}{\partial \zeta}, \tag{9.16}$$

and using the Hamiltonian equations, we obtain

$$\dot{\zeta} = M\dot{\eta} = M\Omega\frac{\partial H}{\partial \eta} \, , \qquad (9.17)$$

so, finally,

$$\dot{\zeta} = M\Omega M^T\frac{\partial H}{\partial \zeta} = \Omega\frac{\partial H}{\partial \zeta}. \qquad (9.18)$$

The two forms for $\dot{\zeta}$ imply a condition on the matrix M,

$$M\Omega M^T = \Omega \, , \qquad (9.19)$$

i.e., the matrix is symplectic. In conclusion, canonical transformations with $H = \tilde{H}$ are symplectic.

9.2 Generalization to Fields

We now generalize to field theory, like we learned how to do in Chapters 6 and 8, by using discretization of space. The usual definition of the momentum conjugate, $p_i \equiv \partial L/\partial \dot{q}_i$, becomes, when i is meant to label discretized space $\vec{x}_i$,

$$p(\vec{x}) = \frac{\partial L}{\partial \dot{\phi}(\vec{x})} = \frac{\partial}{\partial \dot{\phi}(\vec{x})}\int d^3y \mathcal{L}(\phi(\vec{y},t), \partial_\mu\phi(\vec{y},t)). \qquad (9.20)$$

Under the assumption of only an infinitesimal integration domain, we obtain

$$p(\vec{x}) = d^3x\frac{\partial \mathcal{L}(\vec{x},t)}{\partial \dot{\phi}(\vec{x},t)} \equiv \pi(\vec{x})d^3x. \qquad (9.21)$$

Here, $\pi(\vec{x})$ is a canonical momentum conjugate *density* and is written as

$$\pi(\vec{x}) = \frac{\partial \mathcal{L}}{\partial \dot{\phi}(\vec{x})} = \frac{\delta S}{\delta \dot{\phi}(\vec{x})} \, , \qquad (9.22)$$

where the symbol δ refers to functional differentiation, i.e., – when we have a functional (integral over the domain of a function f of a functional of f) – we remove the integral and then use normal differentiation.

Now discretizing, as we learned in Chapter 6 (with a unit of volume ΔV), just being careful about normalization, we write

$$\sqrt{\Delta V}\phi_i(t) \equiv q_i(t)$$
$$\sqrt{\Delta V}\pi_i(t) \equiv p_i(t). \qquad (9.23)$$

Here $\phi_i(t) \equiv \phi(\vec{x}_i, t)$ is the discretized field. Then, moreover, we have the map

$$\frac{\delta f(\phi(\vec{x},t), \pi(\vec{x},t))}{\delta\phi(\vec{x},t)} \rightarrow \frac{1}{\Delta V}\frac{\partial f_i(t)}{\partial \phi_j(t)}$$
$$d^3x \rightarrow \Delta v. \qquad (9.24)$$

As an example of functional differentiation, for a quadratic term in the potential,

$$V_1(t) = \int d^3x m^2 \frac{\phi^2(\vec{x}, t)}{2} , \qquad (9.25)$$

we have

$$\frac{\delta V_1(t)}{\delta \phi(\vec{x}, t)} = m^2 \phi(\vec{x}, t). \qquad (9.26)$$

The map to classical mechanics suggest a generalization of the Poisson brackets for fields, namely for functionals f, g of the field phase space,

$$\{f, g\}_{P.B.} = \int d^3x \left[\frac{\delta f}{\delta \phi(\vec{x}, t)} \frac{\delta g}{\delta \pi(\vec{x}, t)} - \frac{\delta f}{\delta \pi(\vec{x}, t)} \frac{\delta g}{\delta \phi(\vec{x}, t)} \right]. \qquad (9.27)$$

In this case, taking into account that

$$\frac{\delta \phi(\vec{y}, t)}{\delta \phi(\vec{x}, t)} = \delta^3(\vec{x} - \vec{y}); \quad \frac{\delta \pi(\vec{y}, t)}{\delta \pi(\vec{x}, t)} = \delta^3(\vec{x} - \vec{y}) , \qquad (9.28)$$

we obtain for the fundamental Poisson brackets

$$\{\phi(\vec{x}, t), \pi(\vec{x}', t\}_{P.B.} = \delta^3(\vec{x} - \vec{x}') = -\{\pi(\vec{x}', t), \phi(\vec{x}, t)\}_{P.B.}$$
$$\{\phi(\vec{x}, t), \phi(\vec{x}', t)\}_{P.B.} = 0 = \{\pi(\vec{x}, t), \pi(\vec{x}', t\}_{P.B.}. \qquad (9.29)$$

The symplectic formulation is obtained by defining the phase space variables

$$\eta(\vec{x}, t) = (\phi(\vec{x}, t), \pi(\vec{x}, t)), \qquad (9.30)$$

in terms of which the Poisson brackets of two functions of phase space are

$$\{f, g\}_{P.B.} = \int d^3x \left(\frac{\delta f}{\delta \eta} \right)^T \Omega \frac{\delta g}{\delta \eta}. \qquad (9.31)$$

The fundamental Poisson brackets are then written as

$$\{\eta, \eta\}_{P.B.} = \Omega. \qquad (9.32)$$

Moreover, the Hamiltonian equations of motion,

$$\dot{\phi}(\vec{x}, t) = \{\phi(\vec{x}, t), H(t)\}_{P.B.}$$
$$\dot{\pi}(\vec{x}, t) = \{\pi(\vec{x}, t), H(t)\}_{P.B.} , \qquad (9.33)$$

can be written in symplectic form as

$$\dot{\eta}(\vec{x}, t) = \{\eta(\vec{x}, t), H(t)\}_{P.B.}. \qquad (9.34)$$

Again, as before, the canonical transformation is a symplectic transformation on $\eta(\vec{x}, t)$.

The Hamiltonian is now an integral over space of a Hamiltonian density. Indeed, the discretized expression

$$H(t) = \sum_{\vec{x}} p(\vec{x}, t)\dot{\phi}(\vec{x}, t) - L(t) \qquad (9.35)$$

can be generalized to a field theory integral form as

$$H(t) = \int d^3x[\pi(\vec{x},t)\dot{\phi}(\vec{x},t) - \mathcal{L}(\vec{x},t)] \equiv \int d^3x \mathcal{H} \,, \qquad (9.36)$$

where $\mathcal{H}$ is a Hamiltonian (energy) density.

9.3 Examples

We now test our understanding in two examples.

Example 1

Consider the real scalar field Lagrangean

$$\mathcal{L} = -\frac{1}{2}(\partial_\mu\phi)(\partial_\nu\phi)\eta^{\mu\nu} - \frac{1}{2}m^2\phi^2 - \Delta V(\phi) \,, \qquad (9.37)$$

where

$$(\partial_\mu\phi)(\partial_\nu\phi)\eta^{\mu\nu} \equiv (\partial_\mu\phi)^2 = \frac{\dot{\phi}^2}{2} - \frac{(\vec{\nabla}\phi)^2}{2}. \qquad (9.38)$$

This is the most common Lagrangean for a scalar that we can have: a canonical kinetic term, a mass term, and an interaction potential.

We find the canonical conjugate momentum density to $\phi(\vec{x},t)$ as

$$\pi(\vec{x},t) = \dot{\phi}(\vec{x},t), \qquad (9.39)$$

which means that the Hamiltonian density of the system is

$$\mathcal{H} = \pi\dot{\phi} - \mathcal{L} = +\frac{\dot{\phi}^2}{2} + \frac{(\vec{\nabla}\phi)^2}{2} + \frac{m^2\phi^2}{2} + \Delta V(\phi). \qquad (9.40)$$

The Hamiltonian equations of motion are as follows: the ϕ equation,

$$\dot{\phi}(\vec{x},t) = \{\phi(\vec{x},t), H(t)\}_{P.B.} = \int d^3y \frac{\delta\phi(\vec{x},t)}{\delta\phi(\vec{y},t)} \frac{\delta \int d^3z \mathcal{H}(\vec{z},t)}{\delta\pi(\vec{y},t)} - 0$$

$$= \int d^3y \delta^3(\vec{x}-\vec{y})\pi(\vec{y},t) = \pi(\vec{x},t), \qquad (9.41)$$

and the π equation,

$$\dot{\pi}(\vec{x},t) = \{\pi(\vec{x},t), H(t)\}_{P.B.}$$

$$= 0 - \int d^3y \frac{\delta\pi(\vec{x},t)}{\delta\pi(\vec{y},t)} \int d^3z \frac{\delta}{\delta\phi(\vec{y},t)} \left[\frac{\pi^2(\vec{z},t)}{2} - \frac{\phi(\vec{z},t)\vec{\nabla}^2\phi(\vec{z},t)}{2} \right.$$

$$\left. + \frac{m^2\phi^2(\vec{z},t)}{2} + \Delta V(\phi(\vec{z},t)) \right]$$

$$= +\vec{\nabla}^2\phi(\vec{x},t) - m^2\phi(\vec{x},t) - \Delta V'(\phi(\vec{z},t)). \qquad (9.42)$$

The two equations are, thus,

$$\dot{\phi} = \pi$$
$$\dot{\pi} = +\vec{\nabla}^2\phi - m^2\phi - \Delta V'(\phi) \,, \qquad (9.43)$$

which combine to

$$\Box\phi \equiv -\ddot{\phi} + \vec{\nabla}^2\phi = m^2\phi + \Delta V'(\phi) \,, \qquad (9.44)$$

which is the the equation of motion from the Lagrangean formalism.

But note that there is another way to calculate, thinking of the Poisson brackets as derivative operators and using directly the fundamental Poisson brackets, $\{\phi, \pi\}_{P.B.} = -\{\pi, \phi\}_{P.B.} = \delta^3$, to write

$$\{\phi, H\}_{P.B.} = \int d^3y \{\phi, \pi\}_{P.B.} \frac{\partial \mathcal{H}}{\partial \pi} = \pi$$
$$\{\pi, H\}_{P.B.} = \int d^3y \{\pi, \phi\}_{P.B.} \frac{\partial \mathcal{H}}{\partial \phi} = \vec{\nabla}^2\phi - m^2\phi - \Delta V'(\phi). \qquad (9.45)$$

Example 2

We now move to a case with a nontrivial kinetic term for a set of real scalar fields ϕ^I, the nonlinear sigma model

$$\mathcal{L} = -\frac{1}{2}g_{IJ}(\phi)(\partial_\mu\phi^I)(\partial_\nu\phi^J)\eta^{\mu\nu} - \frac{1}{2}g_{IJ}(\phi)\dot{\phi}^I\dot{\phi}^J - \frac{1}{2}g_{IJ}(\phi)\vec{\nabla}\phi^I\vec{\nabla}\phi^J. \qquad (9.46)$$

Then the canonical momentum density conjugate to $\phi^I(\vec{x}, t)$ is

$$\pi_I(\vec{x}, t) = \frac{\partial\mathcal{L}}{\partial\dot{\phi}^I(\vec{x}, t)} = g_{IJ}(\phi)\dot{\phi}^J. \qquad (9.47)$$

Note that we put the index I down on π, according with how it appears on the right-hand side. Define the "inverse metric" g^{IJ} as the inverse matrix to g_{IJ},

$$g^{IK}g_{KJ} = \delta^I_J. \qquad (9.48)$$

Then the Hamiltonian density is

$$\mathcal{H} = \pi_I\dot{\phi}^I - \mathcal{L}$$
$$= \frac{1}{2}g_{IJ}(\phi)(\dot{\phi}^I\dot{\phi}^J + \vec{\nabla}\phi^I\vec{\nabla}\phi^J)$$
$$= \frac{1}{2}g^{IJ}(\phi)\pi_I\pi_J + g_{IJ}(\phi)\vec{\nabla}\phi^I\vec{\nabla}\phi^J. \qquad (9.49)$$

The fundamental Poisson brackets are now

$$\{\phi^I(\vec{x}, t), \pi_J(\vec{x}', t)\}_{P.B.} = \delta^I_J\delta^3(\vec{x} - \vec{x}') = -\{\pi_J(\vec{x}', t), \phi(\vec{x}, t)\}_{P.B.}$$
$$\{\phi^I(\vec{x}, t), \phi^J(\vec{x}', t)\}_{P.B.} = 0 = \{\pi^I(\vec{x}, t), \pi^J(\vec{x}', t)\}_{P.B.}. \qquad (9.50)$$

The Hamiltonian equations of motion are as follows:

• the ϕ^I equation of motion,

$$\dot{\phi}^I(\vec{x}, t) = \left\{\phi^I(\vec{x}, t), \int d^3z\mathcal{H}(\vec{z}, t)\right\}_{P.B.} = \int d^3y\{\phi^I, \pi_J\}\frac{\partial\mathcal{H}}{\partial\pi_J}$$
$$= g^{IJ}(\phi)\pi_J(\vec{x}, t). \qquad (9.51)$$

Inverting this relation, we reobtain $\pi_I = g_{IJ}\dot{\phi}^J$.

- the π_I equation of motion,

$$\dot{\pi}_I(\vec{x}, t) = \{\pi_I(\vec{x}, t), d^3z\mathcal{H}(\vec{z}, t)\}_{P.B.} = \int d^3y\{\pi_I, \phi^J\}_{P.B.}\frac{\partial\mathcal{H}}{\partial\phi^J}$$

$$= -\frac{1}{2}(\partial_I g^{JK}(\phi))\pi_J\pi_K - \frac{1}{2}(\partial_I g_{JK}(\phi))\vec{\nabla}\phi^J\vec{\nabla}\phi^K + \vec{\nabla}(g_{IJ}(\phi)\vec{\nabla}\phi^J), \quad (9.52)$$

where in the last term we used a partial integration. Here we have used the notation

$$\partial_I \equiv \frac{\partial}{\partial\phi^I}. \quad (9.53)$$

Combining the two equations, we obtain

$$\dot{\pi}_I(\vec{x}, t) = \partial_0[g_{IJ}(\phi)\dot{\phi}^J]$$

$$= \vec{\nabla}[g_{IJ}(\phi)\vec{\nabla}\phi^J] - \frac{1}{2}(\partial_I g^{JK}(\phi))\pi_J\pi_K - \frac{1}{2}(\partial_I g_{JK}(\phi))\vec{\nabla}\phi^J\vec{\nabla}\phi^K. \quad (9.54)$$

Taking into account that (since $MM^{-1} = \mathbb{1}$, implying $(\partial M)M^{-1} + M\partial M^{-1} = 0$ for any matrix M)

$$\partial g^{-1} = -g^{-1}\partial g g^{-1} \Rightarrow \partial g^{IJ} = -g^{IK}\partial g_{KL}g^{LJ}, \quad (9.55)$$

we obtain

$$\partial_0(g_{IJ}\dot{\phi}^J) - \vec{\nabla}(g_{IJ}\vec{\nabla}\phi^J) = +\frac{1}{2}(\partial_I g_{JK})\dot{\phi}^J\dot{\phi}^K - \frac{1}{2}(\partial_I g_{JK})(\vec{\nabla}\phi^J)(\vec{\nabla}\phi^K), \quad (9.56)$$

which is the same equation of motion as obtained in the Lagrangean formalism.

Important Concepts to Remember

- The symplectic group $Sp(2N, \mathbb{R})$ is the group of real matrices satisfying $M^T\Omega M = \Omega$, and the unitary symplectic group $USp(2N)$ is the group of complex unitary symplectic matrices.
- The Hamiltonian equations of motion in the symplectic formulation are $\dot{\eta} = \Omega\partial H/\partial\eta = \{\eta, H\}_{P.B.}$, and the fundamental Poisson brackets are $\{\eta, \eta\}_{P.B.} = \Omega$.
- Canonical transformations are symplectic.
- For fields, we define a canonical momentum density $\pi = \partial\mathcal{L}/\partial\dot{\phi}$ and a Hamiltonian density $\mathcal{H} = \pi\dot{\phi} - \mathcal{L}$.
- The Poisson brackets for fields are defined using an integral and functional derivatives, as $\{f, g\}_{P.B.} = \int d^3y[(\delta f/\delta\phi)\delta g/\delta\pi - (\delta f/\delta\pi)\delta g/\delta\phi]$.

Further Reading

For more on Poisson brackets and the symplectic formulation in classical mechanics, see Goldstein's book [1].

Exercises

(1)   Consider the DBI Lagrangean

$$\mathcal{L} = -\frac{1}{L^4}\sqrt{1 + L^4[(\partial_\mu\phi)^2 + m^2\phi^2]}. \tag{9.57}$$

(a) Calculate the Hamiltonian.
(b) Using the Poisson brackets, calculate the Hamiltonian equations of motion.

(2)   Using Poisson brackets, calculate the equations of motion of the Hamiltonian

$$\mathcal{H} = \frac{\pi^2}{2m} + \alpha\pi + \frac{1}{2}(\vec{\nabla}\phi)^2 + \beta[1 - \cos(a\phi)]. \tag{9.58}$$

(3)   Consider the $O(N)$ model, first the linear one (unconstrained), and then the nonlinear case (constrained – i.e., nonlinear sigma model). Using the Poisson brackets, calculate the Hamiltonian equations of motion.

(4)   Calculate the Poisson bracket of the functions of fields

$$f = p^n + [(\vec{\nabla}\phi)^2]^2 + \alpha e^\phi , \quad g = p^2\phi^2 + \phi^2(\vec{\nabla}\phi)^2. \tag{9.59}$$

10 Classical Perturbation Theory and Formal Solutions to the Equations of Motion

In this chapter, we will see how to obtain a perturbative solution to a classical problem, for the solution of the equations of motion in the presence of a source. The general formalism can only take us so far, so after defining it, we specialize to the case of a polynomial potential so that we can be more explicit. We then define a diagrammatic procedure, which is usually defined in quantum field theory as Feynman diagrams but can be specialized at the classical level for solving the equations of motion perturbatively. We end with an application to a nonstandard theory, the case of the sigma model, with noncanonical kinetic term.

10.1 General Formalism

We start with the standard scalar Lagrangean treated so far with a canonical kinetic term, a mass term, and an interaction $\Delta V(\phi)$,

$$
\begin{aligned}
S = \int d^4 x \mathcal{L} &= \int d^4 x \left[-\frac{1}{2}(\partial_\mu \phi)^2 - \frac{1}{2}m^2 \phi^2 - \Delta V(\phi) \right] \\
&= \int d^4 x \left[\frac{1}{2}\phi(\Box - m^2)\phi - \Delta V(\phi) \right],
\end{aligned} \tag{10.1}
$$

where in the second line we did a partial integration.

The equations of motion of this Lagrangean are, as we know and as we can easily re-check from (10.1) by $\delta S/\delta \phi = 0$,

$$
(\Box - m^2)\phi = \Delta V'(\phi). \tag{10.2}
$$

Note that $\Delta V(\phi)$ contains terms at least cubic in the fields, which leads to nonlinear terms in the equation of motion (10.2) (at least quadratic), or interaction terms.

But, moreover, the more interesting classical problem is the (nonlinear) response to a classical source – i.e., the driven solution. Therefore, we can introduce a coupling to an external "currrent" (or source) $J(x)$, completely similarly to what we did for electromagnetism. There we wrote an action

$$
S = \int d^4 x \left[-\frac{1}{4}F_{\mu\nu}^2 + J^\mu A_\mu \right], \tag{10.3}
$$

where J^μ was a current (source) composed of particle (electron) delta functions. Now, similarly, we add a source term, so we write

$$S = \int d^4x[\mathcal{L} + J(x)\phi(x)]. \qquad (10.4)$$

Here J is an external source for ϕ. For instance, if $\phi(x)$ stands for the pion field – which, as we said, is the effective field describing the strong force at intermediate distances – then $J(x)$ is a source for it, which can be a nucleon source or a sum of nucleon delta functions (the nucleons source the pion field like the electron sources the electromagnetic field).

The equations of motion of the modified action are

$$(-\Box + m^2)\phi(x^\rho) = J(x^\rho) - \Delta V'(\phi(x^\rho)). \qquad (10.5)$$

By acting from the left with the inverse of the operator $-\Box + m^2$, we can solve *formally* the equation of motion (10.5), *if we consider that $\Delta V(\phi)$ is a small perturbation.* The formal solution is

$$\phi(x^\rho) = (-\Box + m^2)^{-1}J(x^\rho) - (-\Box + m^2)^{-1}\Delta V'(\phi). \qquad (10.6)$$

Of course, this is a formal solution, for two reasons:

- The inverse of the operator $\Box + m^2$ is defined formally, but we have not defined its explicit action on the terms on the right-hand side.
- More importantly, ϕ appears on the right-hand side as well, in $\Delta V'(\phi)$, so it is actually an *equation*. We intend to solve it iteratively. But that can only be made if we assume that the second term, $\Delta V'(\phi)$, is small.

To fix the first reason, we do a Fourier transformation,

$$\phi(x) = \int \frac{d^4p}{(2\pi)^4} e^{ip \cdot x} \phi(p). \qquad (10.7)$$

Here, $p \cdot x = p_\mu x^\mu = p^\mu x^\nu \eta_{\mu\nu}$. Substituting in the action, we obtain

$$
\begin{aligned}
S &= \int d^4x \left[\frac{1}{2}\phi(x)(\Box - m^2)\phi(x) - \Delta V(\phi(x)) \right] \\
&= \int d^4x \left[\frac{1}{2} \int \frac{d^4p}{(2\pi)^4} \int \frac{d^4q}{(2\pi)^4} \phi(p)e^{ip \cdot x}(-q^2 - m^2)e^{iq \cdot x}\phi(q) \right. \\
&\quad \left. - \Delta V\left(\int \frac{d^4p}{(2\pi)^4} e^{ip \cdot x}\phi(p) \right) \right].
\end{aligned}
\qquad (10.8)
$$

Doing the integral over x using

$$
\int \frac{d^4x}{(2\pi)^4} e^{i(p+q) \cdot x} = \int \frac{dx^0}{2\pi} e^{-i(p+q)^0 x^0} \int \frac{dx^1}{2\pi} e^{+i(p+q)^1 x^1} \int \frac{dx^2}{2\pi} e^{+i(p+q)^2 x^2} \int \frac{dx^3}{2\pi} e^{+i(p+q)^3 x^3}
$$
$$
= \delta^4(p + q) , \qquad (10.9)
$$

and then doing the q integral, we obtain

$$S = -\int \frac{d^4p}{(2\pi)^4} \left[\frac{1}{2}\phi(p)[p^2 + m^2]\phi(-p) \right] - \int d^4x \Delta V(\phi). \qquad (10.10)$$

10.2 Polynomial Potential

In order to proceed, we now specialize to the case most common, a potential given by a power law in fields,

$$\Delta V(\phi) = \frac{\lambda_n}{n!}\phi^n. \tag{10.11}$$

Then we can rewrite the momentum space potential as

$$
\begin{aligned}
\int d^4x \Delta V(\phi) &= \int d^4x \frac{\lambda_n}{n!}\phi^n \\
&= \frac{\lambda_n}{n!} \int \frac{d^4p_1}{(2\pi)^4} \cdots \int \frac{d^4p_n}{(2\pi)^4} \int d^4x e^{i(p_1+\cdots p_n)\cdot x}\phi(p_1)\cdots\phi(p_n) \\
&= \frac{\lambda_n}{n!} \int \frac{d^4p_1}{(2\pi)^4} \cdots \int \frac{d^4p_n}{(2\pi)^4}\phi(p_1)\cdots\phi(p_n)(2\pi)^4\delta^4(p_1+\cdots p_n). \quad (10.12)
\end{aligned}
$$

Moreover, defining also the Fourier transform of the current,

$$J(x) = \int \frac{d^4p}{(2\pi)^4}e^{ip\cdot x}J(p), \tag{10.13}$$

the source term is

$$\int d^4x J(x)\phi(x) = \int \frac{d^4p}{(2\pi)^4}\phi(p)J(-p). \tag{10.14}$$

Finally, now we can write the full action in p space as

$$
\begin{aligned}
S = &-\int \frac{d^4p}{(2\pi)^4}\left[\frac{1}{2}\phi(p)(p^2+m^2)\phi(-p) - \phi(p)J(-p)\right] \\
&-\frac{\lambda_n}{n!}\int \frac{d^4p_1}{(2\pi)^4}\cdots\int \frac{d^4p_n}{(2\pi)^4}\phi(p_1)\cdots\phi(p_n)(2\pi)^4\delta^4(p_1+\cdots p_n), \quad (10.15)
\end{aligned}
$$

leading to the equation of motion

$$
\begin{aligned}
&(p^2+m^2)\phi(-p) \\
&= J(-p) - \frac{\lambda_n}{(n-1)!}\int \frac{d^4p_1}{(2\pi)^4}\cdots\int \frac{d^4p_{n-1}}{(2\pi)^4}\phi(p_1)\cdots\phi(p_{n-1})(2\pi)^4\delta^4(p_1+\cdots p_{n-1}+p),
\end{aligned}
$$
$$\tag{10.16}$$

with the formal solution (changing $p \to -p$ in the argument of the fields)

$$
\begin{aligned}
\phi(p) = &\frac{1}{p^2+m^2}J(p) \\
&-\frac{1}{p^2+m^2}\frac{\lambda_n}{(n-1)!}\int \frac{d^4p_1}{(2\pi)^4}\cdots\int \frac{d^4p_{n-1}}{(2\pi)^4}\phi(p_1)\cdots\phi(p_{n-1})(2\pi)^4\delta^4(p_1+\cdots p_{n-1}-p).
\end{aligned}
$$
$$\tag{10.17}$$

Now we have defined properly the inverse of the kinetic operator, but we still have an *equation* for $\phi(p)$, which needs to be solved iteratively, in a *perturbation theory* in λ_n. We need to assume that λ_n is small enough that the term on the right-hand side proportional to it is small compared with the first.

But before we do that, we will formally discretize space, in its x and p varieties, in order to write the configuration space and momentum space actions (10.1) and (10.10) together as

$$S = -\sum_{ij} \frac{1}{2}\phi_i K_{ij}\phi_j - \frac{\lambda_n}{n!}\sum_i \phi_i^n + \sum_i \phi_i J_i. \qquad (10.18)$$

Here, of course, the matrix K_{ij} represents the kinetic operator, $-\Box + m^2$ or $p^2 + m^2$, thought of as a matrix in discretized space. Note also that $\sum_i \phi_i^n$ strictly speaking only corresponds to the x space formula, but the more complicated p space one $(\sum_{i_1, \ldots, i_n} V_{i_1, \ldots, i_n}\phi_{i_1} \cdots \phi_{i_n})$ could be treated in a similar, if not identical, manner.

Its equation of motion is

$$K_{ij}\phi_j + \frac{\lambda_n}{(n-1)!}\phi_i^{n-1} = J_i, \qquad (10.19)$$

which has the formal solution (multiplying with the inverse of the kinetic operator, K_{ij}^{-1}, from the left,

$$\phi_i = K_{ij}^{-1}J_j - K_{ij}^{-1}\frac{\lambda_n}{(n-1)!}\phi_j^{n-1}. \qquad (10.20)$$

10.3 Diagrammatic Procedure

In order to proceed, we introduce a diagrammatic procedure. We define, as in Fig. 10.1:

- To represent ϕ_i, we write a line starting at point i and ending on a full blob.
- To represent K_{ij}^{-1}, we write a line between points i and j.
- To represent J_j, we write a cross at point j.
- To represent the connection V_n of n ϕ_i's with factor $\lambda_n/(n-1)!$, we write a vertex at point j with n lines coming out of it.

Then the equation can be represented diagramatically as in Figure 10.2.

We will solve this equation self-consistently in perturbation theory in the parameter λ_n, up to a certain order k – i.e., keeping only terms up to λ_n^k. The procedure is to solve the equation at order k; then, in order to find the solution for ϕ at order $k+1$, we substitute on right-hand hand side of the self-consistency equation.

That means that the solution at order zero is to simply to put λ_n to zero, obtaining

$$\phi_i^{(0)} = K_{ij}^{-1}J_j. \qquad (10.21)$$

Diagrammatically, we write that the solution is a line between point i and point j with a cross on it, as in Figure 10.3a.

The first-order solution means replacing $\phi_i^{(0)}$ on the right-hand side of the equation, replacing the ϕ_i line+blob in the vertex. The result is

$$\phi_i^{(1)} = K_{ij}^{-1}J_j - K_{ij}^{-1}\frac{\lambda_n}{(n-1)!}(K_{jk}^{-1}J_k)^{n-1}. \qquad (10.22)$$

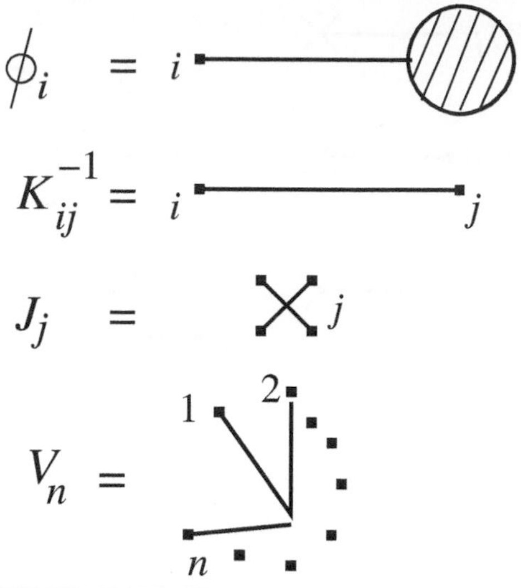

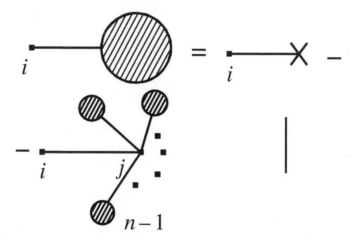

Figure 10.1 Diagrammatic representation of ϕ_i, K_{ij}^{-1}, J_j and $V_n = \frac{\lambda_n}{n!} \phi_i^n$.

Figure 10.2 Equation satisfied by ϕ_i represented diagrammatically.

Diagramatically, we simply replace the blobs by crosses in the vertex term of the self-consistency equation, as in Figure 10.3b.

For the second-order solution, we replace the ϕ on the right-hand side with $\phi_i^{(1)}$. That means diagrammatically that we add a third diagram, where we replace *one* of the legs of the vertex with a full vertex. One might ask why not add a vertex in all the legs, but that would be a diagram of order λ_n^n, whereas we only consider the order λ_n^2. We have, of course, $(n-1)$ ways of replacing a leg by a vertex, and because there are no indices (since the theory is for a *single* scalar field), all the diagrams are identical, simply giving a factor $(n-1)$ to the diagram. The result in the formula is

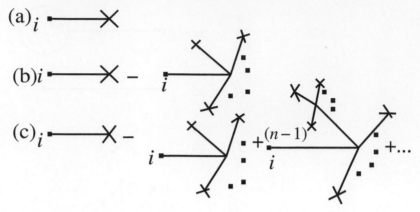

Figure 10.3 Iterative (perturbative) solution to the self-consistent equation for ϕ: (a) Zeroth-order (no iteration) solution. (b) First-order (first iteration) solution. (c) Second-order (second iteration) solution.

$$\phi_i^{(2)} = K_{ij}^{-1} J_j - K_{ij}^{-1} \frac{\lambda_n}{(n-1)!} (K_{jk}^{-1} J_k)^{n-1}$$

$$+ K_{ij}^{-1} \frac{\lambda_n}{(n-2)!} (K_{jk}^{-1} J_k)^{n-2} \left(K_{jk}^{-1} \frac{\lambda_n}{(n-1)!} (K_{kl}^{-1} J_l)^{n-1} \right) + \cdots \quad (10.23)$$

The diagrammatic solution is in Figure 10.3c.

Now we try to understand what this perturbative solution does. It is the solution for the *nonlinear* field generated by source $J(x)$ (the nonlinear response). The field is classical, not quantum, which means that the intensity of the field, measured by $|\phi|^2$, is large. In the electromagnetic case, we would say that $F_{\mu\nu}^2$ and $\epsilon^{\mu\nu\rho\sigma} F_{\mu\nu} F_{\rho\sigma}$ are large. In terms of the quantum theory, what this condition means is that the number of particles (photons, in the case of the electromagnetic field, pions, in the case of the pion field, etc.) is very large.

Note that the diagrammatic procedure developed here is the classical limit of the diagrammatic procedure of quantum field theory, "Feynman diagrams." In the quantum case, the external lines can be, besides the classical field ϕ_i used here – which represents, as we said, a state of *many* quantum particles – also individual quantum (particle) states. Moreover, the diagrams here are only *trees* – i.e., diagrams that have only branchings. But in the quantum case, we could also have *loops*, closed contours.

However, we didn't need here to reference at all the quantum theory, and, moreover, we can think of the procedure simply as a mathematical trick to solve a nonlinear equation, without needing to reference the physics it refers to.

10.4 Noncanonical Case: Nonlinear Sigma Model

The case studied until now was the most common, of a canonical kinetic term and a polynomial potential. But the solution can be written in a more general context for any theory at all. In particular, we will study here the case of a noncanonical kinetic term (though still with two derivatives only), of the nonlinear sigma model, for N scalar fields $\phi^I, I = 1, .., N$, coupled with a source term,

$$S = \int d^4x \left[-\frac{1}{2} g_{IJ}(\phi) \partial_\mu \phi^I \partial_\nu \phi^J \eta^{\mu\nu} + J_I(x) \phi^I(x) \right].$$ (10.24)

The equations of motion are

$$\partial_\mu (g_{IJ}(\phi) \partial^\mu \phi^J) - \frac{1}{2} (\partial_I g_{JK}) \partial_\mu \phi^J \partial_\nu \phi^K \eta^{\mu\nu} + J_I = 0.$$ (10.25)

We rewrite them as

$$g_{IJ}(\phi)(-\Box)\phi^J - (\partial_K g_{IJ}(\phi)) \partial_\mu \phi^K \partial_\nu \phi^J \eta^{\mu\nu} + \frac{1}{2} (\partial_I g_{JK}(\phi)) \partial_\mu \phi^J \partial_\nu \phi^K \eta^{\mu\nu} = J_I.$$ (10.26)

In order to write a formal solution, we need to consider a perturbation theory. The obvious one is to consider that the metric $g_{IJ}(\phi)$ only varies little – i.e., that $\partial_I g_{JK}$ is small. Then, multiplying by $(-\Box)^{-1} g^{IJ}$ from the left (g^{IJ} is the inverse matrix to g_{IJ}), we obtain the formal solution

$$\phi^I = (-\Box)^{-1} g^{IJ}(\phi) J_J + (-\Box)^{-1} g^{IJ}(\phi) (\partial_K g_{JL}) \partial_\mu \phi^K \partial_\nu \phi^L \eta^{\mu\nu}$$
$$- (-\Box)^{-1} g^{IJ}(\phi) \frac{1}{2} (\partial_J g_{KL}(\phi)) \partial_\mu \phi^K \partial_\nu \phi^L \eta^{\mu\nu}.$$ (10.27)

We see that, indeed, if $\partial_I g_{JK}$ is small, we can treat the second and third terms on the right-hand side as perturbations and apply the same procedure as before to find an interative solution. It will just be more involved than the case just analyzed, so we will not describe it further here.

Important Concepts to Remember

- The classical problem we want to solve is the classical response (solution of the equations of motion) in the case of an external source $J(x)$ for the field ϕ – i.e., an extra term $\int d^4x J(x) \phi(x)$ in the action.
- The equation of motion is $(-\Box + m^2)\phi = J - \Delta V'$, and we can write a formal solution as $\phi = (-\Box + m^2)^{-1} J - (-\Box + m^2)^{-1} \Delta V'$.
- To make concrete the inverse of the kinetic operator, we go to momentum space (Fourier transform).
- For a polynomial potential, we can write a concrete equation (formal solution) for ϕ.
- Writing the x and p cases together by discretizing, we get the formal solution (self-consistent equation) $\phi_i = K_{ij}^{-1} J_j - K_{ij}^{-1} (\lambda_n/(n-1)!) \phi_j^{n-1}$.
- We can solve it iteratively via a diagrammatic procedure by replacing the previous step solution back into the equation.
- The diagrammatic procedure is a mathematical trick, but it is based on the reduction of the Feynman diagrams of quantum field theory to the classical case, with external fields (states with large number of particles) and no loops, only trees.
- For the sigma model, we can make a peturbation procedure in $\partial_I g_{JK}$.

Further Reading

There usually is not much done at the classical level for the formal perturbative expansion. In quantum field theory texts, one finds the method and the associated Feynman diagram procedure only at the quantum level; see, for instance, Peskin and Schroeder's book [4], chapter 11.

Exercises

(1) Write the formal solution of the equations of motion for the sine-Gordon model.
(2) Write the perturbative diagrammatic solution for the above found equation for the sine-Gordon model, up to second order.
(3) For the ϕ^3 model ($V = \lambda_3 \phi^3/3!$), write the diagrammatic solution up to $\phi_i^{(3)}$ (included).
(4) Write a formal solution of the equations of motion for the real DBI scalar model.

11   Representations of the Lorentz Group

In this chapter and the next, we will describe the only issues related to quantum theory in this book (however, we will also describe a bit of quantum mechanics when talking about solitons, in order to define their quantization in the collective coordinate method). These are traditionally described in a quantum field theory book and are the reason usually classical field theory is described together with quantum field theory. However, we will not need any notions of intrinsically quantum *field* theory to describe them, only of basic quantum theory.

In this chapter, then, we will consider a classification of the possible fields (functions of spacetime) that we can have as possible *representations of the Lorentz group*. We have already seen the most common example, the vector field, or gauge field A_μ of electromagnetism, as well as the simplest example, the scalar field $\phi(x)$. Now we will consider a more thorough classification.

11.1   Characterization of the Lie Algebra of the Lorentz Group

We have seen in Chapter 4 that the Lorentz group is $SO(3,1)$, which is a noncompact orthogonal Lie group that contains as a subgroup the spatial rotation group $SO(3)$ (special orthogonal compact group). It is composed of the matrices $\Lambda^\mu{}_\nu$, acting on differences in spacetime coordinates,

$$\Delta x'^\mu = \Lambda^\mu{}_\nu \Delta x^\nu, \tag{11.1}$$

such that they satisfy the orthogonality relation

$$\Lambda^T M \Lambda = M, \tag{11.2}$$

where M is the Minkowski metric matrix,

$$M = \eta_{\mu\nu} = \begin{pmatrix} -1 & 0 & 0 & 0 \\ 0 & 1 & 0 & 0 \\ 0 & 0 & 1 & 0 \\ 0 & 0 & 0 & 1 \end{pmatrix}. \tag{11.3}$$

This is a generalization to the Minkowski metric of the orthogonal group of rotations, acting on differences in spatial coordinates,

$$\Delta x'^i = R^i{}_j \Delta x^j, \tag{11.4}$$

such that they satisfy orthogonality,

$$R^T \, \mathbb{1} R = \mathbb{1}. \tag{11.5}$$

As we see, $SO(3,1)$ is obtained by replacing the Euclidean metric δ_{ij} with the Minkowski one $\eta_{\mu\nu}$.

Moreover, we saw in Chapter 3 that up to global issues, we have an equivalence between $SO(3)$ and $SU(2)$ – i.e., the relation is valid in the Lie algebra, $su(2) = so(3)$. As a result, we can write the Lie algebra of $SO(3)$ as the one of $SU(2)$ that we learned in quantum mechanics,

$$[J_i, J_j] = i\epsilon_{ijk}J_k. \tag{11.6}$$

However, we must write it in terms of the usual operators associated with angular momentum, J_{ij}, defined by

$$J_{ij} \equiv \epsilon_{ijk}J_k. \tag{11.7}$$

Then the algebra becomes

$$\begin{aligned} [J_{ij}, J_{kl}] &= \epsilon_{ijm}\epsilon_{klp}[J_m, J_p] = i\epsilon_{ijm}\epsilon_{klp}\epsilon_{mpq}J_q \\ &= i\epsilon_{ijm}\epsilon_{klp}J_{mp}. \end{aligned} \tag{11.8}$$

But we can rewrite this further. Noting that the ϵ_{ijk} symbol has only three indices (there are three spatial dimensions), and since J_{mp} is antisymmetric, $m \neq p$, but also $m \neq i,j$ and $p \neq k,l$, it means that $(ij) \neq (kl)$. Yet they cannot be all different; rather, two of the indices must match, and the others not. One possibility is that $i = k$ and then $m = l, p = j$. But that means that $\epsilon_{klp} = \epsilon_{imj} = -\epsilon_{ijm}$, giving a minus sign, so a possible term on the right-hand side is $-\delta_{ik}J_{lj}$. Another possibility is that $i = l$, and then $m = k, p = j$, and then $\epsilon_{klp} = \epsilon_{mij} = +\epsilon_{ijm}$, so we get a plus sign, and the term is $+\delta_{il}J_{kj}$. The other two terms are obtained from the antisymmetry in (ij) and (kl) that should exist, finally obtaining

$$[J_{ij}, J_{kl}] = i(-\delta_{ik}J_{lj} + \delta_{il}J_{kj} + \delta_{jk}J_{li} - \delta_{jl}J_{ki}). \tag{11.9}$$

Multiplying the generators by a minus sign and rearranging the order of indices, we obtain

$$[J_{ij}, J_{kl}] = i(-\delta_{ik}J_{jl} + \delta_{il}J_{jk} + \delta_{jk}J_{il} - \delta_{jl}J_{ik}). \tag{11.10}$$

But we note that while the original form of the algebra depended on being in three Euclidean dimensions, as it involved the Levi-Civitta symbol ϵ_{ijk}, this form doesn't. It is, in fact, the general Lie algebra of $SO(n)$ for any n (for any n Euclidean dimensions).

We can then easily obtain the Lie algebra of the Lorentz group $SO(3,1)$ by just replacing the Euclidean metric δ_{ij} with the Minkowski metric $\eta_{\mu\nu}$, as we saw earlier that we should (from the defining relation of the group). The result is

$$[J_{\mu\nu}, J_{\rho\sigma}] = i(-\eta_{\mu\rho}J_{\nu\sigma} + \eta_{\mu\sigma}J_{\nu\rho} + \eta_{\nu\rho}J_{\mu\sigma} - \eta_{\nu\sigma}J_{\mu\rho}). \tag{11.11}$$

Inside the Lie algebra of the Lorentz group we have six generators: the three generators of rotations J_{ij} (for $\mu, \nu = i, j$) and three generators for boosts, J_{0i}. Defining in an obvious manner two rotation vectors

$$J_{ij} = \epsilon_{ijk}J_k, \quad J_{0i} = K_i, \tag{11.12}$$

and constructing from them the combinations

$$M_i = \frac{J_i + iK_i}{2}; \quad N_i = \frac{J_i - iK_i}{2},$$
(11.13)

we can prove that M_i, N_i become two independent $SU(2)$ Lie algebras, i.e., that

$$[M_i, M_j] = i\epsilon_{ijk}M_k; \quad [N_i, N_j] = i\epsilon_{ijk}N_k; \quad [M_i, N_j] = 0.$$
(11.14)

We leave the proof of this statement as an exercise.

In conclusion, we have seen that the Lie algebra $so(3,1)$ is the same as $su(2) \times su(2)$. But since the general representation of $SU(2)$ is a spin j representation, with j half integer – i.e., $j = 0, 1/2, 1, 3/2, \ldots$ – the general representation of the Lorentz group should be in terms of two spins, one for each group – i.e., a

$$j \otimes j'$$
(11.15)

representation.

11.2 The Poincaré Group

Note that even though we have described the Lorentz generators as an antisymmetric tensor of generators, $J_{\mu\nu}$, since these are abstract generators (that will themselves act on any representation), we cannot consider the indices μ, ν as transforming under the Lorentz group as vector indices. We can only say that *if we act on some fields*, and *representing the generators on the fields*, we can think of the indices as Lorentz indices. This happens in the case when we represent $J_{\mu\nu}$ as a derivative operator.

This representation is related to the extended group of symmetries of Minkowski spacetime, the *Poincaré group*. Besides the Lorentz group of $J_{\mu\nu}$, the full symmetries of Minkowski spacetime include spacetime translations, $x^\mu \to x^\mu + \alpha^\mu$, generated by the generators P_μ. To obtain their algebra – i.e., the Poincaré algebra – we represent it on fields as a derivative operator. For translations, we have the usual (generalized from the nonrelativistic quantum mechanics, where momentum is i times the spatial derivative, to the relativistic case)

$$P_\mu = i\partial_\mu,$$
(11.16)

such that the action of P_μ on a function is just the Taylor expansion written as a formal series,

$$e^{-i\alpha^\mu P_\mu}f(x^\mu) = \sum_{n \geq 0} \frac{(\alpha^\mu)^n (\partial_\mu)^n f(x)}{n!} = f(x^\mu + \alpha^\mu).$$
(11.17)

We can immediately check that then

$$[P_\mu, P_\nu] = 0.$$
(11.18)

For the Lorentz generator, we can write the "four-angular momentum" representation

$$J_{\mu\nu} = i(x_\mu \partial_\nu - x_\nu \partial_\mu) = x_\mu P_\nu - x_\nu P_\mu; \quad x_\mu \equiv \eta_{\mu\nu} x^\nu.$$
(11.19)

This is just the Lorentz (four-index tensor) generalization of the representation of the angular momenta J_{ij} on wavefunctions in quantum mechanics, and we can check that it satisfies the Lorentz algebra (11.11). We can now calculate the final commutator of the Poincaré algebra, $[P_\mu, J_{\nu\rho}]$, using the relation

$$[P_\mu, x_\nu] = i[\partial_\mu, x_\nu] = i\eta_{\mu\nu}. \tag{11.20}$$

We obtain

$$[P_\mu, J_{\nu\rho}] = i(\eta_{\mu\nu}P_\rho - \eta_{\mu\rho}P_\nu). \tag{11.21}$$

The Poincaré group is denoted by $ISO(3,1)$.

11.3 The Universal Cover of the Lorentz Group

In the case of the rotation group, we have seen in Chapter 3 that there is an isomorphism between the Lie algebras of $SO(3)$ and $SU(2)$, which can be made concrete in a mapping of the elements of the two groups,

$$g = \exp\left[i\frac{\vec{\alpha}\cdot\vec{\sigma}}{2}\right] = \cos\left|\frac{\vec{\alpha}}{2}\right| + \frac{i\vec{\alpha}\cdot\vec{\sigma}}{|\vec{\alpha}|}\sin\left|\frac{\vec{\alpha}}{2}\right| \tag{11.22}$$

where $\vec{\sigma}$ are the Pauli matrices for the $SU(2)$ group and

$$g = e^{i\vec{\theta}\cdot\vec{J}} = g(\theta, \phi, \psi) \tag{11.23}$$

for the $SO(3)$ group element in terms of the Euler angles.

A similar mapping exists for the case of the Lorentz group $SO(3,1)$ into the group $Sl(2,\mathbb{C})$, the group of 2×2 complex matrices with determinant one, which also has six generators (since 2×2 matrices have four complex elements, and the determinant one leaves three complex, or six real, components). In fact, global issues, which we will study later on in the chapter, mean that $Sl(2,\mathbb{C})$ is actually larger than $SO(3,1)$, so it is called the *universal cover of the Lorentz group.*

11.4 The Wigner Method and the Little Group

However, Wigner realized that the correct way to describe fields is not in terms of the representations of the Lorentz group but, rather, as representations of the full Poincaré group $ISO(3,1)$. If we consider just the Lorentz group, we run into some inconsistencies, like non-unitarity, but imposing the Poincaré group is more restrictive.

It is also more physical. If we want to describe representations for *fields*, we have to remember that in a quantum theory, they correspond to a collection of *particles*. So, in particular, a single-particle state of the field is in a representation of the full Poincaré group: the generator P_μ has an eigenvalue on the state, the momentum k_μ,

$$P_\mu |k_\mu\rangle = k_\mu |k_\mu\rangle. \tag{11.24}$$

Indeed, as we saw, we can represent $P_\mu = i\partial_\mu$ on fields, and these have a $e^{-ip\cdot x}$ factor, leading to a $P_\mu = p_\mu$ on states of given momentum, as before. That means that in order to consider a representation of $ISO(3, 1)$, we need to consider not only a representation of the Lorentz group (described by generators J_i and K_i) but also a representation of the subgroup that leaves this momentum (eigenvalue of P_μ in the single particle state) invariant, $W \in L$

$$W^\mu{}_\nu k^\nu = k^\mu. \tag{11.25}$$

The group L is called the *little group*, and we need to seek representations for it. This is the Wigner method.

We see that there is a difference between massless and massive particles.

Massive Particles, $M^2 \neq 0$

In this case, we can go to the rest frame, where

$$k^\mu = (M, 0, 0, 0). \tag{11.26}$$

Obviously, that means that the little group L is simply the group of spatial rotations, with $J_i \in so(3) = su(2)$, which leaves k^μ invariant.

But since

$$J_i = M_i + N_i \tag{11.27}$$

and $J_{0i} = K_i = (M_i - N_i)/i = 0$, the little group is the *diagonal* $SU(2)$ (i.e., with $M_i = N_i$, acting on both $SU(2)$ simultaneously), so we need to seek representations of the diagonal reduction of the Lorentz group to the little group,

$$SU(2) \times SU(2) \to SU(2)_{\text{diag}}. \tag{11.28}$$

Massles Particles $M^2 = 0$

In this case, there is no rest frame, and, at most, we can align the momentum to the x^1 direction and put it in the form

$$k^\mu = (p, p, 0, 0). \tag{11.29}$$

In this form, it is obvious that the group is $ISO(2)$, the Poincaré group in two dimensions. It is composed of the $SO(2)$ group of rotations of the x^2, x^3 directions, plus the $SO(1, 1)$ group of Lorentz rotations in the x^1 direction. Both are Abelian generators.

The analysis is more complicated and will not be described further, but the endpoint of the analysis is, in fact, the same as in the massive case.

11.5 Representations in Terms of Fields

Massive Particles $\longrightarrow$ Fields

We will only consider in detail the case of massive particles.

As we already saw, in order to consider representations of the Poincaré group $ISO(3, 1)$, we need to consider representations of the little group $SU(2)_{\text{diag}}$, but at the same time representations of the Lorentz group. That is, we need to *decompose* the representations of the Lorentz group, which as we saw were representations of spins $j \otimes j'$ for $SU(2) \times SU(2)$, into representations of the little group $SU(2)_{\text{diag}}$.

More precisely, a single field must be in an *irreducible representation* (irrep) of the little group, since it must transform into itself (representation) but must not be possible to subdivide it (irreducible). That is, we must decompose $j \otimes j'$ into the irreps of $SU(2)$ – namely, the spin j'' representations. But this decomposition,

$$j \otimes j' = \oplus_{j''} j'', \tag{11.30}$$

if we identify the $SU(2)$ groups, is nothing but the usual *addition of angular momenta* that we learned in quantum mechanics.

1. The simplest case is when $j = j' = 0$, in which case we simply have

$$0 \otimes 0 = 0. \tag{11.31}$$

Here, 0 is the spin 0 representation – i.e., the scalar field. It means that there is a single function (or component), $\phi(x)$. As we have already said, this is the simplest case of a representation of the Lorentz group, where the numerical value of the field doesn't change under the transformation

$$x'^{\mu} = \Lambda^{\mu}{}_{\nu} x^{\nu} , \tag{11.32}$$

i.e.,

$$\phi'(x') = \phi(x). \tag{11.33}$$

2. The next possibility is when either j or j' is 1/2 and the other is still 0. We obtain two representations that are equal from the point of view of the little group but not from the point of view of the full Lorentz group,

$$1/2 \otimes 0 = 1/2_a \quad \text{and} \quad 0 \otimes 1/2 = 1/2_b. \tag{11.34}$$

Note that 0 means an invariant component, so multiplying by 0 in (11.34) means really the addition of the the angular momentum 0 to the 1/2 one, leading to just 1/2. Indeed, the first case corresponds to $M_i = \sigma_i/2$ (the spin 1/2 representation), but $N_i = 0$ (the spin 0 representation), which leads to

$$J_i = \frac{\sigma_i}{2}; \quad K_i = -\frac{i\sigma_i}{2} , \tag{11.35}$$

whereas the second case corresponds to $M_i = 0$ (the spin 0 representation), but $N_i = \sigma_i/2$ (the spin 1/2 representation), leading to

$$J_i = \frac{\sigma_i}{2}; \quad K_i = +\frac{i\sigma_i}{2}. \tag{11.36}$$

These spin 1/2 representations are called *spinors*, and there are several ways to define them, as we will see later in the book. Here they have been defined as the basic spinor representations of the little group $SO(d-1)$ (which, in 3+1 dimensions, happens to be $= SU(2)$, but in other dimensions, this is not true), having a more general definition.

The two representations are called *Weyl spinor* representations.

Note that the two representations are related by *spatial parity*, which acts on the spatial coordinates by reversal, $\vec{x} \to -\vec{x}$. That means that, considering the (derivative representation) action of $J_{\mu\nu}$ on fields, where the indices μ, ν are vector indices for Lorentz transformations, the action on a spatial index i is with a minus, and on a time index 0 without. Therefore, they transform as

$$K_i \equiv J_{0i} \to -J_{0i} = -K_i; \quad J_{ij} \to J_{ij} \Rightarrow J_i \to J_i. \tag{11.37}$$

By putting together the two irreducible spinor representations $1/2_a$ and $1/2_b$ into a single reducible representation, we obtain a *Dirac spinor*. Because of (11.37), the Dirac spinor is invariant under parity (exchanging its two halves) – i.e., it is a representation of both the Lorentz group and parity.

3. The next possibility is that $j = j' = 1/2$. In this case, the rules of addition of angular momenta decomposition give all the possibilities from $j + j' = 1/2 + 1/2 = 1$, down to $j - j' = 1/2 - 1/2 = 0$, in steps of one – i.e.,

$$1/2 \otimes 1/2 = 1 \oplus 0. \tag{11.38}$$

This represents a decomposition of the fields carrying two spin 1/2 indices, α and $\dot{\alpha}$ (each taking two values, spin-up and spin-down), into irreducible representations of the little group $SO(3)$. Since 2×2 matrices can be decomposed into the complete set of four matrices σ_i and $\mathbb{1}$, we write

$$\psi_{\alpha\dot{\alpha}} = A_i(\sigma_i)_{\alpha\dot{\alpha}} + \phi\delta_{\alpha\dot{\alpha}}. \tag{11.39}$$

Here A_i are the physical components of a vector field A_μ of spin 1, whereas ϕ is a scalar field of spin 0.

As we saw before, a vector field (of spin 1) is either a:

• covariant vector, which transforms as ∂_μ,

$$A'_\mu(x') = \frac{\partial x^\nu}{\partial x'_\mu} A_\nu, \tag{11.40}$$

• or a contravariant vector, which transforms as dx^μ,

$$V'^\mu(x') = \frac{\partial x'^\mu}{\partial x^\nu} V^\nu(x). \tag{11.41}$$

4. Another possibility is to have either j or j' being 1, and the other 0, giving 1 by the rules of addition of angular momenta – i.e.,

$$1 \otimes 0 = 1 \quad \text{or} \quad 0 \otimes 1 = 1. \tag{11.42}$$

This gives a vector field A_μ again.

5. The next possibility (except for the case $j = 1, j' = 1/2$, which is only of interest in supersymmetric theories, more precisely in supergravity) is $j = j' = 1$. According to the rules of addition of angular momenta, we start with $1 + 1 = 2$, and go down to $1 - 1 = 0$ in units of 1 – i.e.,

$$1 \otimes 1 = 2 \oplus 1 \oplus 0. \tag{11.43}$$

In terms of fields, we decompose a field with two Lorentz vector indices – i.e., a matrix A_{ij} – in terms of a symmetric traceless matrix $g_{((ij))}$, an antisymmetric matrix $B_{[ij]}$, and a trace ϕ,

$$A_{ij} = g_{((ij))} + B_{[ij]} + \phi \delta_{ij}. \tag{11.44}$$

The symmetric traceless field corresponds to a "graviton" g_{ij} – i.e., a spin 2 field – the antisymmetric field to a spin 1 field B_{ij}, and the trace to a scalar field ϕ.

Note that, more generally, a (p, q) Lorentz tensor with only spatial components,

$$A^{i_1 \ldots i_p}_{j_1 \ldots j_q} \tag{11.45}$$

decomposes into irreducible representations of the little group. It so happens that the totally antisymmetric representation will always have spin 1, as we will argue later in the book. It could be proven rigorously, but we will not do it here.

This ends the analysis of massive particles.

Massless Particles

But for massless particles, even though we have a more complicated story – since, as we saw, the little group is $ISO(2) \simeq SO(2) \times SO(1, 1)$ – the final result is the same. It seems naively that $SO(2) \simeq U(1)$ gives a priori arbitrary helicity (projection of the angular momentum, or spin, onto the momentum direction), since there doesn't seem to be any restriction on the $SO(2)$ charge = helicity (since $SO(2)$ rotates the directions transverse to the momentum – i.e., angular momentum along the momentum direction). Yet in fact, due to topological issues that we will not describe here, the final result is the same as for the massive case: quantized helicity, equivalent to the massive case.

11.6 The Double Cover of *SO*(3,1) as *Sl*(2,ℂ) and the Definition of Spinors

We have remarked earlier that there is an isomorphism between the Lie algebras of $SO(3,1)$ and $Sl(2,\mathbb{C})$, which means that the two groups are related, modulo global issues. Here we try to make that statement precise.

The $Sl(2,\mathbb{C})$ group is composed of 2×2 matrices V with complex components (the group of general such matrices is called $Gl(2,\mathbb{C})$), but with determinant 1 (which defines $Sl(2,\mathbb{C})$).

Consider a general 2×2 matrix, decomposed into the complete set of matrices $\sigma_\mu = (\mathbb{1}, \vec{\sigma})$ (σ_i are the Pauli matrices), as

$$V = V^\mu \sigma_\mu = \begin{pmatrix} V^0 + V^3 & V^1 - iV^2 \\ V^1 + iV^2 & V^0 - V^3 \end{pmatrix}. \tag{11.46}$$

Since as we can easily see

$$\det V = -V_\mu V^\mu , \tag{11.47}$$

which is a Lorentz invariant if V^μ is a vector, we consider matrix transformations of V, i.e., of the type

$$V \to \lambda V \lambda^\dagger, \tag{11.48}$$

such that $\det V$ is invariant. We can easily check that (since $\det(\lambda V \lambda^\dagger) = \det V |\det \lambda|^2$) for that, we need (if we are to have a Lie group, continously connected with the identity)

$$\det \lambda = +1. \tag{11.49}$$

But that means that $\lambda \in Sl(2,\mathbb{C})$, and since $V^\mu V_\mu$ is invariant, λ must define an $SO(3,1)$ Lorentz transformation on V^μ – i.e., we must have

$$\lambda(V^\mu \sigma_\mu)\lambda^\dagger = (\Lambda^\mu{}_\nu V^\nu)\sigma_\mu. \tag{11.50}$$

That means that, indeed, as a Lie algebra, $so(3,1) = sl(2,\mathbb{C})$. But we need still to understand it at the level of the group itself. There, we note that a set of $Sl(2,\mathbb{C})$ elements (complex matrices of determinant one) is provided by (for $g = e^{i\theta^3 \sigma_3/2}$)

$$g(\theta) = \begin{pmatrix} e^{i\frac{\theta}{2}} & 0 \\ 0 & e^{-i\frac{\theta}{2}} \end{pmatrix} = R_3(\theta) , \tag{11.51}$$

which, as we saw in Chapter 3, is the $SU(2)$ element corresponding to rotation around the third axis. But for $\theta = 2\pi$, we obtain $g = -\mathbb{1}$, so, in fact, we have a difference of $\mathbb{Z}_2 = (\mathbb{1}, -\mathbb{1})$ between the groups, and, therefore,

$$SO(3,1) = \frac{Sl(2,\mathbb{C})}{\mathbb{Z}_2}. \tag{11.52}$$

There is a way to describe this statement that gives a definition for spinors. A general matrix can, by a general theorem, be decomposed into an unitary one, u ($u^\dagger u = \mathbb{1}$) times an exponent of a hermitian one, h ($h^\dagger = h$),

$$\lambda = ue^h. \tag{11.53}$$

But since, as we already mentioned,

$$\det e^h = e^{\operatorname{Tr} h}, \tag{11.54}$$

for $Sl(2, \mathbb{C})$ matrices, $\det \lambda = 1$, we need to have

$$\det u = 1; \quad \operatorname{Tr} h = 0. \tag{11.55}$$

But a traceless hermitian matrix can be decomposed into the Pauli matrices σ_i which are traceless and hermitian,

$$h = \begin{pmatrix} p & q - ir \\ q + ir & -p \end{pmatrix}. \tag{11.56}$$

Since p, q, r are arbitrary ($\in \mathbb{R}$), the e^h piece is parametrized by $\mathbb{R}^3$.

A unitary matrix of determinant one (that is, an $SU(2)$ matrix) u on the other hand is parametrized as ($a, b, c, d \in \mathbb{R}$)

$$u = \begin{pmatrix} a + ib & c + id \\ -c + id & a - ib \end{pmatrix}, \tag{11.57}$$

if its determinant is one – i.e., if

$$a^2 + b^2 + c^2 + d^2 = 1. \tag{11.58}$$

But this is just the definition of S^3. That means that the space $Sl(2, \mathbb{C})$ is topologically the same as $\mathbb{R}^3 \times S^3$. But this is a simply connected space – i.e., all loops on it are contractible to a point.

On the other hand, because of dividing by $\mathbb{Z}_2$, $SO(3, 1) = Sl(2, \mathbb{C})/\mathbb{Z}_2$ is not simply connected, which means that there are loops in it that are not contractible in $SO(3, 1)$ but are in $Sl(2, \mathbb{C})$, its double cover. A 2π rotation around a direction x^3 is defined in quantum mechanics by the helicity (i.e., projection of the spin, or angular momentum, on the direction of the momentum) m_3, as the phase

$$e^{2\pi i m_3}. \tag{11.59}$$

This phase doesn't need to be 1, but the 2π rotation in the double cover, i.e., $Sl(2, \mathbb{C})$, must be 1 – i.e., a 4π rotation in the group must be equivalent to the identity,

$$e^{4\pi i m_3} = 1 \Rightarrow m_3 \in \frac{\mathbb{Z}}{2}. \tag{11.60}$$

Then the Lorentz spin must be half-integer, as we saw, and a rotation by 2π for a half-integer representation gives a minus sign. That is an alternative definition of a spinor.

Important Concepts to Remember

- The Lorentz group reduces to two copies of $SU(2)$, at least in the Lie algebra – i.e., $so(3, 1) = su(2) \times su(2)$.
- Then the general representation of the Lorentz group is $j \otimes j'$ for spin j and spin j' representations for the two $SU(2)$.

- Fields are classified by the representations of the full Poincaré group $ISO(3, 1)$.
- The universal cover of $SO(3, 1)$ is $Sl(2, \mathbb{C})$.
- For invariance of the quantum theory under $ISO(3, 1)$, fields must belong to representations of the liitle group as well, which for massive particles is $SO(d - 1) = SO(3) \simeq SU(2)$.
- Thus, the Wigner method decomposes representations of spin $j \otimes j'$ into irreps of the diagonal $SU(2)$, just like for addition of angular momenta in quantum mechanics.
- We have a scalar representation, two inequivalent (in $SO(3, 1)$) Weyl spinor representations of spin 1/2, a vector representation of spin 1, a symmetric traceless tensor (graviton) of spin 2, etc.
- A Dirac spinor is the sum of the two Weyl spinor representations, which are related by parity, so is a representation of the Lorentz group and parity.
- Globally we have $SO(3, 1) = Sl(2, \mathbb{C})/\mathbb{Z}_2$, and topologically $Sl(2, \mathbb{C})$ is $\simeq \mathbb{R}^3 \times S^3$, which is simply connected.
- Only the double cover of $SO(3, 1)$ needs to be simply connected, which leads to the fact that spinors (half-integer spin fields) give a minus sign under rotations by 2π around an axis.

Further Reading

See, for instance, chapters 8.1 and 10 in [9], chapter 6 in [10], chapter 2 (section 5 and appendix A) in [11].

Exercises

(1) Show that defining $J_{ij} = \epsilon_{ijk}J_k$ and $J_{0i} = K_i$, and, moreover,

$$M_i = \frac{J_i + iK_i}{2}; \quad N_i = \frac{J_i - iK_i}{2}, \tag{11.61}$$

the Lie algebra of the Lorentz group decomposes into two independent $su(2)$,

$$[M_i, M_j] = i\epsilon_{ijk}M_k; \quad [N_i, N_j] = i\epsilon_{ijk}N_k; \quad [M_i, N_j] = 0. \tag{11.62}$$

(2) Decompose into irreducible representations of the little group the case with $j = 2$ and $j' = 1$ for the two $su(2)$ factors, and guess the Lorentz index structure of the decomposition.

(3) Show that the Poincaré algebra $ISO(3, 1)$ can be obtained by a so-called Wigner-Inönü contraction of the "de Sitter algebra" $SO(3, 2)$, written in the same way as the Lorentz algebra (11.11), just with $\eta_{\mu\nu}$ now having two minuses and three pluses on the diagonal. The contraction amounts rescaling some of the generators by an R^2 quantity that is then taken to infinity in the resulting rescaled algebra.

(4) Show explicitly, by calculating the Lie algebras (commutators of generators), that the Lie algebras of $su(2)$ and $sl(2, \mathbb{C})$ are the same.

Statistics, Symmetry, and the Spin-Statistics Theorem

In this chapter, also, we will deal with some concepts of quantum theory, but concepts that don't need the full power of quantum field theory. Specifically, we will address statistics and its relation to spin in the form of the spin-statistics theorem, which relates integer spin with bosons and half-integer spin with fermions, and then we will try to understand symmetries from a more general point of view.

12.1 Statistics

Whereas in classical physics, particles are distinguishable and we can put labels on them (for instance, by following up a particle trajectory as it passes through a specific hole or slit in a panel), in quantum mechanics, that is not true anymore. Particles are *identical*, and, moreover, they are indistinguishable. For instance, in the standard quantum mechanical double slit experiment, we know that particles going through the two slits interfere, and in particular, we can even have a single particle "passing through both slits" and interfering with itself.

But that, in turn, means that if we write down a wave function for a system of particles or, more generally, a quantum (ket) state (such that when multiplying with a basis in a bra state, we get a wave function), the interchange of two particles in the wave function can only give a phase factor. In this way, the probabilities found by absolute values squared of the wave function are unaffected by the permutation. Thus, we write

$$|\ldots(s_1\vec{p}_1)\ldots(s_2\vec{p}_2)\ldots\rangle = \alpha|\ldots(s_2\vec{p}_2)\ldots(s_1\vec{p}_1)\ldots\rangle; \quad \alpha = e^{i\phi}. \quad (12.1)$$

Moreover, since we can permute the particles twice and (normally) obtain the same state,[†] it means that we must have

$$\alpha^2 = 1 \Rightarrow \alpha = \pm1. \quad (12.2)$$

- The case $\alpha = +1$ is called the *Bose-Einstein (BE) statistics*, and the particles are called bosons. It was discovered by Einstein for the photon case, but Bose made the general statements.
- The case $\alpha = -1$ is called the *Fermi–Dirac (FD) statistics*, and the particles are called fermions. It was discovered by Fermi, and Dirac treated the relativistic case and better understood its quantization.

[†] We will see later in the book that this is not necessarily true; we can have "anyons" and also "nonabelions," but for the moment, we will consider only the standard cases.

Note that the assumption of $\alpha^2 = 1$ that permuting the particles twice gives the same result is not necessarily always true. There is a generalization, which can be thought of as having some flux at a particle position, so that permutation is more nontrivial and is better thought of as a rotation, in which case the phase $\alpha = e^{i\phi}$ is general. This is the case of *anyons*, which will be considered later in the book.

Just like one does for the harmonic oscillator, when the photon (or rather, phonon) states are obtained by acting with a creation operator $a^\dagger$ on the vacuum of the theory, we can do a similar thing in the case of any particles. That means that the ket particle state is written as

$$|\ldots(s_1\vec{p}_1)\ldots(s_2\vec{p}_2)\ldots\rangle = N\ldots a^\dagger_{s_1\vec{p}_1} a^\dagger_{s_2\vec{p}_2}\ldots|0\rangle, \quad (12.3)$$

where N is a normalization constant. We can now differentiate between the two possible statistics.

BE Statistics

In this case, we write the defining relation, permuting the particles and obtaining a factor $\alpha = +1$ as an action of operators on states,

$$\ldots a^\dagger_{s_1\vec{p}_1} a^\dagger_{s_2\vec{p}_2}\ldots|0\rangle = \ldots a^\dagger_{s_2\vec{p}_2} a^\dagger_{s_1\vec{p}_1}\ldots|0\rangle \Rightarrow a^\dagger_{s_1\vec{p}_1} a^\dagger_{s_2\vec{p}_2}|\psi\rangle = a^\dagger_{s_2\vec{p}_2} a^\dagger_{s_1\vec{p}_1}|\psi\rangle, \ \forall|\psi\rangle, \quad (12.4)$$

which means that we can write as an commutation operatorial relation

$$[a^\dagger_{s_1\vec{p}_1}, a^\dagger_{s_2\vec{p}_2}] = 0, \quad (12.5)$$

and, taking the Hermitian conjugate, we find also the relation for the annihilation operators,

$$[a_{s_1\vec{p}_1}, a_{s_2\vec{p}_2}] = 0. \quad (12.6)$$

Moreover, since $\langle 0|0\rangle = 1$ (the vacuum is normalized), $(a^\dagger|0\rangle)^\dagger = \langle 0|a$, and $a|0\rangle = 0$, and the momentum state is normalized by a delta function, we obtain

$$\langle s_1\vec{p}_1|s_2\vec{p}_2\rangle = \delta_{s_1s_2}\langle\vec{p}_1|\vec{p}_2\rangle = \delta_{s_1s_2}(2\pi)^3\delta^3(\vec{p}_1-\vec{p}_2)$$
$$= \langle 0|a_{s_1\vec{p}_1} a^\dagger_{s_2\vec{p}_2}|0\rangle = \langle 0|[a_{s_1\vec{p}_1}, a^\dagger_{s_2\vec{p}_2}]|0\rangle, \quad (12.7)$$

where, in the last equality, we have added a term equal to zero in such a way as to obtain the commutator (and we want the commutator, since as we saw earlier, that is what we get for bosons). Comparing the two results, we see that we have

$$[a_{s_1\vec{p}_1}, a^\dagger_{s_2\vec{p}_2}] = (2\pi)^3\delta^3(\vec{p}_1-\vec{p}_2)\delta_{s_1s_2}. \quad (12.8)$$

Fermi–Dirac Statistics

We can repeat the logic above, now with an extra minus sign, so we obtain

$$a^\dagger_{s_1\vec{p}_1} a^\dagger_{s_2\vec{p}_2}|\psi\rangle = -a^\dagger_{s_2\vec{p}_2} a^\dagger_{s_1\vec{p}_1}|\psi\rangle, \ \forall|\psi\rangle, \quad (12.9)$$

which means that we can write anticommutation operatorial relations instead of commutation ones,

$$\{a^\dagger_{s_1\vec{p}_1}, a^\dagger_{s_2\vec{p}_2}\} = 0; \quad \{a_{s_1\vec{p}_1}, a_{s_2\vec{p}_2}\} = 0. \quad (12.10)$$

and we can write also the product of states as

$$
\langle s_1 \vec{p}_1 | s_2 \vec{p}_2 \rangle = \delta_{s_1 s_2} \langle \vec{p}_1 | \vec{p}_2 \rangle = \delta_{s_1 s_2} (2\pi)^2 \delta^3 (\vec{p}_1 - \vec{p}_2)
$$
$$
= \langle 0 | a_{s_1 \vec{p}_1} a^\dagger_{s_2 \vec{p}_2} | 0 \rangle = \langle 0 | \{ a_{s_1 \vec{p}_1}, a^\dagger_{s_2 \vec{p}_2} \} | 0 \rangle, \tag{12.11}
$$

where the only new factor is that we added the term equal to zero in such a way as to form the anticommutator, as expected from the previous anticommutation relations. Thus, we now also obtain

$$
\{ a_{s_1 \vec{p}_1}, a^\dagger_{s_2 \vec{p}_2} \} = (2\pi)^3 \delta^3 (\vec{p}_1 - \vec{p}_2) \delta_{s_1 s_2}. \tag{12.12}
$$

But what decides the statistics of particles? We will see that it is the spin of the particles, as found from the *spin-statistics theorem*. Under this generic title fall several ways to prove the same connection. We will sketch some of them and describe others in words.

12.2 Rotation Matrices in Different Lorentz Representations

But in order to do that, we will first establish some needed formalism to represent Lorentz generators in various representations.

We had seen that the Lorentz generators split into boosts $J_{0i} \equiv K_i$ and spatial rotations, defined by

$$
J_{ij} = \epsilon_{ijk} J_k. \tag{12.13}
$$

For instance, we are interested in rotations around the third axis – i.e., around the direction z – which rotate coordinates in the (12) plane, thus $J_{12} = J_3$ (or J_z). For such a simple rotation, the Lorentz group element is

$$
\lambda^\mu{}_\nu \equiv g = \begin{pmatrix} 1 & 0 & 0 & 0 \\ 0 & \cos\theta_z & \sin\theta_z & 0 \\ 0 & -\sin\theta_z & \cos\theta_z & 0 \\ 0 & 0 & 0 & 1 \end{pmatrix}. \tag{12.14}
$$

But a general Lorentz group transformation is written as

$$
g = e^{i\theta^{\mu\nu} J_{\mu\nu}} = e^{i\theta^i J_i + \beta^i K_i}, \tag{12.15}
$$

where we have redefined the parameters $\theta^{\mu\nu}$ into θ^i (rotation angles) and β^i (boosts) in order to match the decomposition of the generators.

Then the infinitesimal transformation becomes

$$
\delta\lambda^\mu{}_\nu = \delta g = i\theta^i (J_i)^\mu{}_\nu, \tag{12.16}
$$

and it acts on contravariant vectors V^μ, similar to dx^μ. Identifying with the infinitesimal variation of the matrix $\lambda^\mu{}_\nu$, we obtain

$$
\begin{pmatrix} 0 & 0 & 0 & 0 \\ 0 & 0 & \theta_z & 0 \\ 0 & -\theta_z & 0 & 0 \\ 0 & 0 & 0 & 0 \end{pmatrix} = i\theta_z J_z \Rightarrow J_z = -i \begin{pmatrix} 0 & 0 & 0 & 0 \\ 0 & 0 & 1 & 0 \\ 0 & -1 & 0 & 0 \\ 0 & 0 & 0 & 0 \end{pmatrix}. \tag{12.17}
$$

We can similarly calculate the boost generators in this representation. The action of a boost by a parameter $\beta_x = \cosh^{-1}(\gamma)$ on the coordinates is

$$
\begin{pmatrix} t' \\ x' \\ y' \\ z' \end{pmatrix} = \begin{pmatrix} \cosh\beta_x & \sinh\beta_x & 0 & 0 \\ \sinh\beta_x & \cosh\beta_x & 0 & 0 \\ 0 & 0 & 0 & 0 \\ 0 & 0 & 0 & 0 \end{pmatrix} \begin{pmatrix} t \\ x \\ y \\ z \end{pmatrix} \equiv \lambda^\mu{}_\nu \begin{pmatrix} t \\ x \\ y \\ z \end{pmatrix},
\tag{12.18}
$$

which gives the infinitesimal transformation

$$
\delta\lambda^\mu{}_\nu = \begin{pmatrix} 0 & \beta_x & 0 & 0 \\ \beta_x & 0 & 0 & 0 \\ 0 & 0 & 0 & 0 \\ 0 & 0 & 0 & 0 \end{pmatrix} = i\beta_x J_z \Rightarrow K_x = -i \begin{pmatrix} 0 & 1 & 0 & 0 \\ 1 & 0 & 0 & 0 \\ 0 & 0 & 0 & 0 \\ 0 & 0 & 0 & 0 \end{pmatrix}.
\tag{12.19}
$$

We can calculate the eigenvalues of J_z in the usual way, from

$$
\det(J_z - \lambda\,\mathbb{1}) = 0 \Rightarrow \lambda^2(\lambda^2 - 1) - 0 \Rightarrow \lambda = 0, +1, -1.
\tag{12.20}
$$

This is as expected, since we are in a *spin 1* – i.e., (as we saw in the last chapter) a *vector* representation of the little group of the Lorentz group – and so the projection of the spin on any direction should belong to the set $(-1, 0, +1)$. But since the eigenvalues of the matrix J_z are $0, 0, +1, -1$, that means (by a theorem of matrices) that we can diagonalize it by a similarity transformation and have these eigenvalues on the diagonal – i.e., that we have

$$
J_z = U^{-1} \begin{pmatrix} 0 & 0 & 0 & 0 \\ 0 & -1 & 0 & 0 \\ 0 & 0 & 1 & 0 \\ 0 & 0 & 0 & 0 \end{pmatrix} U.
\tag{12.21}
$$

When writing powers of J_z, there will still be a single U^{-1} on the right and one on the left (for instance, $(U^{-1}MU)^2 = U^{-1}MU^{-1}UMU = U^{-1}M^2U$, etc.), and then the same will happen for the exponential (which is an infinite sum of power laws). Moreover, since the matrix in the middle is diagonal, its exponential will be given by the exponentials of the eigenvalues on the diagonal, so, finally, the Lorentz group element in the vector representation corresponding to a rotation by θ_z around the z direction is

$$
\Lambda_{\text{vector}}(\theta_z) = e^{i\theta_z J_z} = U^{-1} \begin{pmatrix} 1 & 0 & 0 & 0 \\ 0 & e^{-i\theta_z} & 0 & 0 \\ 0 & 0 & e^{i\theta_z} & 0 \\ 0 & 0 & 0 & 1 \end{pmatrix} U.
\tag{12.22}
$$

On the other hand, in a *Dirac spinor* representation, which as we saw is the sum of two basic (Weyl spinor) representations, composed each of a spin-up and a spin-down state, so $\begin{pmatrix} \psi \\ \eta \end{pmatrix}$, we must have J_3 represented independently in each spinor state (ψ and η) by the Pauli matrix over two ($J_i = \sigma_i/2$ in each, as we saw in Chapter 3 that we must, to obtain the basic two dimensional representation of $SU(2)$), so

$$J_3 = \begin{pmatrix} \frac{\sigma_3}{2} & \mathbf{0} \\ \mathbf{0} & \frac{\sigma_3}{2} \end{pmatrix} = \begin{pmatrix} +1/2 & 0 & 0 & 0 \\ 0 & -1/2 & 0 & 0 \\ 0 & 0 & +1/2 & 0 \\ 0 & 0 & 0 & -1/2 \end{pmatrix}. \tag{12.23}$$

That means that for the group element, we obtain

$$\Lambda_{\text{spinor}}(\theta_z) = e^{i\theta_z J_3} = \begin{pmatrix} e^{i\theta_z/2} & 0 & 0 & 0 \\ 0 & e^{-i\theta_z/2} & 0 & 0 \\ 0 & 0 & e^{i\theta_z/2} & 0 \\ 0 & 0 & 0 & e^{-i\theta_z/2} \end{pmatrix}. \tag{12.24}$$

In particular, we reobtain the important results, mentioned at the end of the last chapter, that a rotation by 2π around an axis gives a minus for a spinor, since we can verify that

$$\Lambda_{\text{spinor}}(2\pi) = -\mathbb{1}. \tag{12.25}$$

Another observation is that the spinor representation is *projective* – i.e., it can change the phase of a state. Unlike a usual representation, when we multiply two representations, we can get a minus sign or, more generally, a phase,

$$R[g_1]R[g_2] = e^{i\phi(g_1, g_2)} R[g_1 g_2]. \tag{12.26}$$

12.3 The Spin-Statistics Theorem

The spin-statistics theorem says that the integer spin particles obey BE statistics, and the half-integer spin particles obey FD statistics.

As we said, there are several ways to prove it, and we will sketch several of them.

Interchange as a Rotation

The first way to prove it is related to the obsevation that a rotation by π in the spatial plane (plus a translation, depending on the position of the axis of rotation) amounts to an interchange of particles. But since this is a rotation by π of *each* of the two particles of spin s, it is the same as a rotation by 2π of one particle of spin s – which, as we said in the last chapter, gives a factor of

$$e^{2\pi i s} = \pm 1, \tag{12.27}$$

depending on the spin. For half-integer spin, we obtain a -1, thus FD statistics, whereas for integer spin we obtain a $+1$, thus a BE statistics.

We can verify this statement for a spinor representation. We have seen that $\Lambda_{\text{spinor}}(\theta_z)$ is given by (12.24). For a π rotation, we obtain the matrix

$$\Lambda_{\text{spinor}}(\pi) = \begin{pmatrix} i & 0 & 0 & 0 \\ 0 & -i & 0 & 0 \\ 0 & 0 & i & 0 \\ 0 & 0 & 0 & -i \end{pmatrix}. \tag{12.28}$$

But when acting only on spin up states in the z direction, i.e., on

$$\psi_1 = \begin{pmatrix} 1 \\ 0 \\ 1 \\ 0 \end{pmatrix}, \tag{12.29}$$

we obtain

$$\Lambda_{\text{spinor}}(\pi)\psi_1 = i\psi_1 \Rightarrow \Lambda_{\text{spinor}}(\pi)|\psi_1\psi_2\rangle = i^2|\psi_1\psi_2\rangle = -|\psi_1\psi_2\rangle. \tag{12.30}$$

But perhaps the previous proof seems too simple.

Causality Criterion

Another way to prove the theorem is to consider the effect of causality on fields. In a relativistic theory, we know that nothing propagates faster than light. That means that for space-like separation of two events, $(x - y)^2 > 0$, which means that they can be made to be simultaneous (equal time) in some reference frame but cannot be connected by light or any other signal; they cannot be causally connected (they cannot be cause and effect).

In the quantum theory, that means that any observables measured at x and y must be *independent*, since they can't influence one another. In the quantum theory, that means that they must commute,

$$[\mathcal{O}(x), \mathcal{O}(y)] = 0. \tag{12.31}$$

This criterion turns out to suffice to ensure the correct relation between spin and statistics.

We will check this for bosons – more precisely, for a *free* real scalar field $\phi(\vec{x}, t)$ – and we will consider the observable operator to be the field itself (in the classical limit).

Decompose the field in Fourier modes, with normalization N_k and coefficient function $a(\vec{k})$,

$$\phi(\vec{x}, t) = \int d^D k \left[N_k\, a(\vec{k}) e^{-i\omega t + i\vec{k}\cdot\vec{x}} + \bar{N}_k a^\dagger(\vec{k}) e^{+i\omega t - i\vec{k}\cdot\vec{x}} \right]. \tag{12.32}$$

Note that the second term is just the hermitian conjugate of the first, which is needed in order to make the field real.

Since the field is free and classical, it should obey the equation of motion – namely, the KG equation –

$$(\Box - m^2)\phi(\vec{x}, t) = 0, \tag{12.33}$$

which gives

$$\omega^2 - \vec{k}^2 - m^2 = 0 \Rightarrow \omega = \pm\sqrt{\vec{k}^2 + m^2}; \qquad (12.34)$$

therefore, $\omega = \omega(\vec{k})$ is a dependent quantity.

If we now impose the BE statistics when we quantize the mode numbers $a(\vec{k})$ and, moreover, if we quantize them according to the rules derived earlier, we obtain for the commutator of two fields both located at $t = 0$ (so that we have a space-like separation, $(x - y)^2 = (\vec{x} - \vec{y})^2 > 0$),

$$[\phi(\vec{x}, t = 0), \phi(\vec{y}, t = 0)] = \int d^D k |N_k|^2 \left[e^{i\vec{k}\cdot(\vec{x}-\vec{y})} - e^{-i\vec{k}\cdot(\vec{x}-\vec{y})} \right] = 0! \qquad (12.35)$$

In the last equality, we have rescaled $k \to -k$ to obtain zero. Notice that the cancellation was only possible because the first term came from the $[a, a^\dagger]$ terms, whereas the second came from the $[a^\dagger, a]$ terms, and by the BE commutation relations, they are equal up to a minus sign.

But if we impose the FD statistics instead, we obtain

$$\{\phi(\vec{x}, t = 0), \phi(\vec{y}, t = 0)\} = \int d^D k \, |N_k|^2 \left[e^{i\vec{k}\cdot(\vec{x}-\vec{y})} + e^{-i\vec{k}\cdot(\vec{x}-\vec{y})} \right] \neq 0. \qquad (12.36)$$

Note that now the reason the two terms add instead of canceling is that the second term has $\{a^\dagger, a\} = \{a, a^\dagger\}$, which gives the same result as the first term.

For other fields (in particular for fermions; also for vectors – i.e., gauge fields), the analysis is more complicated, and we will not do it here.

But this proof is case by case, so perhaps one still needs more ways to prove the theorem.

Stability of the Vacuum

If we consider the wrong statistics, we find that there are *negative energy contributions* to the total energy from the quantum theory, which means that the potential can grow negative without bound – i.e., the vacuum is unstable. However, to understand this proof, we would need the full quantum theory of fields, so we will not pursue it further.

Lorentz Invariance of the S-Matrix

Finally, the proof most commonly quoted for the spin-statistics theorem is obtained by analyzing the Lorentz invariance properties of the S-matrix for scattering. We find that Lorentz invariance is broken if we don't have the spin-statistics connection. However, this proof also needs the full power of quantum field theory, so we will not describe it further here.

12.4 Symmetries

We now try to understand the way general symmetries act on fields.

We have seen that the action of Lorentz – i.e., spacetime – symmetries on fields (in general with spin – i.e., in representations of the Lorentz group) is given by the general form (for a transformation $x'^\mu = \Lambda^\mu{}_\nu x^\nu$)

$$\phi'^i(x') = R^i{}_j(\Lambda)\phi^j(x). \tag{12.37}$$

Here $R^i{}_j(\Lambda)$ is the matrix of the Lorentz transformation in the representation R. Considering the Lorentz generators in the representation R, $J_{\mu\nu}(R)$, the matrix of the Lorentz transformation in the representation R is

$$R^i{}_j(\Lambda) = \left(e^{i\theta^{\mu\nu}J_{\mu\nu}(R)}\right)^i{}_j. \tag{12.38}$$

For an infinitesimal transformation, we obtain the variation

$$\delta\phi^i(x) = \delta R^i{}_j(\Lambda)\phi^j(x) \simeq i\theta^{\mu\nu}(J_{\mu\nu}(R))^i{}_j\phi^j(x). \tag{12.39}$$

We have just seen the example of the variation in the vector (fundamental) representation – i.e., the action of $\lambda^\mu{}_\nu$ on V^μ (contravariant vector, transforming like dx^μ) – but the principle is the same.

But spacetime symmetries are only the Lorentz transformations noted earlier, and the translations generated by P_μ – i.e., in total the Poincaré group $ISO(3,1)$.

Moreover, there is a theorem, known as the *Coleman–Mandula theorem*, which says that there are no other symmetries – i.e., *internal (non-spacetime)* symmetries – such that they *don't* commute with $ISO(3,1)$. More precisely, the statement is that if there would be internal symmetries that don't commute with $ISO(3,1)$, then scattering would be trivial – i.e., there would be no interactions.

However, as always, a theorem is only as good as its assumptions, and it was not realized that the assumption of a Lie algebra can be broken. In fact, supersymmetry breaks the theorem by breaking this assumption: we can have super-Lie algebras, which have some anticommutators instead of the commutators of the Lie algebras. In supersymmetry, the susy generator Q relates bosons with fermions; thus, it must be a fermion itself, which means it is natural for it to anticommute instead of commuting.

But excluding supersymmetry and super-Lie algebras, we have

$$[J_{\mu\nu}, T_a] = [P_\mu, T_a] = 0 , \tag{12.40}$$

where T_a are the internal symmetry generators. That means that we can consider the internal symmetries independently from the spacetime symmetries.

Consider, then, an internal symmetry that doesn't affect coordinates (we have the same x on both sides, unlike the Lorentz transformation case),

$$\phi'^i(x) = R^i{}_j(\theta^a)\phi^j(x) , \tag{12.41}$$

where the matrix in representations R, depending on the parameters θ^a, is

$$R^i{}_j(\theta^a) = \left(e^{i\theta^a T_a(R)}\right)^i{}_j. \tag{12.42}$$

Then the infinitesimal transformation is

$$\delta\phi^i(x) = \delta R^i{}_j(\theta^a)\phi^j(x) \simeq i\theta^a[T_a(R)]^i{}_j\phi^j(x). \tag{12.43}$$

We have already seen an example, the $O(N)$ model, with N real scalars ϕ^I, $I = 1, .., N$, transforming as

$$\phi^I(x) \rightarrow \phi'^I(x) = R^I{}_J \phi^J(x), \tag{12.44}$$

where the matrix R is orthogonal ($O(N)$ matrix), i.e.,

$$R^T R = \mathbb{1}. \tag{12.45}$$

But we can have more complicated cases, where the action of the internal group is also on fields with spin (in nontrivial representations of the Lorentz group, spinor, vector, etc.), but, because of the Coleman–Mandula theorem, they must act *independently of spacetime*. The classic example is *Yang–Mills theory*, with vector fields A_μ^a in the adjoing representation of a symmetry group G. However, this case is more complicated and will be analyzed later.

A simpler case is of a vector fields in a fundamental representation of a group – i.e., of the same $O(N)$ group – e.g., consider N vector fields $A_\mu^I(x)$. Thus, the transformation is the same as for the scalar $O(N)$ model, despite the extra μ index,

$$A_\mu^I(x) \rightarrow A_\mu'^I(x) = R^I{}_J A_\mu^J(x). \tag{12.46}$$

The infinitesimal transformation is then

$$\delta A_\mu^I(x) = i\theta^a (T_a(R))^I{}_J A_\mu^J(x). \tag{12.47}$$

These symmetry transformations must leave the action invariant. In most cases, that would mean that the Lagrangean density $\mathcal{L}$ is invariant, but that is not necessary: it could be that we have a variation up to a total derivative $\delta\mathcal{L} = \partial_\mu j^\mu$, which gives only a boundary term in the action that vanishes due to the boundary conditions.

The general recipe for infinitesimal transformations (with parameter ϵ) on fields can be put in the form

$$\delta_\epsilon \phi = \{\epsilon Q, \phi\}_{P.B.}, \quad \forall \phi. \tag{12.48}$$

This relation, when considered for *all* the fields in the action, defines the generator Q of the symmetry transformation as an integral over spacetime of functions of the fields, $Q = \int d^4x f(\phi)$, such that the Poisson bracket with each of the fields gives the corresponding symmetry transformation. We will see an example of this fact in Chapter 14, when discussing the energy-momentum tensor. We can view the relation above in two ways: if we know Q, we find all $\delta_\epsilon \phi$, or if we know all $\delta_\epsilon \phi$, we can calculate Q.

Important Concepts to Remember

- Quantum particles are indistinguishable and obey statistics – i.e., properties under interchange, defined by a phase $\alpha = e^{i\phi} = \pm 1$. For $\alpha = +1$, we have Bose-Einstein (BE) statistics and bosons, for $\alpha = -1$ we have Fermi–Dirac (FD) statistics and fermions.
- BE statistics leads to commutation relations $[a, a] = [a^\dagger, a^\dagger] = 0$ and $[a, a^\dagger] = \delta$, and FD statistics leads to anticommutation relations $\{a, a\} = \{a^\dagger, a^\dagger\} = 0$ and $\{a, a^\dagger\} = \delta$.

- A spinor has a Lorentz matrix for rotations around an axis given by $\Lambda_{\text{spinor}}(2\pi) = -\mathbb{1}$, and the corresponding representation is projective – i.e., only up to a phase.
- The spin-statistics theorem says that half-integer spins are fermions and integer spins are bosons.
- We can prove the spin-statistics theorem from the interchange being equal to rotation by π, leading to a minus sign for fermions, from the fact that space-like separated points $(x - y)^2 > 0$ must have commuting operators, $[\mathcal{O}(x), \mathcal{O}(y)] = 0$, from the stability of the vacuum, or from the Lorentz invariance of the S-matrix.
- If we quantize integer spins by FD relations or half-integer spins by BE, we find that fields at space-like separations don't commute.
- Spacetime symmetries are only $ISO(3,1)$, and by the Coleman–Mandula theorem, all other internal symmetries must commute with them if they are based on Lie algebras (so supersymmetry is an exception).
- Internal symmetries thus are independent of spacetime, and defined as $\delta_\epsilon \phi = \{\epsilon Q, \phi\}_{P.B.}$.

Further Reading

See chapters 3.5 in Peskin and Schroeder's book [4] and chapter 12 in Schwartz's [9].

Exercises

(1) Show that
$$[\dot\phi(\vec{x}, t = 0), \dot\phi(\vec{y}, t = 0)] = 0 \tag{12.49}$$
only for the BE, but not for the FD, statistics.

(2) Calculate
$$g = e^{i(\theta^3 J_3 + \theta^2 J_2)}, \tag{12.50}$$
for θ^3, θ^2 finite, in the vector representation, and diagonalize it.

(3) Represent the generator of $SO(N)$ rotations on N scalar fields $\phi^I(x)$ on the fields, analogous to the way we represented the Lorentz generators on them. Then explicitly show that on the fields $\phi^I(x)$, we have
$$[J_{\mu\nu}, T_a] = 0. \tag{12.51}$$

(4) Show explicitly that for BE statistics we also have
$$[\phi^2(\vec{x}, t), \phi^2(\vec{y}, t)] = 0, \tag{12.52}$$
and not only $[\phi(\vec{x}, t), \phi(\vec{y}, t)] = 0$, paralleling the proof in the text.

13 Electromagnetism and the Maxwell Equation; Abelian Vector Fields; Proca Field

In this chapter, we will further describe electromagnetism – in particular, its properties as a gauge theory and a formulation in terms of p-forms, which we then we will generalize to Abelian vector and antisymmetric tensor fields as p-forms. Finally, we will add a mass to the gauge field of electromagnetism to describe the massive vector field, the Proca field.

13.1 Electromagnetism as a $U(1)$ Gauge Field

We have already seen that the Maxwell equations are written as

$$\partial_\mu F^{\nu\mu} = j^\nu \tag{13.1}$$

for the equation of motion, while the Bianchi identity is

$$\partial_{[\mu} F_{\nu\rho]} = 0, \tag{13.2}$$

which allows us to write the field strength as the antisymmetric derivative of a gauge (potential) field,

$$F_{\mu\nu} = \partial_\mu A_\nu - \partial_\nu A_\mu \equiv 2\partial_{[\mu} A_{\nu]}. \tag{13.3}$$

In the last equality, we have written an expression using the antisymmetrization with strength expression (i.e., such that when multiplying by an antisymmetric object, we can just drop the antisymmetrization symbol).

We have also seen that a good Lagrangean density for electromagnetism is

$$\mathcal{L} = -\frac{1}{4} F_{\mu\nu} F^{\mu\nu} = -\frac{1}{4} F_{\mu\nu} F_{\rho\sigma} \eta^{\mu\rho} \eta^{\nu\sigma}, \tag{13.4}$$

to which we add the coupling to a source (current) j^μ, $j^\mu A_\mu$. The Lagrangean density is a relativistic invariant, and it can be written in terms of electric and magnetic field as

$$\mathcal{L} = -\frac{1}{2} F_{0i} F^{0i} - \frac{1}{4} F_{ij} F^{ij} = \frac{\vec{E}^2}{2} - \frac{\vec{B}^2}{2}. \tag{13.5}$$

Since the Lagrangean (for zero source) is written in terms of the field strength $F_{\mu\nu}$ only, without having an explicit A_μ, it means that we have a *local* symmetry, known as *gauge invariance*,

$$A_\mu(x) \to A'_\mu(x) = A_\mu(x) + \partial_\mu \alpha(x). \tag{13.6}$$

The variation, both infinitesimal and finite, is

$$\delta A_\mu(x) = \partial_\mu \alpha(x), \tag{13.7}$$

and under it, the field strength is invariant,

$$\delta F_{\mu\nu} = 2\partial_{[\mu}\delta A_{\nu]} = 2\partial_{[\mu}\partial_{\nu]}\alpha = 0, \tag{13.8}$$

so the Lagrangean is also explicitly invariant.

Note that this is a *redundancy* in the description, since it means that not all A_μ values are independent (we can use the function $\alpha(x)$ to fix one component of A_μ to something). In terms of the nonrelativistic components (the electric potential ϕ and the vector potential $\vec{A}$), $A_\mu = (-\phi, \vec{A})$, the gauge invariance is

$$\delta\phi = -\partial_t \alpha; \quad \delta\vec{A} = \vec{\nabla}\alpha. \tag{13.9}$$

The redundancy of gauge invariance means that the one function $\alpha(x)$ can be used to fix one component of $A_\mu(x)$ to anything – for instance, zero – using a *gauge condition*.

We can have gauge conditions that are Lorentz covariant, or invariant. For instance, there is the *Lorenz gauge* condition,

$$\partial^\mu A_\mu = \eta^{\mu\nu}\partial_\nu A_\mu = 0. \tag{13.10}$$

Note that (Ludvig) Lorenz of the gauge condition is a different person than (Hendrik) Lorentz of Lorentz invariance.

Even though we have used a single general function $\alpha(x)$ to fix another function, $\partial^\mu A_\mu(x)$, we have, in fact, not completely fixed the gauge (removed gauge invariance). We have a residual gauge invariance, $\delta A_\mu(x) = \partial_\mu \alpha$, such that

$$\partial^\mu \delta A_\mu = \partial^\mu \partial_\mu \alpha = \Box\alpha = 0. \tag{13.11}$$

That is, we cannot make a gauge transformation by an arbitrary function (since in that case, the counting above says we have fixed the gauge) but, rather, by function with a restricted dependence on x. We can make this more evident in the Fourier representation, since, then,

$$\alpha(x) = \int \frac{d^4k}{(2\pi)^4} e^{ik\cdot x}\alpha(k) \Rightarrow k^2 = 0. \tag{13.12}$$

Thus, only the coefficients $\alpha(k)$ for $k^2 = 0$ are nonzero.

The equation of motion (Maxwell equation) for zero source, $j^\mu = 0$, is

$$\partial_\mu F^{\mu\nu} = 0 = \partial_\mu \partial^\mu A^\nu - \partial_\mu \partial^\nu A^\mu \Rightarrow \Box A^\nu - \partial^\nu(\partial_\mu A^\mu) = 0. \tag{13.13}$$

In Lorenz gauge, then, we have just the Klein-Gordon (KG) equation for A_μ, like for a massless scalar field,

$$\Box A^\nu = 0. \tag{13.14}$$

In this case of $j^\mu = 0$, we can use the residual gauge invariance to also fix the gauge condition $A_0 = 0$, by a transformation by an $\alpha(x)$ (with $\Box\alpha(x) = 0$) such that

$$\partial_0 \alpha = -A_0 \Rightarrow A_0' = 0. \tag{13.15}$$

To check that this is possible, construct the equation of motion for the right-hand side, which in the Lorenz gauge is $\Box A_0 = 0$. Then we must have the same on the left-hand side – i.e., $\partial_0 \Box \alpha = 0$, which is true, since $\Box \alpha = 0$.

Note that the $A_0 = 0$ gauge is called the *Coulomb gauge*, and together with the Lorenz gauge, implies also $\vec{\nabla} \cdot \vec{A} = 0$. The combined gauge is sometimes called the *radiation gauge* since it is used in describing radiation (which is for $j^\mu = 0$, and satisfies $\Box A_\mu = 0$).

The gauge symmetry of A_μ is an Abelian ($U(1)$) symmetry that is *local* – i.e., it is parametrized by $\alpha = \alpha(x)$ – and the group element is

$$U(x) = e^{i\alpha(x)}. \tag{13.16}$$

But the A_μ gauge field transforms in a way related to the exponent, $\delta A_\mu = \partial_\mu \alpha(x)$, which means that the gauge field is *not* quite in the fundamental representation. In fact, in the non-Abelian case, which will be studied later, the field $A_\mu^a(x)$ transforms in the *adjoint* representation of the non-Abelian gauge group. Of course, in the $U(1)$ case, there is no adjoint representation since $f_{ab}{}^c = 0$ now, so it is a special case. We say that A_μ is an *Abelian vector field*.

13.2 Electromagnetism in p-Form Language

We now represent electromagnetism in a more abstract way, which admits an interesting generalization – namely, representing it in p-form language.

A p-form ω_p is defined by the relation

$$\omega_p \equiv \frac{1}{p!} \omega_{\mu_1 \ldots \mu_p} dx^{\mu_1} \wedge \ldots \wedge dx^{\mu_p}. \tag{13.17}$$

Here the "wedge" or "exterior" product of the differentials is defined as a totally antisymmetric product, so, for instance,

$$dx^\mu \wedge dx^\nu = -dx^\nu \wedge dx^\mu. \tag{13.18}$$

That means that $\omega_{\mu_1 \ldots \mu_p}$ is totally antisymmetric, so the $1/p!$ normalization is so that it can be written as a single term (with strength 1),

$$\omega_p = \omega_{01 \ldots p-1} dx^0 \wedge dx^1 \wedge \ldots \wedge dx^{p-1}. \tag{13.19}$$

We can also define the *exterior derivative* basically like $d \equiv \partial_\nu dx^\nu \wedge$, so

$$d\omega_p = \frac{1}{p!} \partial_\nu \omega_{\mu_1 \ldots \mu_p} dx^\nu \wedge dx^{\mu_1} \wedge \ldots \wedge dx^{\mu_p}. \tag{13.20}$$

Then, from this definition, if we write

$$F_{p+1} = dA_p \equiv \frac{1}{(p+1)!} F_{\mu_1 \ldots \mu_{p+1}} dx^{\mu_1} \wedge \ldots \wedge dx^{\mu_{p+1}}, \tag{13.21}$$

we obtain

$$F_{\mu_1 \ldots \mu_{p+1}} = (p+1) \partial_{[\mu_1} A_{\mu_2 \ldots \mu_{p+1}]}, \tag{13.22}$$

where the total antisymmetrization symbol is, as before, with strength 1 – i.e., such that when we multiply by a totally antisymmetric tensor, we can just drop the antisymmetrization symbol.

We can also define the *wedge product* of a p-form and a q-form in the natural way, by just "wedging" everything together – i.e.,

$$\omega_q \wedge \omega_r = \frac{1}{p!\,q!}\omega_{\mu_1...\mu_q}\omega_{\mu_{q+1}...\mu_{q+r}}dx^{\mu_1} \wedge ... \wedge dx^{\mu_{q+r}}. \tag{13.23}$$

Using this formalism for $p = 1$, we get exactly Maxwell electromagnetism. Indeed, define the gauge field one-form and the field strength two-form,

$$A = A_\mu dx^\mu; \quad F = \frac{1}{2}F_{\mu\nu}dx^\mu \wedge dx^\nu, \tag{13.24}$$

and then they are related by

$$F = dA. \tag{13.25}$$

In components, we obtain

$$F_{\mu\nu} = 2\partial_{[\mu}A_{\nu]} = \partial_\mu A_\nu - \partial_\nu A_\mu. \tag{13.26}$$

The gauge invariance can also be written in form language as

$$\delta A = d\lambda, \tag{13.27}$$

where λ is a 0-form – i.e., a scalar function – so in components we have the usual

$$\delta A_\mu = \partial_\mu \lambda. \tag{13.28}$$

The Bianchi identity $\partial_{[\mu}F_{\nu\rho]} = 0$ is written in form language (by multiplying by $dx^\mu \wedge dx^\nu \wedge dx^\rho$) as

$$dF = 0. \tag{13.29}$$

It implies that $F = dA$, since $d \wedge d = 0$ (derivatives commute), which in turn means that we have gauge invariance $\delta A = d\lambda$.

We can define form integration, which means integrating a p-form over a p-volume,

$$\int_{V_p} \omega_p = \int_{V_p} \frac{1}{p!}\omega_{\mu_1...\mu_p}dx^{\mu_1} \wedge ... \wedge dx^{\mu_p} = \int_{V_p} (dx^0 \wedge ... \wedge dx^{p-1})\omega_{01...p-1}. \tag{13.30}$$

But since $dx^0 \wedge dx^1 \wedge ... \wedge dx^{p-1} = dx^0 dx^1 ... dx^{p-1} = d^p x$, the previous integration coincides with the usual integration as

$$\int_{V_p} \omega_p = \int_{V_p} d^p x \omega_{01...p-1}. \tag{13.31}$$

The p-form integration formalism allows one to collect together many theorems under the guise of a single one, Stokes's theorem,

$$\int_{V_{p+1}} d\omega_p = \int_{(\partial V)_p} \omega_p. \tag{13.32}$$

We next define the *Hodge dual* of a p-form. We start by defining the Levi-Civitta totally antisymmetric tensor in n dimensions with the convention that

$$\epsilon^{01...n-1} = +1. \tag{13.33}$$

In four Minkowski dimensions, we define $\epsilon^{0123} = +1$, and $\epsilon^{\mu\nu\rho\sigma}$ is totally antisymmetric. We can lower its indices, defining

$$\epsilon_{\mu_1\mu_2\mu_3\mu_4} = \eta_{\mu_1\nu_1}\eta_{\mu_2\nu_2}\eta_{\mu_3\nu_3}\eta_{\mu_4\nu_4}\epsilon^{\nu_1\nu_2\nu_3\nu_4} = \det\eta\epsilon^{\mu_1\mu_2\mu_3\mu_4} = -\epsilon^{\mu_1\mu_2\mu_3\mu_4}. \tag{13.34}$$

In the last equality, we have used the fact that the determinant of any $n \times n$ matrix is defined by

$$\det M = \frac{1}{n!}\epsilon^{\mu_1...\mu_n}\epsilon^{\nu_1...\nu_n}M_{\mu_1\nu_1}...M_{\mu_n\nu_n}, \tag{13.35}$$

so, in particular,

$$\det\eta = \frac{1}{4!}\epsilon^{\mu_1\mu_2\mu_3\mu_4}\epsilon^{\nu_1\nu_2\nu_3\nu_4}\eta_{\mu_1\nu_1}...\eta_{\mu_4\nu_4}, \tag{13.36}$$

from which we can derive also (by multiplication by $\epsilon^{\mu_1\mu_2\mu_3\mu_4}$)

$$\eta_{\mu_1\nu_1}...\eta_{\mu_4\nu_4}\epsilon^{\nu_1\nu_2\nu_3\nu_4} = \det\eta\epsilon^{\mu_1\mu_2\mu_3\mu_4}. \tag{13.37}$$

Note that we have in general

$$\epsilon^{\mu_1...\mu_p\nu_1...\nu_q}\epsilon_{\mu_1...\mu_p\rho_1...\rho_q} = -p!\,q!\,\delta^{\nu_1...\nu_q}_{\rho_1...\rho_q}. \tag{13.38}$$

This is proven as follows. Since we have the $\mu_1...\mu_p$ in both epsilons, and since the indices on the epsilons are all different and comprise all the possible index values, it means that as a set, $(\nu_1...\nu_q) = (\rho_1...\rho_q)$, but not necessarily in the same order. We could have $\delta^{\nu_1}_{\rho_1}...\delta^{\nu_q}_{\rho_q}$ or any of the $q!$ permutations of the indices, giving in total $q!\,\delta^{\nu_1...\nu_q}_{\rho_1...\rho_q}$, where the last symbol is totally antisymmetrized with strength 1 (and multiplied by $q!$ since there are $q!$ different terms). Moreover, the sum over $\mu_1...\mu_p$, comprising $p!$ permutations of indices, gives another $p!$ factor. Finally, the single remaining term is $\epsilon^{01...n-1}\epsilon_{01...n-1} = -1$, hence the overall minus sign.

Finally, we define the Hodge dual, or star operation, first on the basis differentials of the p-forms in m-dimensional space,

$$*\left(dx^{\mu_1} \wedge ... \wedge dx^{\mu_p}\right) \equiv \frac{1}{(m-p)!}\epsilon^{\mu_1...\mu_p}{}_{\nu_{p+1}...\nu_m}dx^{\nu_{p+1}} \wedge ... \wedge dx^{\nu_m}, \tag{13.39}$$

thus as we see, it results in a $(m-p)$ form (the $1/(m-p)!$ factor is such as to cancel the factor coming from the equal $(m-p)!$ terms in the sum, the permutations of the indices $\nu_{p+1}...\nu_m$). Then, for a general p-form ω_p,

$$*\left(\omega_p\right) = \frac{1}{(m-p)!}\left[\frac{1}{p!}\omega_{\mu_1...\mu_p}\epsilon^{\mu_1...\mu_p}{}_{\nu_{p+1}...\nu_m}dx^{\nu_{p+1}} \wedge ... \wedge dx^{\nu_m}\right]. \tag{13.40}$$

We can then calculate the wedge (exterior) product

$$\omega_p \wedge *(\omega_p) = \frac{1}{p!\,(m-p)!\,p!} \omega_{\rho_1\ldots\rho_p} \omega_{\mu_1\ldots\mu_p} \epsilon^{\mu_1\ldots\mu_p}{}_{\nu_{p+1}\ldots\nu_m} dx^{\rho_1} \wedge \ldots \wedge dx^{\rho_p} \wedge dx^{\nu_{p+1}} \wedge \ldots \wedge dx^{\nu_m}$$

$$= \frac{1}{p!\,(m-p)!\,p!} \omega_{\rho_1\ldots\rho_p} \omega_{\mu_1\ldots\mu_p} \epsilon^{\mu_1\ldots\mu_p}{}_{\nu_{p+1}\ldots\nu_m} \epsilon^{\rho_1\ldots\rho_p\nu_{p+1}\ldots\nu_m} d^m x$$

$$= -\frac{1}{p!} \omega_{\rho_1\ldots\rho_p} \omega^{\mu_1\ldots\mu_p} \delta^{\mu_1\ldots\mu_p}_{\rho_1\ldots\rho_p} d^m x$$

$$= -\frac{1}{p!} \omega_{\mu_1\ldots\mu_p} \omega^{\mu_1\ldots\mu_p} d^m x. \tag{13.41}$$

Here, in the first equality, we just used the definition of the exterior product; in the second equality, we have used the fact that $dx^{\mu_1} \wedge \ldots \wedge dx^{\mu_m} = d^m x \epsilon^{\mu_1\ldots\mu_m}$ on an m-dimensional space (since the left-hand side is totally antisymmetric but contains the product of all the dx^i values, $dx^0 \wedge dx^1 \wedge \ldots \wedge dx^{m-1} = d^m x$), and in the third, we have used the relation (13.38).

Finally, then, we can define the notation

$$\int d^m x |F_p|^2 \equiv \int F_p \wedge *F_p = -\int dx^0 \wedge dx^1 \ldots \wedge dx^{m-1} \frac{1}{p!} F_{\mu_1\ldots\mu_p} F^{\mu_1\ldots\mu_p}$$

$$= -\int d^m x \frac{1}{p!} F_{\mu_1\ldots\mu_p} F^{\mu_1\ldots\mu_p}. \tag{13.42}$$

In particular, for $m = 4$ (four-dimensional Minkowski space) and $p = 1$ (gauge field for Maxwell electromagnetism), we obtain the Maxwell action,

$$S = \frac{1}{2} \int d^4 x |F_2|^2 = \frac{1}{2} \int F_2 \wedge *F_2 = -\frac{1}{4} \int d^4 x F_{\mu\nu} F^{\mu\nu}. \tag{13.43}$$

It is easy to see in this formalism that, since the $*$ operation can be thought to act on either of the two F values (its action is symmetric), when finding the equations of motion of this action, we can partially integrate either of the two d values on A values to the other side, obtaining two equal terms, for the (Maxwell) equation of motion

$$d * F = 0. \tag{13.44}$$

13.3 General p-Form Fields

One reason to introduce this p-form formalism is that it can be easily generalized. We have already noted that $B_{\mu\nu}$, the antisymmetric tensor obtained in the product of two spin 1 representations, is also a spin 1 representation, like the vector. In fact, as we also said, any totally antisymmetric tensor $A_{\mu_1\ldots\mu_p}$ is in an irreducible spin 1 representation (or spin zero, if it can be Hodge dualized to a scalar), so it can be thought of as a gauge field, like in the Maxwell case. In fact, one can define a field strength $F_{\mu_1\ldots\mu_{p+1}}$, the p-form A_p, and the $(p+1)$-form A_{p+1} with

$$F_{p+1} = dA_p, \tag{13.45}$$

or, in components,

$$F_{\mu_1\ldots\mu_{p+1}} = (p+1)\partial_{[\mu_1}A_{\mu_2\ldots\mu_{p+1}]}. \tag{13.46}$$

The field strength satisfies the Bianchi identity

$$dF_{p+1} = 0, \tag{13.47}$$

or, in components,

$$\partial_{[\mu_1}F_{\mu_2\ldots\mu_{p+1}]} = 0, \tag{13.48}$$

which, in fact, implies that $F_{p+1} = dA_p$. If we write an action in terms of only F_{p+1}, that means there is a gauge invariance, which is, as before, an Abelian local symmetry, now of the form

$$\delta A_p = d\lambda_{p-1}, \tag{13.49}$$

where λ is a $(p-1)$–form symmetry parameter, or, in components,

$$\delta A_{\mu_1\ldots\mu_p} = p\partial_{[\mu_1}\lambda_{\mu_2\ldots\mu_p]}. \tag{13.50}$$

The Maxwell-type action is the simple generalization suggested by the formalism

$$\frac{1}{2}\int d^4x |F_{p+1}|^2 = \frac{1}{2}\int F_{p+1}\wedge *(F_{p+1}) = -\frac{1}{2(p+1)!}\int d^mx F_{\mu_1\ldots\mu_{p+1}}F^{\mu_1\ldots\mu_{p+1}}. \tag{13.51}$$

As before, we can find the equation of motion

$$d*F = 0, \tag{13.52}$$

or, in components,

$$\partial_{\mu_1}F^{\mu_1\ldots\mu_{p+1}} = 0, \tag{13.53}$$

and the details are left as an exercise. If we add a source term to the action for $A_{\mu_1\ldots mu_p}$, of the type $j^{\mu_1\cdots\mu_p}A_{\mu_1\ldots\mu_p}$, the equation of motion becomes

$$\partial_{\mu_1}F^{\mu_1\ldots\mu_{p+1}} + j^{\mu_2\cdots\mu_{p+1}} = 0. \tag{13.54}$$

Note that now it is even less obvious that the Abelian (in the sense that two symmetry transformations commute, their order is irrelevant) local symmetry parametrized by $\lambda_{\mu_1\ldots\mu_{p-1}}(x)$ is related to a symmetry group, but since we have a continuous (local) and commuting transformation, we can infer it. There is no known way to make a p-form gauge invariance non-Abelian despite trying (there are also some no-go theorems about it).

The p-form fields couple to extended objects (their currents $j^{\mu_1\cdots\mu_p}$ are for objects with spatial extension) and are more important in theories in dimensions higher than four.

13.4 The Proca Field

Until now, we have considered a gauge field (with a gauge symmetry), which is by necessity a massless field. Indeed, the Lagrangean must be written in terms of the field strength F_{p+1} only, not of the gauge field A_p, and a mass term would only involve A_p.

But we can also consider massive vector fields A_μ – like, for instance, for the W bosons – appearing in the Standard Model after electroweak symmetry breaking. Such fields are called *Proca fields* and have the Lagrangean

$$\mathcal{L} = -\frac{1}{4} F_{\mu\nu} F^{\mu\nu} - \frac{m^2}{2} A_\mu A^\mu. \tag{13.55}$$

As we said, now there is no gauge invariance, since we have explicity A_μ values in the action, so it changes under this local transformation.

The equations of motion with source j^μ are now

$$\partial_\mu F^{\mu\nu} + j^\nu - m^2 A^\nu = 0, \tag{13.56}$$

which are rewritten as

$$(\Box - m^2) A^\nu + \partial^\nu (\partial_\mu A^\mu) + j^\nu = 0, \tag{13.57}$$

but unlike the gauge field case, we cannot put to zero the second term by a Lorenz gauge condition, so the resulting equation is not just the KG equation. Thus, the massive vector (Proca) field is truly different from a scalar field, even at the classical level.

Important Concepts to Remember

- The gauge field A_μ has gauge invariance $\delta A_\mu = \partial_\mu \alpha(x)$, which is a local symmetry.
- We can fix the covariant Lorenz gauge $\partial^\mu A_\mu = 0$, but there is a residual gauge invariance, allowing us also to fix $A_0 = 0$ (Coulomb gauge), giving together the radiation gauge.
- In p-form language, $F = dA$, the Maxwell action is $-1/2 \int |F|^2$, the Bianchi identiy is $dF = 0$, the equation of motion is $d * F = 0$, and the gauge invariance is $\delta A = d\lambda$.
- We can generalize to totally antisymmetric tensor fields A_p : $A_{\mu_1 \ldots \mu_p}$, which still have spin 1. Then $F_{p+1} = dA_p$, the Bianchi identity is $dF_{p+1} = 0$, the gauge invariance is $\delta A_p = d\lambda_{p-1}$, the action is $-1/2 \int |F_{p+1}|^2$ and the equation of motion is $d * F_{p+1} = 0$.
- A Proca field has a mass term $-m^2 A_\mu A^\mu$ in the action and has no gauge invariance, and its equation of motion cannot be reduced to KG.

Further Reading

For the electromagnetism as a $U(1)$ gauge theory part, see any QFT book – for instance, [9].

Exercises

(1)   Consider the three-dimensional action

$$S_{2+1} = \int d^3x \left[\frac{k}{4\pi} \epsilon^{\mu\nu\rho} A_\mu \partial_\nu A_\rho + m A_\mu A^\mu \right]. \tag{13.58}$$

    (a) Calculate the equations of motion.

    (b) Write it in form language.

(2)   Show that the equation of motion of the action

$$S = \frac{1}{2} \int d^n x \, |F_{p+1}|^2 \tag{13.59}$$

is, in form language,

$$d * F_{p+1} = 0. \tag{13.60}$$

(3)   Consider a p-form field A_p in $2p + 2$ dimensions, and a term in the action

$$S_p = \frac{\theta}{2} \int F_{p+1} \wedge F_{p+1}. \tag{13.61}$$

Show that, if $F_{p+1} = dA_p$, then the term in (13.61) doesn't contribute to the equation of motion for A_p.

(4)   Consider a scalar field $\phi(x)$ in 3+1 dimensions. Calculate the determinant

$$\det(\eta_{\mu\nu} + \partial_\mu \phi \partial_\nu \phi) \tag{13.62}$$

by doing all the contractions of the $\epsilon^{\mu\nu\rho\sigma}$ symbols.

14 The Energy-Momentum Tensor

In this chapter, we will introduce an important concept in classical field theory, the concept of the energy-momentum tensor. This is a relativistic tensor that has the Hamiltonian (energy) density among its components and describes many properties of the system. In particular, it encapsulates some conservation equations but contains some ambiguity, which can be fixed by considering the coupling of the system to gravity, leading to the Belinfante tensor.

14.1 Defining the Energy-Momentum Tensor

We have seen in Chapter 9 that the Hamiltonian is given as an integral of a Hamiltonian density, $H = \int d^3x \mathcal{H}$. In a general theory, it is

$$H = \int d^3x \mathcal{H} = \int d^3x \left[\pi_I(\vec{x}, t)\dot{\phi}^I(\vec{x}, t) - \mathcal{L} \right], \qquad (14.1)$$

where $\pi_I = \partial\mathcal{L}/\partial\dot{\phi}^I$ is the canonical momentum conjugate to ϕ^I. In a theory with a canonical kinetic term, we obtain

$$\mathcal{H} = \frac{(\dot{\phi}^I)^2}{2} + \frac{(\vec{\nabla}\phi^I)^2}{2} + \frac{m^2(\phi^I)^2}{2} + \Delta V(\phi^I), \qquad (14.2)$$

where $\pi_I = \dot{\phi}^I$. But in a relativistic theory, the energy and energy density are not Lorentz invariant concepts, so we want $\mathcal{H}$ to be a part of a relativistically covariant Lorentz four-tensor (i.e., that transforms appropriately under Lorentz transformations).

Consider the standard-type action (without higher derivatives of the fields),

$$S = \int d^4x \mathcal{L}(\phi^I, \partial_\mu\phi^I), \qquad (14.3)$$

with Euler–Lagrange equations of motion

$$\partial_\mu \left(\frac{\partial\mathcal{L}}{\partial(\partial^\mu\phi^I)} \right) - \frac{\partial\mathcal{L}}{\partial\phi^I} = 0. \qquad (14.4)$$

Calculate $\partial_\mu\mathcal{L}$ by the chain rule,

$$\partial_\mu \mathcal{L} = \frac{\partial \mathcal{L}}{\partial \phi^I} \partial_\mu \phi^I + \frac{\partial \mathcal{L}}{\partial(\partial_\nu \phi^I)} \partial_\mu \partial_\nu \phi^I$$

$$= \partial_\nu \left(\frac{\partial \mathcal{L}}{\partial(\partial_\nu \phi^I)} \right) \partial_\mu \phi^I + \frac{\partial \mathcal{L}}{\partial(\partial_\nu \phi^I)} \partial_\mu \partial_\nu \phi^I$$

$$= \partial_\nu \left(\frac{\partial \mathcal{L}}{\partial(\partial_\nu \phi^I)} \partial_\mu \phi^I \right), \tag{14.5}$$

where in the second equality we have used the Euler–Lagrange equations of motion. Finally, we can write the equation as

$$\partial_\nu \left[-\frac{\partial \mathcal{L}}{\partial(\partial_\nu \phi^I)} \partial_\mu \phi^I + \delta_\mu^\nu \mathcal{L} \right] = 0. \tag{14.6}$$

Then, defining the tensor

$$T_\mu{}^\nu \equiv -\frac{\partial \mathcal{L}}{\partial(\partial_\nu \phi^I)} \partial_\mu \phi^I + \delta_\mu^\nu \mathcal{L}, \tag{14.7}$$

called the *energy-momentum tensor*, we can write the equation we derived as

$$\partial_\nu T_\mu{}^\nu = 0. \tag{14.8}$$

14.2 Conservation Equations

This is a (set of) *conservation equation(s)*, as we can now show. Indeed, writing a relativistic four-current as the union of a charge density ρ and a vector current density $\vec{j}$, $j^\mu = (\rho, \vec{j})$, we show that

$$\partial_\mu j^\mu = 0 \tag{14.9}$$

means that charge is conserved.

Conservation of some charge means that the total quantity must be fixed, so the decrease in charge Q in a volume is given by the current flowing out from the surface of the volume. Since the current flowing out is defined by $\vec{j} \cdot d\vec{S}$, conservation means

$$\frac{\partial Q}{\partial t} = -\oint_{S=\partial V} \vec{j} \cdot d\vec{S}. \tag{14.10}$$

Expressing also Q as an integral, we obtain

$$\frac{\partial}{\partial t} \int_V \rho dV = -\oint_{S=\partial V} \rho \vec{v} \cdot d\vec{S} \equiv -\oint_{S=\partial V} \vec{j} \cdot d\vec{S} = -\int_V \vec{\nabla} \cdot \vec{j} dV. \tag{14.11}$$

In the last equality, we have used the Stokes's theorem (the generalization of the Gauss's law), which in form integration is written as $\int_M df = \int_{\partial M} f$. Finally, that means

$$\int_V dV [\partial_0 \rho + \vec{\nabla} \cdot \vec{j}] = 0, \quad \forall V, \tag{14.12}$$

which allows us to write the infinitesimal form

$$\partial_0 \rho + \vec{\nabla} \cdot \vec{j} = 0, \tag{14.13}$$

or in Lorentz covariant form

$$\partial_\mu j^\mu = 0. \tag{14.14}$$

Integrating over a four-manifold (our Minkowski space) the Lorentz covariant form we obtain, using Stokes's theorem,

$$\int_{M^4} d^4x \, \partial_\mu j^\mu = \oint_{\partial M^4} j^\mu d\Sigma_\mu, \tag{14.15}$$

where $d\Sigma_\mu$ is the surface element of the three-dimensional surface ∂M^4 at the boundary of the four-dimensional manifold.

In our case, $T_\mu{}^\nu$ is the current, with an extra index μ that labels the charges. The integral form of the conservation equation $\partial_\nu T_\mu{}^\nu = 0$ is

$$\oint_{\partial M^4} T^{\mu\nu} d\Sigma_\nu = 0, \tag{14.16}$$

which states that the quantities

$$P^\mu = \int_{\Sigma_3 \equiv V} T^{\mu\nu} d\Sigma_\nu \tag{14.17}$$

are conserved and, as we will see, coincide with the four-momentum. Note that, in this case, $\Sigma_3 = V$ is a surface with unit element $d\Sigma_\mu$ pointing in the time direction – i.e., with only nonzero component $d\Sigma_0 = dV$. This defines a time direction, the same way as a 2-dimensional surface element $d\vec{S}$ in three-dimensional Euclidean space defines the direction normal to the plane of of the surface element.

Indeed, in general for a current j^μ, consider the boundary ∂M^4 of the four-dimensional space as composed of a boundary S_∞ at spatial infinity, together with two surfaces S_{t_1} at constant time t_1 and S_{t_2} at constant time $t_2 > t_1$ (we can consider $t_1 = -\infty$ and $t_2 = +\infty$, but it is not necessary), as in Figure 14.1. We can consider that fields and currents go to

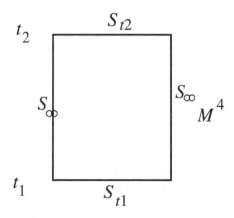

The surface needed for the conservation equation for charge.

zero at spatial infinity, so the integral over S_∞ vanishes. Then the conservation equation is

$$0 = \oint_{\partial M^4} j^\mu d\Sigma_\mu = \int_{t=t_2} d^3x j^0 - \int_{t=t_1} d^3x j^0 = Q(t_2) - Q(t_1), \qquad (14.18)$$

which means that $Q(t_2) = Q(t_1)$ – i.e., the charge Q is conserved.

In our case, then, the energy P^0 would be given by (note that $T^{00} = -T_0{}^0$)

$$P^0 = \int_{S_t} T^{0\mu} d\Sigma_\mu = \int_V dV T^{00} = \int_V dV \left[\frac{\partial \mathcal{L}}{\partial \dot{\phi}^I} \dot{\phi}^I - \mathcal{L} \right] = \int_V dV [\pi_I \dot{\phi}^I - \mathcal{L}] = \int_V dV \mathcal{H}, \qquad (14.19)$$

which is indeed correct.

Then, indeed, the quantity

$$P^\mu = \oint_{\Sigma_3} T^{\mu\nu} d\Sigma_\nu = \int_V dV T^{\mu 0} \qquad (14.20)$$

is the four-momentum, by Lorentz invariance (P^0 is part of the Lorentz vector P^μ), which also means that

$$P^i = \int_V dV T^{i0} \qquad (14.21)$$

is the three-momentum. Then it also means that T^{i0} is the momentum density (per unit volume).

14.3 An Ambiguity in $T_{\mu\nu}$ and Ways to Fix It

Before we continue, we consider a subtlety: $T_\mu{}^\nu$ is only defined up to a total derivative. Indeed, any current j^μ is only defined up to a total derivative: we can add $\partial_\nu j^{\nu\mu}$ to it where $j^{\nu\mu}$ is antisymmetric in $\mu\nu$, since then $\partial_\mu[\partial_\nu j^{\nu\mu}] = 0$, so the conservation equation $\partial_\mu j^\mu = 0$ is unaffected.

Then, for the energy momentum tensor, we can add a term

$$T^{\mu\nu} \to T^{\mu\nu} + \partial_\rho j^{\mu\nu\rho}, \qquad (14.22)$$

where $j^{\mu\nu\rho}$ is antisymmetric in $\nu\rho$, $j^{\mu\nu\rho} = -j^{\mu\rho\nu}$, without changing the conservation equation $\partial_\nu T_\mu{}^\nu = 0$. Moreover, the charge (momentum) is also unaffected, since

$$P'^\mu - P^\mu = \int_{S_t=V} \partial_\rho j^{\mu\nu\rho} d\Sigma_\rho = \int_{\partial V} j^{\mu\nu\rho} d\Sigma_{\nu\rho} = 0. \qquad (14.23)$$

In the second equality, we have used Stokes's theorem, and in the last equality we have considered that the boundary of the spatial slice, ∂V, is a "surface at spatial infinity" on which we assume all fields and currents vanish.

Given this ambiguity, how do we define $T^{\mu\nu}$ uniquely?

To do so, one possibility is to consider the angular momentum,

$$\vec{L} = \sum_a \vec{r}_a \times \vec{p}_a, \qquad (14.24)$$

or in components

$$L_i = \epsilon_{ijk} \sum_a x_j^a p_k^a. \tag{14.25}$$

Here, the sum over a is a sum over degrees of freedom. Defining the angular momentum tensor operator, we obtain

$$L_{ij} = \epsilon_{ijk} L_k = \sum_a \epsilon_{ijk} \epsilon_{ilm} x_l^a p_m^a = \sum_a (x_i^a p_j^a - x_j^a p_i^a). \tag{14.26}$$

Here we used $\epsilon_{ijk}\epsilon_{klm} = 2\delta_{ij}^{lm} = \delta_i^l \delta_j^m - \delta_i^m \delta_j^l$.

The final result is the same as the (ij) components of the $J_{\mu\nu}$ operator that we already found as a function of the translation operator P_μ in Chapter 11,

$$J_{\mu\nu} = \sum_a (x_\mu^a p_\nu^a - x_\nu^a p_\mu^a). \tag{14.27}$$

The only difference is the sum over a and having p_μ^a instead of the abstract operator P_μ, which amounts to being in a given representation in terms of particles, where we just sum over p_μ^a. If instead of particles, we consider a continuum, we obtain an integral expression,

$$J^{\mu\nu} = \int (x^\mu dP^\nu - x^\nu dP^\mu). \tag{14.28}$$

But the general definition of $P^\mu = \int T^{\mu\nu} d\Sigma_\nu$ implies the differential $dP^\mu = T^{\mu\nu} d\Sigma_\nu$, so, finally,

$$J^{\mu\nu} = \int_{S_t} (x^\mu T^{\nu\rho} - x^\nu T^{\mu\rho}) d\Sigma_\rho = \int_{S_t} J^{\mu\nu\rho} d\Sigma_\rho. \tag{14.29}$$

Considering that this four-angular momentum $J^{\mu\nu}$ is the charge for a current $J^{\mu\nu\rho}$, its conservation (assuming no current flows out of the system) is $\partial_0 J^{\mu\nu 0} = 0$, and the integrated form, over the same M^4 (with the same ∂M^4 composed of S_∞ and S_{t_2}, S_{t_1} in Figure 14.1 as before) is

$$\begin{aligned}
0 &= \int_{\partial M^4} J^{\mu\nu\rho} d\Sigma_\rho = \int_{\partial M^4} (x^\mu T^{\nu\rho} - x^\nu T^{\mu\rho}) d\Sigma_\rho \\
&= \int_{t_2} dV(x^\mu T^{\nu 0} - x^\nu T^{\mu 0}) - \int_{t_1} dV(x^\mu T^{\nu 0} - x^\nu T^{\mu 0}) \\
&= \int_{M^4} d^4 x \, \partial_\rho (x^\mu T^{\nu\rho} - x^\nu T^{\mu\rho}). \tag{14.30}
\end{aligned}$$

In the last equality, we have used Stokes's theorem. Since we have an integral over volume, valid for any volume, we can then impose the local conservation as well,

$$0 = \partial_\rho (x^\mu T^{\nu\rho} - x^\nu T^{\mu\rho}) = \delta_\rho^\mu T^{\nu\rho} - \delta_\rho^\nu T^{\mu\rho} + x^\mu \partial_\rho T^{\nu\rho} - x^\nu \partial_\rho T^{\mu\rho}, \tag{14.31}$$

and, using the conservation of $T^{\mu\nu}$, $\partial_\nu T^{\mu\nu} = 0$, we obtain that

$$T^{\nu\mu} = T^{\mu\nu}, \tag{14.32}$$

i.e., the energy-momentum tensor defined uniquely as above must be symmetric. We can simply symmetrize the original definition $\tilde{T}^{\mu\nu}$ in (14.7), as $T^{\mu\nu} = \tilde{T}^{\mu\nu} + \tilde{T}^{\nu\mu}$. However, we need to remember that we can always add a total derivative term and that we can also symmetrize, so there is still an ambiguity (to add $\partial_\rho j^{\mu\nu\rho}$, where $j^{\mu\nu\rho} = -j^{\mu\rho\nu} = +j^{\nu\mu\rho}$).

14.4 Interpretation of $T^{\mu\nu}$

We have already seen that $T^{00} = T_{00}$ has the interpretation as $\mathcal{H}$, the energy density, and that T^{i0} is momentum density. We write the conservation equations for $T_{\mu\nu}$, in components, as

$$\partial_0 T^{00} + \partial_i T^{0i} = 0$$
$$\partial_0 T^{i0} + \partial_j T^{ij} = 0. \tag{14.33}$$

Integrating the first equation over a spatial volume V, we obtain

$$\partial_0 E \equiv \partial_0 \int_V dV T^{00} = - \int_V \partial_i T^{0i} dV = - \oint_{\partial V} T^{0i} d\Sigma_i, \tag{14.34}$$

where in the last equality we have used Stokes's theorem. The equation then says that the loss of energy in the volume V is given by the flux of T^{0i} through the surface of the volume – that is, T^{0i} is the energy flux density (i.e., the energy loss through the unit surface per unit time). If we use the symmetric energy-momentum tensor, $T^{0i} = T^{i0}$ has the dual interpretation of momentum density and energy flux density.

Integrating the second equation over the same spatial volume V, we obtain in the same way

$$\partial_0 P^i \equiv \partial_0 \int_V dV T^{i0} = - \int_V dV \partial_j T^{ij} = - \oint_{\partial V} T^{ij} d\Sigma_j. \tag{14.35}$$

Since the loss of momentum in the volume V is given by the flux of T^{ij} through the surface of the volume, it follows that T^{ij} is the momentum flux density (momentum loss through the unit surface per unit time). It is also called the "stress tensor."

Consequently, the relativistically covariant object $T_{\mu\nu}$ is also called, besides energy-momentum tensor, the "stress-energy tensor" (though perhaps "stress-energy-momentum tensor" would have been more appropriate).

14.5 The Belinfante Tensor

To obtain the generally used form of the symmetric energy-momentum tensor, there is a specific construction that comes out of the definition of $T_{\mu\nu}$ as the object that couples in the action to perturbations in the metric of spacetime, $\delta g^{\mu\nu}$. We promote the Minkowski metric $\eta_{\mu\nu}$ to a general metric $g_{\mu\nu}(x)$ (an arbitrary function of spacetime) and vary the action with respect to it, so

$$T_{\mu\nu} \equiv -\frac{2}{\sqrt{-g}} \frac{\delta S}{\delta g^{\mu\nu}}. \tag{14.36}$$

Note first the functional derivative; we get rid of the integration in the action, obtaining a local object, as it should. Second, note that when we change the metric, we also change the integration measure, from d^4x to $d^4x\sqrt{-g}$ (which is the general integration measure when we have a nontrivial metric, as we will see later in the book). This definition gives the so-called Belinfante tensor.

In order to derive the Belinfante tensor, we derive a formula. For a general matrix M, $M \cdot M^{-1} = \mathbb{1}$ implies $dMM^{-1} = -MdM^{-1}$. Moreover, we have the matrix formula

$$\det M = e^{\operatorname{Tr} \ln M}, \tag{14.37}$$

which, when differentiated, gives

$$\delta \det M = e^{\operatorname{Tr} \ln M} \operatorname{Tr}(\delta \ln M) = \det M \operatorname{Tr}(M^{-1}\delta M) = -\det M \operatorname{Tr}(M\delta M^{-1}). \tag{14.38}$$

Since $\delta\sqrt{\det M}/\sqrt{\det M} = 1/2\delta \det M/\det M$, applying the formula (14.38) for the matrix $M = g_{\mu\nu}$ (the metric), and since the matrix inverse metric is denoted by $g^{\mu\nu}$, we obtain

$$\delta\sqrt{-g} = -\frac{1}{2}\sqrt{-g}g_{\mu\nu}\delta g^{\mu\nu}. \tag{14.39}$$

We are now ready to derive the Belinfante tensor for scalars and electromagnetism, for instance.

Note that the Belinfante tensor is automatically symmetric since $\delta g^{\mu\nu}$ is symmetric. However, that doesn't mean that it has always the desired properties. For instance, it is sometimes said that we need to have a traceless energy-momentum tensor, $T^\mu{}_\mu = 0$, for a "conformal" theory. We don't need the precise definition of conformal theories here, but it will suffice to say they are theories with no mass (or equivalently energy or length) scales. In particular, a free massless scalar,

$$S = -\frac{1}{2}\int d^4x(\partial_\mu\phi)\partial_\nu\phi\eta^{\mu\nu} \rightarrow -\frac{1}{2}\int d^4x\sqrt{-g}(\partial_\mu\phi)(\partial_\nu\phi)g^{\mu\nu} \tag{14.40}$$

is such a theory.

But while in $d = 2$, spacetime dimensions we would indeed obtain a traceless Belinfante tensor, in $d = 4$, we don't if we use the (naive) Belinfante tensor, which is

$$T^\phi_{\mu\nu} = \partial_\mu\phi\partial_\nu\phi - \frac{1}{2}\eta_{\mu\nu}(\partial_\rho\phi)^2. \tag{14.41}$$

However, we can add a total derivative term $\partial_\rho J^{\mu\nu\rho}$ such that we can have a traceless energy-momentum tensor, a so-called *improved energy-momentum tensor*,

$$T^{\phi,I}_{\mu\nu} = \partial_\mu\phi\partial_\nu\phi - \frac{1}{2}\eta_{\mu\nu}(\partial_\rho\phi)^2 + \frac{1}{6}(\eta_{\mu\nu}\Box - \partial_\mu\partial_\nu)\phi^2. \tag{14.42}$$

In fact, we can obtain such an extra term from the action for the scalar coupled to gravity if we impose that this action is also conformally invariant.[‡]

Then, provided we take into account the symmetries of the system, the use of the Belinfante tensor as the energy-momentum tensor is the correct thing to do.

[‡] In general dimension, we need to add to the Lagrangean the extra term $\frac{(d-2)}{4(d-1)}R\phi^2$, where R is the "Ricci scalar" that will be derived later in the book, becoming $\frac{1}{6}R\phi^2$ in $d = 4$. This, then, gives the extra term in the energy-momentum tensor.

14.6 Example: The Electromagnetic Field

The Lagrangean for pure electromagnetism is

$$\mathcal{L} = -\frac{1}{4}F_{\mu\nu}F^{\mu\nu} = -\frac{1}{2}[(\partial_\mu A_\nu)^2 - (\partial_\mu A_\nu)(\partial_\nu A_\mu)]. \qquad (14.43)$$

That means that the original definition of the energy-momentum tensor in (14.7) gives

$$\begin{aligned}
\tilde{T}_\mu{}^\nu &= -\frac{\partial\mathcal{L}}{\partial(\partial_\nu A_\rho)}\partial_\mu A_\rho + \delta_\mu^\nu\mathcal{L} \\
&= (\partial^\nu A^\rho - \partial^\rho A^\nu)\partial_\mu A_\rho - \frac{1}{4}\delta_\mu^\nu F_{\rho\sigma}F^{\rho\sigma} \\
&= F^{\nu\rho}\partial_\mu A_\rho - \frac{1}{4}\delta_\mu^\nu F_{\rho\sigma}F^{\rho\sigma}.
\end{aligned} \qquad (14.44)$$

To symmetrize it, we need to subtract the total derivative

$$F^{\nu\rho}\partial_\rho A_\mu = \partial_\rho(A_\mu F^{\nu\rho}), \qquad (14.45)$$

where in the equality we have used the equation of motion (valid classically – i.e., "on-shell") $\partial_\rho F^{\nu\rho} = 0$. We obtain the manifestly symmetric result (thus guaranteeing that this is the result of symmetrization)

$$T_\mu{}^\nu = F^{\nu\rho}F_{\mu\rho} - \frac{1}{4}\delta_\mu^\nu F_{\rho\sigma}F^{\rho\sigma}. \qquad (14.46)$$

In fact, we can check that this is also the Belinfante tensor. Moreover, we can check that in this case (free electromagnetism is a conformal theory) we do have $T^\mu{}_\mu = 0$.

In components, since we already calculated $-1/4 F_{\rho\sigma}F^{\rho\sigma} = (\vec{E}^2 - \vec{B}^2)/2$, with the definitions $F^{0i} = E^i$, we obtain first

$$T_0{}^0 = F^{0i}F_{0i} + \frac{\vec{E}^2 - \vec{B}^2}{2} = -\frac{\vec{E}^2 + \vec{B}^2}{2}, \qquad (14.47)$$

which means that indeed we obtain the energy density,

$$T_{00} = \frac{\vec{E}^2 + \vec{B}^2}{2} = \mathcal{H}. \qquad (14.48)$$

Next, we obtain

$$T_0{}^i = F^{ij}F_{0j} = -\epsilon^{ijk}B_k E_j = -(\vec{E} \times \vec{B})^i, \qquad (14.49)$$

which means that indeed we obtain the momentum density

$$T^{0i} = (\vec{E} \times \vec{B})^i. \qquad (14.50)$$

Finally, we obtain the Maxwell stress tensor

$$\begin{aligned}
T_{ij} = T_i{}^j &= F^{j\rho}F_{i\rho} + \frac{1}{2}\delta_i^j(\vec{E}^2 - \vec{B}^2) = F^{j0}F_{i0} + F^{jk}F_{ik} + \frac{1}{2}\delta_i^j(\vec{E}^2 - \vec{B}^2) \\
&= -E_i E_j - B_i B_j + \frac{1}{2}\delta_{ij}(\vec{E}^2 + \vec{B}^2).
\end{aligned} \qquad (14.51)$$

Here we have used $\epsilon^{jkl}\epsilon_{ikm} = \delta_i^j\delta_m^l - \delta_i^l\delta_m^j$.

Further Reading

See, for instance, chapter 4, section 32, in Landau and Lifshitz [3].

Exercises

(1) Calculate the energy-momentum tensor defined as

$$T_\mu^{\ \nu} = -\frac{\partial\mathcal{L}}{\partial(\partial_\nu\phi^I)}\partial_\mu\phi^I + \delta_\mu^\nu\mathcal{L} \tag{14.52}$$

and the symmetric Belinfante tensor for the sine-Gordon model.

(2) Calculate the energy-momentum tensor of "Born–Infeld electromagnetism,"

$$\mathcal{L} = -\frac{1}{L^4}\sqrt{1 + \frac{L^4}{2}F_{\mu\nu}F^{\mu\nu} - L^8\left(\frac{1}{8}\epsilon^{\mu\nu\rho\sigma}F_{\mu\nu}F_{\rho\sigma}\right)^2}. \tag{14.53}$$

(3)   Calculate the Belinfante energy-momentum tensor for the massless DBI scalar action in 3+1 dimensions,

$$\mathcal{L} = -\frac{1}{L^4}\sqrt{1 + L^4(\partial_\mu\phi)^2}, \tag{14.54}$$

as well as the "improved" tensor, which is traceless.

(4)   Calculate the components and their interpretation for an electromagnetic wave, with $\vec{E} \perp \vec{B} \perp \vec{n}$ (as we will see later) and $\vec{E} = \vec{E}_0 \cos\omega t$.

Motion of Charged Particles and Electromagnetic Waves; Maxwell Duality

In this chapter, we will describe the electromagnetic field generated by static and moving charged particles and traveling electromagnetic waves, and we will finish by describing a symmetry of the Maxwell equations in vacuum called Maxwell duality, interchanging electric, and magnetic fields, which will be used later in the book.

We have alread seen that the Maxwell equations are

$$\partial_\mu F^{\nu\mu} = j^\nu \text{ (eqs. of m.)} \quad \partial_{[\mu} F_{\nu\rho]} = 0 \text{ (Bianchi id.)} \Rightarrow F_{\mu\nu} = \partial_\mu A_\nu - \partial_\nu A_\mu. \quad (15.1)$$

Besides them, we have the Lorentz four-force law describing the motion of charged particles in an electromagnetic field,

$$\frac{dp^\mu}{d\tau} = qF^{\mu\nu}u_\nu. \quad (15.2)$$

Together, these equations are enough to describe the whole of classical electromagnetism (they are the equations of motion of the field A_μ and of the particles). However, in the following, we will see more concretely how to use them to calculate the electromagnetic field in various cases of interest. This is a very large subject, treated properly in a course on electromagnetism, so we will show here only the general principles and no specific applications.

15.1 Set of Static Charges

We start our description of the way to deal with electromagnetism with the easiest case, a single static charge. In this case, we have the usual Gauss's law. We surround the charge Q with a surface S, spherically symmetric around the charge, as in Figure 15.1a. Then integrating $\partial_\mu F^{\nu\mu} = j^\nu$ over the volume V inside S (so that $\partial V = S$), for the static charge with $j^0 = Q\delta^3(r), j^i = 0$, and using Gauss's theorem (generalized to Stokes's theorem, in form language $\int_D df = \int_{\partial D} f$), we find

$$\int_{S=\partial V} F^{0i} dS_i = \int_V dV \partial_i F^{0i} = \int_V dV j^0 = Q. \quad (15.3)$$

Taking into account that $F^{0i} = E^i$ and considering S as a sphere centered on Q, we find

$$Q = \int_S \vec{E} \cdot d\vec{S} = 4\pi R^2 E(R) \Rightarrow E(R) = \frac{Q}{4\pi R^2}. \quad (15.4)$$

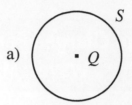

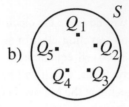

Figure 15.1 (a) Gauss law surface around a single charge Q. (b) Gauss's law surface around a spherically symmetric distribution of charge.

But we can more generally consider a general surface S and various charges Q_i inside it, and then Gauss's law gives the electric flux from the sum of the charges,

$$\int_S \vec{E} \cdot d\vec{S} = \sum_i Q_i. \qquad (15.5)$$

We can find the electric field at a point in space as the sum of the electric fields generated by each charge, $\vec{E} = \sum_i \vec{E}_i$, but if we are interested in just the electric flux through a surface, we can use the above formula. In the case of lack of symmetry of the distribution of charge, we cannot "undo" the integral to find the electric field; we can only do that in the case of spherical symmetry, as in Figure 15.1b in the same way as for a single charge.

15.2 Uniformly Moving Charges

We can next find the result for the electromagnetic field of uniformly moving charges. We start by finding the result for a single uniformly (constant velocity) moving charge by simply doing a Lorentz boost on the previous case of a single static charge.

The gauge field $A^\mu = (\phi, \vec{A})$ is a contravariant vector and so transforms as dx^μ under boosts – that is, as (for motion on z)

$$\begin{aligned} dt &= \gamma(dt' + \beta dz') \\ dz &= \gamma(dz' + \beta dt'), \end{aligned} \qquad (15.6)$$

and dx and dy are invariant.

That means that A^μ transforms as

$$\begin{aligned} \phi &= \gamma(\phi' + \beta A'_z) \\ A_z &= \gamma(A'_z + \beta \phi'). \end{aligned} \qquad (15.7)$$

and A_x, A_y are invariant. As we can see from the previous formula, there is no time dependence in A^μ – that is, the field is *stationary*. Then, since

$$E^i = F^{0i} = \partial^0 A^i - \partial^i A^0$$
$$B^i = \frac{1}{2}\epsilon^{ijk}F_{jk} = \epsilon^{ijk}\partial_j A_k, \qquad (15.8)$$

$B^z = F_{xy} = \partial_x A_y - \partial_y A_x$ is invariant, and so is $E^z = F^{0z} = \partial^0 A^z - \partial^z A^0$, which transforms as $dt \wedge dz$,

$$dt \wedge dz = \gamma^2(1 - \beta^2)dt' \wedge dz' = dt' \wedge dz'. \qquad (15.9)$$

Then E^x transforms as $dt \wedge dx$, with $dz \wedge dx \sim F^{zx} = B_y$, and E^y as $dt \wedge dy$, with $dz \wedge dy \sim F^{zy} = -B_x$, so

$$E_x = \gamma(E'_x + \beta B'_y)$$
$$E_y = \gamma(E'_y - \beta B'_x). \qquad (15.10)$$

Also, B^x transforms as $dy \wedge dz$, where $dy \wedge dt \sim F^{y0} = -E^y$ and B^y transforms as $dz \wedge dx$, where $dt \wedge dx \sim F^{0x} = E^x$, so

$$B_x = \gamma(B'_x - \beta E'_y)$$
$$B_y = \gamma(B'_y + \beta E'_x). \qquad (15.11)$$

In the nonrelativistic case, $\beta \ll 1$, we find ($\gamma \simeq 1$ and $\vec{v} = \beta n_z$)

$$\vec{E} \simeq \vec{E}' + \vec{B} \times \vec{v}$$
$$\vec{B} \simeq \vec{B}' - \vec{E}' \times \vec{v}. \qquad (15.12)$$

We can verify that we have two Lorentz invariants,

$$-\frac{1}{2}F_{\mu\nu}F^{\mu\nu} = \vec{E}^2 - \vec{B}^2 \qquad (15.13)$$

and

$$F_{\mu\nu}F_{\rho\sigma}\epsilon^{\mu\nu\rho\sigma} = 4F_{0i}F_{jk}\epsilon^{ijk} = 8\vec{E}\cdot\vec{B}. \qquad (15.14)$$

Coming back to our case, boosting the static electric charge, with $\vec{B}' = 0$, we obtain the magnetic field

$$\vec{B} \simeq \vec{v} \times \vec{E} \simeq \vec{v} \times \frac{Q\hat{r}}{4\pi R^2} = \frac{(Q\vec{v}) \times \hat{r}}{4\pi R^2}. \qquad (15.15)$$

Since $Q\vec{v}$ is the current of a single moving charge, summing over infinitesimal charges, $dQ\vec{v} = \vec{j}dV$ (see Figure 15.2), we obtain the *Biot-Savart law*,

$$\vec{B} = \int_V dV \frac{\vec{j} \times \hat{r}}{4\pi R^2}. \qquad (15.16)$$

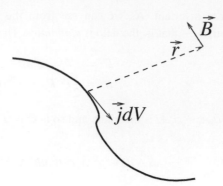

Biot-Savart law: Infinitesimal contribution.

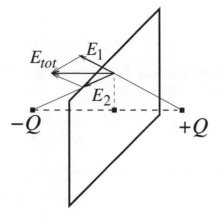

Mirror image charge method. The charges on the surface reorient themselves to have constant V on the suface, which is possible only if there is a mirror charge $-Q$, so that the total electric field $\vec{E}_{tot} = \vec{E}_1 + \vec{E}_2$ is everywhere perpendicular to the surface.

15.3 Electrostatics Methods

For a collection of static charges, together with implicit movable charges inside a *perfect conductor*, we have a system solved with a specific method, the *mirror image method*.

Indeed, a perfect conductor ($\sigma = \infty$) is characterized by a zero electric field inside the conductor (since otherwise $\vec{j} = \sigma\vec{E}$ is infinite for any nonzero $\vec{E}$). For an infinite planar conductor, that means that $\vec{E}$ must be perpendicular to the plane ($\vec{E}$ in the parallel direction is zero) or that (since $\vec{E} = -\vec{\nabla}\phi$) the electric potential ϕ is constant inside the conductor. But the only way for that to be preserved in the presence of an electric charge Q, whose electric field would be radial from the charge and so would have a parallel component to the conductor, is for charges inside it to move in such a way as to simulate the presence of a *mirror charge* $-Q$, situated on the opposite side of the conductor. Indeed, we can easily check, vectorially, that for two opposite charges, Q and $-Q$, on the median plane, the electric field is everywhere perpendicular to it, as in Figure 15.3.

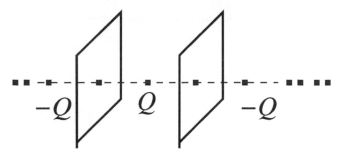

Figure 15.4   Mirror image charge method for two conducting planes: The first two images. Draw the others for the exercise.

That means that for a single charge on a side of a perfect conducting plane, the electric field at the minimum distance position is

$$E(0) = 2\frac{Q}{4\pi R^2},\tag{15.17}$$

twice as much as in the absence of the conducting plane.

If we have more than one conducting plane, we must take care and consider the image of the images as well (the image in one plane has an image in the other, etc.), forming an infinite set of mirror charges. We leave it, then, as an exercise to calculate the electric field at one of the planes, at the minimum distance position from the charge (see Figure 15.4).

15.4 The Electric and Magnetic Fields Away from a Collection of Particles: Multipole Expansions

For a collection of charged particles with charges Q_a at positions $\vec{R}_a$, as we saw, the electric potential is

$$\phi(\vec{r}) = \sum_a \frac{Q_a}{4\pi|\vec{r} - \vec{R}_a|}.\tag{15.18}$$

But the issue is how to describe this electric field far away from the collection of charges, for $|\vec{r}| \gg |\vec{R}_a|$, where we cannot discern the individual charges. For a general function $f(\vec{r} - \vec{R})$, we can consider a Taylor expansion in the multiple coordinates $\vec{r}$, so to first nontrivial term, we have

$$f(\vec{r} - \vec{R}) \simeq f(\vec{r}) - \vec{R} \cdot \vec{\nabla}_{\vec{r}} f(\vec{r}).\tag{15.19}$$

That means that we have

$$\phi \simeq \frac{\sum_a Q_a}{4\pi r} - \left(\sum_a Q_a \vec{R}_a\right) \cdot \vec{\nabla}_{\vec{r}}\frac{1}{4\pi r}$$
$$= \frac{\sum_a Q_a}{4\pi r} + \frac{\vec{d} \cdot \hat{r}}{4\pi r^2},\tag{15.20}$$

where $\hat{r} \equiv \vec{r}/r$ and we have defined the *electric dipole moment*

$$\vec{d} \equiv \sum_a Q_a \vec{R}_a. \tag{15.21}$$

The electric field coming from the above potential is

$$\vec{E} = -\vec{\nabla}\phi = \frac{\sum_a Q_a}{4\pi r^2}\hat{r} - \vec{\nabla}\left(\frac{\vec{d}\cdot\hat{r}}{4\pi r^2}\right)$$

$$= \frac{(\sum_a Q_a)\,\hat{r}}{4\pi r^2} + \frac{3(\hat{r}\cdot\vec{d})\hat{r} - \vec{d}}{4\pi r^3}. \tag{15.22}$$

The first term is the "monopole" term that contains all the electric charge in the origin and the next is the dipole, depending on the dipole moment.

More generally, keeping all the terms in the Taylor expansion, we obtain the so-called *multipole expansion*, taking $\partial_{i_1} \ldots \partial_{i_n}$ on $1/r$, which gives

$$\phi = \phi^{(0)} + \phi^{(1)} + \phi^{(2)} + \cdots, \tag{15.23}$$

so we obtain that

$$\phi^{(n)} \propto \frac{1}{r^{n+1}}. \tag{15.24}$$

Here $\phi^{(0)}$ is the monopole term, $\phi^{(1)}$ is the dipole term, $\phi^{(2)}$ is a "quadrupole" term, etc. ($\phi^{(n)}$ is a "2^n-pole"). For instance, the quadrupole term is, thus,

$$\phi^{(2)} = \left(\sum_a Q_a R_i^a R_j^a\right)\frac{\partial^2}{\partial x_i \partial x_j}\frac{1}{4\pi r}. \tag{15.25}$$

Moving over to a magnetic field, since $\vec{B} = \vec{\nabla}\times\vec{A}$ and

$$\vec{\nabla}\times\vec{B} = \partial_0\vec{E} + \vec{j}, \tag{15.26}$$

for static fields, $\partial_t\vec{E} = 0$, and using

$$[\vec{A}\times(\vec{B}\times\vec{C})]^i = \epsilon^{ijk}A_j\epsilon^{klm}B_l C_m = (\delta_i^l\delta_j^m - \delta_i^m\delta_j^l)A^j B_l C_m = [(A\cdot C)\vec{B} - (A\cdot B)\vec{C}]^i, \tag{15.27}$$

so that

$$\vec{\nabla}\times(\vec{\nabla}\times\vec{A}) = \vec{\nabla}(\vec{\nabla}\cdot\vec{A}) - \Delta\vec{A}, \tag{15.28}$$

in the gauge $\vec{\nabla}\cdot\vec{A} = 0$ we obtain

$$\Delta\vec{A} = -\vec{j}. \tag{15.29}$$

This is the vectorial form of the same electric potential equation $\Delta\phi = -\vec{\nabla}\cdot\vec{E} = -\rho$, so its solution is the same one,

$$\vec{A} = \int_{\vec{R}\in V} dV \frac{\vec{j}_{\vec{R}}}{4\pi|\vec{r} - \vec{R}|}. \tag{15.30}$$

Taking the curl, we obtain the same Biot-Savart law from before,

$$\vec{B} = \int_{\vec{R}\in V} dV \frac{\vec{j}_{\vec{R}}\times\hat{r}}{4\pi|\vec{r} - \vec{R}|^2}, \tag{15.31}$$

as we should.

The discrete form of (15.30), from $\vec{j}dV \to dq_a\vec{v}_a$, is

$$\vec{A} = \sum_a \frac{q_a\vec{v}_a}{4\pi|\vec{r} - \vec{R}_a|}. \tag{15.32}$$

We can obtain a similar *multipole expansion* for the magnetic potential as the Taylor expansion in multiple coordinates. To the first nontrivial order, we obtain

$$\vec{A} \simeq \frac{\sum_a q_a\vec{v}_a}{4\pi r} - \sum_a q_a\vec{v}_a \left(\vec{R}\cdot\vec{\nabla}\frac{1}{4\pi r}\right). \tag{15.33}$$

The same double vector product formula as (15.27) gives

$$-(\vec{R}_a \times \vec{v}_a) \times \hat{r} = -(\vec{v}_a\cdot\hat{r})\vec{R}_a + (\vec{R}_a\cdot\hat{r})\vec{v}_a, \tag{15.34}$$

but the first term is zero when summed over with $\sum_a q_a$ for a closed material, in which also the total current is zero, $\sum_a q_a\vec{v}_a = 0$, so $\sum_a(q_a\vec{v}\cdot\hat{r})\vec{R}_a = 0$. Indeed, $\sum_a q_a\vec{R}_a = 0$ (defining the center of the distribution) implies also $\sum_a q_a\vec{v}_a = 0$ and then also $\sum_a(q_a\vec{v}_a\cdot\hat{r})\vec{R}_a = 0$.
Finally, we obtain

$$\vec{A} = \frac{\sum_a q_a\vec{v}_a}{4\pi r} + \frac{\vec{\mu} \times \hat{r}}{4\pi r^2} + \cdots, \tag{15.35}$$

where we have defined the *magnetic dipole moment*,

$$\vec{\mu} = \sum_a q_a\vec{R}_a \times \vec{v}_a. \tag{15.36}$$

Note also that

$$\sum_a q_a\vec{v}_a = \frac{d}{dt}\sum_a q_a\vec{R}_a = \frac{d}{dt}\vec{d}, \tag{15.37}$$

where $\vec{d}$ is the electric dipole moment, so we have

$$\vec{A} = \frac{\dot{\vec{d}}}{4\pi r} + \frac{\vec{\mu} \times \hat{r}}{4\pi r^2} + \cdots \tag{15.38}$$

15.5 Electromagnetic Waves

Having described the electromagnetic field generated by charges, we move on to the free propagating electromagnetic field in the absence of any charges (sources), so for $j^\mu = 0$. We choose to work in the *radiation gauge*, $A_0 = \phi = 0$, and $\vec{\nabla}\cdot\vec{A} = 0$, since, as we already said, it is the most appropriate in the absence of sources. Then the equation of motion, $\partial_\mu\partial^\mu A^\nu - \partial^\nu\partial^\mu A_\mu = 0$ becomes just the KG equation of motion for $\vec{A}$,

$$\Box\vec{A} = (-\partial_0^2 + \vec{\nabla}^2)\vec{A} = 0. \tag{15.39}$$

In this context, the equation is also called the d'Alembert equation (and $\Box$ the d'Alembertian) or the wave equation. Note that we actually only need to be in the Lorentz gauge to have $\Box\vec{A} = 0$, $A_0 = 0$ is only used so as to eliminate also the remaining A_0 equation.

Plane Wave Solutions

We consider a solution in the type of a plane wave – i.e., a solution that is independent on y and z. Then the wave equation for a function f is

$$\partial_0^2 f - \partial_x^2 f = (\partial_0 - \partial_x)(\partial_0 + \partial_x)f = 0, \tag{15.40}$$

and its most general solution is the sum of a term that depends only on $t + x$ and one that depends only on $t - x$,

$$f = f_1(t - x) + f_2(t + x), \tag{15.41}$$

since $(\partial_0 - \partial_x)f_2 = 0$ and $(\partial_0 + \partial_x)f_1 = 0$. Here f_1 is a forward-moving wave, and f_2 is backward-moving one.

That means that in our case, choosing only the forward-moving wave, we obtain

$$\vec{A} = \vec{A}(t - x). \tag{15.42}$$

Since now $\vec{E} = -\partial_0\vec{A}$ and $\vec{B} = \vec{\nabla} \times \vec{A}$, by denoting $u = t - x$ and d/du as a prime, we obtain

$$\begin{aligned} \vec{E} &= -\vec{A}' \\ \vec{B} &= \vec{\nabla}(t - x) \times \vec{A}' = -\vec{n} \times \vec{A}', \end{aligned} \tag{15.43}$$

where we have used the fact that $\vec{\nabla}x = \vec{n}_x$, the unit vector in the direction of propagation of the wave – i.e., the x direction. We finally obtain

$$\vec{B} = \vec{n} \times \vec{E}. \tag{15.44}$$

But the radiation gauge condition $\vec{\nabla} \cdot \vec{A} = 0$ becomes on our solution

$$\partial_x A_x = -A'_x = 0 \Rightarrow \partial_t A_x = F_{0x} = -E_x = 0, \tag{15.45}$$

or rather

$$\vec{n} \cdot \vec{E} = 0, \tag{15.46}$$

which says that the electromagnetic wave is transverse to the direction of its propagation. The two conditions (15.44) and (15.46) have been derived in the radiation gauge, but since they are written in terms of gauge invariant objects, they are valid in any gauge.

The three vectors $(\vec{n}, \vec{E}, \vec{B})$ are therefore mutually orthogonal and, moreover, then (15.44) implies $|\vec{E}| = |\vec{B}|$, or $\vec{E}^2 - \vec{B}^2 = 0$ for the planar wave.

Energy Flux

We have seen in the last chapter that the energy flux of electromagnetism is

$$T_0{}^i = T^i{}_0 = (\vec{E} \times \vec{B})^i \equiv S^i. \tag{15.47}$$

Here $\vec{S} = \vec{E} \times \vec{B}$ is called the *Poynting vector*, and from (15.47), it describes the energy flux. Using (15.44) and (15.46), (15.27) and $\vec{E}^2 = \vec{B}^2$, we find

$$\vec{S} = \vec{E} \times \vec{B} = \vec{E} \times (\vec{n} \times \vec{E}) = \vec{E}^2\vec{n} - (\vec{n} \cdot \vec{E})\vec{E} = \vec{E}^2\vec{n} = \vec{B}^2\vec{n}. \tag{15.48}$$

We can, in fact, also write a symmetric form,

$$\vec{S} = \frac{\vec{E}^2 + \vec{B}^2}{2}\vec{n}, \tag{15.49}$$

which emphasizes its connection with the energy density $(\vec{E}^2 + \vec{B}^2)/2$.

15.6 Arbitrary Moving Charges

Until now, we have considered only stationary (time-independent fields) moving charges. But if we consider arbitrary charges – for instance, a *single* moving charge that starts moving at a finite time – we cannot have time-independent fields. Then the perturbations in the fields move at a finite speed, c, which means that potentials are *retarded* – i.e., they depend on charge distributions at a previous time – for the *electric potential*

$$\phi(\vec{r}, t) = \phi_0 + \int_V dV' \frac{1}{4\pi R}\rho(\vec{r}', t - R/c) \equiv \phi_0 + \int_V dV \frac{1}{4\pi R}\rho_{t-R/c}. \tag{15.50}$$

Here $\vec{R} \equiv \vec{r} - \vec{r}'$ and $dV' \equiv d^3\vec{r}'$, and we have put $\epsilon_0 = 1$.

Similarly, for the *vector potential*, we find the formula depending on retarded time,

$$\vec{A}(\vec{r}, t) = \vec{A}_0 + \int_V dV' \frac{1}{4\pi R}\vec{j}(\vec{r}', t - R/c) \equiv \vec{A}_0 + \int_V dV \frac{\vec{j}_{t-R/c}}{4\pi R}. \tag{15.51}$$

Derivation

We can be more formal in the derivation of (15.50), the formula for the electric potential and consider first the field of an infinitesimal moving charge, $\rho = dq(t)\delta(\vec{R})$, where $\vec{R}$ is the position from the origin of coordinates, and then the equation of motion in spherical coordinates is

$$\Delta\phi - \frac{1}{c^2}\frac{\partial^2\phi}{\partial t^2} = \frac{1}{R^2}\frac{\partial}{\partial R}\left(R^2\frac{\partial\phi}{\partial R}\right) - \frac{1}{c^2}\frac{\partial^2\phi}{\partial t^2} = dq(t)\delta(R). \tag{15.52}$$

Substituting $\phi = \psi/R$, we obtain the equation

$$\frac{\partial^2\psi}{\partial R^2} - \frac{1}{c^2}\frac{\partial^2\psi}{\partial t^2} = R\,dq(t)\delta(R), \tag{15.53}$$

whose solution at $R \neq 0$ is $\psi = f(t - R/c) + f(t + R/c)$. Choosing the physical retarded form, we get

$$\phi = \frac{f(t - R/c)}{R} \tag{15.54}$$

for $R \neq 0$. At $R \to 0$, then, the $\Delta\phi$ terms dominates over $\partial^2\phi/\partial t^2$, and the equation becomes the usual Coulomb equation, just with a time dependent source,

$$\Delta\phi = dq(t)\delta(\vec{R}), \tag{15.55}$$

with solution $\phi = dq(t)/4\pi R$. Comparing then at $R \to 0$ with the general form, we find $f(t) = dq(t)/(4\pi)$, so finally

$$\phi = \frac{dq(t - R/c)}{4\pi R}. \tag{15.56}$$

We can now integrate over a volume and write $dq = \rho dV$ and translate from the origin to $\vec{r}$, understanding $\vec{R}$ as $\vec{r} - \vec{r}'$, with $dV \to d^3\vec{r}' \equiv dV'$, and obtain (15.50).

Now there is still a subtlety that we glossed over – namely that R itself, in the retarded formula $t' = t - R/c$, depends on time – so we should correct that omission.

Lienard–Wiechert Potentials

We want now to specialize the formulas (15.50) and (15.51) to the case of an single arbitrary moving charge q and to take into account the said subtlety. Consider then $\vec{r}' = \vec{r}_0(t)$, and then consider that the length R that is relevant for retardation is at time t' (generation of field) itself so, really,

$$t = t' + \frac{R(t')}{c}. \tag{15.57}$$

Defining the four-vector $R^\mu = r^\mu - r'^\mu = (c(t - t'), \vec{r} - \vec{r}')$, the covariant form of the relation is null propagation, $R_\mu R^\mu = 0$.

We first consider the case when we are in a reference system where the charge is at rest at t', so we have a Coulomb potential,

$$\phi = \frac{q}{4\pi R(t')} = \frac{q}{4\pi c(t - t')}, \quad \vec{A} = 0. \tag{15.58}$$

This then matches with the formulas (15.50) and (15.51) in the case $\vec{v}(t') = 0$, so $\vec{j}(t') = 0$. It is easily seen then that the correct four-vector relation that is valid in any reference frame is

$$A^\mu = \frac{qu^\mu}{R_\nu u^\nu}, \tag{15.59}$$

which reduces to (15.58) and also is consistent with (15.50) and (15.51), considering that the four-velocity is $u^\mu = \gamma(1, \vec{v})$ and that $qu^\mu = \gamma(q, q\vec{v}) = \gamma(q, \vec{j})$.

Writing it in three-dimensional notation, we obtain the *Lienard–Wiechert potentials*,

$$\phi = \frac{q}{4\pi \left(r - \frac{\vec{v} \cdot \vec{r}}{c}\right)}$$

$$\vec{A} = \frac{q\vec{v}}{4\pi \left(r - \frac{\vec{v} \cdot \vec{r}}{c}\right)} = \vec{v}\phi, \tag{15.60}$$

and remember that the times on the right-hand side are t', defined in (15.57).

Note that a shortcut was to just replace

$$\vec{R} = \vec{r} - \vec{r}' \to \vec{R} - (t - t')\vec{v} = \vec{R} - \frac{R}{c} \cdot \vec{v}, \tag{15.61}$$

i.e., the distance with a time-corrected distance, with the motion in between emission and observation, so that by multiplying by $\hat{r}$, we obtain (15.60).

15.7 Generation of Electromagnetic Waves by Dipoles

We expect that if the motion of a charge is sinusoidal (oscillatory), it will generate fields that are sinusoidal, so perhaps it will generate electromagnetic waves? The oscillatory behavior is certainly right, but a single charge is not.

One can prove that electromagnetic waves can only be generated by dipole moments or higher, not by a monopole, so we need at least two charges to generate them (a single charge has no dipole or higher moments).

We will not present the full proof of this statement, though it can be found in any book on classical electrodynamics: Landau and Lifshitz have a more thorough explanation, JD Jackson provides even more detail. We will only sketch the main idea.

We need to combine the multipole expansion with the idea of the retarded potentials. One can, for instance, use the Lienard–Wiechert form. We are looking at a large distance from the distribution of charges so that the multipole expansion is good. Then the wave component of the electromagnetic field appears, as we saw in both ϕ and $\vec{A}$, which means that it is enough to look at $\vec{A}$ to find the wave component.

Moreover, we saw in (15.38) that the first terms in $\vec{A}$ depend on $\ddot{\vec{d}}$ and $\dot{\vec{\mu}}$ (now understood as retarded potentials) but not on the total charge ("electric monopole" term) Q. The leading term depends on the time derivative of the electric dipole moment and is, in fact, the *wave term generated by the dipole*,

$$\vec{A} = \frac{1}{4\pi c R_0}\dot{\vec{d}},\tag{15.62}$$

where we have used the Lienard–Wiechert notation for $R = R_0$ as the center of the distribution in the multipole expansion. But then, for an electromagnetic wave, we have the relations (15.43), which can be rewritten as

$$\vec{B} = \frac{1}{c}\dot{\vec{A}} \times \vec{n}\,, \vec{E} = \vec{B} \times \vec{n} = \left(\frac{1}{c}\dot{\vec{A}} \times \vec{n}\right) \times \vec{n},\tag{15.63}$$

finally obtaining for the electric and magnetic fields,

$$\vec{E} = -\frac{1}{4\pi c^2 R_0}(\ddot{\vec{d}} \times \vec{n}) \times \vec{n}, \vec{B} = \frac{1}{4\pi c^2 R_0}\ddot{\vec{d}} \times \vec{n},\tag{15.64}$$

and the energy flux vector (energy per unit time and transverse surface, $dE/dtdS_\perp$) is

$$\vec{S} = \frac{1}{(4\pi R_0)^2 c^4}|\ddot{\vec{d}} \times \vec{n}|^2 \vec{n}.\tag{15.65}$$

We can also calculate the intensity of emitted radiation (energy per unit time) in the unit solid angle,

$$\frac{dI}{d\Omega} = \frac{dE}{dt\,dS_\perp}\frac{dS_\perp}{d\Omega} = |\vec{S}|\frac{R_0^2 d\Omega}{d\Omega} = \frac{1}{(4\pi)^2 c^4}|\ddot{\vec{d}}|\sin^2\theta,\tag{15.66}$$

where θ is the angle between $\ddot{\vec{d}}$ and $\vec{n}$. Integrating over $d\Omega = 2\pi \sin\theta d\theta$, we find

$$I = \frac{2}{3c^4}|\ddot{\vec{d}}|^2.\tag{15.67}$$

15.8 Maxwell Duality

The source-free Maxwell's equations have a symmetry called Maxwell duality. Indeed, they are

$$\partial_\mu F^{\mu\nu} = 0; \quad \partial_{[\mu} F_{\nu\rho]} = 0, \tag{15.68}$$

or, in form language,

$$d * F = 0; \quad dF = 0, \tag{15.69}$$

which means that we have the symmetry $F \to *F$, which leaves them invariant. Note that $*^2 = -1$, since

$$\frac{1}{2}\epsilon^{\mu\nu\rho\sigma}\frac{1}{2}\epsilon_{\rho\sigma\lambda\tau} = -\delta^{\mu\nu}_{\lambda\tau}, \tag{15.70}$$

as we already saw. Then

$$F_{\mu\nu} \to *F_{\mu\nu} = \frac{1}{2}\epsilon_{\mu\nu\rho\sigma}F^{\rho\sigma} \tag{15.71}$$

gives ($\epsilon^{0ijk} = \epsilon^{ijk} = -\epsilon_{0ijk}$)

$$\epsilon_{ijk}B_k = F_{ij} \to -\epsilon_{ijk}F^{0k} = -\epsilon_{ijk}E_k$$
$$-E_i = -F^{0i} = F_{0i} \to -\frac{1}{2}\epsilon^{ijk}F_{jk} = -B_i, \tag{15.72}$$

so, in total,

$$\vec{B} \to -\vec{E}; \quad \vec{E} \to \vec{B}. \tag{15.73}$$

This duality symmetry will be useful for us later, when discussing magnetic monopoles.

Important Concepts to Remember

- Static charges are completely characterized by Gauss's law for ϕ.
- Uniformly moving charges obey the Biot-Savart law for $\vec{B}$, obtained by boosting a static charge.
- In the presence of perfectly conducting planes, we can use the mirror image method, when a charge Q (even a virtual charge, an image) has an image of $-Q$ at an equidistant position on the other side of the plane.
- A multipole expansion is a Taylor expansion in several components (variables); it has a monopole (total charge), dipole moment (first moment), quadrupole moment (second moment), etc.
- The multipole expansion for ϕ gives the electric dipole $\vec{d} = \sum_a Q_a \vec{R}_a$, and the one for $\vec{A}$ gives the magnetic dipole $\vec{\mu} = \sum_a q_a \vec{R}_a \times \vec{v}_a$.
- In the vacuum in the Lorentz gauge, $\vec{A}$ satisfies the wave equation $\Box \vec{A} = 0$ and so has propagating wave solutions.
- The planar electromagnetic wave $\vec{A} = \vec{A}(t - x)$ has transverse fields $\vec{B} = \vec{n} \times \vec{E}$ and $\vec{n} \cdot \vec{E} = 0$ for propagation along $\vec{n}$.
- The energy flux is the Poynting vector $\vec{S} = \vec{E} \times \vec{B}$.

- A set of arbitrary moving charges (not stationary) has retarded potentials, and for a single charge, we obtain the Lienard–Wiechert potentials.
- The Maxwell equations have the duality symmetry $F \rightarrow *F$, which means $\vec{E} \rightarrow \vec{B}$ and $\vec{B} \rightarrow -\vec{E}$.

Further Reading

See parts of chapters 3, 4, 5, 6, 8, and 9 in Landau and Lifshitz [3].

Exercises

(1) Consider a charge Q in between two equidistant perfectly conducting plates, situated each at a distance R from the charge, as in Figure 15.4. Calculate the electric field on the conducting plate at the minimum distance point from the charge.

(2) Consider a homogenous spherical distribution of charge. What multipoles does it have? Can it radiate electromagnetic waves?

(3) Consider three charges $-Q, +3Q, -2Q$ situated on a line, with the $+3Q$ in the center and the other two equidistantly at distance R. Calculate the electric field, in the multipole expansion, at a distance $r \gg R$, in the direction perpendicular to the line, calculating it from the monopole, dipole, and quadrupole moments.

(4) For the configuration at exercise 3, with $+3Q$ fixed and the distance R to the other two charges varying as $R = R_0 \cos \omega t$, calculate the intensity of the emitted radiation.

16 The Hopfion Solution and the Hopf Map

Field theories often have solutions with special properties, either topological or of stability, that make them take particle-like properties. We have previewed this fact when discussing the solitonic water wave (solution of the KdV equation). We will discuss these nonlinear solutions at length in the second part of the book, but here we will start with a simple theory: our favorite linear theory, Maxwell's electromagnetism. It is less known that Maxwell's electromagnetism also admits solitonic-like structures, here in the simple sense of having an associated topology. The solutions are not stable, nor do they keep their shape, so they are not on par with particles in this case. The basic solution is called a "Hopfion," and we will describe it in this chapter.

16.1 Sourceless Maxwell's Equations and Conserved "Helicities"

The Hopfion is a solution of the Maxwell's equations in vacuum – i.e., without sources. In the nonrelativistic notation, they are

$$\vec{\nabla} \times \vec{E} = -\frac{\partial \vec{B}}{\partial t}; \quad \vec{\nabla} \cdot \vec{E} = 0$$

$$\vec{\nabla} \times \vec{B} = +\frac{1}{c^2} \frac{\partial \vec{E}}{\partial t}; \quad \vec{\nabla} \cdot \vec{B} = 0, \tag{16.1}$$

and we will put $c = 1$ in the following. In the absence of sources, we can work in the gauge $A_0 = \phi = 0$, so only the vector potential $\vec{A}$ is nonzero and still have the (residual) gauge invariance $\delta \vec{A} = \vec{\nabla} \alpha(\vec{x})$. But, moreover, just like $\vec{\nabla} \cdot \vec{B} = 0$ implies the existence of $\vec{A}$, now so does $\vec{\nabla} \cdot \vec{E} = 0$ imply the existence also of an electric-magnetic dual vector potential $\vec{C}$, such that

$$\vec{E} = \vec{\nabla} \times \vec{C}; \quad \vec{B} = \vec{\nabla} \times \vec{A}. \tag{16.2}$$

Of course, this is a redundant description, since $\vec{A}$ is enough to describe the system, and

$$\vec{E} = -\frac{\partial}{\partial t} \vec{A} \equiv \vec{\nabla} \times \vec{C}, \quad \vec{B} = \vec{\nabla} \times \vec{A} = -\partial_t \vec{C}. \tag{16.3}$$

But the purpose of introducing $\vec{C}$ is that now we can define integrals over space, "Chern–Simons forms," in terms of $\vec{B}$ and $\vec{C}$.

First, the *electric helicity* is the Chern–Simons form of $\vec{C}$,

$$H_{ee} = \int d^3x \vec{C} \cdot \vec{E} = \int d^3x \vec{C} \cdot \vec{\nabla} \times \vec{C} = \int d^3x \epsilon^{ijk} C_i \partial_j C_k. \tag{16.4}$$

Then, its electromagnetic dual, the *magnetic helicity*, is the Chern–Simons form of $\vec{A}$,

$$H_{mm} = \int d^3x \vec{A} \cdot \vec{B} = \int d^3x \epsilon^{ijk} A_i \partial_j A_k, \tag{16.5}$$

Then the *electric-magnetic cross helicity* is the BF form for $\vec{C}$ and $\vec{A}$,

$$H_{em} = \int d^3x \vec{C} \cdot \vec{B} = \int d^3x \epsilon^{ijk} C_i \partial_j A_k, \tag{16.6}$$

and then its electric-magnetic dual, the *magnetic-electric cross helicity*, is the BF form for $\vec{A}$ and $\vec{C}$,

$$H_{me} = \int d^3x \vec{A} \cdot \vec{E} = \int d^3x \epsilon^{ijk} A_i \partial_k C_k. \tag{16.7}$$

When performing (residual) gauge transformations with gauge parameter $\alpha(\vec{x})$ that go to zero at infinity, the previous helicities are invariant, which is left as an exercise to check. But, a priori, these "helicities" could evolve in time. Their time variation is

$$\partial_t H_{mm} = \int d^3x (\partial_t \vec{A} \cdot \vec{B} + \vec{A} \cdot \partial_t \vec{B}) = -\int d^3x (\vec{E} \cdot \vec{B} + \vec{A} \cdot (\vec{\nabla} \times \vec{E}))$$

$$= -2 \int d^3x \vec{E} \cdot \vec{B}$$

$$\partial_t H_{ee} = \int d^3x (\partial_t \vec{C} \cdot \vec{E} + \vec{C} \cdot \partial_t \vec{E}) = -\int d^3x (\vec{B} \cdot \vec{E} + \vec{C} \cdot (\vec{\nabla} \times \vec{B}))$$

$$= -2 \int d^3x \vec{E} \cdot \vec{B}$$

$$\partial_t H_{me} = \int d^3x (\partial_t \vec{A} \cdot \vec{E} + \vec{A} \cdot \partial_t \vec{E}) = \int d^3x (-\vec{E} \cdot \vec{E} + \vec{A} \cdot (\vec{\nabla} \times \vec{B}))$$

$$= -\int d^3x (\vec{E}^2 - \vec{B}^2) ,$$

$$\partial_t H_{em} = \int d^3x (\partial_t \vec{C} \cdot \vec{B} + \vec{C} \cdot \partial_t \vec{B}) = \int d^3x (-\vec{B} \cdot \vec{B} + \vec{C} \cdot (\vec{\nabla} \times \vec{E}))$$

$$= \int d^3x (\vec{E}^2 - \vec{B}^2). \tag{16.8}$$

In (16.8), we have used partial integration and the Maxwell's equations.

From the explicit expressions, we note that conservation of H_{mm} and H_{ee} is equivalent to $\vec{E} \cdot \vec{B} = 0$ (at least as an integral), and conservation of H_{em} and H_{me} is equivalent to $\vec{E}^2 - \vec{B}^2 = 0$. That means that if we restrict to configurations with $\vec{E} \cdot \vec{B} = 0$ and $\vec{E}^2 - \vec{B}^2 = 0$, or in complex notation $(\vec{E} + i\vec{B})^2 = 0$, or in relativistic notation

$$\epsilon^{\mu\nu\rho\sigma} F_{\mu\nu} F_{\rho\sigma} \propto \vec{E}^2 - \vec{B}^2 = 0, \quad F_{\mu\nu} F^{\mu\nu} \propto \vec{E}^2 - \vec{B}^2 = 0 , \tag{16.9}$$

the helicities will be conserved. Then, for such solutions, it is to be expected that the helicities will be proportional to some sort of topological number.

16.2 Hopf Map and Hopf Index

Indeed, it will be the case, and the corresponding number will be the Hopf index, associated with the Hopf map. The Hopf maps are famous maps in mathematics. We will be concerned with just the first Hopf map, the best known of them. It is a map: $S^3 \to S^2 = \mathbb{CP}^1$. If we consider the $\mathbb{CP}^1$ representation, the map is also known as the Hopf fibration and is defined as follows.

Consider complex coordinates on $\mathbb{C}^2$, $Z^1 = X^1 + iX^2$ and $Z^2 = X^3 + iX^4$. We can define an S^3 by

$$\sum_{\alpha=1,2} Z^\alpha Z^\dagger_\alpha = |Z^1|^2 + |Z^2|^2 = 1. \tag{16.10}$$

The relation is invariant under multiplication of the coordinates by a phase, $Z^\alpha \to e^{i\theta} Z^\alpha$. The phase $e^{i\theta} \in U(1) = S^1$, since it is part of the S^3, defines a "fiber" for the "fibration" of S^3. The remaining two-dimensional space is a sphere S^2,

$$\sum_{i=1}^{3} x_i x^i = 1, \tag{16.11}$$

and the map between x^i and Z^α is the Hopf map, defining the Hopf fibration,

$$x_i = Z^\dagger_\alpha (\tilde{\sigma}_i)^\alpha{}_\beta Z^\beta \equiv Z^\beta (\sigma_i)_\beta{}^\alpha Z^\dagger_\alpha. \tag{16.12}$$

where σ_i are the Pauli matrices and $\tilde{\sigma}_i$ their transposes. We see that, indeed, the map is invariant under $Z^\alpha \to e^{i\alpha} Z^\alpha$, so the $U(1) = S^1$ "fiber" is transverse to the S^2. More precisely, the Hopf fibration is the statement that at each point on the S^2, there is an S^1 of a different size (it is "fibered" over S^2), and the resulting three-dimensional manifold is the S^3, see Figure 16.1.

The Hopf fibration is a Hopf map of "winding number" 1 – i.e., of Hopf index one – meaning that the S^2 wraps the S^3 exactly once.

A Hopf map of Hopf index 1 that will be more useful for us is constructed as follows. Consider a complex field in three-dimensional space, $\phi(x_1, x_2, x_3) = \phi_1(x_1, x_2, x_3) + i\phi_2(x_1, x_2, x_3)$. If the field goes to a constant (which could be zero) at infinity, $\phi \to \phi_0$ as

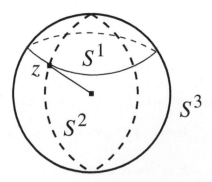

Figure 16.1 The Hopf fibration: At each position z on S^2 there is an S^1 of a different size; thus, S^3 is "fibered" over S^2.

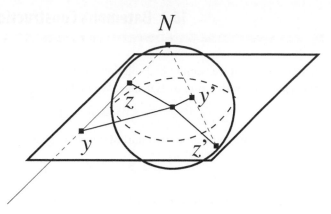

Figure 16.2 The stereographic projection, here represented for S^2, though it is the same for S^3 (except we can't draw it on a two-dimensional plane): A line from the North Pole (N) passes through the sphere at z and then through the plane at y if z is the in Northern Hemisphere, and first through the plane at y' and then through the sphere at z' if z' is in the Southern Hemisphere.

$|\vec{r}| \to \infty$, we can identify all points at infinity and say that the map is defined on $\mathbb{R}^3 \cup \{\infty\}$, which is topologically the same as S^3. Indeed, the $\mathbb{R}^3 \cup \{\infty\}$ can be identified with S^3 via the usual stereographic projection: consider an S^3 and an $\mathbb{R}^3$ intersecting it over an Equator. Then consider a line from the North Pole intersecing the sphere at $z \in S^3$, and then the $\mathbb{R}^3$ "plane," at y, as in Figure 16.2. This defines the stereographic map. The complex field can also take any value, including ∞, which is the same in all directions, so the scalar field maps S^3 to $\mathbb{C} \cup \{\infty\} \simeq S^2$, where the topological equality is defined now by the usual stereographic projection (S^2 to $\mathbb{C} \cup \{\infty\}$).

Thus, $\phi(x_1, x_2, x_3)\colon S^3 \to S^2$ must be a Hopf map. A Hopf map of Hopf index 1 is

$$\phi_H(x,y,z) = \frac{2(x+iz)}{2z + i(r^2-1)}; \quad r^2 = x^2 + y^2 + z^2. \tag{16.13}$$

To see that, we must define the Hopf index better. Note that a map from S^3 to S^2 can be defined by the antisymmetric object

$$F_{ij} = \frac{1}{2\pi i}\frac{\partial_i \phi^* \partial_j \phi - \partial_i \phi \partial_j \phi^*}{(1+|\phi|^2)^2}, \tag{16.14}$$

since this object takes value on the S^2. In fact, this is written as a $F_{ij} = \partial_i A_j - \partial_j A_i$, where

$$A_i = \frac{1}{4\pi i}\frac{\partial_i \ln \phi - \partial_i \ln \phi^*}{(1+|\phi|^2)}. \tag{16.15}$$

The integral of the Chern–Simons form of A_i on S^3 is then a topological number,

$$N = \int_{S^3} \epsilon^{ijk} A_i F_{jk}, \tag{16.16}$$

the Hopf index. For $\phi = \phi_H$ defined in (16.13), we obtain a Hopf index of 1.

16.3 Bateman's Construction

We have seen that we want to find solutions that have $(\vec{E} + i\vec{B})^2 = 0$. In order to do so, we have a simple construction by Bateman. We first define the *Riemann-Silberstein vector* as

$$\vec{F} = \vec{E} + i\vec{B}. \tag{16.17}$$

In terms of it, the sourceless (vacuum) Maxwell's equations are

$$\vec{\nabla} \times \vec{F} = i\frac{\partial}{\partial t}\vec{F}; \quad \vec{\nabla} \cdot \vec{F} = 0. \tag{16.18}$$

We can now introduce an ansatz that automatically satisfies half of the equations in (16.18), namely $\vec{\nabla} \cdot \vec{F} = 0$, by

$$\vec{F} = \vec{\nabla}\alpha \times \vec{\nabla}\beta, \quad \alpha, \beta \in \mathbb{C}. \tag{16.19}$$

The remaining Maxwell's equations are now written as

$$i\vec{\nabla} \times (\partial_t \alpha \vec{\nabla}\beta - \partial_t \beta \vec{\nabla}\alpha) = \vec{\nabla} \times \vec{F}. \tag{16.20}$$

We can solve them by considering that the equation without $\vec{\nabla} \times$ is valid – i.e.,

$$i(\partial_t \alpha \vec{\nabla}\beta - \partial_t \beta \vec{\nabla}\alpha) = \vec{F} = \vec{\nabla}\alpha \times \vec{\nabla}\beta. \tag{16.21}$$

But, then, taking one $\vec{F}$ as the left-hand side expression and the other as the right-hand side expression, we find

$$\vec{F}^2 = i(\partial_t \alpha \vec{\nabla}\beta - \partial_t \beta \vec{\nabla}\alpha) \cdot (\vec{\nabla}\alpha \times \vec{\nabla}\beta) = 0 \Rightarrow$$
$$\vec{E}^2 - \vec{B}^2 = 0, \quad \vec{E} \cdot \vec{B} = 0, \tag{16.22}$$

so Bateman's construction leads to null fields, $\vec{F}^2 = 0$, as we wanted.

16.4 The Hopfion Solution

The basic "Hopfion" solution of electromagnetism is found by giving a specific form for α, β in Bateman's construction,

$$\alpha = \frac{A - 1 + iz}{A + it};$$
$$\beta = \frac{x - iy}{A + it},$$
$$A \equiv \frac{1}{2}(x^2 + y^2 + z^2 - t^2 + 1). \tag{16.23}$$

Since $\vec{F}^2 = 0$ by construction, all the helicities are conserved in time, but we find that

$$H_{ee} = H_{mm} \neq 0; \quad H_{em} = H_{me} = 0. \tag{16.24}$$

It is to be expected then that the solution will have a conserved topological number, and it is indeed true.

To see that, we note there is another way to obtain the same solution that emphasizes the connection with the Hopf index. Consider the antisymmetric object in (16.14), and define it as the spatial components of the electromagnetic field tensor, thus (since $F_{ij} = \epsilon_{ijk}B_k$)

$$\vec{B} = \frac{1}{2\pi i}\frac{\vec{\nabla}\phi \times \vec{\nabla}\phi^*}{(1+\phi^*\phi)^2}. \tag{16.25}$$

Then consider also a similar formula for the electric-magnetic dual tensor,

$$\tilde{F}_{\mu\nu} \equiv \frac{1}{2}\epsilon_{\mu\nu\rho\sigma}F^{\rho\sigma}. \tag{16.26}$$

Then $\tilde{F}_{ij}$ is related in the same way with another complex scalar field θ, so (since $\tilde{F}_{ij} = \epsilon_{ijk}E_k$),

$$\vec{E} = \frac{1}{2\pi i}\frac{\vec{\nabla}\theta \times \vec{\nabla}\theta^*}{(1+\theta^*\theta)^2}. \tag{16.27}$$

We choose two Hopf maps of winding number 1 (16.13) at $t = 0$ for both ϕ and θ,

$$\phi(t=0,x,y,x) = \frac{2(z-iy)}{2x+i(r^2-1)} = \phi_H(z,-y,x)$$
$$\theta(t=0,x,y,z) = \frac{2(x+iz)}{-2y+i(r^2-1)} = \phi_H(x,z,-y). \tag{16.28}$$

We can find the solution of the Maxwell's equations for any time t, which is

$$\phi = \frac{Az+t(A-1)+i(tx-Ay)}{Ax+ty+i(A(A-1)-tz)}$$
$$\theta = \frac{Ax+ty+i(Az+t(A-1))}{tx-Ay+i(A(A-1)-tz)}, \tag{16.29}$$

by choosing it to satisfy Lorentz covariant forms,

$$F_{\mu\nu} = \frac{1}{2\pi i}\frac{\partial_\mu\phi^*\partial_\nu\phi - \partial_\mu\phi\partial_\nu\phi^*}{(1+\phi^*\phi)^2}$$
$$\tilde{F}_{\mu\nu} = \frac{1}{2\pi i}\frac{\partial_\mu\theta^*\partial_\nu\theta - \partial_\mu\theta\partial_\nu\theta^*}{(1+\theta^*\theta)^2}. \tag{16.30}$$

These ensure that $\partial_{[\mu}F_{\nu\rho]} = 0$ (the covariant version of $\vec{\nabla}\cdot\vec{B} = 0$, satisfied by (16.25), as we can easily check), as well as its electric-magnetic dual, $\partial_\mu F^{\mu\nu} = \partial_{[\mu}\tilde{F}_{\nu\rho]} = 0$, so the full Maxwell's equations are satisfied.

Note that now we have a nontrivial check, since the same fields are given in terms of either ϕ or θ (if we restrict to $t = 0$ as before, it is not true: $\vec{B}$ depends on the spatial derivatives of ϕ, but $\vec{E}$ depends on its time derivatives, which are not known unless we have the full t dependent formula; reversely, $\vec{E}$ depends on the spatial derivatives of θ, but $\vec{B}$ depends on its time derivatives).

Calculating $\vec{E}$ and $\vec{B}$ explicitly, we find then that the same electric and magnetic fields are obtained as in Bateman's construction.

16.5 More General Solutions and Properties

More general solutions can be obtained in several ways. One simple way is if we replace α and β by holomorphic functions of them, $f(\alpha, \beta)$ and $g(\alpha, \beta)$. Then

$$\vec{F} = \vec{\nabla} f \times \vec{\nabla} g = (\partial_\alpha f \partial_\beta g - \partial_\beta f \partial_\alpha g)\vec{\nabla}\alpha \times \vec{\nabla}\beta \equiv h(\alpha, \beta)(\vec{\nabla}\alpha \times \vec{\nabla}\beta). \qquad (16.31)$$

I will not show it here, but this leads to another solution of the Maxwell's equations (16.18), as we can explicitly check.

Then we can choose, in particular, $f = \alpha^p$ and $g = \beta^q$ and find that the electric and magnetic fields are nested tori, forming (p, q)−torus knots at the center. In particular, the original Hopfion has electric and magnetic fields that are *linked*, like two rungs in a chain. We say we have a *linking number 1*.

The time evolution of the Hopfion solution is found to be the linked tori of electric and magnetic field lines evolve radially inward from infinity, until reaching a minimum at $t = 0$, then move outward for positive t. Thus, the *size and shape* of the solution is time depedent, but the topological numbers characterizing it (Hopf index, linking number, helicities) aren't. Note that this fact might seem trivial, since Maxwell's theory is linear, so there are no interactions, and the sum of two solutions is a new solution. However, the solution is defined on the subspace of null solutions, which imposes a nonlinear constraint on the fields. In particular, the sum of two null solutions is not necessarily a null solution (only in very particular cases it is). That means that the conservation of topological numbers is a nontrivial statement about the solution.

Important Concepts to Remember

- One can define helicities H_{ee}, H_{mm}, H_{em} and H_{me} as the CS and BF forms for $\vec{A}$ and its electric-magnetic dual, $\vec{C}$.
- The helicies are conserved for null fields, $\vec{E} \cdot \vec{B} = 0$ and $\vec{E}^2 - \vec{B}^2 = 0$.
- The (first) Hopf map is a map from S^3 to $S^2 = \mathbb{CP}^1$, defining a fibration of winding number 1.
- Considering a complex field in three-dimensional space, $\phi(x_1, x_2, x_3)$, it defines a Hopf map $S^3 \to S^2$ if it tends to a constant at infinity because of the stereographic projection.
- One can define an A_i and F_{ij} from ϕ that defines the Hopf index as the CS form on S^3, $\int_{S^3} \epsilon^{ijk} A_i F_{jk}$.
- Bateman's construction for null fields is $\vec{F} = \vec{E} + i\vec{B} = \vec{\nabla}\alpha \times \vec{\nabla}\beta, \alpha, \beta \in \mathbb{C}$, giving $\vec{F}^2 = 0$.
- The Hopfion solution has nonzero and conserved H_{ee} and H_{mm}, Hopf index 1 and linking number 1 for electric and magnetic fields.
- In Bateman's construction, replacing α, β by holomorphic functions $f(\alpha, \beta), g(\alpha, \beta)$, we obtain new solutions. If $f = \alpha^p, g = \beta^q$, we obtain a (p, q)−knotted solution.

Further Reading

See, for instance, the paper [12].

Exercises

(1) Check that the helicities $H_{ee}, H_{mm}, H_{em}, H_{me}$ are invariant under residual gauge transformations with $\alpha(\vec{x})$ that go to zero at infinity.

(2) Check that the electric and magnetic fields for the Hopfion in Bateman's construction match with the ones from ϕ, θ in (16.29).

(3) Calculate the electric and magnetic fields for the (p, q)–knotted solution, with $f = \alpha^p$ and $g = \beta^q$, where α, β are the quantities for the Hopfion solution.

(4) Check that the Hopf index for the Hopfion solution is 1.

17 Complex Scalar Field and Electric Current: Gauging a Global Symmetry

In this chapter, we will better describe complex scalar fields and their associated electric currents, and we will know how to "gauge" (or to make local) a global symmetry.

17.1 Complex Scalar Field

We can construct in the obvious way a complex scalar field out of two real scalars ϕ_1, ϕ_2, as

$$\phi(x) = \phi_1(x) + i\phi_2(x) \Rightarrow \phi^*(x) = \phi_1(x) - i\phi_2(x). \tag{17.1}$$

Then the kinetic term for the combination can also be rewritten in complex notation as

$$\mathcal{L}_{\text{kin}} = -\frac{1}{2}(\partial_\mu \phi_1)^2 - \frac{1}{2}(\partial_\mu \phi_2)^2 = -\frac{1}{2}(\partial_\mu \phi)(\partial^\mu \phi^*). \tag{17.2}$$

Note that varying with respect to ϕ_1 (real part of ϕ) and ϕ_2 (imaginary part of ϕ) independently gives the equations of motion $\Box\phi_1 = \Box\phi_2 = 0$.

But there is a subtlety. If we want to treat both ϕ and ϕ^* (instead of the real and imaginary parts) as independent variables, we must drop the 1/2 factor to write

$$\mathcal{L}_{\text{kin}} = -(\partial_\mu \phi)(\partial^\mu \phi^*). \tag{17.3}$$

Indeed, in this case, the equation of motion for ϕ^* is

$$\frac{\delta S}{\delta \phi^*} = +\partial_\mu \partial^\mu \phi = 0, \tag{17.4}$$

i.e., the sum of the two equations of motion for ϕ_1 and ϕ_2, without an extra factor (keeping the 1/2 would give an overall 1/2 in the equation of motion). Another way of saying this is that because of the Jacobian of the transformation from ϕ_1, ϕ_2 to ϕ, ϕ^*, we need to multiply the action by 2.

Thus, a general Lagrangean for a complex scalar with a canonical kinetic term, but which is manifestly real, is

$$\mathcal{L} = -(\partial_\mu \phi)(\partial^\mu \phi^*) - m^2 \phi^* \phi - \Delta V(\phi^* \phi). \tag{17.5}$$

The same comment applies to the mass term (which is $m^2(\phi_1^2 + \phi_2^2)/2$ for the real case) as for the kinetic term. Note that this form is actually restricted, simply on the basis of reality we could have written $V(\phi) + V(\phi^*)$, for instance, but we chose $V(\phi^* \phi)$ instead.

In the case of a free massive field ($\Delta V = 0$), we obtain the KG equation,

$$(\Box - m^2)\phi = 0. \tag{17.6}$$

It is worth writing the solution of the (free) KG equation in the Fourier expansion – i.e., in momentum space. We write a Fourier decomposition over the spatial coordinates, but written in terms of $p \cdot x = \vec{p} \cdot \vec{x} - p^0 t$, namely

$$\phi = \int \frac{d^3 p}{(2\pi)^3} \left(a_{\vec{p}} e^{ip \cdot x} + \tilde{a}_{\vec{p}}^\dagger e^{-ip \cdot x} \right). \tag{17.7}$$

Note that we have introduced arbitrary coefficients $a_{\vec{p}}, \tilde{a}_{\vec{p}}^\dagger$ for the Fourier modes. For a real field, we would have needed $\tilde{a}_{\vec{p}} = a_{\vec{p}}$, but for a complex field the two coefficients of $e^{ip \cdot x}$ and $e^{-ip \cdot x}$ are independent. The KG equation is then on the Fourier modes

$$- (p^2 + m)(a_{\vec{p}} e^{ip \cdot x} + \tilde{a}_{\vec{p}}^\dagger e^{-ip \cdot x}) = 0 , \tag{17.8}$$

i.e., the energy is given in terms of the momentum,

$$- (p^0)^2 + \vec{p}^2 = -m^2 \Rightarrow p^0 = \sqrt{\vec{p}^2 + m^2}. \tag{17.9}$$

Another way to say this is that the two independent solutions of the KG equation are $e^{ip \cdot x}$ and $e^{-ip \cdot x}$, for $p^2 + m^2 = 0$, and their coefficients are arbitrary functions of the independent $\vec{p}$.

The choice of the potential as a function of $\phi^* \phi$ was such that we have a global $U(1)$ symmetry,

$$\phi(x) \to e^{i\alpha}\phi(x); \quad \phi^*(x) \to e^{-i\alpha}\phi^*(x). \tag{17.10}$$

It is obvious then that $\mathcal{L}$ is invariant, since both $\phi^* \phi$ and $(\partial_\mu \phi^*)(\partial^\mu \phi)$ are invariant under it. Note that the kinetic term would not be invariant under a local transformation, with $\alpha = \alpha(x)$ (the derivative could act on it), but only in the global case.

At the infinitesimal level, we have the $U(1)$ symmetry transformation

$$\delta\phi(x) = i\alpha\phi(x); \quad \delta\phi^*(x) = -i\alpha\phi^*(x). \tag{17.11}$$

17.2 Electric Current and Charge

Consider the quantity

$$j^\mu \equiv \frac{\delta S}{\delta(\partial_\mu \phi_i)} \frac{\delta_\alpha \phi_i}{\alpha} , \tag{17.12}$$

called a current, where ϕ_i are all the fields in the theory. In our case, those are ϕ and ϕ^*, so

$$j^\mu = \frac{\partial \mathcal{L}}{\partial(\partial_\mu \phi)} \frac{\delta_\alpha \phi}{\alpha} + \frac{\partial \mathcal{L}}{\partial(\partial_\mu \phi^*)} \frac{\delta_\alpha \phi^*}{\alpha} = i(\phi^* \partial^\mu \phi - \phi \partial^\mu \phi^*). \tag{17.13}$$

We show that this quantity is *conserved on-shell* (when using the equations of motion) – i.e., $\partial_\mu j^\mu = 0$. We have, in general,

$$\partial_\mu j^\mu = i\partial_\mu(\phi^*\partial^\mu\phi - \phi\partial^\mu\phi^*) = i(\phi^*\Box\phi - \phi\Box\phi^*). \tag{17.14}$$

To show that this is zero, consider the equations of motion for ϕ^* and ϕ,

$$(\Box - m^2)\phi - \Delta V'(\phi^*\phi)\phi = 0; \quad (\Box - m^2)\phi^* - \Delta V'(\phi^*\phi)\phi^* = 0, \tag{17.15}$$

and multiply the first relation by ϕ^* and the second by ϕ, and then subtract them, obtaining

$$\phi^*\Box\phi - \phi\Box\phi^* = 0. \tag{17.16}$$

The relation $\partial_\mu j^\mu = 0$ is the local form corresponding to a conservation of a charge (so we say that the current is conserved), as we saw in Chapter 14, when talking about the energy-momentum tensor. Reviewing the argument, we can use Stokes's theorem (or Gauss's theorem, in this case) to write

$$\oint_{\partial M_4 = \Sigma} j^\mu d\Sigma_\mu = \int_{M_4} dV\partial^\mu j_\mu = 0. \tag{17.17}$$

Choosing a closed three-surface Σ that has a Σ_{t_1} (spatial surface at time t_1), a Σ_{t_2} (spatial surface at time $t_2 > t_1$), and a surface S_∞ connecting them, a surface at spatial infinity plus the time direction, and assuming that all fields vanish at infinity (on S_∞), so $\int_{S_\infty} j^\mu d\Sigma_\mu = 0$, we find that

$$\int_{\Sigma_{t_1}} j^\mu d\Sigma_\mu = \int_{\Sigma_{t_2}} j^\mu d\Sigma_\mu, \tag{17.18}$$

i.e., the quantity

$$Q \equiv \int_{S_t = V} j^\mu d\Sigma_\mu = \int_{S_t = V} j^0 d^3x, \tag{17.19}$$

the charge of the current j^μ, is conserved (constant in time).

The interpretation of Q is of *electric charge*. But to see that, we must couple the scalar ϕ with electromagnetism. Indeed, electric charge is the strength of the coupling of a field to electromagnetism, since, as we saw when talking about particles interacting with electromagnetism, this coupling is $\int d^4x j^\mu(x^\rho)A_\mu(x^\rho)$, becoming $\int d\tau Q A_0(x^\rho(\tau))$ for a particle world line.

In $j^\mu A_\mu$, j^μ was the electric current of the particles. We extend this idea to the notion of a current generated by a field, coupling to the same electromagnetism through $j^\mu A_\mu$. Moreover, we want to make electromagnetism dynamical, so we consider its kinetic term. In all, we replace

$$\mathcal{L}_0 \to \mathcal{L} = \mathcal{L}_0 + j^\mu A_\mu - \frac{1}{4}F_{\mu\nu}F^{\mu\nu}. \tag{17.20}$$

It would seem that if we do this, we can now interpret j^μ as the electric current of the complex field and Q as its electric charge.

17.3 Gauging a Global Symmetry

But that is still not so, since we are missing a crucial ingredient. We know that in electromagnetism, we have a local symmetry called gauge invariance. Under the transformation
by $\alpha(x)$,

$$\delta A_\mu = -\partial_\mu \alpha(x), \tag{17.21}$$

the kinetic term for A_μ is invariant. This is an Abelian ($U(1)$) invariance, and a local one.

But note that $\mathcal{L}$ was only globally invariant under $U(1)$. Did we make it local by adding
the coupling to A_μ, $j^\mu A_\mu$? Under a local transformation

$$\phi(x) \to e^{i\alpha(x)} \phi(x), \tag{17.22}$$

the globally symmetric Lagrangean $\mathcal{L}_0$ changes as

$$\mathcal{L}_0 \to \mathcal{L}_0 - \partial_\mu(e^{i\alpha(x)}\phi(x))\partial^\mu(e^{-i\alpha(x)}\phi^*) + (\partial_\mu\phi(x))(\partial^\mu\phi^*(x)). \tag{17.23}$$

Under an infinitesimal transformation, this becomes

$$\delta\mathcal{L} = -\partial_\mu\alpha(x)i(\phi\partial^\mu\phi^* - \phi^*\partial^\mu\phi) = j^\mu\partial_\mu\alpha(x). \tag{17.24}$$

The coupling term transforms as

$$\delta(j^\mu A_\mu) = j^\mu \delta A_\mu + A_\mu \delta j^\mu = -j^\mu\partial_\mu\alpha(x) + A_\mu\delta j^\mu. \tag{17.25}$$

On the other hand, we have

$$\begin{aligned}
\delta j^\mu &= i(\delta(\phi^*\partial^\mu\phi) - \delta(\phi\partial^\mu\phi^*)) \\
&= i\delta\phi^*\partial^\mu\phi + i\phi^*\partial^\mu\delta\phi - i\delta\phi\partial^\mu\phi^* - i\phi\partial^\mu\delta\phi^* \\
&= -2(\partial^\mu\alpha(x))|\phi|^2.
\end{aligned} \tag{17.26}$$

That means that by coupling to electromagnetism, we obtain

$$\delta\left(\mathcal{L} + j^\mu A_\mu - \frac{1}{4}F_{\mu\nu}F^{\mu\nu}\right) = -2A_\mu(\partial^\mu\alpha(x))|\phi|^2 = 2A_\mu\delta A^\mu|\phi|^2. \tag{17.27}$$

But that means that if we also add a term quadratic in A_μ (quartic in the fields; the term
$j^\mu A_\mu$ was cubic in the fields) to the action – namely, if we write

$$\mathcal{L}' = \mathcal{L}_0 + j^\mu A_\mu - \frac{1}{4}F_{\mu\nu}F^{\mu\nu} - A_\mu A^\mu|\phi|^2, \tag{17.28}$$

then, since

$$\delta(-A_\mu A^\mu|\phi|^2) = +2A_\mu(\partial^\mu\alpha(x))|\phi|^2, \tag{17.29}$$

the extra term in the local variation cancels, and the new action is *locally* $U(1)$ invariant –
i.e., is gauge invariant.

The procedure is a simple example of the general *Noether procedure* to *gauge (make
local) a global symmetry*. We have added a coupling of the current j^μ to a gauge field A_μ,
with its own kinetic term, and then added a term of higher nonlinearity in the fields to the
action (quartic vs. cubic), and we obtained invariance.

But, in general, if invariance was not obtained, we would have needed to add terms of the needed (highest) nonlinearity in the fields to the transformation laws for the fields ($\delta\phi_i$) and if even then we didn't succeed, we would continue with even higher nonlinearity terms in the action and then in the transformation laws, etc., until either we obtain invariance or we continue to infinity.

This Noether procedure is a "brute force" one. But if we have some ideas about what to expect, we can improve the procedure at any stage. For instance, in this case, we see that we can rewrite the Lagrangean as

$$\mathcal{L}' = -(D_\mu\phi)(D^\mu\phi)^* - m^2|\phi|^2 - \frac{1}{4}F_{\mu\nu}F^{\mu\nu} - \Delta V(\phi^*\phi), \qquad (17.30)$$

where we have defined the *covariant derivative*

$$D_\mu\phi = \partial_\mu\phi - iA_\mu\phi, \qquad (17.31)$$

or, more generally, for a coupling with electric charge q between the field and electromagnetism, which means a transformation law for the scalar of $\phi \to e^{iq\alpha(x)}\phi$,

$$D_\mu\phi = \partial_\mu\phi - iqA_\mu\phi. \qquad (17.32)$$

The name covariant is used in many ways: we have covariant vectors for Lorentz transformations, we will see later in the book covariance with respect to general coordinate transformations (general covariance), etc. But here it refers to covariance with respect to gauge transformations, so we have a *gauge covariant derivative*.

In this instance, the name comes from the fact that the transformation law is gauge covariant – i.e., that $D_\mu\phi$ transforms in the same way as ϕ itself, as we can check:

$$\begin{aligned} D_\mu = \partial_\mu\phi - iqA_\mu\phi &\to (\partial_\mu - iqA'_\mu)\phi' = (\partial_\mu - iq(A_\mu - \partial_\mu\alpha))e^{iq\alpha(x)}\phi \\ &= e^{iq\alpha(x)}(\partial_\mu\phi - iqA_\mu\phi) = e^{iq\alpha(x)}D_\mu. \end{aligned} \qquad (17.33)$$

We can now also check that indeed the covariant derivative terms give all the electromagnetic-scalar field coupling terms,

$$\begin{aligned} -(D_\mu\phi)(D^\mu\phi)^* &= -(\partial_\mu\phi - iqA_\mu\phi)(\partial^\mu\phi^* + iqA_\mu\phi^*) \\ &= -(\partial_\mu\phi)\partial^\mu\phi + iqA_\mu(\phi\partial^\mu\phi^* - \phi^*\partial_\mu\phi) - q^2A_\mu^2|\phi|^2 \\ &= -(\partial_\mu\phi)\partial^\mu\phi + qj^\mu A_\mu - q^2A_\mu^2|\phi|^2. \end{aligned} \qquad (17.34)$$

We note that the term quadratic in fields is the original one, the $j^\mu A_\mu$ is cubic, and the last term is quartic.

In the Noether procedure of making a global symmetry local (gauging it), one thing that we almost always must use (hence, it can be used to simplify the procedure) is to replace normal derivatives with gauge covariant ones, $\partial_\mu \to D_\mu$. This is called the *minimal coupling to the gauge field*.

Note that we have used the example of the scalar field, but any field can be used for gauging. We also showed only the example of an Abelian gauge group, since we will study non-Abelian gauge invariance (Yang–Mills theory) only later, but the same gauging procedure can be performed in the case of a non-Abelian global symmetry group. We can also gauge just a subgroup of the total global symmetry group.

- A complex scalar field with canonical kinetic term and potential depending on $\phi^* \phi$ has a global $U(1)$ symmetry.
- There is a current, $j^\mu = i(\phi^* \partial^\mu \phi - \phi \partial^\mu \phi^*)$, which is conserved on-shell, $\partial_\mu j^\mu = 0$.
- Its charge $Q = \int j^0 d^3 x$ is interpreted as electric charge, but for that, we must couple to electromagnetism via a $j^\mu A_\mu$ term and a kinetic term for A_μ.
- Moreover, we need to respect the gauge invariance of electromagnetism – i.e., to "gauge," or make local – the global $U(1)$ symmetry.
- For gauging, we have the Noether procedure of adding $j^\mu A_\mu$ and a kinetic term and then adding terms nonlinear in fields of increasing nonlinearity to both the action and the transformation laws untill we find invariance.
- In the case of gauge theories, the net effect is often that we replace ∂_μ with the (gauge) covariant derivative $D_\mu = \partial_\mu - iqA_\mu$, called minimal coupling to the gauge field.
- The gauge covariant derivative transforms covariantly, $D_\mu \phi \rightarrow e^{i\alpha(x)} D_\mu \phi$.
- Gauging and the Noether procedure can be applied to a subgroup of the global symmetry group and can be done for any field and any symmetry group.

Further Reading

See for instance chapter 9 in Schwartz [9].

Exercises

(1)   Consider the $O(2)$ model

$$\mathcal{L} = -\frac{1}{2}(\partial_\mu \phi^I)\partial^\mu \phi^I - V(\phi^I \phi^I), \qquad (17.35)$$

where $I = 1, 2$ and $\phi^I \in \mathbb{R}$. Gauge the (continuous part of the) global symmetry.

(2)   Consider the model

$$\mathcal{L} = -\frac{1}{2}(\partial_\mu \phi^I)^\dagger (\partial^\mu \phi^I) - V(\phi_I^\dagger \phi^I), \qquad (17.36)$$

where $I = 1, 2$ and $\phi^I \in \mathbb{C}$.

(a) What is its global symmetry?
(b) Gauge an Abelian subgroup of the global symmetry.

(3)   Consider the nonlinear sigma model obtained from the $O(N)$ model ($N \geq 3$),

$$\mathcal{L} = -\frac{1}{2}(\partial_\mu \phi^I)(\partial^\mu \phi^I), \qquad (17.37)$$

by imposing the constraint

$$\sum_{I=1}^{N} \phi^I \phi^I = R^2.$$

(17.38)

(a) Can one gauge a subset of the symmetry group?

(b) If yes, then do it for an Abelian subgroup.

(4)   For the case at exercise 3, but for $N = 2$, can one gauge the continous part of the symmetry group? If so, do it.

18 The Noether Theorem and Applications

In this chapter, we will describe one of the most important and influential theorems in modern theoretical physics, Noether's theorem, relating symmetries and conserved charges. It was found by Emmy Noether in 1915, and it changed the way we approach theoretical physics. We will then apply it to several systems of interest.

18.1 Setup

The theorem associates a conserved current and a conserved charge to any global symmetry. We, therefore, consider a general global symmetry. While it is not always the case, we mostly consider linearly realized symmetries – i.e., symmetries that are linear in the fields,

$$\delta\phi^i = \epsilon^a (iT_a)^i{}_j \phi^j. \tag{18.1}$$

Note that there are also nonlinearly realized symmetries, which usually is a sign of some spontaneously broken linear symmetry, in which case we can have nonlinear terms (nonlinear in ϕ^i; a constant is also nonlinear, for example) on the right-hand side. But in most of this chapter, we will stick to linear symmetries.

Varying the action for the standard Lagrangean depending only on first derivatives, $\mathcal{L}(\phi^i, \partial_\mu \phi^i)$, we obtain

$$
\begin{aligned}
\delta S &= \int d^4 x \left[\frac{\partial \mathcal{L}}{\partial \phi^i} \delta\phi^i + \frac{\partial \mathcal{L}}{\partial(\partial_\mu \phi^i)} \partial_\mu \delta\phi^i \right] \\
&= \int d^4 x \left[\left(\frac{\partial \mathcal{L}}{\partial \phi^i} - \partial_\mu \left(\frac{\partial \mathcal{L}}{\partial(\partial_\mu \phi^i)} \right) \right) \delta\phi^i + \partial_\mu \left(\frac{\partial \mathcal{L}}{\partial(\partial_\mu \phi^i)} \delta\phi^i \right) \right],
\end{aligned} \tag{18.2}
$$

where we have used, as usual, $\partial_\mu \delta\phi^i = \delta \partial_\mu \phi^i$ and partially integrated in the second equality so as to obtain the Euler–Lagrange equations in the first term.

The statement of invariance under the global symmetry is $\delta S = 0$ *for the variation* $\delta\phi^i$ *under the symmetry transformation*; in the linear case (18.1) and *on-shell* – i.e., considering that the Euler–Lagrange equations of motion are satisfied – we obtain

$$\int_{M^4} d^4 x \partial_\mu \left(\frac{\partial \mathcal{L}}{\partial(\partial_\mu \phi^i)} \delta\phi^i \right) = \int_{M^4} d^4 x \partial_\mu \left(\frac{\partial \mathcal{L}}{\partial(\partial_\mu \phi^i)} i\epsilon^a (T_a)^i{}_j \phi^j \right) = 0. \tag{18.3}$$

That means that we can define the *Noether current* associated with the global symmetry as

$$j_\mu^a = \sum_{i,j} \frac{\partial \mathcal{L}}{\partial(\partial_\mu \phi^i)} i(T_a)^i{}_j \phi^j(x), \tag{18.4}$$

in which case the statement of invariance under the global symmetry becomes

$$\epsilon_a \int_{M^4} \partial^\mu j_\mu^a = 0, \ \forall \epsilon^a \Rightarrow \partial^\mu j_\mu^a = 0. \tag{18.5}$$

18.2 Noether's Theorem

The statement of the theorem is thus: For every continuous global symmetry under a group G with generators T_a, there is an associated conserved current j_μ^a, defined earlier and thus a conserved charge

$$Q^a(t) = \int_{S_t} d^3x j^{0a} \Rightarrow \frac{dQ^a(t)}{dt} = 0. \tag{18.6}$$

As we already showed in Chapters 14 and 17, this follows from the current conservation, since by the use of the Stokes theorem (or Gauss's law) $\int_M df = \int_{\partial M} f$,

$$0 = \int_{M^4} d^4x \partial_\mu j^{\mu a} = \oint_{\Sigma = \partial M^4} d\Sigma^\mu j_\mu^a = \int_{S_{t_2}} d^3x j^{0a} - \int_{S_{t_1}} d^3x j^{0a} \equiv Q^a(t_2) - Q^a(t_1). \tag{18.7}$$

Here we have used a surface at the boundary of four-dimensional spacetime region M^4 of the type $\Sigma = \partial M^4 = S_{t_2} \cup S_{t_1} \cup S_\infty$, and we have assumed that on the surface at spatial infinity (times time) S_∞, fields vanish, so also the current vanishes, whereas on S_{t_2} and S_{t_1} the normals have opposite directions (they go out of the volume M^4), hence the scalar product $d\Sigma^\mu j_\mu^a$ comes with opposite signs.

Thus, the invariance under the symmetry means the existence of a *conserved charge*,

$$Q^a(t_2) = Q^a(t_1) \Rightarrow \frac{d}{dt}Q^a = 0. \tag{18.8}$$

Note that this is a classical field theory book, so we work always *on-shell* (using the Euler–Lagrange equations of motion), but it is useful to remember that in the quantum theory – i.e., *off-shell* – we need to consider the more general result,

$$0 = \delta S = \int d^4x i \epsilon^a (T_a)^i{}_j \left[\left(\frac{\partial \mathcal{L}}{\partial \phi^i} - \partial_\mu \left(\frac{\partial \mathcal{L}}{\partial(\partial_\mu \phi^i)} \right) \right) \phi^j + \partial_\mu \left(\frac{\partial \mathcal{L}}{\partial(\partial_\mu \phi^i)} \phi^j \right) \right]. \tag{18.9}$$

18.3 The Noether Procedure

We can also consider promoting the global transformation parameter to a local one, $\epsilon^a(x)$. But in that case, the transformation is not a symmetry anymore. The previous variation of

S is then zero, but there is now an extra term, coming from the fact that $\epsilon^a(x)$ was under the derivative ∂_μ in the total derivative term. Thus, in this case, we obtain

$$\delta S = 0 + i \int d^4x \sum_{a,i,j} (\partial_\mu \epsilon^a)(T_a)^i_j \left[\frac{\partial \mathcal{L}}{\partial(\partial_\mu \phi^i(x))} \phi^j(x) \right] = i \sum_a \int d^4x (\partial^\mu \epsilon^a(x)) j^a_\mu(x)$$

$$= -i \sum_a \int d^4x \epsilon^a(x)(\partial^\mu j^a_\mu(x)). \tag{18.10}$$

In the last equality, we used partial integration and the fact that we consider $j^a_\mu(x)\epsilon_a(x)$ to vanish on ∂M^4.

Therefore, in order to obtain invariance under this *local* transformation – i.e., in order to gauge the global symmetry – we must add a gauge field A^a_μ to the theory with a transformation law,

$$\delta A^a_\mu = \partial_\mu \epsilon^a + \text{nonlinear} \tag{18.11}$$

(we will see this more precisely later in the book, when dealing with Yang–Mills theories), and then add a coupling

$$\int d^4x A^a_\mu j^{\mu a} \tag{18.12}$$

to the action, then more nonlinear terms to the transformation laws, then to the action, etc. This is the Noether procedure for gauging, first described in Chapter 17.

Note that the Noether theorem is associated only with a *global symmetry*. If we have a local symmetry, the equivalent of the Noether current will not be conserved ($\partial_\mu j^{\mu a} = 0$) but rather will be covariantly conserved ($D_\mu j^{\mu a} = 0$). This doesn't lead, by integration over an M_4, to a conserved charge in the usual sense described earlier.

18.4 A Subtlety and a General Form: Extra Terms in the Current

In the previous section, we implicitly assumed that the Lagrangean was invariant under the symmetry, $\delta \mathcal{L} = 0$, but that is not necessarily so, since all we need is $\delta S = 0$, and we can have

$$\mathcal{L} \to \mathcal{L} + \partial_\mu(\epsilon^a J^{\mu a}). \tag{18.13}$$

Indeed, then

$$\delta S = \int_{M^4} d^4x \epsilon^a \partial_\mu J^{\mu a} = \oint_{\partial M^4 = \Sigma} \epsilon^a J^{\mu a} d\Sigma_\mu, \tag{18.14}$$

which means that we find

$$\int_{M^4} d^4x \partial_\mu \left(\frac{\partial \mathcal{L}}{\partial(\partial_\mu \phi^i)} \Delta^a \phi^i - J^{\mu a} \right) = 0 \tag{18.15}$$

instead. Thus, the current in this more general case is

$$j^{\mu a}(x) = \frac{\partial \mathcal{L}}{\partial(\partial_\mu \phi^i)} \Delta^a \phi^i - J^{\mu a}. \tag{18.16}$$

Here we have also considered a more general transformation law, $\delta\phi^i = \epsilon^a \Delta^a \phi^i$ (thus, $\Delta^a \phi^i$ is the transformation law divided by the parameter), not necessarily a linear one. In the linear case, we have

$$(\epsilon^a \Delta^a \phi^i) = i\epsilon^a (T_a)^i_{\ j}\phi^j. \tag{18.17}$$

18.5 Applications

We now turn to applying the Noether theorem to specific symmetries.

1. Translations and the Energy-Momentum Tensor

An important example, and the most relevant case of a $J^{\mu a}$ term, was already considered: the energy-momentum tensor $T_\mu^{\ \nu}$ – which, as we saw, was a current associated with the "charge" of momentum P_μ.

But the momentum P_μ can be thought of as a generator associated with (symmetry under) translations. Indeed, we represent these generators as

$$P_\mu = -i\partial_\mu, \tag{18.18}$$

by which we mean that its group action on a field $\phi(x^\rho)$,

$$e^{ia^\mu P_\mu}\phi(x^\rho) = e^{a^\mu \partial_\mu}\phi(x^\rho) \equiv \sum_{n\geq 0} \frac{(a^\mu \partial_\mu)^n}{n!}\phi(x^\rho) \tag{18.19}$$

is the four-dimensional generalization of the formal expression for the Taylor expansion,

$$e^{a\partial_x}f(x) = \sum_{n\geq 0}\frac{a^n}{n!}\frac{d^n}{dx^n}f(x) = f(x+a). \tag{18.20}$$

That means that

$$e^{ia^\mu P_\mu}\phi(x^\rho) = \phi(x^\rho + a^\rho), \tag{18.21}$$

so, indeed, P_μ is the generator of translations.

In this case, the symmetry parameter is $\epsilon^a \to a^\nu$, so under the symmetry transformation = translation, the Lagrangean varies by

$$\delta\mathcal{L} = a^\nu \partial_\nu \mathcal{L} = a^\nu \partial_\mu(\mathcal{L}\delta^\mu_\nu) \to \epsilon^a \partial^\mu J^a_\mu, \tag{18.22}$$

which implies that

$$J^a_\mu \to \mathcal{L}\delta^\nu_\mu. \tag{18.23}$$

We also have for the symmetry transformation law

$$(\epsilon^a \Delta^a)\phi^i = a^\nu \partial_\nu \phi^i \Rightarrow \epsilon^a \to a^\nu. \tag{18.24}$$

Finally, we obtain for the current associated with translations

$$j^\mu_a \to T^\mu_{\ \nu} = \frac{\partial\mathcal{L}}{\partial(\partial_\mu\phi^i)}\partial_\nu\phi^i - \mathcal{L}\delta^\mu_\nu. \tag{18.25}$$

This is the conventional energy-momentum tensor (the nonsymmetric tensor), and it was obtained via Noether's theorem from invariance under translations.

2. $U(1)$ (Abelian) Symmetry

Another example that we have considered before is the case of the $U(1)$ Abelian symmetry, which, as we saw already in Chapter 17, is associated with the conserved electric charge.

For an $U(1)$ Abelian symmetry, the transformation law is

$$\phi(x) \to e^{i\alpha}\phi(x) \Rightarrow \delta\phi(x) = i\alpha\phi(x); \quad \delta\phi^*(x) = -i\alpha\phi^*(x). \tag{18.26}$$

The fields $\phi^i(x)$ are (ϕ, ϕ^*), and $\delta\mathcal{L} = 0$, so $J^\mu = 0$, and we obtain for the current

$$j^\mu(x) = \frac{\partial\mathcal{L}}{\partial(\partial_\mu\phi)}\frac{\delta\phi}{\alpha} + \frac{\partial\mathcal{L}}{\partial(\partial_\mu\phi^*)}\frac{\delta\phi^*}{\alpha}. \tag{18.27}$$

For the Lagrangean

$$\mathcal{L} = -(\partial_\mu\phi)(\partial^\mu\phi^*) - V(\phi\phi^*), \tag{18.28}$$

we obtain

$$j^\mu(x) = -(\partial_\mu\phi^*)i\phi + (\partial_\mu\phi)i\phi^* = i(\phi^*\partial_\mu\phi - \phi\partial_\mu\phi^*), \tag{18.29}$$

which is the electric current we obtained in the last chapter for the complex scalar.

3. $O(N)$ Model with $O(N)$ Symmetry

We consider the $O(N)$ model Lagrangean

$$\mathcal{L} = -\frac{1}{2}(\partial_\mu\phi^I)(\partial^\mu\phi^I) - V(\phi^I\phi^I), \tag{18.30}$$

invariant under the $O(N)$ transformation

$$\phi^I \to (e^{i\alpha^a T_a})^I{}_J\phi^J = M^I{}_J\phi^J, \tag{18.31}$$

which infinitesimally becomes

$$\delta\phi^I = i\alpha^a(T_a)^I{}_J\phi^J. \tag{18.32}$$

Then the current is

$$j^{\mu a}(x) = \frac{\partial\mathcal{L}}{\partial(\partial_\mu\phi^I)}\frac{\delta\phi^I}{\alpha} = -\partial^\mu\phi^I(iT_a)^I{}_J\phi^J. \tag{18.33}$$

4. Vector Fields in the Fundamental Representation of $O(N)$

Consider, as in Chapter 12, vector fields transforming under a *global* $O(N)$ symmetry in the fundamental representation, A_μ^I. Note that we don't consider non-Abelian gauge fields – i.e., Yang–Mills fields – which would transform in the adjoint representation (these will be treated later in the book).

We consider the Lagrangean

$$\mathcal{L} = -\frac{1}{4}F_{\mu\nu}^I F^{\mu\nu I} - \frac{m^2}{2}A_\mu^I A^{\mu I}$$
$$= -\frac{1}{2}(\partial_\mu A_\nu^I)^2 + \frac{1}{2}(\partial_\mu A_\nu^I)(\partial^\nu A^{\mu I}) - \frac{m^2}{2}A_\mu^I A^{\mu I}. \tag{18.34}$$

The first term is the kinetic term for N Abelian gauge fields, and the second is a mass term that kills the local invariance, leaving only the global $O(N)$ invariance.

The global $O(N)$ transformation law is

$$A^I_\mu(x) \rightarrow M^I{}_J A^J_\mu(x) \Rightarrow \delta A^I_\mu(x) = i\theta^a (T_a)^I{}_J A^J_\mu(x). \tag{18.35}$$

Then the current is

$$j^{\mu a}{}_\nu = \frac{\partial \mathcal{L}}{\partial(\partial_\mu A^I_\nu)} \frac{\delta A^I_\nu}{\theta^a} = -F^I_{\mu\nu} i(T_a)^I{}_J A^J_\mu. \tag{18.36}$$

18.6 The Charge as a Function of Fields

Finally, we explain the formula described in Chapter 12, in which the action of the generator T_a of a symmetry transformation law, represented on the fields, is given by a charge Q_a, expressed in terms of fields, through the formula

$$\delta_\epsilon \phi^I = \{\epsilon^a Q_a, \phi^I\}_{P.B.}. \tag{18.37}$$

We understand this formula as saying that for the action on the fields of the generators, we represent T_a via Q_a, and the action is given via the Poisson brackets. But, moreover, Q_a is obtained as the Noether charge associated with the Noether current $j^{\mu a}$,

$$Q_a(t) = \int_{S_t} j^\mu_a d\Sigma_\mu = \int_{S_t} d^3 x j^0_a$$
$$= \int_{S_t} d^3 x \left[\frac{\partial \mathcal{L}}{\partial(\dot{\phi}^J)} i(T_a)^J{}_K \phi^K(x) \right], \tag{18.38}$$

and then re-expressed in terms of the canonically conjugate momenta

$$\pi_J(x) \equiv \frac{\partial \mathcal{L}}{\partial \dot{\phi}^J}, \tag{18.39}$$

as

$$Q_a = \int_{S_t} d^3 x \, \pi_J (iT_a)^J{}_K \phi^K. \tag{18.40}$$

Then, using the fundamental Poisson brackets,

$$\{\phi^I(x), \phi^J(y)\}_{P.B.} = 0; \quad \{\phi^I(x), \pi_J(x)\}_{P.B.} = \delta^I_J \delta^3(x - y), \tag{18.41}$$

we finally obtain

$$\{\epsilon^a Q_a, \phi^I\}_{P.B.} = -i\epsilon^a (T_a)^I{}_K \phi^K, \tag{18.42}$$

which is the correct transformation law.

Important Concepts to Remember

- The Noether theorem associates with every continuous global symmetry a conserved current and a conserved charge.
- For a linear global symmetry, $\delta\phi^i = i\epsilon^a(T_a)^i{}_j\phi^j$, the current is $j^a_\mu = \partial\mathcal{L}/\partial(\partial\phi^i)(iT_a)^i{}_j\phi^j$, $\partial_\mu j^{\mu a} = 0$ and the charge is $Q^a = \int_{S_t} d^3 x j^{0a}$.

- For a general symmetry, with $\delta\phi^i = \epsilon^a \Delta^a \phi^i$ and a $\delta\mathcal{L} = \epsilon^a \partial_\mu J_a^\mu$, we have $j^{\mu a} = \partial\mathcal{L}/\partial\phi^i \Delta^a \phi^i - J^{\mu a}$.
- The charge associated with translation invariance is the four-momentum P_μ and the current is the energy-momentum tensor $T_\mu{}^\nu$.
- The charge associated with $U(1)$ invariance is electric charge.
- The transformation laws on the fields for the group with generators T_a are $\delta\phi^i = \{\epsilon^a Q_a, \phi^i\}_{P.B.}$, where Q_a, the representative of T_a on the fields, is the charge associated with the symmetry.

Further Reading

See, for instance, chapter 3 in Schwartz [9].

Exercises

(1) Write down the Noether current and charge associated with the $SU(N)$ global symmetry of the model with Lagrangean

$$\mathcal{L} = -\frac{1}{2}(\partial_\mu \phi^J)^\dagger \partial^\mu \phi^I g_{I\bar J}(\phi), \tag{18.43}$$

assuming that there is such a symmetry.

(2) Consider the Noether charge

$$Q = \int d^3x f^{AB}{}_C \pi_A \pi_B \phi^C, \tag{18.44}$$

where $f^{AB}{}_C$ are constants. Calculate the transformation laws for the fields.

(3) Consider Noether charges

$$Q^A = \int d^3x f^{AB}{}_C \pi_B \phi^C. \tag{18.45}$$

Calculate the commutator of two Noether charges by calculating the Poisson brackets

$$\{\epsilon_A Q^A, \epsilon_B Q^B\}_{P.B.}. \tag{18.46}$$

(4) Consider two complex scalar fields, ϕ_1, ϕ_2, with Lagrangean

$$\mathcal{L} = -\sum_{I=1,2}\frac{1}{2}(\partial_\mu\phi^I)(\partial^\mu\phi^I) - V(|\phi_1^2 + \phi_2^2|, |\phi_1^2 - \phi_2^2|). \tag{18.47}$$

(a) Write the Noether current and the charge associated with the global symmetry present in general.
(b) Is there a symmetry in a special case for the potential? What is the Noether current in this case?

Nonrelativistic and Relativistic Fluid Dynamics: Fluid Vortices and Knots

In this chapter, we will describe fluid dynamics in both its usual, nonrelativistic form, leading to the Navier–Stokes equations, and its more general relativistic form. We will then describe some solutions of the fluid equations that have vorticity and nontrivial knottedness. While fluid vortices are the reason knot theory was invented in the nineteenth century and are easy to visualize, explicit knotted solutions were unknown until recently.

For a general *interacting* quantum system, we will see that the fluid expansion is a general *perturbative* description in derivatives of the fields: The ideal fluid is the leading term in the expansion, which doesn't have any dissipative interactions. The next terms is a viscous term, encoding the first order of the dissipative interactions and having an extra derivative on the velocity field, etc.

We have summarized the description of fluids in Section 7.2, without much detail and derivations, but we will give now a more in-depth treatment.

19.1 Ideal Fluid Equations

A fluid is a collection of interacting particles, with a density that is approximately constant (to zeroth order). Fluid dynamics is defined by a fluid velocity $\vec{v}(\vec{r}, t)$ (the velocity of the particles inside the fluid), density $\rho(\vec{r}, t)$, and pressure $p(\vec{r}, t)$ fields.

Consider a volume V bounded by a surface ∂V and fluid passing through it. Then the increase in the total mass inside the volume is equal to minus the flux of particles through the surface ∂V – i.e.,

$$\frac{\partial}{\partial t} \int_V \rho dV = - \oint_{\partial V} \rho \vec{v} \cdot d\vec{A}. \tag{19.1}$$

Using Stokes's law or, more precisely, Gauss's law, $\oint_{\partial V} \vec{F} \cdot d\vec{A} = \int_V \vec{\nabla} \cdot \vec{F}$, we obtain

$$\int_V dV \left[\frac{\partial \rho}{\partial t} + \vec{\nabla} \cdot (\rho \vec{v}) \right] = 0. \tag{19.2}$$

Since the volume is arbitrary, we obtain the local *continuity equation*,

$$\frac{\partial \rho}{\partial t} + \vec{\nabla} \cdot (\rho \vec{v}) = 0. \tag{19.3}$$

The dynamics equation is based on Newton's law $\vec{F} = m\vec{a}$ for the particles of the fluid, integrated over a volume. The force through the boundary equals the integral of

$d\vec{F} = -pd\vec{A}$, plus any external force per unit mass $\vec{f}_{ext}$ we have, and $dm\,\vec{a} = \rho\,dV\,d\vec{v}/dt$, so, finally,

$$\int_V \rho \frac{d\vec{v}}{dt} dV = + \int_V \rho dV \vec{f}_{ext} - \oint_{\partial V} p d\vec{A} = \int_V dV(\rho\vec{f}_{ext} - \vec{\nabla}p)\,, \tag{19.4}$$

where in the last equality we have used Stokes's law. Since $\vec{v} = \vec{v}(\vec{r}, t)$, by the chain rule, we obtain

$$\frac{d\vec{v}}{dt} = \frac{\partial\vec{v}}{\partial t} + (\vec{v} \cdot \vec{\nabla})\vec{v}. \tag{19.5}$$

Since the volume of integration is arbitrary, we also have the local equation – which by dividing with ρ, is

$$\frac{\partial\vec{v}}{\partial t} + (\vec{v} \cdot \vec{\nabla})\vec{v} = \vec{f}_{ext} - \frac{\vec{\nabla}p}{\rho}. \tag{19.6}$$

This is *Euler's equation* for ideal fluids – i.e., nonviscous (non-dissipative) fluids. This equation will be generalized by introducing a dissipative viscous force term on the right-hand side, leading to the Navier–Stokes equation.

The equations of the ideal fluid are then the continuity and Euler equations, which completely determine the ideal fluid, as we have seen in Chapter 7.

But we can combine the equations to find another one as well: first we write

$$\begin{aligned}\frac{\partial}{\partial t}(\rho v_i) &= \rho\frac{\partial v_i}{\partial t} + v_i\frac{\partial\rho}{\partial t}\\ &= -\rho(\vec{v} \cdot \vec{\nabla})v_i - \vec{\nabla}_i p + \rho\dot{f}_{ext} - v_i(\vec{\nabla} \cdot (\rho\vec{v})),\end{aligned} \tag{19.7}$$

where we have used the Euler equation in the first term and the continuity equation in the second. Then, finally, we obtain, in the case of zero external force,

$$\frac{\partial}{\partial t}(\rho v_i) = -\partial_j(\rho v_i v_j + p\delta_{ij}) \equiv -\partial_j\Pi_{ij}. \tag{19.8}$$

But $\rho v_i v_j + p\delta_{ij} = \Pi_{ij}$ is momentum (in i direction) flux density in the j direction, since ρv_i is momentum density, and the pressure p is also momentum flux density. But the momentum flux density Π_{ij} is a component T_{ij} of the energy-momentum tensor $T_{\mu\nu}$, as seen in Chapter 14. Since $T_{00} = T^{00} = \rho$ is energy density, and $-T_{0i} = T^{0i} = \rho v_i$ is momentum density, then the continuity equation together with (19.8) are written as the conservation of the relativistic energy-momentum tensor,

$$\partial_\mu T^{\mu\nu} = 0. \tag{19.9}$$

19.2 Viscous Fluid and Navier–Stokes Equation

But we don't need to consider the relativistic generalization yet. At the nonrelativistic level, the next order in the derivative expansion – i.e., in the description of the fluid as an expansion in derivatives on $\vec{v}(\vec{r}, t)$ – is to add a *viscous term* to the stress tensor Π_{ij},

$$\Pi_{ij} = p\delta_{ij} + \rho v_i v_j + \sigma_{ij}, \tag{19.10}$$

where σ_{ij} is the new term, the viscous stress tensor. This term, being the first-order approximation, must contain first derivatives of the velocity, and contain them linearly. Moreover, we consider on physical grounds that for a fluid in uniform rotation, with velocity $\vec{v} = \vec{\Omega} \times \vec{r}$ ($v^i = \epsilon^{ijk}\Omega^j x^k$), there must be no viscous term, which restricts the dependence on the velocity to the symmetric combination $\partial_i v_j + \partial_j v_i$.

Finally, then, expanding σ_{ij} in irreducible representations of the rotation group $SO(3)$, namely symmetric traceless and the trace, we write

$$\sigma_{ij} = -\eta \left(\partial_i v_j + \partial_j v_i - \frac{2}{3}\delta_{ij}\partial_k v_k \right) - \zeta \delta_{ij}\partial_k v_k. \tag{19.11}$$

The constant η is called *shear viscosity* and the constant ζ is called *bulk viscosity*. Thermodynamic stability will, in fact, require $\eta \geq 0, \zeta \geq 0$. It is clear that the sign must be defined by thermodynamics, since viscosity means friction, which generates heat, and heat cannot be absorbed back. The signs were chosen such that the heat is only generated.

We substitute (19.10), with the σ_{ij} given previously into the conservation equation, and obtain

$$\frac{\partial}{\partial t}(\rho v_i) = -\frac{\partial}{\partial x_j}\left[p\delta_{ij} + \rho v_i v_j - \eta \left(\partial_i v_j + \partial_j v_i - \frac{2}{3}\delta_{ij}\partial_k v_k \right) - \zeta \delta_{ij}\partial_k v_k \right]. \tag{19.12}$$

We put the $-\partial_j(v_j \rho v_i)$ term on the left-hand side, use the continuity equation to cancel some terms, and thus write the equation in the form of Euler's equation plus corrections,

$$\rho \left[\frac{\partial v_i}{\partial t} + (\vec{v} \cdot \vec{\nabla})v_i \right] = -\partial_i p + \partial_j \left[\eta \left(\partial_j v_i + \partial_i v_j - \frac{2}{3}\delta_{ij}\partial_k v_k \right) + \zeta \delta_{ij}\partial_k v_k \right]. \tag{19.13}$$

In all generality, η and ζ are thermodynamic coefficients and so can depend on T and p, but if they depend only slightly, we can write

$$\rho \left[\frac{\partial \vec{v}}{\partial t} + (\vec{v} \cdot \vec{\nabla})\vec{v} \right] = -\vec{\nabla}p + \eta \Delta \vec{v} + \left(\zeta + \frac{\eta}{3} \right) \vec{\nabla}(\vec{\nabla} \cdot \vec{v}). \tag{19.14}$$

This is the *Navier–Stokes equation*.

19.3 Relativistic Generalization of Fluid Dynamics

Until now, we have presented norelativistic, everyday, fluids. But we saw how to write a relativistic generalization. We just need to write a covariant expansion for $T_{\mu\nu}$ that satisfies the conservation equation $\partial_\mu T^{\mu\nu} = 0$, in terms of a comoving (moving with the fluid) particle 4-velocity $u^\mu = dx^\mu/d\tau$. More precisely, we must also add whatever other currents J_I^μ we have and write the conservation laws together,

$$\partial_\mu T^{\mu\nu} = 0, \quad \partial_\mu J_I^\mu = 0. \tag{19.15}$$

The fields are now $u^\mu(x^\rho)$, the pressure $p(x^\rho)$ and energy density $\rho(x^\rho)$, and the ideal fluid has

$$T_{\mu\nu} = \rho u^\mu u^\nu + p(\eta^{\mu\nu} + u^\mu u^\nu). \tag{19.16}$$

We can define the projector onto states normal to u^μ,

$$P^{\mu\nu} \equiv \eta^{\mu\nu} + u^\mu u^\nu. \tag{19.17}$$

It satisfies the projector conditions

$$P^{\mu\nu} u_\mu = 0 \;\text{ and }\; P^{\mu\rho} P_{\rho\nu} = P^\mu{}_\nu \equiv P^{\mu\rho} \eta_{\rho\nu}. \tag{19.18}$$

The ansatz for the charge currents is

$$J_I^\mu = q_I u^\mu. \tag{19.19}$$

In the nonrelativistic limit, $u^\mu \simeq (1, \vec{v})$, we have

$$T_{00} = \rho; \quad T_{ij} = \rho v_i v_j + p\delta_{ij}; \quad J_I^\mu = (q_I, q_I \vec{v}). \tag{19.20}$$

For an ideal fluid, we have also a conserved entropy current, since we have no dissipation. In the viscous case, the entropy current will not be conserved, because of dissipation. The entropy current is, as a function of the entropy density s,

$$J_{s,\text{ideal}}^\mu = s\, u^\mu, \tag{19.21}$$

and is conserved,

$$\partial_\mu (J_s^\mu)_{\text{ideal}} = 0. \tag{19.22}$$

In the nonrelativistic limit, $u^\mu \simeq (1, v^i)$, its conservation becomes just

$$\partial_t s + \vec{v} \cdot \vec{\nabla} s = 0. \tag{19.23}$$

Viscous (Dissipative) Fluid

Since fluid dynamics is an expansion in derivatives, we can again write the viscous fluid, the first-order approximation in the description of the fluid, by adding terms linear in the first derivative on u^μ, both to the energy-momentum tensor and to the charge currents,

$$\begin{aligned}
(T_{\mu\nu})_{\text{dissipative}} &= \rho u^\mu u^\nu + p P^{\mu\nu} + \Pi_{(1)}^{\mu\nu} \\
(J_I^\mu)_{\text{dissipative}} &= q_I u^\mu + \Upsilon^\mu.
\end{aligned} \tag{19.24}$$

Expanding $\Pi_{(1)}^{\mu\nu}$ and Υ^μ in $\partial_\nu u^\rho$ and substituting in the conservation equations will lead to the relativistic generalization of the Navier–Stokes equations.

To fix the ambiguity in the definition of the energy-momentum tensor (which we discussed in Chapter 14), we impose the condition of the *Landau frame*, which is

$$\Pi_{(1)}^{\mu\nu} u_\mu = 0; \quad \Upsilon^\mu u_\mu = 0, \tag{19.25}$$

We decompose the derivatives on the velocity, $\partial^\nu u^\mu$, into a part parallel to the velocity u^μ, and a part transverse to it, which can be further decomposed into a trace, the *divergence* θ, and a traceless part, containing a symmetric traceless part, the *shear* $\sigma^{\mu\nu}$, and an antisymmetric part, the *vorticity* $\omega^{\mu\nu}$, so that, finally,

$$\partial^\nu u^\mu = -a^\mu u^\nu + \sigma^{\mu\nu} + \omega^{\mu\nu} + \frac{1}{d-1}\theta P^{\mu\nu}, \tag{19.26}$$

where a^μ is the acceleration, and the various quantities are defined by

$$a^\mu = u^\nu \partial_\nu u^\mu$$
$$\theta = \partial_\mu u^\mu = P^{\mu\nu} \partial_\mu u_\nu$$
$$\sigma^{\mu\nu} = \partial^{(\mu} u^{\nu)} + u^{(\mu} a^{\nu)} - \frac{1}{d-1} \theta P^{\mu\nu}$$
$$\omega^{\mu\nu} = \partial^{[\mu} u^{\nu]} + u^{[\mu} a^{\nu]}. \tag{19.27}$$

Since $\Pi^{\mu\nu}_{(1)}$ is a symmetric tensor (and since, as we argued in the nonrelativistic case, the stress tensor must only contain the symmetric part of ∂v), it can only be composed of the irreducible parts $\sigma^{\mu\nu}$ and $\theta P^{\mu\nu}$, whose coefficients are the shear and bulk viscosities, respectively (by comparison with the nonrelativistic case),

$$\Pi^{\mu\nu}_{(1)} = -2\eta \sigma^{\mu\nu} - \zeta \theta P^{\mu\nu}. \tag{19.28}$$

Substituting in the relativistic conservation equation $\partial_\mu T^{\mu\nu} = 0$, we obtain the relativistic form of the Navier–Stokes equation, which we leave as an exercise to do.

To construct a first-order ansatz for the conserved currents, we first construct the only vector made up of a general $\partial_\mu u_\nu$ term,

$$l^\mu = \epsilon_{\nu\rho\sigma}{}^\mu u^\nu \partial^\rho u^\sigma. \tag{19.29}$$

Then we consider that the current must depend on the derivatives of the chemical potentials μ_J for the charges q_J and of the temperature, for a final ansatz,

$$\Upsilon^\mu_{I(1)} = -\tilde{\kappa}_{IJ} P^{\mu\nu} \partial_\nu \left(\frac{\mu_J}{T} \right) - \mathcal{U}_I l^\mu - \gamma_I P^{\mu\nu} \partial_\nu T. \tag{19.30}$$

Here, $\tilde{\kappa}_{IJ}$, $\mathcal{U}_I$ and γ_I are thermodynamic coefficients, just like the shear and bulk viscosities.

19.4 Vorticity and Helicity of Ideal Fluid

As we saw in Chapter 7*, we can define a *vorticity* of the fluid as

$$\vec{\omega} = \vec{\nabla} \times \vec{v}. \tag{19.31}$$

Taking the curl $\vec{\nabla} \times$ of the Euler equation, and writing the external force per unit mass as a potential one, $\vec{f}_{\text{ext}} = -\vec{\nabla}\pi$, we obtain the equation for the vorticity,

$$\partial_t \vec{\omega} + \vec{\nabla} \times (\vec{\omega} \times \vec{v}) = 0. \tag{19.32}$$

Moreover, from the same Euler equation we can obtain the equation

$$\partial_t (\vec{v} \cdot \vec{\omega}) - \vec{\nabla} \cdot \left[\vec{\omega} \left(\frac{v^2}{2} - \int \frac{dp}{\rho} - \pi \right) - \vec{v}(\vec{v} \cdot \vec{\omega}) \right] = 0, \tag{19.33}$$

which we leave as an exercise to prove.

We can now define the *fluid helicity*

$$H = \int d^3x \, \vec{v} \cdot \vec{\omega} = \int d^3x \, \vec{v} \cdot (\vec{\nabla} \times \vec{v}) = \int d^3x \, \epsilon^{ijk} v_i \partial_j v_k \,, \qquad (19.34)$$

analogous to the magnetic helicity defined in Chapter 16*, being also written like a Chern–Simons term, this time for v_i instead of A_i. Then integrating (19.33) over space and considering vanishing boundary conditions for the fields at infinity, by the Stokes law, the integral $\int_V \vec{\nabla} \cdot [\ldots]$ turns into $\int_{\partial V_\infty} d\vec{S} \cdot [\ldots]$, which vanishes. Therefore, we obtain that the fluid helicity is conserved (constant in time).

19.5 Small Fluctuations and Fluid Waves

We can rewrite the ideal fluid equations using the *enthalpy per unit mass* $h = H/M$. The enthalphy $H = U + PV$ has differential derived from $dU - TdS - PdV + d\Pi$ (where Π is the potential), giving

$$dh = \frac{dp}{\rho} + Tds + d\pi, \qquad (19.35)$$

so for an adiabatic (isentropic) flow, $ds = 0$, we have

$$\vec{\nabla}h = \frac{1}{\rho}\vec{\nabla}p + \vec{\nabla}\pi \qquad (19.36)$$

or, more generally,

$$\delta h = \frac{\delta p}{\rho} + \delta\pi = c_s^2 \frac{\delta\rho}{\rho} + \delta\pi, \qquad (19.37)$$

where $c_s^2 = \partial p/\partial\rho$ is the sound speed squared. The Euler equation then can be rewritten as

$$\partial_t \vec{v} + (\vec{v} \cdot \vec{\nabla})\vec{v} + \vec{\nabla}h = 0. \qquad (19.38)$$

If $\partial_t \pi = 0$, similarly to (19.37), we obtain

$$\frac{1}{\rho}\partial_t\rho = \frac{1}{c_s^2}\partial_t h \,, \qquad (19.39)$$

and, if $\vec{\nabla}\pi = 0$, we have also

$$\frac{1}{\rho}\vec{\nabla}\rho = \frac{1}{c_s^2}\vec{\nabla}h \,, \qquad (19.40)$$

and, thus, the continuity equation can be rewritten as

$$\partial_t h + \vec{v} \cdot \vec{\nabla}h + c_s^2 \vec{\nabla} \cdot \vec{v} = 0. \qquad (19.41)$$

Moreover, using the relation (valid for any vector)

$$(\vec{v} \cdot \vec{\nabla})\vec{v} = -\vec{v} \times (\vec{\nabla} \times \vec{v}) + \vec{\nabla}(\vec{v}^2/2), \qquad (19.42)$$

and the vorticity definition $\vec{\omega} = \vec{\nabla} \times \vec{v}$, the Euler equation can also be rewritten in the form

$$\partial_t \vec{v} + \vec{\omega} \times \vec{v} + \vec{\nabla}\left(\frac{\vec{v}^2}{2} + h\right) = 0. \tag{19.43}$$

The ideal fluid equations are now (19.38) and (19.41). For small fluctuations, we can neglect the nonlinear (in fields) terms in them and write

$$\begin{aligned}\partial_t \vec{v} + \vec{\nabla} h &= 0 \\ \partial_t h + c_s^2 \vec{\nabla} \cdot \vec{v} &= 0.\end{aligned} \tag{19.44}$$

Taking $\vec{\nabla}\cdot$ of the first equation and ∂_t of the second, and subtracting them, we obtain

$$[-\partial_t^2 + c_s^2 \vec{\nabla}^2]h = 0 , \tag{19.45}$$

that is, the wave equation with velocity c_s for the enthalpy per unit mass, h. That means that small (enthalpy) fluctuations propagate with velocity c_s in the fluid.

19.6 Fluid Vortices and Knots

Since the fluid has a conserved helicity H, it is plausible that it can support solutions with nontrivial knottedness. In fact, knot theory was actually invented to deal with fluid knots. After the original work by Helmholtz in 1858, Lord Kelvin found the topological robustness of fluid knots and started the area of mathematics now known as knot theory. But knotted solutions are quite difficult to obtain, and there were no explicit solutions until the 1960s.

Here we will present a specific solution that appears just from rewriting pure electromagnetism as a fluid and importing the "Hopfion" solution of electromagnetism that we described before.

We first note that the relativistic conservation equation for the energy-momentum tensor, $\partial_\mu T^{\mu\nu} = 0$, for an ideal fluid becomes

$$u_\mu u_\nu \partial^\mu (\rho + P) + (\rho + P)(\partial^\mu u_\mu)u_\nu + (\rho + P)u_\mu \partial^\mu u_\nu + \partial_\nu P = 0. \tag{19.46}$$

In the nonrelativistic limit, $u^\mu \simeq (1, \vec{v})$, $|\vec{v}|^2 \ll 1$, $P \ll \rho$, the $\nu = 0$ component of equation (19.46) gives the continuity equation,

$$\partial_t \rho + \vec{\nabla} \cdot (\rho \vec{v}) = 0, \tag{19.47}$$

whereas the $\nu = i$ component gives the equation

$$\vec{v} \cdot \left(\partial_t \rho + \vec{\nabla} \cdot (\rho \vec{v})\right) + \rho \frac{\partial \vec{v}}{\partial t} + \rho(\vec{v} \cdot \vec{\nabla})\vec{v} + \vec{\nabla} P = 0, \tag{19.48}$$

which, after using the continuity equation, reduces to the Euler equation.

But the same $\nu = 0$ and $\nu = i$ equations are obtained not just in the nonrelativistic limit but also for a more unusual type of fluid, a "null dust," with $u^\mu u_\mu = 0$ (null velocity) and $P = 0$ (no pressure). It is no surprise that under certain conditions, we can write

Maxwell's electromagnetism as the same kind of null fluid. We can indeed rewrite the energy-momentum tensor as the one of the null fluid,

$$T_{00} = \frac{1}{2}(\vec{E}^2 + \vec{B}^2)^2 \equiv \rho$$
$$T_{0i} = [\vec{E} \times \vec{B}]_i \equiv \rho v_i$$
$$T_{ij} = -\left[E_i E_j + B_i B_j - \frac{1}{2}(\vec{E}^2 + \vec{B}^2)\delta_{ij} \right] \equiv \rho v_i v_j. \qquad (19.49)$$

Then it follows that $P = 0$ and

$$\rho \leftrightarrow \frac{1}{2}(\vec{E}^2 + \vec{B}^2); \quad v_i \leftrightarrow \frac{[\vec{E} \times \vec{B}]_i}{\frac{1}{2}(\vec{E}^2 + \vec{B}^2)}. \qquad (19.50)$$

In order for T_{ij} to match as well, we need some consistencty conditions. Since, as we can check, for electromagnetism $T^\mu{}_\mu = 0$, so $T_{00} = T_{ij}\delta^{ij}$, we need $\vec{v}^2 = 1$, so the fluid must be null.

But replacing the velocity in (19.50) in this condition, we find

$$(\vec{E}^2 - \vec{B}^2)^2 + 4(\vec{E} \cdot \vec{B})^2 = 0 \Rightarrow \vec{E} \cdot \vec{B} = 0 \quad \text{and} \quad \vec{E}^2 - \vec{B}^2 = 0, \qquad (19.51)$$

so the electromagnetic configuration must be null, $\vec{F}^2 \equiv (\vec{E} + i\vec{B})^2 = 0$, which means that the Hopfion solution can indeed be imported to the null fluid, as it is null.

After some algebra, one can find that the energy density is

$$\rho = \frac{16\left((t - z)^2 + x^2 + y^2 + 1\right)^2}{\left(t^4 - 2t^2(x^2 + y^2 + z^2 - 1) + (x^2 + y^2 + z^2 + 1)^2\right)^3}, \qquad (19.52)$$

and the velocity is

$$v_x = \frac{2(y + x(t - z))}{1 + x^2 + y^2 + (t - z)^2},$$
$$v_y = \frac{-2(x - y(t - z))}{1 + x^2 + y^2 + (t - z)^2},$$
$$v_z^2 = 1 - v_x^2 - v_y^2. \qquad (19.53)$$

Moreover, one can also calculate and find that the helicity is nonzero, though it diverges. This is, then, an example of a knotted fluid solution. To see that we have a linked toroidal (thus knotted) structure, see the representation of the velocity field Figure 19.1.

Important Concepts to Remember

- The fluid description of an interacting system is an expansion in derivatives.
- For an ideal fluid, there is no dissipation, and the equations are the continuity equation (mass conservation) and the Euler equation (dynamics equation). We also obtain the conservation of the stress tensor Π_{ij}.
- For a first-order dissipative fluid, we add terms to Π_{ij} that are linear in the first derivatives of the velocity, proportional to the shear viscosity η and the bulk viscosity ζ.
- The conservation of the viscous energy-momentum tensor gives the Navier–Stokes equation.

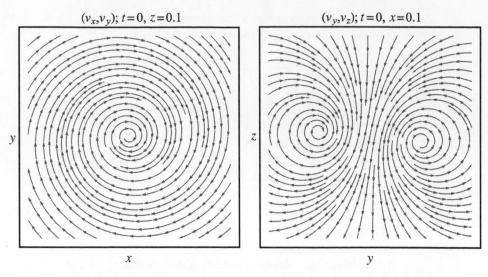

(v_x, v_y); $t = 0$, $z = 0.1$　　　　　　　(v_y, v_z); $t = 0$, $x = 0.1$

Figure 19.1 Two sections of the velocity field of the Hopfion solution in orthogonal directions: velocities in the (x, y) plane (left) and (y, z) plane (right). Rotating the right figure in the (x, y) directions (because of the evident rotational symmetry in the left figure), the linked torus structure is apparent. The figures are slightly shifted from the $x = 0, z = 0$ center.

- In the relativistic case, the equations of motion are the conservation laws for the energy-momentum tensor, $\partial_\mu T^{\mu\nu} = 0$, as well as the conservations of any currents (including the entropy current for the ideal fluid).
- For the ideal fluid, we have $T_{\mu\nu} = \rho u^\mu u^\nu + p(\eta^{\mu\nu} + u^\mu u^\nu)$.
- For an ideal fluid, the helicity $H = \int d^3x \, \vec{v} \cdot \vec{\omega}$ is conserved, where $\vec{\omega} = \vec{\nabla} \times \vec{v}$ is the vorticity.
- Small fluctuations in the fluid propagate at the speed of sound c_s, $c_s^2 = \partial p / \partial \rho$.
- An ideal null fluid, $u^\mu u_\mu = 0$, with $p = 0$, satisfies the same equations as the ideal nonrelativistic fluid.
- Electromagnetism is a null fluid if $\vec{F}^2 = (\vec{E} + i\vec{B})^2 = 0$, so the Hopfion solution can be imported to the fluid.

Further Reading

See, for instance, [12].

Exercises

(1) Find the relativistic form of the Navier–Stokes equation by substituting the relativistic ansatz for $T^{\mu\nu}$ in its conservation equation.

(2)   Prove equation (19.33) from the original Euler equation.
(3)   Calculate the helicity of the fluid Hopfion solution.
(4)   Repeat the procedure that led to the wave equation for small enthalpy per unit mass, h, fluctuations, to find the fluctuation equation for the velocity $\vec{v}$. Is it (or can it be reduced to) the wave equation with velocity c_s?
(5)   Show that the ansatz (19.24), (19.30) satisfies the conservation equation for the current.

PART II

SOLITONS AND TOPOLOGY;
NON-ABELIAN THEORY

Kink Solutions in ϕ^4 and Sine-Gordon, Domain Walls and Topology

In this chapter, we will begin the analysis of classical solutions of field theories character-ized by topological numbers – i.e., solitons. We start with solitons of scalar field theories in 1+1 dimensions, so-called kink solutions in ϕ^4 theories and in the sine-Gordon model. Their trivial extension to 3+1 dimensions will lead to domain walls.

20.1 Setup

We consider real scalar field theories with a canonical kinetic term, i.e.,

$$S = \int d^2x\mathcal{L} = \int dx\, dt\left[-\frac{1}{2}(\partial_\mu\phi)^2 - V(\phi)\right], \qquad (20.1)$$

where $(\partial_\mu\phi)^2 = -\dot{\phi}^2 + \phi'^2$, whose equation of motion, as we saw, is the modified KG equation,

$$\Box\phi = V'(\phi). \qquad (20.2)$$

The canonical conjugate momentum to ϕ is

$$p_\phi = \frac{\partial\mathcal{L}}{\partial\dot{\phi}} = \dot{\phi}. \qquad (20.3)$$

Then the Hamiltonian density is

$$\mathcal{E} = \mathcal{H} = p_\phi\dot{\phi} - \mathcal{L} = \frac{\dot{\phi}^2}{2} + \frac{\phi'^2}{2} + V(\phi). \qquad (20.4)$$

We have expressed it in terms of $\dot{\phi}$ instead of p_ϕ since we want to use it as giving the energy density of a classical solution – i.e., the soliton.

We will also consider a complex scalar field version – where, as we mentioned, we will treat ϕ and ϕ^* as independent – so we have no factor of 1/2 in the action – i.e.,

$$S = \int d^2x\left[-(\partial_\mu\phi)(\partial^\mu\phi^*) - V(\phi,\phi^*)\right]. \qquad (20.5)$$

The canonical conjugate momenta to the variables ϕ, ϕ^* are

$$p_\phi = \frac{\partial\mathcal{L}}{\partial\dot{\phi}} = \dot{\phi}^*; \quad p_{\phi^*} = \frac{\partial\mathcal{L}}{\partial\dot{\phi}^*} = \dot{\phi}, \qquad (20.6)$$

so the Hamiltonian density is

$$\mathcal{E} = \mathcal{H} = p_\phi\dot{\phi} + p_{\phi^*}\dot{\phi}^* - \mathcal{L} = |\dot{\phi}|^2 + |\phi'|^2 + V(\phi,\phi^*). \qquad (20.7)$$

20.2 Analysis of Classical Solutions

The theory has mass scales, so there should be scalar field configurations that move at a speed less than c, so there can be a center of mass frame where these soliton solutions are static. Therefore, we consider static solutions, $\phi = \phi(x)$ only (independent of t), which implies that the modified KG equation becomes

$$\frac{d^2\phi}{dx^2} = \frac{dV}{d\phi}. \tag{20.8}$$

We now observe that this equation is formally the same as the equation for classical motion of a particle in the inverted potential

$$U_{cl} = -V, \tag{20.9}$$

if we replace $\phi(x)$ with the particle trajectory $x(t)$ in $U_{cl}(x)$ – i.e.,

$$\frac{d^2x}{dt^2} = -U_{cl}(x). \tag{20.10}$$

The energy of the static solutions is then

$$E = \int dx \mathcal{E} = \int dx \left[\frac{1}{2}\left(\frac{d\phi}{dx}\right)^2 + V(\phi) \right]. \tag{20.11}$$

Multiplying the static KG equation (20.8) by ϕ' and integrating over x, we obtain

$$\int dx \frac{d}{dx}\left(\frac{\phi'^2}{2}\right) = \int dx \phi'\phi'' = \int dx \frac{d\phi}{dx}\frac{dV}{d\phi} = \int d\phi \frac{dV}{d\phi}. \tag{20.12}$$

But we are interested in solutions that can be interpreted as some states in the theory – i.e., with solutions of *finite energy*, which means that necessarily at $x \to \pm\infty$, $\phi' \to 0$, and $V \to 0$. That means that when integrating the above equation, there is no integration constant, and we obtain

$$\frac{\phi'^2}{2} = V. \tag{20.13}$$

In terms of the classical particle in the inverted potential, this would be the virial theorem, equating the kinetic energy with (minus) the potential energy. Moreover, we can integrate the virial theorem and find an implicit solution, since we get

$$\frac{d\phi}{dx} = \pm\sqrt{2V(\phi)} \Rightarrow x - x_0 = \pm\int_{\phi(x_0)}^{\phi(x)} \frac{d\tilde{\phi}}{\sqrt{2V(\tilde{\phi})}}. \tag{20.14}$$

This is an implicit solution, since performing the integral in a particular potential gives $x = x(\phi)$, which can be inverted to obtain $\phi = \phi(x)$.

The Newtonian particle analogy defined earlier is useful in figuring out whether there exists a soliton solution.

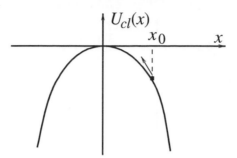

Figure 20.1 Inverted $V = \lambda\phi^4/4!$, now the classical Newtonian problem for $U_{cl}(x) = -\lambda x^4/4!$. If we start at x_0, we can only move a finite time to the top of the potential.

$V(\phi) = \lambda\phi^4/4$ Potential

For instance, consider the case of the ϕ^4 theory, with $\lambda > 0$, needed in order to have a stable theory (otherwise, the potential is negative definite and unbounded from below, so the theory is unstable to perturbations which drive the energy down the potential). The inverted potential then has a maximum at $x = 0$ and is unbounded from below.

A nontrivial – i.e., finite energy – solution corresponds in the particle picture to a particle that ends at rest (zero velocity), $dx/dt \rightarrow 0$ ($d\phi/dx \rightarrow 0$). This is only possible if the particle starts off at some point x_0 (down the potential), with a nonzero velocity such that the virial theorem is satisfied (thus it must start at a finite time t_0), in the direction of $x = 0$, and asymptotically (at $t \rightarrow \infty$) ends up at $x = 0$. But this would correspond to a solution that starts at some finite x_0 and goes to $x \rightarrow +\infty$; see Figure 20.1. That is not quite a soliton, which would be a solution over the whole real line, from $x = -\infty$ to $x = +\infty$.

Higgs Potential

To obtain a soliton solution, defined over the whole real line, we must modify the potential to the so-called Higgs, or Mexican hat (or wine bottle), potential,

$$V(\phi) = \frac{\lambda}{4}\left(\phi^2 - \frac{\mu^2}{\lambda}\right)^2 = \frac{\lambda\phi^4}{4} - \frac{\mu^2\phi^2}{2} + \frac{m^4}{4\lambda}. \tag{20.15}$$

This potential is positive definite (again if $\lambda > 0$), has two minima at $V = 0$, in between them a local maximum at $\phi = 0$, and grows without bound at $\phi \rightarrow \pm\infty$. We observe that $m^2 = -\mu^2 < 0$, but this only means that the theory is unstable around $\phi = 0$. We can define the mass squared locally, as $V''(\phi)$, and $m^2 = -\mu^2 < 0$ only at $\phi = 0$, since the potential has a local maximum there.

The theory has two vacua,

$$\phi_{1,2} = \pm\phi_0 = \pm\frac{\mu}{\sqrt{\lambda}}, \tag{20.16}$$

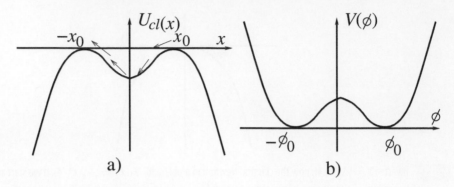

Figure 20.2 Higgs potential: (a) Inverted potential for the analog particle motion. The soliton corresponds to a particle rolling off one maximum with zero initial velocity (at $t = -\infty$), and reaching the other maximum also with zero velocity (at $t = +\infty$. (b) The Higgs potential, with two minima. The soliton corresponds to a field starting at one minimum at $x = -\infty$ and reaching the other at $x = +\infty$.

of zero energy, $V(\phi_{1,2}) = 0$. Around these vacua, $V''(\phi_{1,2}) \equiv m^2(\phi_{1,2}) > 0$, so the theory is stable here, see Figure 20.2b. In the inverted potential, $U_{cl}(x)$ has two maxima at $x_{1,2} \leftrightarrow \phi_{1,2}$ and a local minimum in between at $x = 0$, and it is unbounded from below at $x \to \pm\infty$.

We see that there is a simple way to obtain a soliton: in the particle picture, we have a particle starting at $t = -\infty$ at one of the maxima with zero velocity, but with an infinitesimal perturbation, after which we let it roll until the other maxima, where it ends at $t = +\infty$ again with zero velocity, see Figure 20.2a. Then in the soliton picture, the solution starts at $x = -\infty$ in one vacuum and goes at $x = +\infty$ to the other vacuum.

The soliton solution is found using the general formula (20.14) – which, for the Higgs potential, is

$$x - x_0 = \pm\frac{\sqrt{2}}{\sqrt{\lambda}} \int_{\phi(x_0)}^{\phi(x)} \frac{d\tilde\phi}{\tilde\phi^2 - \mu^2/\lambda} = \pm\frac{\sqrt{2}}{\mu} \int_0^{\phi(x)} \frac{d(\tilde\phi\sqrt{\lambda}/\mu)}{\frac{\lambda\tilde\phi^2}{\mu^2} - 1}$$

$$= \pm\frac{\sqrt{2}}{\mu} \tanh^{-1}\left(\pm\frac{\sqrt{\lambda}\phi}{\mu}\right). \tag{20.17}$$

Here we have chosen $\phi(x_0) = 0$, so the integral is from 0 to $\phi(x)$, and we choose the + sign in front, since arctanh is an odd function, so it is enough to consider the $\pm$ inside it. Then, inverting the formula, we find

$$\phi(x) = \pm\frac{\mu}{\sqrt{\lambda}} \tanh\left[\frac{\mu}{\sqrt{2}}(x - x_0)\right]. \tag{20.18}$$

The solution with a + in front is called the *kink* solution (see Figure 20.3) – which, as we see, is a soliton interpolating between the two vacua, ϕ_1 at $x = -\infty$ and ϕ_2 at $x = +\infty$. The *antikink* is the one with a minus, since we have

$$\phi_{antikink}(x) = -\phi(x) = -\phi_{antikink}(-x). \tag{20.19}$$

We have used the term "soliton" very loosely until now, but we should now define it. Solitons were first observed by John Scott Russel, who observed a soliton water wave in the Union Canal in Scotland, tracking it for a few kilometers and seeing it maintain its shape. The technical definition of solitons is that they keep their shape even when

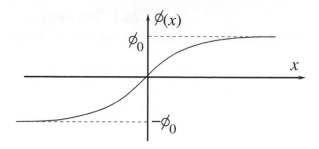

Figure 20.3 The kink solution.

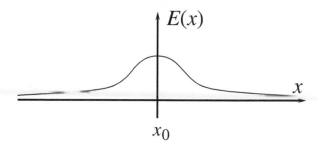

Figure 20.4 The energy density of the kink solution.

interacting with each other, and emerge from a collision unscathed. From this technical point of view, the kink described earlier is not a soliton, since scattering of two solitons changes their shape in general.

But in this book, we will deal with *topological* solitons, which means objects that are characterized by a nonzero topological charge (topological number). Such a number cannot change; hence, these soliton may change their shape, but they still possess the same charge, characterizing them as a nontrivial object in the theory.

The energy density of the kink is

$$\mathcal{E}(x) = \frac{\phi'^2}{2} + V(\phi) = 2V(\phi) = \frac{\lambda}{2}\left(\frac{\mu^2}{\lambda}\tanh^2\left(\frac{\mu}{\sqrt{2}}(x-x_0)\right) - \frac{\mu^2}{\lambda}\right)^2$$
$$= \frac{\mu^4}{2\lambda}\frac{1}{\cosh^4\frac{\mu(x-x_0)}{\sqrt{2}}}, \tag{20.20}$$

where in the second equality we have used the virial theorem proved. The energy density is peaked at x_0, the position of the kink; see Figure 20.4.

The total energy of the solution, which is the classical mass of the static soliton, is the spatial integral of (20.20),

$$M_{\rm cl} = \int_{-\infty}^{+\infty} dx \mathcal{E}(x) = \frac{\mu^3}{\sqrt{2}\lambda}\int_{-\infty}^{+\infty}\frac{d\left(\frac{\mu(x-x_0)}{\sqrt{2}}\right)}{\cosh^4\left(\frac{\mu(x-x_0)}{\sqrt{2}}\right)}$$
$$= \frac{2\sqrt{2}\mu^3}{3\lambda}. \tag{20.21}$$

Note that the numerical integral above equals 4/3, as we can check.

20.3 Topology

The classical solitons we describe now and later are, as we said, defined by topology in field space. Topology refers to quantities (topological numbers) that cannot be changed by continuous processes – i.e., by small deformations. The standard example is the genus of two-dimensional (Euclidean) surfaces, defined by Euler. A sphere or its deformations have genus 0, a torus has genus 1, etc.

Unlike the case of Noether's theorem from Chapter 18, where we have a conserved current and corresponding conserved charge associated with a global symmetry, in the case of topological charges, there is a conserved current that is *automatically* conserved so that its charges cannot be changed by small perturbations.

To construct a conserved current in scalar theories in 1+1 dimensions is easy. With a particular normalization, defined for the specific theory we are describing, the current is

$$j^\mu = \frac{\sqrt{\lambda}}{2\mu} \epsilon^{\mu\nu} \partial_\nu \phi = \epsilon^{\mu\nu} \partial_\nu \frac{\phi}{2\phi_0}. \tag{20.22}$$

Here, the Levi-Civitta tensor in 1+1 dimensions is defined as usual by $\epsilon^{01} = +1$. The current is automatically conserved,

$$\partial_\mu j^\mu = 0, \tag{20.23}$$

because derivatives commute ($\partial_{[\mu}\partial_{\nu]} = 0$).

The charge associated with this current is

$$Q = \int_{S_t} j^\mu d\Sigma_\mu = \int_{-\infty}^{+\infty} dx j^0(x) = \frac{\sqrt{\lambda}}{2\mu} \int_{-\infty}^{+\infty} dx\, \phi'(x)$$

$$= \frac{1}{2}\left[\frac{\phi(x = +\infty)}{\phi_0} - \frac{\phi(x = -\infty)}{\phi_0}\right]. \tag{20.24}$$

As we have already discussed, the energy being finite means that at $x = \pm\infty$, $d\phi/dx \to 0$ and $V \to 0$, which means

$$\phi(x \to \pm\infty) = \pm\frac{\mu}{\sqrt{\lambda}}. \tag{20.25}$$

That means that Q can take only integer values, and more precisely the values $-1, 0, +1$.

- We have $Q = 0$ if $\phi(x \to +\infty) = \phi(x \to -\infty)$ – i.e., trivial vacuum solutions.
- We have $Q = +1$ for the kink solution.
- We have $Q = -1$ for the antikink solution.

Note that this Q is a topological charge – i.e., a topological number in field space. The values for ϕ form an interval, and Q is a kind of "winding number" for this interval, meaning that it counts how many times the field winds on the interval.

A topological number cannot change under small perturbations, which means that a topological soliton, containing a definite charge Q, cannot decay by changing Q. Quantum mechanical decays can occur, if the mass of the soliton equals the sum of masses of other particles or solitons in the theory (for instance, when moving on the "moduli space," the space of parameters, of a supersymmetric theory, when crossing so-called walls in this space) and if the topological charge is conserved.

That is, the topological charge divides classical and quantum states of the theory into "topological sectors" characterized by a given value of the topological charge, and no continuous classical processes or quantum processes can change that. If a decay occurs, the sum of the topological charges of the decay products must equal the topological charge of the original soliton.

It is this stability property of topological solitons that makes them so important in field theory: they act very much as particles, and this is not a coincidence. In many theories, there are *dualities* that exchange particles and solitons, and when one description is perturbative, the other is non-perturbative, and vice versa.

20.4 Domain Walls

Until now, we have discussed solutions in 1+1 dimensions, but we live in 3+1 dimensions. We can trivially lift the kink solution to 3+1 dimensions, with action

$$S = \int d^3x \, dt \, \mathcal{L}, \tag{20.26}$$

and $\mathcal{L}$ is the same Lagrangean, now with ∂_μ in 3+1 dimensions, but where nothing depends on x_2 and x_3. Then we have the same vacua, and the kink solution has

$$\phi(x_1 \to \pm\infty) \to \pm\frac{\mu}{\sqrt{\lambda}}. \tag{20.27}$$

Then kink thus becomes a "domain wall," a wall in space, located at $x_1 = x_{1,0}$ and extended in x_2, x_3, that interpolates between the two different minima on each side of the wall, see Figure 20.5.

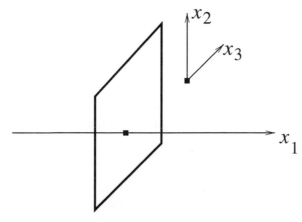

Figure 20.5   The domain wall solution looks like a wall in spacetime.

Of course, such a domain wall now has only a finite energy *density*, $d^2E/(dx_2 dx_3)$, and if the directions x_2, x_3 are truly infinite, the total energy would be infinite as well. But, of course, the infinite domain wall is an abstraction; in reality (for instance, in cosmology), there is always a finite surface. For example, the domain wall could form a closed surface or could end on another type of soliton (for instance, a string). In cosmology, where one actually looks for domain walls that were created close to the Big Bang, the space also expands, making the analysis more complicated.

20.5 Sine-Gordon System

We now consider a more complicated but more interesting system, the sine-Gordon model in 1+1 dimensions, with Lagrangean

$$\mathcal{L} = -\frac{1}{2}(\partial_\mu \phi)^2 + \frac{m^4}{\lambda}\left[\cos\left(\frac{\sqrt{\lambda}}{m}\phi\right) - 1\right]. \tag{20.28}$$

The potential is

$$V(\phi) = \frac{m^4}{\lambda}\left[1 - \cos\left(\frac{\sqrt{\lambda}}{m}\phi\right)\right] = \frac{2m^4}{\lambda}\sin^2\left(\frac{\sqrt{\lambda}}{m}\frac{\phi}{2}\right) \simeq \frac{m^2\phi^2}{2} + \frac{\lambda\phi^4}{4!}, \tag{20.29}$$

so m and λ have the appropriate definitions for mass and four-point coupling. Note that compared with the definition in Chapter 6,

$$V(\phi) = \frac{\alpha}{a^2}[1 - \cos(a\phi)], \tag{20.30}$$

we have $m^2 = \alpha$, $\lambda = \alpha a^2$. The equation of motion,

$$\Box\phi = -\ddot{\phi} + \phi'' = \frac{m^3}{\sqrt{\lambda}}\sin\frac{\sqrt{\lambda}}{m}\phi, \tag{20.31}$$

is a sine-modified KG equation, or "sine-Gordon," hence the name. The minima of the theory, or vacua, for which $V(\phi_N) = 0$, are at

$$\phi_N = \frac{m}{\sqrt{\lambda}}2\pi N \equiv \phi_1 N, \tag{20.32}$$

so, unlike the Higgs potential case, we have an infinite number of vacua.

As before, we can define a topological current, which is actually the same, except the different normalization – used for this particular theory (sine-Gordon),

$$j^\mu = \frac{\sqrt{\lambda}}{4\pi m}\epsilon^{\mu\nu}\partial_\nu\phi = \epsilon^{\mu\nu}\partial_\nu\frac{\phi}{2\phi_0}. \tag{20.33}$$

The topological charge is then

$$Q = \int_{-\infty}^{+\infty} dx\, j^0(x) = \frac{\sqrt{\lambda}}{4\pi m}[\phi(x = +\infty) - \phi(x = -\infty)] = \frac{N_{+\infty} - N_{-\infty}}{2}, \tag{20.34}$$

where $N_{+\infty}$ is the value of N for the vacuum at $x = +\infty$, and $N_{-\infty}$ is the value of N for the vacuum at $x = -\infty$.

We can now construct the static (and localized) soliton, also called the *sine-Gordon soliton*, using the general formula (20.14),

$$x - x_0 = \pm \int_{\phi(x_0)}^{\phi(x)} \frac{d\tilde{\phi}}{\sqrt{2V(\tilde{\phi})}} = \pm \frac{1}{m} \int_{\phi(x_0)}^{\phi(x)} \frac{d(\sqrt{\lambda}\tilde{\phi}/2m)}{\sin\left(\frac{\sqrt{\lambda}}{2m}\tilde{\phi}\right)}$$

$$= \pm \frac{1}{m} \ln \tan \frac{\sqrt{\lambda}\phi}{4m}. \tag{20.35}$$

Here we have chosen $\sqrt{\lambda}\phi(x_0)/4m = \pi/4$, so that $\ln\tan$ gives zero at x_0. Inverting the solution, we find (for the plus sign) the sine-Gordon soliton

$$\phi(x) = \frac{4m}{\sqrt{\lambda}} \tan^{-1}[\exp(m(x - x_0))]. \tag{20.36}$$

The solution interpolates betwen ϕ_{-1} at $x = -\infty$ and ϕ_1 at $x = +\infty$, which means that it has the topological charge $Q = +1$. The antisoliton has the minus sign in front and, thus, $Q = -1$.

But in this case, unlike the Higgs case, the charge Q can take any integer value, $Q \in \mathbb{Z}$. In fact, due to the periodicity of the potential, we can think of ϕ space as being compactified on an S^1. In this case, Q is really the winding number of the field ϕ around this circle.

The classical mass of the soliton is given as the total energy of the fields configuration,

$$M_{\mathrm{cl}} = E = \int dx\, 2V(\phi) = \frac{2m^3}{\lambda} \int_{-\infty}^{+\infty} d(mx)[1 - \cos(4\tan^{-1}(\exp(m(x - x_0))))]$$

$$= \frac{8m^3}{\lambda}. \tag{20.37}$$

Here we have used that the numerical integral equals 4, as we can easily check.

Note also that the sine-Gordon model is *integrable* – i.e., it has an infinite number of integrals of motion, which allows one to solve it completely – in principle: any solution can be written in terms of the whole set of integrals of motion. Note that for a classical system with a finite number of degrees of freedom, integrability means that there is the same number of integrals of motion. For a field theory, there is an infinite number of degrees of freedom, so an infinite number of integrals of motion. In fact, the sine-Gordon model is actually integrable even at the quantum level, which involves a nontrivial extension of the definition of integrability. In particular, it means that one can, in principle, completely solve the scattering of any particles of solitons in the theory. This was done for sine-Gordon in a series of papers from the 1970s and 1980s.

Complex Sine-Gordon Model

There is actually a complex version of the sine-Gordon model, which will be useful later. We consider a complex scalar ϕ, and the potential

$$V(\phi, \phi^*) = |U(\phi)|^2, \tag{20.38}$$

where

$$U(\phi) = m^2 \sqrt{\frac{2}{\lambda}} \sin \frac{\sqrt{\lambda}}{m} \frac{\phi}{2}. \tag{20.39}$$

We see that keeping only the real part of ϕ, we obtain the real sine-Gordon model.

Important Concepts to Remember

- For real scalar field theories in 1+1 dimensions with potential $V(\phi)$, the static solutions $\phi(x)$ formally have the same form as for a classical particle moving in an inverted potential $U_{\rm d} = -V$, if we map $\phi(x) \to x(t)$.
- For finite energy solutions, we have the virial theorem $\phi'^2/2 = V$, which can be integrated to give the classical solution $\phi(x)$.
- $\lambda\phi^4$ theory doesn't have a soliton, but the Higgs potential has a "kink" soliton solution that interpolates between the two vacua of the theory and has a finite mass.
- Topological solitons are characterized by a topological charge Q for a current, in 1+1 dimensions $j^\mu \propto \epsilon^{\mu\nu}\partial_\nu\phi$, that is automatically conserved and that divides the solutions in topological sectors.
- The topology is in field space, and it defines a winding number.
- Kinks lift up to 3+1 dimensions into domain walls, with trivial extension in the two extra spatial dimensions.
- The sine-Gordon model has an infinite number of vacua, $\phi_N = \phi_1 N$, and topological charge $Q \in \mathbb{Z}$.

Further Reading

See chapter 2 in [13].

Exercises

(1)   Consider a potential for a real scalar field ϕ in 1+1 dimensions,

$$V(\phi) = \lambda\phi^4 + \frac{m^2}{2}\phi^2 + D. \tag{20.40}$$

Considering the inverted potential for a classical particle analogy, $U_{\rm cl}(x)$, classify the possible cases for ranges of λ and m^2 that give soliton solutions of different types (starting at $x = -\infty/t = -\infty$ and reaching $x = +\infty(t = +\infty)$).

(2)   Repeat the exercise for *compacton* solutions, which start at $x = x_1 (t = t_1)$ and reach $x = x_2 (t = t_2)$; both are finite.

(3)   Consider a complex sine-Gordon model. Write the topological current and the electric charge current associated with the *kinetic* term. Are they related?

(4)   Classify compacton solutions (see exercise 2) for the (real) sine-Gordon model.

21 The Skyrmion Scalar Field Solution and Topology

In this chapter, we will describe an important topological soliton solution of a $3 + 1$ dimensional scalar theory with scalars living in an $SU(2)$ group, of relevance to particle physics, the Skyrmion. The model is for low-energy QCD, and the soliton has been identified with a baryon – i.e., a proton or neutron.

21.1 Defining the Skyrme Model

QCD, the theory of gluons and quarks, has an approximate $SU(2)_L \times SU(2)_R$ global symmetry acting on the lightest quarks, u and d, and one knows that the theory at low energy is described by three scalar particles, the pions, transforming under the group $SU(2)_V$, which is diagonal subgroup of $SU(2)_L \times SU(2)_R$. The other $SU(2)$ group that is broken is a chiral symmetry (we will not explain this term here), so at low energy we have the important physics of *chiral symmetry breaking*. The exact (as opposed to phenomenologically described) mechanism for that is an important unsolved problem of theoretical physics.

To model this system, we construct a 2×2 scalar matrix Σ transforming under $SU(2)_L \times SU(2)_R$ bi-fundamentally – i.e., the scalars transform under one group from the left and under the other from the right (hence the L, R indices). A complete set of 2×2 matrices is $(\mathbb{1}, \sigma^a)$, where σ^i are the Pauli matrices. Then we can expand Σ in this basis as

$$\Sigma(x) = \sigma(x)\,\mathbb{1} + i\sigma^a \pi^a(x). \tag{21.1}$$

It will turn out that the π^a values will be related to the pions, but let's ignore that for now and consider (σ, π^a) to be just a set of real scalars. Note that $SO(4) \simeq SU(2) \times SU(2)$, so, in fact, the matrix element can be thought of as a Lie algebra element of $SO(4)$ acting on the two-dimensional representation (Weyl spinor, already encountered, and to be explained better later in the book). The transformation law under $SU(2)_L \times SU(2)_R$ is

$$\Sigma \to g_L \Sigma g_R^\dagger. \tag{21.2}$$

A phenomenological model for the chiral symmetry breaking is with a simple kinetic term for Σ and a potential that fixes the modulus of Σ,

$$\mathcal{L}_L = -\frac{1}{4}\,\mathrm{Tr}[\partial_\mu \Sigma \partial^\mu \Sigma^\dagger] - V(\Sigma^\dagger \Sigma). \tag{21.3}$$

This model is called the *linear sigma model*, since it has a canonical kinetic term. The potential V usually chosen is a Higgs potential, studied in more detail in the next chapters,

$$V = -\frac{\mu^2}{4}\,\mathrm{Tr}(\Sigma^\dagger\Sigma) + \frac{\lambda}{16}[\mathrm{Tr}(\Sigma^\dagger\Sigma)]^2. \tag{21.4}$$

What is important is that the minimum of the potential is at a constant value v for σ, so up to an element in a diagonal $SU(2)_V$ group, Σ is constant on this minimum:

$$\Sigma = vU(x), \quad U(x) = \exp\left[\frac{i\sigma^a\pi'^a(x)}{v}\right]. \tag{21.5}$$

Being at the minimum of the potential amounts to a *spontaneous* breaking of the symmetry from $SU(2)_L \times SU(2)_R$ to $SU(2)_V$, whose matrix element is $U(x)$. Indeed, now the transformation law is $U \to g_V U g_V^\dagger$, for $g_V \in SU(2)_V$. Now $\pi'^a(x)$ are massless fields and are identified with the pions of low-energy QCD in the massless limit. Spontaneous symmetry breaking will be studied in more detail later in the book, but for the moment we just say that it means that the vacuum of a theory has less symmetry than the full theory – i.e., choosing a vacuum breaks the symmetry spontaneously.

Replacing Σ at the minimum in the Lagrangean of the linear sigma model, we find the low energy Lagrangean of QCD,

$$\mathcal{L}_{NL} = -\frac{v^2}{4}\,\mathrm{Tr}[\partial_\mu U \partial^\mu U^\dagger]. \tag{21.6}$$

This is called the *nonlinear sigma model* since, while the kinetic term might look canonical, we have the constraint $U^\dagger U = 1$. If we solve it in terms of $\pi'^a(x)$ as presented earlier, the kinetic term for $\pi'^a(x)$ is nonlinear and takes the form of a sigma model, hence the name. We will drop the prime on $\pi^a(x)$ from now on, since the fields describe the pions. The expanded Lagrangean is

$$\mathcal{L}_{NL} \simeq -\frac{1}{2}\partial_\mu\pi^a\partial^\mu\pi^a + \frac{v^{-2}}{6}(\pi^a\pi^a\partial_\mu\pi^b\partial^\mu\pi^b - \pi^a\pi^b\partial_\mu\pi^a\partial^\mu\pi^b) + \cdots \tag{21.7}$$

In fact, from this Lagrangean, we can find that $v = f_\pi$, the pion decay constant. Since in the real life pions have a mass, we can add a mass term to the nonlinear sigma model, of the type

$$\mathcal{L}_{\mathrm{mass}} = f_\pi\,\mathrm{Tr}[MU + M^\dagger U^\dagger], \quad M = \begin{pmatrix} m_u & 0 \\ 0 & m_d \end{pmatrix}. \tag{21.8}$$

In the case in which we consider the masses of the up and down quarks to be approximately the same, we can find the pion mass to be $m_\pi^2 = 2m_u/f_\pi$, and, after subtracting a constant, the mass term becomes

$$\mathcal{L}_{\mathrm{mass}} = \frac{m_\pi^2 f_\pi^2}{2}\,\mathrm{Tr}[U + U^\dagger - 2]. \tag{21.9}$$

The Lagrangean of low energy QCD is then $\mathcal{L}_{NL} + \mathcal{L}_{\mathrm{mass}}$.

But one can ask what kind of nonlinear term we can add to the model such that we obtain nontrivial interactions that can stabilize a solution with topological charge. In fact, it turns out that there are many possibilities, but the one considered by Skyrme is as follows:

$$\mathcal{L}_{\text{Skyrme}} = -\frac{f_\pi^2}{4} \, \text{Tr}[\partial_\mu U \partial^\mu U^\dagger] + \frac{1}{32e^2} \, \text{Tr}([\partial_\mu U U^\dagger, \partial_\nu U U^\dagger][\partial^\mu U U^\dagger, \partial^\nu U U^\dagger]). \quad (21.10)$$

The nonlinear term is called a Skyrme term. We can define the current in the Lie algebra of $SU(2)_V$,

$$R_\mu \equiv \partial_\mu U U^\dagger = \partial_\mu U U^{-1}, \quad (21.11)$$

in terms of which the Lagrangean is (using the cyclicity of the trace, and the fact that $\partial_\mu U^{-1} = -U^{-1}\partial_\mu U U^{-1}$)

$$\mathcal{L}_{\text{Skyrme}} = +\frac{f_\pi^2}{4} \, \text{Tr}[R_\mu R^\mu] + \frac{1}{32e^2} \, \text{Tr}([R_\mu, R_\nu][R^\mu, R^\nu]). \quad (21.12)$$

Note that we can also use $L_\mu = U^{-1}\partial_\mu U$ to write the Skyrme model, and then we obtain the same Lagrangean above, with R_μ replaced by L_μ.

21.2 Analysis of the Model

For simplicity, we choose to scale away the $f_\pi^2/2$ and $1/(2e^2)$ terms to give

$$\mathcal{L}_{\text{Skyrme}} = +\frac{1}{2} \, \text{Tr}[R_\mu R^\mu] + \frac{1}{16} \, \text{Tr}([R_\mu, R_\nu][R^\mu, R^\nu]). \quad (21.13)$$

The equations of motion of the model (with respect to U) are

$$-\partial_\mu \frac{\partial \mathcal{L}}{\partial(\partial_\mu U)} + \frac{\partial \mathcal{L}}{\partial U^{-1}} \frac{\partial U^{-1}}{\partial U} = 0 \Rightarrow$$

$$-\partial_\mu \left(U^{-1} \left(R^\mu + \frac{1}{4}[R^\nu, [R^\mu, R_\nu]] \right) \right) + \partial_\mu U^{-1} \left(R^\mu + \frac{1}{4}[R^\nu, [R^\mu, R_\nu]] \right)$$

$$= -U^{-1}\partial_\mu \left(R^\mu + \frac{1}{4}[R^\nu, [R^\mu, R_\nu]] \right) = 0, \quad (21.14)$$

where we have used $\partial R_\mu / \partial(\partial_\mu U) = U^{-1}$, $\partial R / \partial U^{-1} = \partial_\mu U$ and $dU^{-1} = -U^{-1}dUU^{-1}$. Thus we obtain the Skyrme field equation,

$$\partial_\mu \left(R^\mu + \frac{1}{4}[R^\nu[R^\mu, R_\nu]] \right) = 0. \quad (21.15)$$

Note that this is written in the form of a current conservation,

$$\partial_\mu \tilde{R}^\mu = 0, \quad \tilde{R}^\mu = R^\mu + \frac{1}{4}[R^\nu[R^\mu, R_\nu]]. \quad (21.16)$$

The energy of the system in the static case ($R_0 = \partial_0 UU^{-1} = 0$) is

$$E = \int d^3x \left[-\frac{1}{2} \operatorname{Tr}[R_i R_i] - \frac{1}{16} \operatorname{Tr}([R_i, R_j][R_i, R_j]) \right]$$
$$= \int d^3x \left[-\frac{1}{2} \operatorname{Tr}[L_i L_i] - \frac{1}{16} \operatorname{Tr}([L_i, L_j][L_i, L_j]) \right]. \tag{21.17}$$

In the same static case, the equation of motion is $\partial_i \tilde{R}^i = 0$. Integrating over space and using the Stokes's law,

$$0 = \int_V \partial_i \tilde{R}^i = \oint_{\partial V} \tilde{R}^i dS^i. \tag{21.18}$$

But if the fields go to zero at infinity, $\pi^a(x) \to 0$, $U(x) \to 1$ and $R_i = \partial_i UU^{-1} \to i\partial_i \pi^a \sigma^a \to 0$. Then, moreover,

$$\tilde{R}_i \simeq R_i \simeq i\partial_i \pi^a \sigma^a. \tag{21.19}$$

That means that there is no monopole field at infinity, $\pi^a = c^a/r$, since if there were, then we would get

$$0 = \oint_{S^2_\infty} \frac{c^a \sigma^a n^i}{r^2} n^i dS = 4\pi c^a \sigma^a. \tag{21.20}$$

Thus a static field configuration cannot have a monopole part. But the configuration must be more complicated, as it carries a topological charge.

21.3 Topological Numbers

When describing the Hopfion, we have described a Hopf index, defined in terms of a complex scalar field $\phi(x_1, x_2, x_3)$. We mentioned that if ϕ tends to a constant at infinity, we have effectively compactified the spatial $\mathbb{R}^3$ to S^3 by the addition of the point at infinity. Moreover, the complex plane for ϕ also can be compactified by the stereographic projection $\mathbb{C} \cup \{\infty\} \to S^2$, so the Hopf map is a map from S^3 to S^2. But the (inverse) stereographic projection defines a unit vector $\vec{n}$, $\vec{n}^2 = 1$, from the complex scalar ϕ, so we have a map $\vec{n}(x_1, x_2, x_3): S^3 \to S^2$. Then the Hopf index can be equivalently written as the wrapping number for this map,

$$N = \frac{-i}{24\pi^2} \int d^3x \epsilon^{ijk} \operatorname{Tr}[(\partial_i \vec{n} \cdot \vec{\sigma})(\partial_j \vec{n} \cdot \vec{\sigma})(\partial_k \vec{n} \cdot \vec{\sigma})]. \tag{21.21}$$

Using the fact that $\sigma_i \sigma_j = \delta_{ij} + i\epsilon_{ijk}\sigma_k$, which implies that $Tr[\sigma_{[i}\sigma_j\sigma_{k]}] = 2i\epsilon_{ijk}$, we find

$$N = \frac{1}{12\pi^2} \int d^3x \epsilon^{ijk} \epsilon_{abc} (\partial_i n^a)(\partial_j n^b)(\partial_k n^c). \tag{21.22}$$

But now, a similar topological number, a *winding number*, is found for the $SU(2)-$ valued matrix $U(x)$. Indeed, it is now a map from $\mathbb{R}^3$ to $SU(2)$. But $SU(2) = (SU(2) \times SU(2))/SU(2) \simeq SO(4)/SO(3) = S^3$, and if $U(x)$ is constant at infinity – for instance if it equals 1 ($\pi^a(x) = 0$ at infinity) – then we can again compactify space, $\mathbb{R}^3 \cup \{\infty\} = S^3$, so the $U(x)$ is effectively a map between S^3 and S^3. As such, we can define a winding number, defining how many times one sphere wraps around the other. The winding number is similar to the above Hopf index, and can be defined formally in the same way as the Hopf index is in (21.21), as (since at infinity $R_i \simeq i\partial_i\pi^a\sigma^a$)

$$B = -\frac{1}{24\pi^2}\int d^3x \epsilon^{ijk} \operatorname{Tr}[R_i R_j R_k] = -\frac{1}{24\pi^2}\int d^3x \epsilon^{ijk} \operatorname{Tr}[L_i L_j L_k]. \tag{21.23}$$

This winding number will be identified with the baryon number of QCD, hence the notation B.

21.4 Hedgehog Configuration and Skyrmion Solution

The ansatz for the Skyrmion solution involves a "hedgehog" configuration,

$$U(x) = \exp[if(r)\vec{n} \cdot \vec{\sigma}]; \quad \vec{n} \equiv \frac{\vec{r}}{r}. \tag{21.24}$$

Then we have (by the formula proven in Chapter 3), when discussing the $SU(2)$ group,

$$U(x) = \cos f(r) + i\vec{n} \cdot \vec{\sigma} \sin f(r). \tag{21.25}$$

We then find

$$L_i = U^{-1}\partial_i U = in_i(\vec{n} \cdot \vec{\sigma})\left(f' - \frac{\sin f \cos f}{r}\right) + i\sigma_i \frac{\sin f \cos f}{r} - i\frac{\sin^2 f}{r}\epsilon_{ijk}n_j\sigma_k, \tag{21.26}$$

and, from it,

$$L_i L_i = -f'^2 - 2\frac{\sin^2 f}{r^2}$$

$$\operatorname{Tr}([L_i, L_j][L_i, L_j]) = -16\left[2f'^2\frac{\sin^2 f}{r^2} + \frac{\sin^4 f}{r^4}\right], \tag{21.27}$$

which we leave as an exercise to prove.

Then the energy functional on the ansatz is

$$E = 4\pi\int_0^\infty dr\left[r^2 f'^2 + 2\sin^2 f(1 + f'^2) + \frac{\sin^4 f}{r^4}\right], \tag{21.28}$$

and the equation of motion becomes

$$(r^2 + 2\sin^2 f)f'' + 2rf' + \sin 2f\left(f'^2 - 1 - \frac{\sin^2 f}{r^2}\right) = 0. \tag{21.29}$$

One must solve this equation of motion with some appropriate boundary conditions. It turns out that there is no analytic solution, and one must resort to numerical methods. The boundary condition at infinity is obvious: we must have $f(r) \to 0$ at $r \to \infty$. The boundary at $r = 0$ is found from imposing that the baryon number is an integer. Substituting the ansatz into the baryon number formula, and doing the contractions, out of which only four are nonzero,

$$\epsilon^{ijk}\epsilon_{abc}\delta_i^a\delta_j^b\delta_k^c = 6$$
$$\epsilon^{ijk}\epsilon_{abc}n_i n^a \delta_j^b \delta_k^c = 2$$
$$\epsilon^{ijk}\epsilon_{abc}n_i n^a \epsilon_{jbm}n^m \epsilon_{kcp}n^p = 2$$
$$\epsilon^{ijk}\epsilon_{abc}\delta_i^a \epsilon_{jbm}n^m \epsilon_{kcp}n^p = 2 \,, \tag{21.30}$$

and the other six are zero, we obtain (after some algebra left as an exercise)

$$B = -\frac{2}{\pi}\int_0^\infty dr f' \sin^2 f = -\frac{1}{\pi}\int_0^\infty df(1 - \cos 2f) = -\frac{1}{\pi}f\Big|_0^\infty + \frac{1}{2\pi}\sin 2f\Big|_0^\infty . \tag{21.31}$$

Using the boundary condition at infinity, $f(\infty) = 0$, we find

$$B = \frac{1}{\pi}f(0) - \frac{\sin 2f(0)}{2\pi}. \tag{21.32}$$

We see that in order to obtain a baryon number $B = N$, the boundary condition at 0 must be

$$f(0) = N\pi. \tag{21.33}$$

That is the same condition as from the fact that at $r = 0$, the vector $\hat{n} = \vec{r}/r$ is undefined, so in

$$U(r = 0) = U^{f(0)\hat{n}\cdot\vec{\sigma}} = \cos f(0) + i\hat{n}\cdot\vec{\sigma}\sin f(0) \tag{21.34}$$

we need to cancel the term proportional to $\hat{n}$. So, indeed, we have consistency: for a consistent ansatz, we have an integer baryon number.

Solving asymptotically (imposing an ansatz $f(r) \sim C/r^n$) eq. (21.29), we find that at large r,

$$f(r) \simeq \frac{C}{r^2} + \cdots \tag{21.35}$$

Solving numerically with this boundary condition and $f(0) = \pi$, we find $C \simeq 2.16$.

21.5 Generalizations of the Skyrme Model

An obvious generalization is to consider the group $SU(N)$ instead of $SU(2)$ (here $N = N_f$ is a number of flavors of quarks), i.e., including more than the u, d quarks – for instance, s and perhaps c. Indeed, the Lagrangean (21.10) doesn't know about $SU(2)$ as it is written, so it can be trivially generalized. Then the $SU(2)$ Skyrmion can be embedded into it, but there are also more complicated solitons that will not be shown here.

Another generalization is to add a topological term, a "Wess-Zumino-Witten" term, written as an integral over a five-dimensional space whose boundary is our four-dimensional space, $\partial M_5 = M_4$,

$$S_{WZW} = -\frac{iN_C}{240\pi^2}\int_{M_5} d^5x \epsilon^{\mu\nu\rho\sigma\tau} \text{Tr}[R_\mu R_\nu R_\rho R_\sigma R_\tau] = \int_{M_4 = \partial M_5}(\cdots). \tag{21.36}$$

Here N_C is a number of *colors* of quarks. This term doesn't contribute to the *classical energy* but is important for the quantization of the model. As such, we will not explain anything more about it.

Another "generalization," or rather a specialization, is to restrict in the $SU(2) = S^3$ group to a S^2 equator of the sphere, parametrized by a unit vector $\vec{n}$, $\vec{n}^2 = 1$. The Skyrme field then becomes $U = i\vec{n} \cdot \vec{\sigma}$, and we recover the Hopf map we started from, $\mathbb{R}^3 \cup \{\infty\} \to S^2$, with the Hopf index (21.21).

The Lagrangean of this model, called the *Skyrme-Faddeev model* (proposed by Faddeev in 1975), is then

$$\mathcal{L}_{S-F} = \int d^4x \left[-\partial_\mu \vec{n} \cdot \partial^\mu \vec{n} - \frac{1}{2} (\partial_\mu \vec{n} \times \partial_\nu \vec{n}) \cdot (\partial^\mu \vec{n} \times \partial^\nu \vec{n}) \right]. \tag{21.37}$$

The model has solitons that have two linked strings, with Hopf charge 1.

It was also realized later that there are many higher-order terms other than the Skyrme term that can stabilize the soliton with baryon charge $B = 1$.

One such interesting example is a $SU(2)$ version of the Dirac-Born-Infeld scalar Lagrangean, what we will call the *Skyrme-DBI model*, namely

$$\mathcal{L}_{\text{Skyrme-DBI}} = f_\pi^2 \beta^2 \, \text{Tr} \left[\sqrt{1 + \frac{1}{2\beta^2} L_\mu L^\mu} - 1 \right]. \tag{21.38}$$

Expanding if for large β, we find

$$\mathcal{L}_{\text{Skyrme-DBI}} \simeq \frac{f_\pi^2}{4} \, \text{Tr}[L_\mu L^\mu] - \frac{f_\pi^2}{32\beta^2} \, \text{Tr}[(L_\mu L^\mu)^2] + \cdots \tag{21.39}$$

Compare this with (21.12), with R_μ replaced by L_μ.

The analysis of the model is similar. Using formulas (21.26) and (21.27), we find that the energy of a static configuration of the hedgehog ansatz (21.24) type (for which then the energy is minus the Lagrangean) is

$$E = 8\pi f_\pi^2 \beta^2 \int_0^\infty (1 - R) r^2 dr; \quad R \equiv \sqrt{1 - \frac{1}{\beta^2} \left(\frac{f'^2}{2} + \frac{\sin^2 f}{r^2} \right)}. \tag{21.40}$$

The equation of motion can be obtained by varying the energy functional (which equals the action, on a static configuration) with respect to f, obtaining

$$\left(r^2 \frac{f'}{R} \right)' = \frac{\sin 2f}{R}. \tag{21.41}$$

Explicitly, this becomes

$$\left(r^2 - \frac{1}{\beta^2} \sin^2 f \right) f'' + (2rf' - \sin 2f)$$
$$- \frac{1}{\beta^2} \left(rf'^3 - f'^2 \sin 2f + \frac{3}{r} \sin^2 f - \frac{1}{r^2} \sin 2f \sin^2 f \right) = 0. \tag{21.42}$$

One can find a numerical solution, the *DBI-Skyrmion*, with the same boundary conditions, $f(0) = \pi$ and $f(\infty) = 0$. At large r, substituting a general power law ansatz $f = a/r^n + b/r^{n+m} + c/r^{n+p} + \cdots$ in the equations of motion, one finds

$$f(r) \simeq \frac{a}{r^2} - \frac{a^3}{21r^6} - \frac{a^3}{3\beta^2 r^8} + \cdots,$$ (21.43)

and numerically we can find that for $f(0) = \pi$, the solution has $a \simeq -0.809466$.

Important Concepts to Remember

- QCD has spontaneous chiral symmetry breaking $SU(2)_L \times SU(2)_R \to SU(2)_V$ at low energies.
- A phenomenological model for it has a 2×2 matrix $\Sigma(x) = \sigma(x)\mathbb{1} + i\pi^a(x)\sigma^a$, transforming as $\Sigma \to g_L \Sigma g_R^\dagger$, fixed at low energies to $\Sigma = vU(x), U(x) = \exp[i\pi^{a'}(x)\sigma^a/v]$.
- The low-energy QCD Lagrangean is then the nonlinear sigma model $\mathcal{L} = -f_\pi^2/4 \operatorname{Tr}[\partial_\mu U \partial^\mu U^\dagger]$, to which we can add a mass term.
- The Skyrme model adds to the low-energy action a Skyrme term, written in terms of $R_\mu = \partial_\mu U U^\dagger$ or $L_\mu = U^\dagger \partial_\mu U$, to stabilize soliton solutions.
- There is a winding number for the Skyrme model, for $S^3 \to S^3$, identified with the baryon number, $B = -1/24\pi^2 \int d^3x \epsilon^{ijk} \operatorname{Tr}[L_i L_j L_k]$.
- The Skyrmion solution has a hedgehog ansatz, $U = \exp[if(r)\hat{n} \cdot \vec{\sigma}]$, with boundary conditions (for consistency) $f(\infty) = 0, f(0) = N\pi$, giving a baryon number $B = N$.
- We can write a $SU(N)$ Skyrme model, and add a WZW term. Restricting to an equatorial S^2 leads to the Skyrme-Faddeev model.
- We can replace the Skyrme term with other terms. One possibility is the Skyrme-DBI model.

Further Reading

For Skyrmions, see chapter 9 in [14]. For DBI Skyrmions, see the original paper [15].

Exercises

(1)   Prove the formulas (21.27) for the hedgehog configuration.
(2)   Prove the formula (21.32) for the baryon number B on the hedgehog ansatz.
(3)   Prove the equation of motion (21.42) of the Skyrme-DBI model.
(4)   Prove that the equation of motion (21.42) has the DBI-Skyrmion solution (21.43) at large r.

22 Field Theory Solitons for Condensed Matter: The XY and Rotor Model, Spins, Superconductivity, and the KT Transition

In this chapter, we consider field theoretic models coming from condensed matter that have solitonic solutions, and their physical implications. We will describe the XY and rotor model, then its vortex soliton. We will then show how vortices drive the Kosterlitz–Thouless (KT) phase transition in a simple way, and finally, we will write the Landau–Ginzburg (LG) model for superconductivity and its associated vortices.

22.1 The XY and Rotor Model

We have seen in Chapter 8* that we can consider a theory for spins and, in a certain continuum limit, obtain a simple field theory action. For completeness, we review it here again.

Consider a two-dimensional Heisenberg model, but with *two-dimensional spins*, also known as the XY model (since the spins belong to the *xy* plane),

$$H = -J \sum_{\langle ij \rangle} \vec{S}_i \cdot \vec{S}_j = -J \sum_{\langle ij \rangle} \cos(\theta_i - \theta_j), \tag{22.1}$$

where θ_i is the angle made by $\vec{S}_i$ in the plane and $\langle ij \rangle$ refers to interactions only between nearest neighbors. We have also put $|\vec{S}^2| = 1$ for simplicity. We can also add a kinetic term for the angles, so

$$H = \frac{1}{2} \sum_i \dot{\theta}_i^2 - J \sum_{\langle ij \rangle} \cos(\theta_i - \theta_j). \tag{22.2}$$

In this form, the model is also known as the rotor model, or the $O(2)$ model (since it has $O(2)$ invariance). At the quantum level, we can also replace the conjugate momentum, $\dot{\theta}_i$, with $\partial/\partial(i\theta_i)$ so that $[\theta_i, \dot{\theta}_j] = i\delta_{ij}$, obtaining the quantum rotor model,

$$H = \frac{1}{2} \sum_i \left(\frac{\partial}{i\partial\theta_i} \right)^2 - J \sum_{\langle ij \rangle} \cos(\theta_i - \theta_j). \tag{22.3}$$

For small fields, as we said in Chapter 8*, we obtain just a free relativistic massless scalar field.

We can obtain the same model from another important Hamiltonian in condensed matter, the *bosonic Hubbard model*. It consists of bosons with creation operator $b_i^\dagger$ at site i, and number operator $\hat{n}_i = b_i^\dagger b_i$, with a quadratic kinetic term for $\hat{n}_i$, minimized for a constant boson density n_0, and an interaction in the form of a "hopping term" with strength t, for the boson to hop from site i to nearest neighbor site j – i.e., $b_j^\dagger b_i$ – so

$$H = E_C \sum_i (\hat{n}_i - n_0)^2 - t \sum_{\langle ij \rangle} (b_i^\dagger b_j + b_j^\dagger b_i). \tag{22.4}$$

Here E_C is the energy required to move a boson from site i to infinity, and t is the hopping energy.

We can rewrite the kinetic term by redefining $E_C = U/2$ and $2E_C n_0 = U/2 + \mu$, with μ a chemical potential, to obtain

$$H = \frac{U}{2} \sum_i \hat{n}_i(\hat{n}_i - 1) - \mu \sum_i \hat{n}_i - t \sum_{\langle ij \rangle} (b_i^\dagger b_j + b_j^\dagger b_i). \tag{22.5}$$

If we represent the commutation relation

$$[\hat{n}_i, b_i^\dagger] = b_i^\dagger \tag{22.6}$$

by the identification

$$\hat{n}_i \equiv \frac{\partial}{i \partial \theta_i}; \quad b_i = \sqrt{n_0} e^{i\theta_i}, \tag{22.7}$$

replacing in the Hamiltonian we obtain the quantum rotor model (22.3).

22.2 Field Theory and Vortices

The simple rotor model (22.2) reduces, under the continuum limit for the spins, $\theta_i - \theta_j \to a\vec{\nabla}\theta$, and as we saw in Chapter 8*, the Hamiltonian reduces to the one of a relativistic massless scalar,

$$H = \frac{1}{2} \int d^2x [\dot{\theta}^2 + J(\vec{\nabla}\theta)^2]. \tag{22.8}$$

This field theory contains vortex solutions. Indeed, for static solutions, the equation of motion is just

$$\vec{\nabla}^2 \theta = 0, \tag{22.9}$$

but we need to remember that θ is a periodic variable, $\theta = \theta + 2\pi$, which means that a generic solution of the equations of motion has

$$\oint_C \vec{\nabla}\theta \cdot d\vec{l} = \oint_C d\theta = 2\pi n, \quad n \in \mathbb{Z}. \tag{22.10}$$

If α is the polar angle in two spatial dimensions, for an n-charge static vortex with $n \neq 0$, we have $\theta = n\alpha$, so, since

$$\vec{\nabla} = \frac{\partial}{\partial z}\hat{e}_z + \frac{\partial}{\partial r}\hat{e}_r + \frac{1}{r}\frac{\partial}{\partial \alpha}\hat{e}_\alpha, \tag{22.11}$$

where z is the third direction, perpendicular to the plane, we obtain

$$\vec{\nabla}\theta = \frac{n}{r}\hat{e}_\alpha = \frac{n}{r}\hat{e}_z \times \hat{e}_r. \tag{22.12}$$

The energy of this vortex is thus, replacing the above in H,

$$E = \frac{J}{2}\int(\vec{\nabla}\theta)^2 = \frac{n^2 J}{2}\int_a^R \frac{2\pi r dr}{r^2} = \pi n^2 J \ln\frac{R}{a}. \tag{22.13}$$

Here, a is a small-distance cutoff, and R is the size of the system. For n well-separated vortices we have n instead of n^2 (n times the result for $n = 1$).

22.3 Kosterlitz–Thouless Phase Transition

The previous simple calculation provides the basis for the famous Kosterlitz–Thouless (KT) phase transition. This is a *quantum phase transition* – i.e., one that is driven by quantum fluctuations, not thermal fluctuations. It is a phase transition between a system of bound vortex–antivortex pairs (when single vortices disappear) at low temperature, and a state with unpaired (single) vortices at high temperature. The phase transition line in the temperature vs. some coupling parameter g (like a doping, for instance) is oblique, as in Figure 22.1, so the KT transition also occurs at $T = 0$, between two states of $g > g_C$ and $g < g_C$.

To see the existence of the transition, consider a simple argument. As we saw earlier, for a system of size R with vortices of core size a, we have

$$E \sim \kappa \ln\frac{R}{a}, \tag{22.14}$$

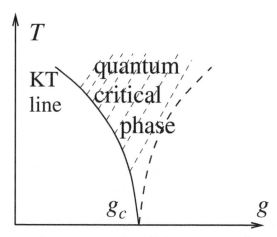

Figure 22.1 The phase diagram of a system with a quantum critical phase, in the (g, T) plane. The Kosterlitz–Thouless (KT) phase transition line is diagonal and starts at $T = 0$ at $g = g_c$.

where κ is a constant depending on the system. If the rotor model is a good description, then $\kappa = \pi J$. The number of possible positions for the vortex is

$$N \sim \frac{R^2}{a^2},\tag{22.15}$$

so the Boltzmann entropy of the vortices is

$$S = k_B \ln N \simeq 2k_B \ln \frac{R}{a}.\tag{22.16}$$

Then the free energy of the system of single vortices is

$$F = E - TS \simeq (\kappa - 2k_B T) \ln \frac{R}{a}.\tag{22.17}$$

That means that there is a phase transition at

$$T_C = \frac{\kappa}{2K_B} \sim \frac{\pi J}{2k_B}.\tag{22.18}$$

The transition is between single vortices at high T (since then the vortex free energy is negative) and bound vortices at low T (since then the vortex free energy is positive, and a smaller free energy is found for bound vortices).

There are many systems that are well described by the rotor or bosonic Hubbard model, and therefore have a KT phase transition, like two-dimensional superfluids and superconductors, two-dimensional crystals, etc.

For a superconductor at nonzero tempetature, Cooper pairs of electrons bind into a scalar field, and the wavefunction becomes macroscopic. Over a macroscopic region, $\psi = |\psi|e^{i\theta}$, and the Hamiltonian density is

$$\mathcal{H} = \left| \left(i\hbar\vec{\nabla} - \frac{e}{c}\vec{A} \right) \frac{\psi}{\sqrt{2m}} \right|^2.\tag{22.19}$$

The macroscopic number density of superconducting Cooper pairs is $n_s^0 = |\psi|^2$, so substituting $\psi = \sqrt{n_s^0}e^{i\theta}$ in (22.19), we obtain *for a two-dimensional model*

$$\frac{F}{k_B T} = \frac{\hbar^2 n_s^0}{2mk_B T} \int d^2r \left(\vec{\nabla}\theta - \frac{e}{\hbar c}\vec{A}(\vec{r}) \right)^2,\tag{22.20}$$

that is, the rotor model coupled to a gauge potential.

The majority of two (spatial) dimensional superconducting models present high T_C superconductivity, and are described as above.

22.4 Landau–Ginzburg Model

In fact, the bosonic Hubbard model in 2+1 dimensions can be shown to lead to a field theory of the form of a *relativistic version of the Landau–Ginzburg model*,

$$S = \int d^{2+1}x \left[-(\partial_t\phi)^2 + v^2|\vec{\nabla}\phi|^2 + (g - g_c)|\phi|^2 + u|\phi|^4 \right],\tag{22.21}$$

where $u = U/2$, though we will not derive it here. We will just mention that the field ϕ here is a continuum limit for a discrete field

$$\phi_i \sim \alpha_i a_i + \beta_i h_i^\dagger, \tag{22.22}$$

and $a_i^\dagger$ is a creation operator for an extra boson around the $\hat{n}_i = n_0$ vacuum, and $h_i^\dagger$ is a creation operator for a "hole" around the same vacuum (an absence of a boson).

However, the standard theory of superconductivity, described by the nonrelativistic (standard) Landau–Ginzburg model in three spatial dimensions, also has vortices.

The Landau–Ginzburg phenomenological model for a type II phase transition with order parameter $|\psi|$ has a free energy density

$$F = F_0 - \alpha|\psi|^2 + \frac{\beta}{2}|\psi|^4, \tag{22.23}$$

and $-\alpha \simeq c(T - T_c)$, so that for $T < T_c$, we are in a phase with nonzero $|\psi|$ at the minimum of F, and for $T > T_C$, we are in a phase with $|\psi| = 0$ at the minimum of F.

For a superconductor, we can write a local free energy density

$$F_s = F_n^0 - \alpha|\psi|^2 + \frac{\beta}{2}|\psi|^4 + \frac{1}{2m}|(i\hbar\vec{\nabla} - ie\vec{A}/c)\psi|^2 - \int \vec{M} \cdot d\vec{B}_a. \tag{22.24}$$

At zero applied magnetic field $\vec{B}_a$ and $\vec{A} = 0$, we can find a "vortex" (or, rather, kink) soliton solution in 1+1 dimensions.

Consider the boundary condition $\psi = 0$ at $x = 0$ so that the system is in the normal (non-superconducting state) at the core, but $|\psi| = |\psi_0| = \sqrt{\alpha/\beta}$, the vacuum solution of the LG model, at $x = \infty$. Then the equation of motion,

$$\left[-\frac{\hbar^2}{2m}\vec{\nabla}^2 - \alpha + \beta|\psi|^2 \right]|\psi| = 0, \tag{22.25}$$

has the solution

$$|\psi| = |\psi_0| \tanh\left(\frac{\sqrt{2}x}{\xi} \right), \quad \xi \equiv \sqrt{\frac{\hbar^2}{2m\alpha}}. \tag{22.26}$$

The solution of the model *coupled to electromagnetism* in 2+1 dimesions is a vortex, which, however, will be described later.

Important Concepts to Remember

- The XY model, or rotor model, or $O(2)$ model, is a model of two-dimensional spins in two dimensions, with nearest neighbor interaction.
- The quantum version of the rotor model can be obtained from the bosonic Hubbard model.
- In the continuum limit, the rotor model gives a free massless relativistic scalar, but with a periodic field variable.
- The model has vortices $\theta = N\alpha$, whose energy is $\pi J \ln R/a$.
- The vortices present a Kosterlitz–Thouless phase transition, between a low-temperature, low-coupling phase of bound (paired) vortices and a high-temperature, high-coupling of single (unpaired) vortices, with transition temperature $T_c = \pi J/sk_B$.

- Two-dimensional superconductors have an effective rotor model coupled to electromagnetism.
- The bosonic Hubbard model reduces to a relativistic Landau–Ginzburg model.
- The nonrelativistic Landau–Ginzburg model for a superconductor has a kink solution $\psi = \psi_0 \tanh$ $(\sqrt{2}x/\xi)$ in 1+1 dimensions, and vortex solutions in 2+1 dimensions.

Further Reading

See, for instance, chapters 9 and 11 in [16].

Exercises

(1) Check that (22.26) is a solution of the equation of motion of the LG model.
(2) Check explicitly that with the identification (22.7), the bosonic Hubbard model (22.5) reduces to the quantum rotor model (22.3).
(3) Define the topological charge for the kink in the LG model. Is it still a topological charge when we introduce a nonzero gauge field?
(4) Calculate approximately the energy of a vortex–antivortex pair, situated at a distance $R \gg a$. Deduce from it that the vortex and antivortex attract each other.

23 Radiation of a Classical Scalar Field: The Heisenberg Model

In this chapter, we will describe the scalar field analog of electromagnetic radiation and radiation from a moving source. We know that electromagnetic radiation is understood classically as a wave carrying away energy and that it can be created by an accelerated source, in particular by an oscillating dipole. Quantum mechanically, of course, the radiation is composed of photons, and in the limit of large number of photons (large bosonic occupation number), we obtain a classical field.

Similarly, the simplest bosonic field, the scalar, can have scalar radiation, a wave carrying away energy, while quantum mechanically it is composed of scalar particles. Again, in the limit of large number of scalar particles (large occupation number), we obtain a classical field.

We will first describe the models we will analyze: the free scalar field, the (polynomially) interacting scalar field, and, finally, the most relevant model, the DBI scalar field. Then we will present the model of Heisenberg for high-energy scattering of nucleons (protons and neutrons), which, as we will see, reduces the scattering of pion field shock waves in a scalar model for the pion. We will define this model to be the DBI model based on the form of the shock waves. Finally, we will show what the radiation from the scattering looks like and its physical interpretation as a wave composed of scalar field particles (pions) in the limit of large numbers of pions.

23.1 Scalar Models

We begin with an analysis of the possible scalar models to consider.

Equations of Motion and Energy Densities

If for an electromagnetic wave the energy density is (for $\mu_0 = \epsilon_0 = 1$)

$$\mathcal{E} = \frac{\vec{E}^2 + \vec{B}^2}{2} ,$$

(23.1)

for a free massive scalar wave, satisfying the KG equation

$$(\Box - m^2)\phi = 0 ,$$

(23.2)

the energy density is

$$\mathcal{E} = \frac{\dot{\phi}^2 + (\vec{\nabla}\phi)^2 + m^2\phi^2}{2}.$$

(23.3)

Considering also a standard $\lambda \phi^4/4$ interaction potential, the equation of motion of the scalar becomes

$$(\Box - m^2 - \lambda \phi^2)\phi = 0 \, , \tag{23.4}$$

and the energy density is

$$\mathcal{E} = \frac{\dot{\phi}^2 + (\vec{\nabla}\phi)^2 + m^2 \phi^2}{2} + \lambda \frac{\phi^4}{4}. \tag{23.5}$$

But finally, we will consider a Lagrangean of the DBI scalar type, with a mass term inside the square root – i.e., with the Lagrangean

$$\mathcal{L} = l^{-4} \left[1 - \sqrt{1 + l^4[(\partial_\mu \phi)^2 + m^2 \phi^2]} \right] . \tag{23.6}$$

Here l is a length scale. The equation of motion coming from this Lagrangean is

$$\frac{-\Box \phi + m^2 \phi}{\sqrt{1 + l^4[(\partial_\mu \phi)^2 + m^2 \phi^2]}} - \partial_\mu \phi \partial_\mu \frac{1}{\sqrt{1 + l^4[(\partial_\mu \phi)^2 + m^2 \phi^2]}} \Rightarrow$$

$$-\Box \phi + m^2 \phi + l^4 \frac{[(\partial_\mu \partial_\nu \phi)(\partial_\mu \phi)\partial_\nu \phi + (\partial_\mu \phi)^2 m^2 \phi]}{1 + l^4[(\partial_\mu \phi)^2 + m^2 \phi^2]} = 0. \tag{23.7}$$

Now the canonical momentum conjugate to ϕ is

$$\pi_\phi = \frac{\partial \mathcal{L}}{\partial \dot{\phi}} = \frac{\dot{\phi}}{\sqrt{1 + l^4[(\partial_\mu \phi)^2 + m^2 \phi^2]}} \, , \tag{23.8}$$

which means that the Hamiltonian density, giving the energy density, is

$$\mathcal{H} = \pi \dot{\phi} - \mathcal{L} = \frac{l^{-4} + (\nabla \phi)^2 + m^2 \phi^2}{\sqrt{1 + l^4[(\partial_\mu \phi)^2 + m^2 \phi^2]}} - l^{-4}. \tag{23.9}$$

Scalar Waves

All these models are relativistic, and the equation of motion starts with $\Box$ acting on ϕ, which means that we can find small fluctuations moving at the speed of light – i.e., scalar waves.

For the free massless KG equation, $\Box \phi = 0$, a planar wave propagating in the positive x direction would be the general solution

$$\phi = \phi(t - x/c). \tag{23.10}$$

Since the $\Box$ operator in cylindrical coordinates is (for $c = 1$)

$$\Box = -\partial_t^2 + \partial_z^2 + \partial_r^2 + \frac{1}{r}\partial_r + \frac{1}{r^2}\partial_\theta^2, \tag{23.11}$$

for a cylindrical wave, propagating in the r direction, when $\Box$ acts only on (r, t), we have the equation of motion

$$\left(-\partial_t^2 + \frac{1}{r}\partial_r + \partial_r^2 \right) \phi(r, t) = 0. \tag{23.12}$$

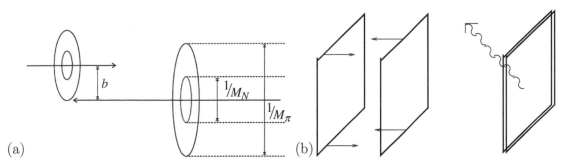

(a) (left) Two nucleons scattering, becoming pancake shaped because of Lorentz contraction. (b) (right) In the limit of high energy, we have two shock waves scattering. When the waves touch, they emit scalar wave radiation.

23.2 Shock Wave Solutions for Relativistic Sources: The Heisenberg Model

In 1952, Heisenberg proposed a simple model for high-energy scattering of nucleons that even now describes well the asymptotics of large energy of the total cross section for scattering in QCD (even though the model was before the definition of QCD). It is a semiclassical model, and we will not describe the particle physics implications that need some quantum mechanics to be analyzed; we will only describe the model's classical aspects.

The basic idea is that a nucleon that interacts at low energy via the pions, which are (pseudo)scalar particles, creates a pion field centered on it. As the nucleon moves with speeds comparable with the speed of light, the pion field moves with it, but, moreover, Lorentz contracts in the direction of motion, becoming pancake shaped. In the limit that the nucleon speed approaches the speed of light, the pancake becomes a shock wave centered on the moving nucleon (see Figure 23.1a). Making the further simplification that the pions can be described by a single real scalar field (even though, in reality, it is at least a triplet of scalars, as in the Skyrme model, and its properties under parity are different: it is a pseudoscalar), we arrive at a shock wave of a real scalar field $\pi(x)$.

Like a shock wave of any kind – fluid, electromagnetic, etc. – this shock wave will generate (classical) radiation, which can be detected. We will calculate the form of the energy spectrum of the radiated wave during the collision (see Figure 23.1b).

The first observation is that for a particle source moving at the speed of light in the x direction, and thus for a shock wave field sourced by it doing the same, π must be a function of the Lorentz invariant

$$s = t^2 - x^2. \tag{23.13}$$

In this case, the solution $\phi = \phi(s)$ is boost invariant: under a boost $x^+ \to e^\beta x^+, x^- \to e^{-\beta} x^-$ ($x^\pm \equiv t \pm x$), so $\phi = \phi(s = x^+ x^-)$ is invariant. But a general classical solution need not be boost invariant. For instance, the electron field of electromagnetism is not. Moreover, the fact that ϕ is a scalar only means that

$$\phi'(x'^+, x'^-) = \phi(x^+, x^-). \tag{23.14}$$

Then, by defining a solution in some reference system, equation (23.14) defines it in another.

What is special about the shock wave solution moving at the speed of light is this: we must have $x^+ = 0$ or $x^- = 0$ be a front of constant ϕ for the solution. In fact, we impose $\phi(x^+ = 0) = a$ or $\phi(x^- = 0) = a$. That means that it must be a power law near the lightcone: for $x^- \sim 0$, we have $\phi - a \propto (x^-)^q$. But then the only way we have the scalar property (23.14) is if we have also the same power of x^+, so

$$\phi(x^+, x^-) = \phi(x^+ x^-) = \phi(s). \tag{23.15}$$

We look for a plane wave shock wave, thus independent of y, z, so $\phi = \phi(s)$ only. In that case,

$$(\partial_\mu \phi)^2 = -(\partial_t \phi(s))^2 + (\partial_x \phi(s))^2 = -4s \left(\frac{d\phi}{ds} \right)^2. \tag{23.16}$$

Now we must decide what is the correct action for the pion in the shock wave limit, but to do so, we must analyze the shape of the resulting shock waves.

Free Scalar Wave

For the free massive scalar field, the Lagrangean becomes

$$\mathcal{L} = \frac{1}{2} \left[4s \left(\frac{d\phi}{ds} \right)^2 - m^2 \phi^2 \right], \tag{23.17}$$

and the KG equation is now

$$4 \frac{d}{ds} \left(s \frac{d\phi}{ds} \right) + m^2 \phi = 0. \tag{23.18}$$

This equation can be put into the form of the Bessel equation of index 0 by the substitution $s = x^2$:

$$\frac{4}{2mx} \frac{d}{mdx} \left(\frac{mx}{2} \frac{d\phi}{mdx} \right) + \phi^2 = 0. \tag{23.19}$$

Then the solution with the boundary condition $\phi(s = 0) = 0$ and, with the causal condition $\phi(s < 0) = 0$, is

$$\phi(s) = a J_0(m\sqrt{s}), \quad s \geq 0$$
$$= 0, \quad s < 0. \tag{23.20}$$

Here, a is an arbitrary constant. Then, expanding near $s = 0$, we have

$$\phi \simeq a(1 - m^2 s + \cdots), \tag{23.21}$$

which means that, at $s > 0$, we have

$$(\partial_\mu \phi)^2 = -4s \left(\frac{d\phi}{ds} \right)^2 \simeq -4s(am^2)^2 \to 0. \tag{23.22}$$

We will argue later that this behavior is not good for the pion shock wave, and we want the result to be finite at the shock.

Interacting Scalar Wave

The next possibility is to add the $\lambda\phi^4/4$ interaction (the only possible *renormalizable* interaction in four dimensions, as one will see and understand what is the relevance in quantum field theory), so the Lagrangean becomes

$$\mathcal{L} = \frac{1}{2}\left[4s\left(\frac{d\phi}{ds}\right)^2 - m^2\phi^2\right] - \frac{\lambda}{4}\phi^4, \tag{23.23}$$

and the equation of motion is now

$$4\frac{d}{ds}\left(d\frac{d\phi}{ds}\right) + m^2\phi + \lambda\phi^3 = 0. \tag{23.24}$$

Now there is no exact solution, but we are mostly interested in the solution near $s = 0$. Solving perturbatively the equation, we obtain the solution

$$\phi = a\left[1 - (m^2 + \lambda a^2)s + \frac{1}{4}(m^2 + 3\lambda a^2)s^2 + \cdots\right], \quad s \geq 0$$
$$= 0, \quad s < 0. \tag{23.25}$$

We leave the proof of this as an exercise. Now, again,

$$(\partial_\mu\phi)^2 = -4s\left(\frac{d\phi}{ds}\right)^2 \simeq -4sa^2(m^2 + \lambda a^2)^2 \to 0. \tag{23.26}$$

DBI Scalar Wave

In fact, Heisenberg understood that any polynomial potential would not work, and in order to get a finite result for $(\partial_\mu\phi)^2$, we need to consider a consistent model with an infinite number of derivative interactions. The model considered is the DBI scalar model, with a mass inside the square root (23.6).

For $\phi = \phi(s)$, the Lagrangean becomes, then,

$$\mathcal{L} = l^{-4}\left[1 - \sqrt{1 + l^4\left(-4s\left(\frac{d\phi}{ds}\right)^2 + m^2\phi^2\right)}\right], \tag{23.27}$$

and the equation of motion (23.7) becomes

$$4\frac{d}{ds}\left(s\frac{d\phi}{ds}\right) + m^2\phi + \frac{8l^4s\left(\frac{d\phi}{ds}\right)^2}{1 + l^4[-4s\left(\frac{d\phi}{ds}\right)^2 + m^2\phi^2]}\left[\frac{d\phi}{ds} - \frac{m^2\phi}{2} + 2s\frac{d^2\phi}{ds^2}\right] = 0. \tag{23.28}$$

But we can simplify this equation. Multiplying by the denominator (which is possible if it doesn't vanish), and canceling and then rewriting some terms, we obtain the simpler form

$$4\frac{d}{ds}\left(s\frac{d\phi}{ds}\right) + m^2\phi = 8sl^4\left(\frac{d\phi}{ds}\right)^2 \frac{\left[\frac{d\phi}{ds} + m^2\phi\right]}{1 + l^4 m^2 \phi^2}. \tag{23.29}$$

For $m = 0$, we can find an exact solution to the equation,

$$\phi = \frac{1}{a}\log\left(1 + \frac{a^2}{2l^4}s + \frac{a}{2l^4}\sqrt{4l^4 s + a^2 s^2}\right), \quad s \ge 0,$$
$$= 0, \quad s < 0, \tag{23.30}$$

as we can check by direct substitution.

For $m \ne 0$, we can again find only a perturbative solution. For small s, we have

$$\phi = \frac{\sqrt{s}}{l^2}(1 + a\, s\, m^2 + \cdots), \quad 0 \le s \ll 1/m^2$$
$$= 0, \quad s < 0, \tag{23.31}$$

whereas for large s, we have a kind of traveling wave solution,

$$\phi \simeq \gamma s^{-1/4} m^{-1/2} \cos(m\sqrt{s} + \delta), \quad s \gg 1/m^2. \tag{23.32}$$

Now however, near $s = 0$, $\phi \simeq \sqrt{s}/s^2$, so

$$(\partial_\mu \phi)^2 = 4s\left(\frac{d\phi}{ds}\right)^2 \simeq -l^{-4} \tag{23.33}$$

is constant, as we wanted (and discontinuous, since at $s < 0$ it vanishes). That means that nonlinearities must play a role at the shock, which is what we expected. As a result, Heisenberg decides that the Lagrangean (23.6) is the correct one to describe the pion field shock wave. In fact, as Heisenberg sketched, and we have described in more detail in [17], the action is quite unique; there are only few possible generalizations.

23.3 Radiation from Scalar Waves

The final assumption made by Heisenberg is that the scalar field shock wave solution corresponds to the classical field of pion radiation created in the collision of two ultrarelativistic (very close to the speed of light) hadron sources. This assumption seems somewhat asymmetrical, since there is a single shock wave, generated by an ultrarelativistic source, but we can assume, for instance, that one source is at rest or that we are in the center of mass, and we consider only the part of the field (given by one hadron) propagating on x^-, not the one (given by the the other) propagating on x^+. In either case, the result is that the pion classical field must be equal to the field of radiation in the collision.

To find the spectrum of radiation, we Fourier transform from x to k the classical solution near the shock (23.31), by

$$\phi(k,t) = l^{-2} \int_0^t dx e^{ikx} \sqrt{t^2 - x^2}(1 + am^2(t^2 - x^2) + \ldots). \qquad (23.34)$$

If we keep only the leading term (put $a = 0$ in (23.34)), we obtain

$$\phi(k,t) \simeq l^{-2} \frac{\pi}{2} \frac{|t|}{|k|} (J_1(|k||t|) + i\mathbf{H_1}(|k||t|)), \qquad (23.35)$$

where J_1 is a Bessel function and $\mathbf{H_1}$ is a Struve function.

This seems perhaps unenlightening, but when expanding at large k, we get

$$\phi - l^{-2} i \frac{|t|}{|k|} \simeq \sqrt{-i} l^{-2} \sqrt{\frac{\pi}{2}} |t|^{1/2} |k|^{-3/2} e^{-i|k||t|} \left(1 + \frac{3}{8|k||t|} e^{2i|k||t|}\right). \qquad (23.36)$$

The term that was subtracted on the left-hand side is a constant (non-oscillatory) piece, which therefore cannot be interpreted as radiation and must be dropped when considering the radiation.

Next we must consider the energy density carried away by the wave. The energy density is (23.9), and we see that it has the square root of the Lagrangean in the denominator. As we saw, near $s = 0$, we have $(\partial_\mu \phi)^2 \sim -l^{-4}$ and $\phi \to 0$, which means that the square root vanishes and the energy density blows up.

But following Heisenberg, we can assume on physical grounds that we have a small perturbation in the shock wave that regulates the divergence, and the denominator becomes a nonvanishing constant. Then, after substituting the form in (23.36), we obtain the spectrum

$$\frac{dE}{dk} \propto k^2 \phi(k)^2 \sim \frac{\text{const.}}{k}. \qquad (23.37)$$

Note that this spectrum is not valid for arbitrary k. Indeed, in reality, the shock wave must have a finite thickness in $\sqrt{s}$ of the order of the Lorentz contracted value of $1/m$ – that is, $\sqrt{s_{0m}} \equiv \sqrt{s_{min}} = \sqrt{1 - v^2}/m$ – which means that (for a sufficiently large time t), $\phi(k,t)$ in (23.36) should be cut off at the relativistic mass of the pion, $k_{0m} = 1/r_{0m} = \gamma m$. Moreover, we have a minimum cutoff in momentum given by the mass of the radiated pion, so the spectrum is valid only for $m \leq l \leq k_{0m} = \gamma m$.

Important Concepts to Remember

- Scalar fields have propagating wave solutions (plane, cylindrical, spherical, ...) just like electromagnetism.
- Nucleons interact at low energies via pions, which form a pion field around them that Lorentz contracts as they move. In the ultrarelativistic limit, we obtain a pion field shock wave.
- The planar pion field shock wave depends only on $s = t^2 - x^2$, $\phi = \phi(s)$ and vanishes at the shock, $\phi(s = 0) = 0$, since by causality $\phi(s < 0) = 0$.
- Both a free and an interacting scalar wave have $\phi(s) \simeq a + \mathcal{O}(s)$, leading to $(\partial_\mu \phi)^2 \to 0$ at the shock.

- A DBI scalar wave has $\phi(s) \simeq \sqrt{s}/l^2$ near the shock, leading to a constant and finite $(\partial_\mu \phi)^2$.
- By going to momentum space, we see that the DBI shock wave has a constant and an oscillatory (radiation) part near the shock.
- The pion field shock wave describes the pion field radiation in a collision of two ultrarelativistic hadrons, and the spectrum is $dE/dk = \text{const.}/k$.

Further Reading

See [17], where the Heisenberg model is generalized, and more details on it are given.

Exercises

(1) Show that the KG equation with $\lambda \phi^3$ interaction has the solution (23.25).
(2) Show that (23.30), (23.31), and (23.32) are solutions of the DBI equations of motion.
(3) Check explicitly by substituting in the equations of motion that the large k (radiative) solution near the shock is (23.36).
(4) If we change the potential from ϕ^4 to ϕ^n in (23.23), find what is the modification to the perturbative solution (23.25).

24   Derrick's Theorem, Bogomolnyi Bound, the Abelian-Higgs System, and Symmetry Breaking

In Chapter 20, we found soliton solutions in the theory of a single real scalar field in $d = 1$ space dimensions. In this chapter, we will consider higher dimensions. First, we will show that there are no solitons (stable, static solutions) for $d > 1$ with a single real scalar field, a statement known as Derrick's theorem. Then, we will describe how to obtain the soliton solution directly from topology, via the so-called Bogomolnyi bound in $d = 1$, a statement that will generalize to $d > 1$. Finally, we move on to the Abelian-Higgs system, in which we will find $d = 2$ solitons in the next chapter.

24.1   Derrick's Theorem

Consider a real scalar field model in d space dimensions, with canonical kinetic term and a potential,

$$\mathcal{L} = -\frac{1}{2}(\partial_\mu \phi)(\partial^\mu \phi) - V(\phi). \tag{24.1}$$

The Hamiltonian (energy) density is

$$\mathcal{H} = \frac{\dot{\phi}^2}{2} + \frac{(\vec{\nabla}\phi)^2}{2} + V(\phi). \tag{24.2}$$

We want to find a soliton, which is a static solution, $\dot{\phi} = 0$, so its energy will be written as

$$E = E_{\text{kin}} + E_{\text{pot}} = \int d^d x \left[\frac{1}{2}(\vec{\nabla}\phi)^2 + V(\phi)\right]. \tag{24.3}$$

Now we modify the shape of the soliton solution by applying a scale transformation on it, $x^i \to \lambda x^i$ – i.e., we consider the rescaled function, or new field configuration,

$$\tilde{\phi}(x) \equiv \phi(\lambda x). \tag{24.4}$$

The energy of the $\tilde{\phi}$ field configuration is

$$E(\lambda) = \int d^d x \left[\frac{1}{2}(\vec{\nabla}\tilde{\phi})^2 + V(\tilde{\phi})\right] \equiv E_{\text{kin}}(\lambda) + E_{\text{pot}}(\lambda). \tag{24.5}$$

But, given that $\tilde{\phi}(x) = \phi(\lambda x)$, by defining $x^i = x'^i/\lambda$, so $\vec{\nabla}_x = \lambda \vec{\nabla}_{x'}$, we obtain

$$E(\lambda) = \lambda^{2-d} \int d^d x' (\vec{\nabla}_{x'}\phi(x'))^2 + \lambda^{-d} \int d^d x' V(\phi(x')) = \lambda^{2-d} E_{\text{kin}} + \lambda^{-d} E_{\text{pot}}. \tag{24.6}$$

217

But our assumption was that the soliton – i.e., stable solution – was at $\lambda = 1$. That means that $\lambda = 1$ must be a minimum of the energy – i.e., that

$$\left. \frac{dE}{d\lambda} \right|_{\lambda=1} = (2 - d)E_{\text{kin}} - dE_{\text{pot}} = 0. \tag{24.7}$$

If $d = 1$, we obtain $E_{\text{kin}} = E_{\text{pot}}$, the virial theorem used in Chapter 20. But for $d = 2$, we obtain $E_{\text{pot}} = 0$, and for $d > 2$, we have that the sum of two negative terms is zero, which implies $E_{\text{kin}} = E_{\text{pot}} = 0$. That means that, indeed, for $d > 1$, we have no static, stable soliton solutions in the theory of a single real scalar field. In order to obtain a soliton solution, we need to add fields to the model.

24.2 Bogomolnyi Bound

But before we do that, we consider a way to obtain the $d = 1$ solitons directly from a topological condition, called the *Bogomolnyi bound*, which we will generalize to $d > 1$.

The energy of a static ($\dot{\phi} = 0$) soliton is

$$E = \int dx \left[\frac{\phi'^2}{2} + V(\phi) \right]. \tag{24.8}$$

While both terms are positive definite ($V(\phi) \geq 0$ for the system to be stable), putting them to zero to minimize the energy just gives the trivial vacuum solution. In order to obtain a nontrivial solution, we will "complete the square" to obtain a square whose vanishing is nontrivial and be left with a term that is topological.

The energy of the static solution can then be written as

$$\begin{aligned} E &= \int dx \left[\frac{1}{2}(\phi' \pm \sqrt{2V(\phi)})^2 \mp \sqrt{2V(\phi)}\phi' \right] \\ &= \int_{-\infty}^{+\infty} dx \left[\frac{1}{2}(\phi' \pm \sqrt{2V(\phi)})^2 \right] \mp \int_{\phi(-\infty)}^{\phi(+\infty)} d\phi \sqrt{2V(\phi)}. \end{aligned} \tag{24.9}$$

We can now define the *superpotential*,

$$W(\phi) \equiv \int d\phi \sqrt{2V(\phi)}, \tag{24.10}$$

so that

$$V(\phi) = \frac{1}{2} \left(\frac{dW}{d\phi} \right)^2. \tag{24.11}$$

That is always possible if $V(\phi) \geq 0$ – i.e., it is positive definite. Note that the term superpotential is borrowed from supersymmetric theories, where we have the same relation between V and W, but here it is defined for an arbitrary theory, not necessarily a supersymmetric one.

Then we have for the energy of a solution (and putting back the term $\dot{\phi}^2/2$ for the case of a time-dependent solution)

$$E = \int_{-\infty}^{+\infty} dx \frac{\dot{\phi}^2}{2} + \int_{-\infty}^{+\infty} dx \left[\frac{1}{2} \left(\phi' \pm \frac{dW}{d\phi} \right)^2 \right] \mp \int_{\phi(-\infty)}^{\phi(+\infty)} dW(\phi(x)). \qquad (24.12)$$

Then the first two terms are positive definite and the last is topological if the solution has finite energy, so it tends to the vacua ϕ_n at infinity, $\phi(\pm\infty) \to \phi_{n,m}$ – i.e., we have the *Bogomolnyi bound* (note that the absolute value is because we have bounds for both signs),

$$E \geq |W(\phi_n(+\infty)) - W(\phi_m(-\infty))|. \qquad (24.13)$$

From it, we can derive directly the soliton solution, by putting the squares to zero, such as to saturate the bound – i.e., $\dot{\phi} = 0$ (static solution) and

$$\phi' = \mp \frac{dW}{d\phi} = \mp\sqrt{2V(\phi)}. \qquad (24.14)$$

Indeed, this is the condition we found in Chapter 20, which we used to solve for the soliton. Note that, unlike the equations of motion, this equation is *first order in derivatives*, another characteristic of equations from the Bogomolnyi bound.

We see that the soliton is the minimum energy solution in a "topological sector," characterized by a topological charge – i.e., given the boundary conditions ϕ_n, ϕ_m at infinity.

Example: The Kink Soliton for the Higgs System

To see how that works, we consider the kink soliton from Chapter 20 for the real scalar Higgs potential

$$V = \frac{\lambda}{4} \left(\phi^2 - \frac{\mu^2}{\lambda} \right)^2 \equiv \frac{1}{2} \left(\frac{dW}{d\phi} \right)^2. \qquad (24.15)$$

Then we find

$$W(\phi) = \frac{\mu^2}{\sqrt{2\lambda}} \phi \left(\frac{\phi^2\lambda/\mu^2}{3} - 1 \right). \qquad (24.16)$$

The energy of the soliton, the topological value, is

$$E_{\text{top}} = |W(\phi_+) - W(\phi_-)| = \frac{\sqrt{2}\mu^2}{3\sqrt{\lambda}} (\phi_+ - \phi_-) = \frac{2\sqrt{2}\mu^3}{3\lambda} Q, \qquad (24.17)$$

where Q is the topological charge that takes the values $-1, 0, +1$, as we showed in Chapter 20. For the kink, $Q_{\text{kink}} = +1$.

We leave the same analysis in the sine-Gordon case as an exercise.

We will see that there exists a Bogomolnyi bound for solitons in higher dimensions too.

As we saw from Derrick's theorem, if we want to find solitons in $d > 1$, we must couple the scalar to other fields, for instance gauge fields A_μ.

24.3 Abelian-Higgs System and Symmetry Breaking

Consider a complex scalar (so as to be possible to couple it to a gauge field; as we saw, a real scalar cannot be coupled) with canonical kinetic term, coupled to a gauge field through the Lagrangean

$$\mathcal{L} = -|D_\mu \phi|^2 - V(|\phi|^2) - \frac{1}{4} F^2_{\mu\nu}, \tag{24.18}$$

where $D_\mu \phi = \partial_\mu \phi - ieA_\mu \phi$. Thus the coupling to A_μ is minimal coupling, and since $V = V(|\phi|)$, the Lagrangean is $U(1)$ invariant. We start with a global $U(1)$ invariance, but by gauging it, we obtain a local $U(1)$ gauge invariance,

$$\phi(x) \to e^{i\alpha(x)}\phi(x); \quad A_\mu(x) \to A_\mu(x) + \frac{1}{e}\partial_\mu \alpha(x). \tag{24.19}$$

We choose a potential that has a continuum of vacua, the *Higgs potential* (see Figure 24.1a),

$$V(\phi) = \frac{\lambda}{2}\left(|\phi|^2 - \frac{\mu^2}{\lambda}\right)^2. \tag{24.20}$$

Note that the factor of $1/2$ difference in the potential (and the kinetic term) is due to the fact that we consider ϕ and ϕ^* as independent variables. We can rewrite the potential as

$$V(\phi) = -\mu^2 \phi^* \phi + \frac{\lambda}{2}|\phi|^4 + \frac{\mu^4}{2\lambda}. \tag{24.21}$$

The first term is needed in order to obtain *symmetry breaking*; the second is needed in order to have $V \geq 0$ at large ϕ, and the last is introduced in order to have $V(\phi) = 0$ at the minima. The vacua – i.e., minima of the potential – are given by

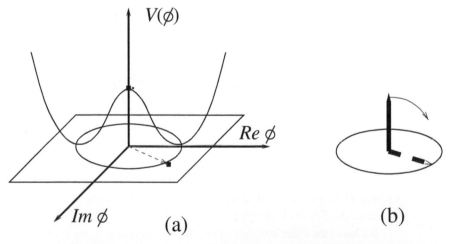

Figure 24.1 (a) The (complex) scalar Higgs potential, in the shape of a Mexican hat or a wine bottle bottom. There is a set of vacua forming a circle, and when choosing one, one spontaneously breaks $U(1)$ symmetry. (b) Symmetry breaking exemplified by a vertical pen falling onto a table, choosing a vacuum out of a circle of possibilities.

$$\phi = \phi_0 e^{i\theta} = \sqrt{\frac{\mu^2}{\lambda}} e^{i\theta} \, , \forall \theta. \tag{24.22}$$

Thus we have a continuum of vacua, continuously connected to one another. Moreover, the global $U(1)$ symmetry, multiplying ϕ by $e^{i\alpha}$, relates the vacua, so they are all equivalent.

The $U(1)$ local symmetry is *broken spontaneously* by the choice of a vacuum. That is, the fact of choosing one of the vacua means that now there is no more $U(1)$ symmetry, though all vacua are equivalent under the symmetry. This is pictorially understood by considering a vertical pen on a table under the action of gravity. When we leave the pen free, any small fluctuation will make it drop in some particular direction on the horizontal plane of the table, thus breaking the rotational symmetry in the plane by choosing a direction, as in Figure 24.1b. This is called spontaneous breaking of symmetry.

Note that all the vacua are equivalent. There is another possibility in principle – to have inequivalent vacua – in which we say we have a *moduli space* or space of parameters for the inequivalent vacua, referenced in previous chapters. In our case, there is no moduli space.

Also note that in the case of a real scalar without a gauge field with the Higgs potential, we had a $\mathbb{Z}_2$ global symmetry, $\phi \leftrightarrow -\phi$, broken spontaneously by the choice of vacuum in between

$$\phi_\pm = \pm\phi_0 = \pm\sqrt{\frac{\mu^2}{\lambda}}. \tag{24.23}$$

In this case also, we have equivalent vacua, but there is no local symmetry, only a global one.

Masses of Fluctuations

The Higgs potential has the shape of a Mexican hat, with a circle of vacua, $\phi = \phi_0 e^{i\theta}$ and a maximum at $\phi = 0$. The fact that the second derivative at 0, which defines the mass squared, is negative,

$$\frac{d^2 V}{d\phi d\phi^*}\bigg|_{\phi=0} \equiv m^2 = -\mu^2, \tag{24.24}$$

only means that $\phi = 0$ is an unstable situation (like the vertical pen on the horizontal table), not that there is some pathology here. Then $\phi = 0$ will roll away in some random direction, choosing one of the possible vacua.

Around the vacuum at $\phi = \phi_0 = \sqrt{\mu^2/\lambda}$, we expand the scalar in small real fluctuations ϕ_1 and ϕ_2 as

$$\phi = \phi_0 + \frac{\phi_1(x) + i\phi_2(x)}{\sqrt{2}} \, , \tag{24.25}$$

so the kinetic term for the complex scalar (at $A_\mu = 0$) becomes the canonical kinetic term for the real scalars ϕ_1 and ϕ_2,

$$-|\partial_\mu \phi|^2 = -\frac{1}{2}[(\partial_\mu \phi_1)^2 + (\partial_\mu \phi_2)^2]. \tag{24.26}$$

The potential term, for small fluctuations ϕ_1, ϕ_2, gives

$$
\begin{aligned}
V(\phi) &= \frac{\lambda}{2}\left[\left|\sqrt{\frac{\mu^2}{\lambda}} + \frac{\phi_1 + i\phi_2}{\sqrt{2}}\right|^2 - \frac{\mu^2}{\lambda}\right]^2 \\
&= \frac{\lambda}{2}\left[\sqrt{\frac{2\mu^2}{\lambda}}\phi_1 + \frac{\phi_1^2 + \phi_2^2}{2}\right]^2 \\
&\simeq \frac{1}{2}(2\mu^2\phi_1^2 + 0 \cdot \phi_2^2) + \mathcal{O}(\phi^3).
\end{aligned}
\tag{24.27}
$$

That means that we can define, around the vacuum at ϕ_0, the masses squared

$$
m_1^2 \equiv \left.\frac{\partial^2 V}{\partial \phi_1^2}\right|_{\phi=\phi_0} = 2\mu^2 > 0, \quad m_2^2 \equiv \left.\frac{\partial^2 V}{\partial \phi_2^2}\right|_{\phi=\phi_0} = 0.
\tag{24.28}
$$

The mass squared m_1^2 of the "Higgs mode," as it is called in 3+1 dimensions, is positive, which means that this is a stable situation, with the potential rising around ϕ_0 in the real direction on both sides.

The fact that the mass squared m_2^2 for the imaginary direction around ϕ_0 is zero means that we have a cost-free motion on the circle in the direction of the continuum of vacua, a so-called Goldstone boson mode. It, indeed, costs no energy to move in the direction of the vacua.

24.4 Unitary Gauge

Gauge invariance is a local symmetry for electromagnetism coupled to the complex scalar, for which we can choose a gauge, thus fixing the symmetry. One useful gauge in the case of the Abelian-Higgs system is the so-called unitary, or physical, gauge. We use the symmetry $\phi(x) \to e^{i\alpha(x)}\phi(x)$ to fix $\phi_2 = 0$ everywhere – i.e., $\phi \in \mathbb{R}$. Then the Lagrangean becomes

$$
\mathcal{L} = -\frac{1}{4}F_{\mu\nu}^2 - (\partial_\mu\phi)^2 - e^2\phi^2 A_\mu A^\mu - V(\phi).
\tag{24.29}
$$

We see that this corresponds to a mass term for the vector, $-\frac{m_V^2}{2}A_\mu A^\mu$, around the vacuum at ϕ_0 – namely,

$$
m_V^2 = 2e^2\phi_0^2.
\tag{24.30}
$$

We can see this in another way as well. If we write $\phi = |\phi|e^{i\theta}$, then

$$
D_\mu\phi = e^{i\theta}(\partial_\mu|\phi| + i\partial_\mu\theta|\phi| - ieA_\mu|\phi|).
\tag{24.31}
$$

We now redefine the gauge field as

$$
A_\mu = A'_\mu + \frac{1}{e}\partial_\mu\theta,
\tag{24.32}
$$

and in this way we get rid of θ in the Lagrangean. Under this transformation,

$$|D_\mu\phi|^2 = (\partial_\mu|\phi|)^2 + e^2 A'_\mu A'^\mu |\phi|^2, \tag{24.33}$$

and in the vacuum the second term gives a vector mass of $m_V^2 = 2e^2\phi_0^2$, and $F'_{\mu\nu} = F_{\mu\nu}$ (since $F_{\mu\nu} = \partial_\mu A_\nu - \partial_\nu A_\mu$), and $V = V(|\phi|)$, so θ disappears from $\mathcal{L}$.

We say that "the vector eats the Goldstone boson mode of the scalar and becomes massive." Note that this is consistent with counting of the degrees of freedom: as we saw, a gauge field (electromagnetism) in 3+1 dimensions has two degrees of freedom, which can be identified with the two directions transverse to the propagation direction $\vec{n}$ of a wave, corresponding to $\vec{E}$ and $\vec{B}$, whereas a massive vector (Proca field) has three degrees of freedom, the extra one being the one made redundant by the local invariance $\delta A_\mu(x) = \partial_\mu \alpha(x)$. Thus the A_μ with two degrees of freedom eats one scalar with one and becomes a Proca field with three degrees of freedom.

The gauge symmetry

$$\delta A_\mu = \partial_\mu \alpha, \quad \delta\phi = |\phi|e^{i\theta+ie\alpha} - |\phi|e^{i\theta} \Rightarrow \delta\theta = e\alpha \tag{24.34}$$

becomes, under the redefinition,

$$\delta A'_\mu = \delta\left(A_\mu - \frac{1}{e}\partial_\mu\theta\right) = 0, \quad \delta|\phi| = 0. \tag{24.35}$$

That means that the gauge invariance is lost – i.e., it acts trivially on the remaining fields – as expected.

24.5 Non-Abelian Higgs System

In the non-Abelian case, one gauges a global non-Abelian symmetry under a Lie group G with Lie algebra generators T_a,

$$\phi_i \rightarrow (e^{i\alpha^a T_a})_{ij}\phi_j, \tag{24.36}$$

by introducing gauge fields A_μ^a, via minimal coupling – i.e., by introducing a covariant derivative

$$D_\mu\phi_i = (\partial_\mu\delta_{ij} - igA_\mu^a(T_a)_{ij})\phi_j. \tag{24.37}$$

Consider that the vacua of the potential $V(\phi)$ are

$$\phi_i = (\phi_0)_i, \tag{24.38}$$

such that, in general, a subgroup $H \subset G$ leaves it invariant. In the Abelian case, the whole group $G = U(1)$ was broken spontaneously, but in general, if H leaves the vacuum invariant, then G is broken spontaneously only to H.

The kinetic term of the scalar becomes, on the vacuum,

$$-\frac{1}{2}|D_\mu\phi_i|^2 = -\frac{1}{2}m_{ab}^2 A_\mu^a A^{b\mu}, \tag{24.39}$$

where the mass matrix is

$$m_{ab}^2 = g^2 (T^a \phi_0)_i (T^b \phi_0)_i. \tag{24.40}$$

The most relevant case is for a gauge group $G = SU(2) \simeq SO(3)$, and a Higgs potential

$$V = \lambda \left(\phi^\dagger \phi - \frac{\mu^2}{\lambda} \right)^2. \tag{24.41}$$

The covariant derivative is written in terms of the generators of the Lie algebra $su(2)$, $\tau^a = \sigma^a / 2$, where σ^a are the Pauli matrices, as

$$D_\mu \phi = (\partial_\mu - ig A_\mu^a \tau^a) \phi, \tag{24.42}$$

where ϕ is in the doublet of $SU(2)$, acted upon by σ^a. We can use the global $SU(2)$ symmetry to fix the vacuum into the form

$$\phi = \frac{1}{\sqrt{2}} \begin{pmatrix} 0 \\ v \end{pmatrix}, \tag{24.43}$$

where $v \in \mathbb{R}$. On this vacuum, the kinetic term for the scalar is

$$|D_\mu \phi|^2 = (D_\mu \phi)^\dagger (D^\mu \phi) = \frac{g^2}{2} \begin{pmatrix} 0 & v \end{pmatrix} \tau^a \tau^b \begin{pmatrix} 0 \\ v \end{pmatrix} A_\mu^a A^{b\mu}. \tag{24.44}$$

Because of the symmetry, we can replace $\tau^a \tau^b$ by $\tau^{(a} \tau^{b)} = \frac{1}{4} \frac{\{\sigma^a, \sigma^b\}}{2} = \delta^{ab}/4$. That means that we obtain a mass term in the Lagrangean,

$$\mathcal{L}_{\text{mass}} = -\frac{g^2 v^2}{8} A_\mu^a A^{a\mu}, \tag{24.45}$$

i.e., a vector mass $m_V = gv/2$.

Important Concepts to Remember

- Derrick's theorem says that there are no purely scalar static, stable solitons in $d > 1$ spatial dimensions, so in order to find solitons, we need more fields.
- The Bogomolnyi bound, obtained by completing squares in the energy in two different ways, says that $E \geq |W(\phi(+\infty)) - W(\phi(+\infty))|$, and the saturation of the bound is for the soliton solution.
- $W(\phi) = \int d\phi \sqrt{2V(\phi)}$ is the superpotential, like in supersymmetric theories, and the vanishing of the other, quadratic, terms gives a set of first-order equations, $\dot{\phi} = 0$ and $\phi' = \mp dW/d\phi$.
- The soliton is the minimum energy solution in a topological sector, defined by $W(\phi_+) - W(\phi_-)$, since $\phi_\pm$ must be vacua in order to have finite energy, so $W(\phi_+) - W(\phi_-) \propto Q_{\text{top}}$.
- For the Higgs potential, the Abelian-Higgs system (complex scalar coupled to Abelian gauge field) has a continuum of vacua, and choosing one breaks spontaneously the $U(1)$ symmetry.
- The vacua are all equivalent under the global symmetry, so we have no moduli space (inequivalent vacua).
- Around $\phi = 0$, the ϕ is unstable ($m^2 < 0$), but around a minimum ϕ_0, we have a stable mode, $m^2 > 0$ ("the Higgs") and a massless mode ("the Goldstone boson").

- The vector A_μ eats the Goldstone boson scalar mode θ and becomes massive, with mass $m_V = \sqrt{2}e\phi_0$, via the redefinition $A_\mu = A'_\mu + \partial_\mu\theta/e$.
- In the non-Abelian case, the symmetry is broken from G to H by the choice of vacua $(\phi_0)_i$.
- In the $SU(2)$ case, $\phi_0 = (0 \quad v)$ and $m_V = gv/2$.

Further Reading

See, for instance, chapter 2 in [13] and chapter 2 in [14].

Exercises

(1) Check that the sine-Gordon soliton saturates the Bogomolnyi bound.

(2) Consider a *complex* version of the scalar model, with

$$V = \frac{1}{2}\left|\frac{dW}{d\phi}\right|^2. \tag{24.46}$$

Rederive the Bogomolnyi bound and check that in the sine-Gordon soliton and kink cases, it is saturated.

(3) Count the number of degrees of freedom to check that it is consistent with the statement that "the vector eats the Goldstone boson mode of the scalar and becomes massive" in the non-Abelian case.

(4) Consider the case $G = SU(2)$ for the non-Abelian Higgs system, and calculate the vacua and mass matrix in a vacuum.

25 The Nielsen-Olesen Vortex, Topology and Applications

After describing in the last chapter the Abelian-Higgs system, in this chapter, we will find the topological soliton in 2+1 dimensions, the vortex. Indeed, a vortex is a solution in a plane, which has field lines that are tangent to circles around a point, and that is what we will find here as well. The solution was found by Holger Bech Nielsen and Poul Olesen; hence, it is usually called the Nielsen-Olesen vortex. Previously, however, there was work in condensed matter by Alexei Abrikosov, who found that there are vortices in type II superconductors, looking like flux tubes from the point of view of 3+1 dimensions, which form a vortex lattice (Abrikosov lattice). Therefore, the vortices are also sometimes called ANO vortices.

We will see that the Abelian-Higgs system also has a Bogomolnyi bound, and the Nielsen-Olesen vortex is the solution that saturates the bound, like in 1+1 dimensions. But unlike the 1+1 dimensional case, the saturation of the bound will occur only in a special limit on the parameters of the theory, called the BPS (after Bogomolnyi, Prasad, and Sommerfield) limit.

25.1 Setup

The action of the Abelian-Higgs system is the same as in the last chapter, considered in 2+1 dimensions,

$$S = \int d^3x \left[-\frac{1}{4}F^2_{\mu\nu} - |D_\mu\phi|^2 - \frac{\lambda}{2}(|\phi|^2 - \phi_0^2)^2 \right], \tag{25.1}$$

where $\phi_0^2 = \mu^2/\lambda$.

We must choose a gauge for electromagnetism, in order to use the Hamiltonian formalism. We choose the Coulomb gauge, $A_0 = 0$, where, therefore,

$$D_0\phi = \dot{\phi}. \tag{25.2}$$

Moreover, we obtain

$$-\frac{1}{4}F^2_{\mu\nu} = -\frac{1}{2}F^2_{12} + \frac{1}{2}F^2_{0i} = -\frac{1}{2}F^2_{12} + \frac{1}{2}\dot{A}_i^2, \tag{25.3}$$

since $F_{0i} = \dot{A}_i - \partial_i A_0 = \dot{A}_i$.

The momenta conjugate to ϕ, ϕ^*, A_i are then

$$p_\phi = \frac{\partial \mathcal{L}}{\partial \dot{\phi}} = \dot{\phi}^*, \quad p_{\phi^*} = \frac{\partial \mathcal{L}}{\partial \dot{\phi}^*} = \dot{\phi}, \quad p_{A_i} = \frac{\partial \mathcal{L}}{\partial \dot{A}_i} = \dot{A}_i. \tag{25.4}$$

The energy is the spatial integral of the Hamiltonian density, so

$$E = \int \mathcal{H} = \int d^2x \left[p_\phi \dot\phi + p_{\phi^*} \dot\phi^* + p_{A_i} \dot{A}_i - \mathcal{L} \right]$$
$$= \int d^2x \left[|\dot\phi|^2 + |D_i\phi|^2 + (\dot{\vec{A}})^2 + \frac{1}{2}F_{12}^2 + \frac{\lambda}{2}(|\phi|^2 - \phi_0^2)^2 \right]. \qquad (25.5)$$

The energy is written as a sum of squares; however, if we were to minimize each term, we would put everything to zero, and we would obtain just a trivial solution.

25.2 Bogomolnyi Bound and BPS Limit

As in the last chapter, we consider static solutions. For the Bogomolnyi bound, the terms $|\dot\phi|^2$ and $(\dot{\vec{A}})^2$ are positive, so the bound will still be true, whereas for finding the vortex solution, we need $\dot\phi = 0 = \dot{\vec{A}}$.

We need to complete squares, like in the last chapter, so that the squares vanishing imply a nontrivial solution, and we are left over with a term that is topological. We complete squares for both F_{12}^2 and $|D_1\phi|^2 + |D_2\phi|^2$. We have the equalities

$$\frac{\lambda}{2}(|\phi|^2 - \phi_0^2)^2 = \frac{\lambda - e^2}{2}(|\phi|^2 - \phi_0^2)^2 + \frac{e^2}{2}(|\phi|^2 - \phi_0^2)^2$$

$$\frac{1}{2}F_{12}^2 + \frac{e^2}{2}(|\phi|^2 - \phi_0^2)^2 = \frac{1}{2}[F_{12} + e(|\phi|^2 - \phi_0^2)]^2 - eF_{12}(|\phi|^2 - \phi_0^2)$$

$$|D_1\phi|^2 + |D_2\phi|^2 = |(D_1 + iD_2)\phi|^2 - i(D_1\phi)^\dagger D_2\phi + i(D_2\phi)^\dagger D_1\phi. \quad (25.6)$$

Moreover, using the fact that

$$2\epsilon^{ij}D_iD_j = \epsilon^{ij}[D_i, D_j] = \epsilon^{ij}[\partial_i - ieA_i, \partial_j - ieA_j] = -ie\epsilon^{ij}F_{ij} = -2ieF_{12}, \qquad (25.7)$$

we have

$$-i\epsilon_{ij}\partial_i(\phi^\dagger D_j\phi) = -i\epsilon_{ij}((D_i\phi)^\dagger - ieA_i\phi^*)D_j\phi - i\phi^\dagger \epsilon^{ij}(D_i + ieA_i)D_j\phi$$
$$= -i\epsilon^{ij}(D_i\phi)^\dagger(D_j\phi) - eF_{12}|\phi|^2$$
$$= -eF_{12}|\phi|^2 - i(D_1\phi)^\dagger D_2\phi + i(D_2\phi)^\dagger D_1\phi. \qquad (25.8)$$

Using this equation in reverse and equations (25.6), and summing them, we obtain

$$|D_1\phi|^2 + |D_2\phi|^2 + \frac{1}{2}F_{12}^2 + \frac{\lambda}{2}(|\phi|^2 - \phi_0^2)^2 - eF_{12}|\phi|^2$$

$$+\frac{e}{2}(|\phi|^2 - \phi_0^2)^2 - i(D_1\phi)^\dagger D_2\phi + i(D_2\phi)^\dagger D_1\phi$$

$$= \frac{\lambda - e^2}{2}(|\phi|^2 - \phi_0^2)^2 + \frac{1}{2}[F_{12} + e(|\phi|^2 - \phi_0^2)]^2 + eF_{12}\phi_0^2$$

$$+|(D_1 + iD_2)\phi|^2 - i\epsilon^{ij}\partial_i(\phi^\dagger D_j\phi)$$

$$-eF_{12}|\phi|^2 + \frac{e}{2}(|\phi|^2 - \phi_0^2)^2 - i(D_1\phi)^\dagger D_2\phi + i(D_2\phi)^\dagger D_1\phi. \qquad (25.9)$$

The last four terms on both sides are the same, so we can cancel them, and then the left-hand side is just the Hamiltonian density for static solutions, which can thus be re-expressed as the right-hand side.

The energy of the solution is, thus,

$$E = \int dx\, dy\, \mathcal{E} = \int dx\, dy\, \left\{ [F_{12} + e(|\phi|^2 - \phi_0^2)]^2 + |(D_1 + iD_2)\phi|^2 \right\}$$

$$+ e\phi_0^2 \int F_{12} dx\, dy - i \int dx\, dy\, \epsilon^{ij} \partial_i (\phi^\dagger D_j \phi) + \frac{\lambda - e^2}{2} \int dx\, dy (|\phi|^2 - \phi_0^2)^2. \quad (25.10)$$

In two spatial dimensions, $F_{ij} = \epsilon_{ij} B$, where B is the magnetic field (since in three spatial dimensions we have $F_{ij} = \epsilon_{ijk} B_k$, and by putting $k = 3$, outside the two dimensions, we obtain the two-dimensional formula), so the first term on the second line of equation (25.10) is the magnetic flux, $\int dx\, dy\, B \equiv \Phi_m$, a topological quantity, which means we are on the right track. The last term will vanish in the BPS limit, so it remains to deal with the boundary term.

But in two dimensions (on the complex plane), we have the Green–Riemann theorem, the two-dimensional version of the general Stokes theorem $\int_D df = \int_{\partial D} f$, namely

$$\int_D dx\, dy \left(\frac{\partial Q}{\partial x} - \frac{\partial P}{\partial y} \right) = \int_{\partial D} (P dx + Q dy), \quad (25.11)$$

or with two-dimensional indices $i, j = 1, 2$,

$$\int_{\partial D} P \equiv \int_{\partial D} P_i dx^i = \int_D \epsilon^{ij} \partial_i P_j\, dx\, dy \equiv \int_D dP, \quad (25.12)$$

where $P_i = (P, Q)$ and $P = P_i dx^i$ is a one-form. Applying it to our boundary term, we find

$$\int_{\mathbb{R}^2} dx\, dy\, \epsilon^{ij} \partial_i (\phi^\dagger D_j \phi) = \int_{\Sigma_\infty} (\phi^\dagger D_j \phi) dx^j. \quad (25.13)$$

But a soliton is a static, finite energy, stable (i.e., minimum of the energy) configuration. As in the previous chapters, the condition of finite energy means that the potential and kinetic terms must vanish at infinity, so ϕ must tend to a vacuum,

$$\phi \to \phi_0 e^{i\theta} \Rightarrow V(\phi) \to 0, \quad (25.14)$$

and the covariant derivative must vanish faster than $1/r$ (so that $\int dx\, dy \simeq 2\pi \int dr\, r$ of $|D_i \phi|^2$ vanishes),

$$|D_i \phi|^2 \to 0 \Rightarrow D_i \phi \to 0. \quad (25.15)$$

Since $D_i \phi \to 0$ faster than $1/r$, we obtain that (the integration element on the circle at infinity is $r d\theta$)

$$\int_{\Sigma_\infty} (\phi^\dagger D_j \phi) dx^j = 0. \quad (25.16)$$

The magnetic flux becomes, using the Green–Riemann formula and polar coordinates

$$z = x + iy = r e^{i\alpha}, \quad (25.17)$$

an integral over the circle at infinity,

$$\frac{\Phi_m}{2\pi} = \frac{1}{2\pi}\int_{\mathbb{R}^2} F_{12}\,dx\,dy = \frac{1}{2\pi}\oint_{C=S_\infty} A_i dx^i = \frac{1}{2\pi}\int_0^{2\pi} A_\alpha(r\to\infty)d\alpha. \tag{25.18}$$

At infinity,

$$\phi = |\phi|e^{i\theta} \to \phi_0 e^{i\theta}, \tag{25.19}$$

so $\partial_i \ln|\phi| \to 0$, and since, as we saw $D_i\phi \to 0$, we obtain also

$$\frac{1}{|\phi|}(\partial_i - ieA_i)|\phi|e^{i\theta} = e^{i\theta}[\partial_i \ln|\phi| - ieA_i + i\partial_i\theta] \to 0, \tag{25.20}$$

so, finally,

$$\partial_i\theta - eA_i \to 0 \Rightarrow \partial_\alpha\theta - eA_\alpha \to 0. \tag{25.21}$$

The magnetic flux of the soliton (finite energy solution) is, therefore,

$$\frac{Q}{2\pi} \equiv \frac{\Phi_m}{2\pi} = \frac{1}{2\pi}\int_{\mathbb{R}^2} F_{12}\,dx\,dy = \frac{1}{2\pi e}\int_0^{2\pi}\partial_\alpha\theta d\alpha = \frac{N}{e}. \tag{25.22}$$

Here, as we see, N is the number of times θ (the phase of the scalar field at infinity) winds around the polar angle α at infinity. But the scalar field ϕ needs to be single valued under a rotation at infinity, so that means that $N \in \mathbb{Z}$, which means that the magnetic flux of the soliton is quantized. In fact, as we will see later, the magnetic flux (or magnetic charge, $\int \vec{B}\cdot d\vec{S} = Q_m$ if $\mu_0 = 1$) is quantized anyway, though here we have obtained it for the soliton only, from the single-valuedness of ϕ.

Finally, then, the energy of the static, finite energy solution becomes

$$E = \int dx\,dy\,\{|(D_1 + iD_2)\phi|^2 + [F_{12} + e(|\phi|^2 - \phi_0^2)]^2\}$$
$$+ 2\pi N\phi_0^2 + \frac{\lambda - e^2}{2}\int dx\,dy(|\phi|^2 - \phi_0^2)^2, \tag{25.23}$$

and, for a non-static solution, we add the positive terms $|\dot\phi|^2$ and $\dot{\vec{A}}^2$. That means that if $\lambda \geq e^2$, we have the Bogomolnyi bound (or BPS bound in this case) for finite energy solutions,

$$E \geq 2\pi N\phi_0^2. \tag{25.24}$$

The right-hand side is a topological quantity. First, N can be interpreted, as we saw, as units of flux, or magnetic charge in 2+1 dimensions, which is quantized and topological (cannot change by small fluctuations). Second, N is also the winding number of the scalar field phase θ at infinity, $\theta = N\alpha$ at $r \to \infty$, which is definitely a topological number.

25.3 BPS Equations and the Vortex Solution

Moreover, in the *BPS limit*, $\lambda \to e^2$, the bound can be saturated by the static solution that puts the perfect square terms on the first line to zero. This will be our vortex solution.

The *BPS equations* putting to zero the perfect squares in this BPS limit $\lambda = e^2$ are

$$(D_1 + iD_2)\phi = 0$$
$$F_{12} + e(|\phi|^2 - \phi_0^2) = 0, \tag{25.25}$$

which are first order in derivatives (as opposed to the equations of motion, which are second order in derivatives), a property shared by the solitons saturating a Bogomolnyi bound, both in 1+1 dimensions and in higher dimensions (the pattern will continue in 3+1 and 4+1 dimensions also).

To find a vortex solution to the BPS equations, we need to consider a *vortex ansatz* for the scalar field. We have seen that at infinity we need $\phi = |\phi|e^{i\theta}$, with $\theta = N\alpha$, but in order to have a minimum energy, it is logical to assume that the same happens *everywhere* – i.e., that we have

$$\phi = |\phi|(r)e^{i\theta} , \quad \theta = N\alpha. \tag{25.26}$$

Then the first BPS equation becomes

$$(\partial_1 + i\partial_2)|\phi| - ie(A_1 + iA_2)|\phi| + i|\phi|(\partial_1 + i\partial_2)\theta = 0, \tag{25.27}$$

and, separating the real and imaginary parts, we obtain

$$A_1 = \frac{1}{e}[\partial_2 \ln|\phi|(r) + \partial_1\theta]$$

$$A_2 = \frac{1}{e}[-\partial_1 \ln|\phi|(r) + \partial_2\theta]. \tag{25.28}$$

Considering that in polar coordinates we have ($\partial_i r = x_i/r$)

$$\partial_1 = \frac{x}{r}\partial_r - \frac{y}{r^2}\partial_\alpha, \quad \partial_2 = \frac{y}{r}\partial_r + \frac{x}{r^2}\partial_\alpha, \tag{25.29}$$

we can write

$$A_1 = \frac{y}{er}\left(\frac{d}{dr}\ln|\phi|(r) - \frac{N}{r}\right)$$

$$A_2 = -\frac{x}{er}\left(\frac{d}{dr}\ln|\phi|(r) - \frac{N}{r}\right), \tag{25.30}$$

or

$$A_i = \frac{\epsilon_{ij}x^j}{r}a(r), \tag{25.31}$$

where

$$ea(r) = \frac{d}{dr}\ln|\phi|(r) - \frac{N}{r}. \tag{25.32}$$

We have thus obtained the gauge fields as a function of the scalar field. But we can then substitute in the second BPS equation in (25.25) the original form (25.28), and obtain a single equation for the scalar field $|\phi|(r)$,

$$-e(|\phi|^2 - \phi_0^2) = F_{12} = \partial_1 A_2 - \partial_2 A_1$$

$$= -\frac{1}{e}(\partial_1^2 + \partial_2^2)\ln|\phi|(r) + \frac{1}{e}\partial_1\partial_2\theta - \frac{1}{e}\partial_2\partial_1\theta$$

$$= -\frac{1}{e}\vec{\nabla}^2 \ln|\phi|(r) + \frac{1}{e}\epsilon^{ij}\partial_i\partial_j\theta, \tag{25.33}$$

or

$$\vec{\nabla}^2 \ln|\phi| = \epsilon^{ij}\partial_i\partial_j\theta + e^2(|\phi|^2 - \phi_0^2). \tag{25.34}$$

One would think that $\epsilon^{ij}\partial_i\partial_j\theta = 0$ because derivatives commute, and that is true everywhere except at $r = 0$. Indeed, since $\theta = N\alpha$ and α (the polar angle) is not well defined at $r = 0$, it can be nonzero there. In other words, it can be proportional to the delta function on the plane,

$$\epsilon^{ij}\partial_i\partial_j\theta = c\delta^2(r). \tag{25.35}$$

To find the proportional constant, we must integrate both sides over a small disk D of vanishing radius $R \to 0$. Then the right-hand side is $= c$, whereas on the left-hand side we use the Green–Riemann formula and obtain

$$c = \int_{C_R}\partial_i\alpha dx^i = \int_0^{2\pi}d\alpha\partial_\alpha\alpha = 2\pi, \tag{25.36}$$

thus,

$$\epsilon^{ij}\partial_i\partial_j\alpha = 2\pi\delta^2(r). \tag{25.37}$$

Then, finally, the equation for the scalar field becomes

$$\vec{\nabla}^2 \ln|\phi|(r) = 2\pi N\delta^2(r) + e^2(|\phi|^2 - \phi_0^2). \tag{25.38}$$

Note that this is an equation derived from the two BPS equations, even though it has two derivatives (because we have substituted one of the BPS equations into the other), so it is only true for the vortex solution (the minimum of the energy). A pictorial representation of the vortex solution is in Figure 25.1.

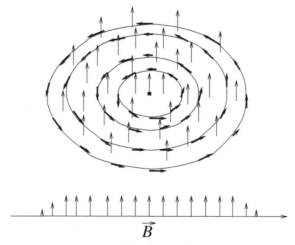

Figure 25.1 Vortex solution: The scalar field is represented together with its direction in field space, making a vortex. The magnetic field is transverse to the plane of the vortex and is limited to the area where the scalar field is nontrivial, as seen in the presentation the follows.

But, since F_{12} is finite (since we saw that the total flux is N) from the second BPS equation in (25.25), we find that $|\phi|^2 - \phi_0^2$ is finite, so if we integrate (25.38) over the small disk D_R of vanishing radius $R \to 0$, we get a vanishing contribution,

$$\int_{D_R} dx\, dy\, \vec{\nabla} \cdot \vec{\nabla} \ln |\phi|(r) = 2\pi N + e^2 \int_{D_R} dx\, dy(|\phi|^2 - \phi_0^2) = 2\pi N + \mathcal{O}(R^2). \quad (25.39)$$

The left-hand side can be calculated using the Green–Riemann formula, as

$$\int_{C_R} dl \frac{d}{dr} \ln |\phi|(r) = 2\pi R \frac{d}{dr} \ln |\phi| \Big|_{r=R}, \quad (25.40)$$

so we obtain

$$r \frac{d}{dr} \ln |\phi| \Big|_{r=R \to 0} = \frac{r}{|\phi|} \frac{d|\phi|}{dr} \Big|_{r \to 0} = N. \quad (25.41)$$

That means that as $r \to 0$, $|\phi| \to 0$ as

$$|\phi| \to A r^N. \quad (25.42)$$

This solution is an N-vortex solution, where the N vortices are coincident. The single vortex solution is for $N = 1$.

One can find the solution of the BPS equations in the next order, and it is, in fact,

$$|\phi| \sim A r e^{-\frac{e^2 \phi_0^2 r^2}{4}} \equiv A r e^{-\frac{m^2 r^2}{8}}, \quad (25.43)$$

where m is the mass of the vector m_V after symmetry breaking (see chapter 24), which for the BPS limit $\lambda = e^2$ equals also the mass of the "Higgs" m_1,

$$m = m_V = \sqrt{2} e \phi_0 = \sqrt{2} e \frac{\mu}{\sqrt{\lambda}} = \sqrt{2} \mu = m_1. \quad (25.44)$$

We leave the proof of the subleading order as $r \to 0$ as an exercise.

To find the behavior at $r \to \infty$, we consider a small fluctuation around the value at infinity $\phi_0 e^{i\theta}$, namely

$$\phi = \phi_0 e^{i\theta} + \frac{\delta\phi}{\sqrt{2}}. \quad (25.45)$$

Substituing this in the action and keeping only the quadratic piece in the small fluctuation $\delta\phi$, we obtain for the potential term

$$-\frac{\lambda}{2}(|\phi|^2 - \phi_0^2)^2 \simeq -\frac{\lambda}{2} 2\delta\phi^2 \phi_0^2 = -\frac{2\mu^2 \delta\phi^2}{2}, \quad (25.46)$$

and for the kinetic term

$$-|D_\mu \phi|^2 \simeq -\left| \frac{\partial_\mu \delta\phi}{\sqrt{2}} - ieA_\mu \left(\phi_0 + \frac{\delta\phi}{\sqrt{2}}\right) \right|^2 \simeq -\frac{(\partial_\mu \delta\phi)^2}{2} + \mathcal{O}(A_\mu^2 \delta\phi), \quad (25.47)$$

for a total quadratic action for $\delta\phi$ of

$$S_\phi \simeq \int d^3x \left[-\frac{1}{2}(\partial_\mu \delta\phi)^2 - \frac{2\mu^2 \delta\phi^2}{2} \right]. \quad (25.48)$$

Its equation of motion is the KG equation, which for a static solution at $r \to \infty$ is ($\vec{\nabla}^2 = \partial_r^2 + 1/r\partial_r$ for spherically symmetric situations in two spatial dimensions)

$$(\Box - m^2)\delta\phi(r) = (\vec{\nabla}^2 - m^2)\delta\phi(r) \simeq (\partial_r^2 - m^2)\delta\phi(r) = 0. \qquad (25.49)$$

Its solution that goes to zero at infinity is $\delta\phi(r) = Be^{-mr}$, so, in total,

$$\phi \to \phi_0 e^{i\theta} + Be^{-mr}. \qquad (25.50)$$

Here, as before, $m = \sqrt{2}\mu = m_1 = m_V = \sqrt{2}e\phi_0$.

We see that the two behaviors at $r \to 0$ and $r \to \infty$ have two free constants, A and B, that were not fixed by the BPS equations in perturbation theory. In fact, we cannot find an analytic solution for the vortex valid over the whole region, so we must resort to numerical methods. One can find the numerical solution, which fixes A and B, by the so-called "shooting method." We impose the behavior at one end – for instance, $r = 0$ – and modify the corresponding coefficient – in this case A – until the numerical solution at the other end – for instance, $r = \infty$ – has the right behavior, and then we also find the other coefficient, B. Thus the numerical solution provides simultaneously A and B.

25.4 Applications

From the behavior at $r = 0$ and the BPS equations, we find that the magnetic field near the core of the vortex is

$$B \equiv F_{12} = -e(|\phi|^2 - \phi_0^2) \to +e\phi_0^2 = \text{const.} \qquad (25.51)$$

Moreover, we see that at $r \sim 1/m$, $|\phi|$ starts to change considerably away from the small r limit, so B also starts to vanish. Therefore, B is nonzero on a radius of order $1/m$ and is zero outside.

As we mentioned at the begining of the chapter, the main application was found before the Nielsen-Olesen analysis, by Abrikosov, and it involves type II superconductors. In general, in superconductors, magnetic fields cannot penetrate too much, since the photon A_μ becomes massive inside it, via the Higgs mechanism, which was described in the previous chapter. They can only penetrate a distance of order of $1/m_{\text{photon}}$. The scalar field of the Abelian-Higgs model is formed of Cooper pairs (two electrons of opposite spins and same momenta pairing up into a scalar) inside the superconductors.

Inside type II superconductors, however, we can have magnetic field penetrating in flux tubes aligned with $\vec{B}$, that in a 2+1 dimensional cross section look like vortices. The parallel flux tubes form a lattice, called Abrikosov lattice, since they interact and repel, so they become equidistant. Note that there are no magnetic monopoles, which in the case of an infinite material would have to be the endpoints of the flux tubes: the magnetic field B of the vortex in the $(1, 2)$ plane is really B_3, perpendicular to the plane, and, as we saw, the solution has a nonzero Φ_m, so a magnetic charge Q_m. But in the case of the finite material *and no external B*, what happens is that the magnetic field lines curve near the surface of the material in such a way that we don't have sources for them (magnetic monopoles); see

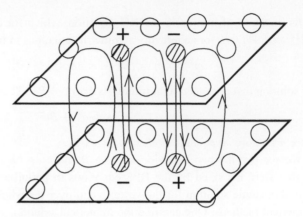

Figure 25.2 Flux tubes in a type II superconductor form an Abrikosov vortex lattice. A cross section of the material would show an array (lattice) of vortices. The field lines in the flux tubes curve near the surfaces of the material, in such a way as not to have enpoints for the flux lines, which would necessarily be magnetic monopoles in the absence of an external B. The resulting constraint in the absence of an external B is that nearby vortices (flux tubes) alternate their charges (positive vs. negative), for a zero total flux.

Figure 25.2. If there is an external magnetic field, we create a vortex lattice of total flux equal to the total flux of the external magnetic field through the material.

Important Concepts to Remember

- At infinity, finite energy solutions have $\phi \to \phi_0 e^{i\theta}$ and $D_i\phi \to 0$, implying $A_\alpha = \partial_\alpha\theta/e$, and θ winds an integer number of times around α at infinity.
- The Abelian-Higgs system has a Bogomolnyi bound for $\lambda - e^2 \geq 0$, $E \geq 2\pi N\phi_0^2$, where N is the magnetic charge, or flux, by $\Phi_m = 2\pi N/e$, and winding number for θ around α at infinity.
- The vortex solution is a static solution that saturates the Bogomolnyi bound in a given topological sector and solves the (first order in derivatives) BPS equations in the BPS limit $\lambda = e^2$.
- The vortex ansatz is $\phi(r,\alpha) = |\phi|(r)e^{iN\alpha}$ for an N-vortex solution, and we find a second-order equation for $|\phi|$ with a delta function source at $r = 0$.
- At $r \to 0$, the N-vortex solution is $\phi \sim Ar^N$, and at $r \to \infty$, $\phi \to \phi_0 e^{i\theta} + Be^{-mr}$, where $m = m_1 = m_v$ is the common mass for the Higgs and vector fields in the BPS limit $\lambda = e^2$.
- The magnetic field and the nontrivial scalar are concentrated in a radius $r \sim 1/m$ around the vortex core.
- In a type II superconductor, there are vortices, extended to flux tubes in 3+1 dimensions, oriented along an external magnetic field $\vec{B}$, and forming an Abrikosov lattice.

Further Reading

See chapter 7 in [14].

Exercises

(1) Prove that near the core of the vortex, $r \to 0$, we have

$$|\phi| \sim A r e^{-\frac{m^2 r^2}{8}}, \tag{25.52}$$

where $m = \sqrt{2} e \phi_0 = \sqrt{2} \mu = m_1 = m_V$.

(2) Find the sub-subleading behavior $f(r)$ of the vortex at $r \to \infty$,

$$\phi \to \phi_0 e^{i\theta} + B e^{-mr} + f(r). \tag{25.53}$$

(3) Use the BPS equations, integrated over the plane, together with the condition for saturation of the BPS bound for the magnetic flux, to estimate the size of vortex.

(4) Consider N BPS vortices separated from each other, so that there are N delta functions on the plane in (25.38). Show how the vortex ansatz is modified and how (25.38) is now consistent with the BPS equations.

26 Non-Abelian Gauge Theory and the Yang–Mills Equation

In this chapter, we will see how to write a gauge theory for a non-Abelian gauge group. The theory was developed by C. N. Yang and Robert Mills in 1954 to explain strong interactions, but simply from a theoretical perspective. There was no way to preserve gauge invariance with a mass term, as was understood must be the case in strong interactions. It took until the development of the Higgs effect in non-Abelian gauge theory that it was understood that the mass comes from spontaneous symmetry breaking, and the theory was considered useful for strong interactions. At the beginning, it was just an interesting theoretical problem being solved: how to gauge a non-Abelian gauge theory and find the corresponding pure non-Abelian gauge theory.

26.1 Non-Abelian Gauge Groups

We consider a non-Abelian symmetry Lie (continuous) group G, with generators T_a, $a = 1, \ldots, N_G$. The group element is

$$U = e^{i\alpha^a T_a} \in G. \tag{26.1}$$

The basic step that is not entirely obvious from the Abelian case but can be guessed is that the gauge field must belong to the Lie algebra, $A_\mu = A_\mu^a T_a$. In the Abelian case, the group element is $U = e^{i\alpha(x)}$, but A_μ transforms as $\delta A_\mu = \partial_\mu \alpha$. Strictly speaking, in this case, there is no adjoint representation, since $f_{abc} = 0$ ($[A_\mu, A_\nu] = 0$).

But in the non-Abelian case, the gauge field is in the adjoint representation – i.e., the representation has an index a – defined by

$$(T_a)_{bc} = i f_{abc}. \tag{26.2}$$

Here, f_{abc} are the structure constants, appearing in the commutation relations for the group (the Lie algebra),

$$[T_a, T_b] = i f_{ab}{}^c T_c. \tag{26.3}$$

As seen from (26.3), we use the convention with Hermitian generators, $(T_a)^\dagger = T_a$. To fix the normalization of the generators, and thus in particular to fix f_{abc}, we use the trace in the representation R of the product of generators, which must be, by invariance arguments, proportional to δ_{ab},

$$\text{Tr}_R[T_a T_b] = T_R \delta_{ab}. \tag{26.4}$$

Here, as denoted, T_R depends on the representation. That means that fixing the normalization just means fixing it in some representation, in which case T_R is fixed in any other representation.

A useful convention that we will adopt since we will mostly deal with gauge fields is $T_R = 1$ for $R =$ the adjoint representation for $SU(N)$, so $\mathrm{Tr}_{\mathrm{adj.}}(T_a T_b) = \delta_{ab}$.

Note that f_{abc} is totally antisymmetric. Indeed, besides the obvious $f_{abc} = -f_{bac}$ from its definition, given the relation $\mathrm{Tr}_{\mathrm{adj}}(T_a T_b) = \delta_{ab}$, we can also write

$$
\begin{aligned}
f_{abc} &= \mathrm{Tr}([T_a, T_b]T_c) = \mathrm{Tr}(T_a T_b T_c) - \mathrm{Tr}(T_b T_a T_c) = \mathrm{Tr}(T_c T_a T_b) - \mathrm{Tr}(T_a T_c T_b) \\
&= \mathrm{Tr}([T_c, T_a]T_b) = f_{cab},
\end{aligned}
\tag{26.5}
$$

where we have used the cyclicity of the trace. We will be using this relation ($f_{abc} = f_{cab}$) implicitly many times later.

Another important number characterizing the representation is the "Casimir of the representation R," C_R, the quadratic form made up of T_a, which is constant as a matrix – i.e.,

$$
\sum_a (T_R^a)^2 = C_R \, \mathbb{1}.
\tag{26.6}
$$

Also, defining N_R as the dimension of the representation R, we have a relation between the constants T_R, C_R, N_G and N_R,

$$
T_R N_G = C_R N_R.
\tag{26.7}
$$

In the adjoint representation, $N_R = N_G$, and

$$
T_R = C_R \equiv C_2(G),
\tag{26.8}
$$

the second Casimir of the group.

We will mostly use the group $SU(N)$, as it is the most relevant for the real world. Indeed, in the Standard Model of particle physics, we have the gauge group $SU(3) \times SU(2) \times U(1)$. In this case, since for $U(N)$ the Lie algebra is $N \times N$ matrices, for $SU(N)$ (which has one constraint on the Lie algebra), we have $N_G = N^2 - 1$. For the fundamental representation of $SU(N)$, $N_R = N$.

26.2 Coupling the Gauge Field to Other Fields

In order to have a nontrivial theory, we need to couple the gauge fields to other fields, since the pure gauge theory is not observable. The real world is made up of fermions (and, to a much lesser degree, scalars), so, in order to observe something, we need to couple the gauge fields to them.

Consider fields ϕ_i, transforming in a representation R of the gauge group G as

$$
\phi_i(x) \to U_{ij}\phi_j(x).
\tag{26.9}
$$

At the infinitesimal level, we have

$$U = e^{ig\alpha^a T_a} \simeq 1 + ig\alpha^a T_a + \cdots \tag{26.10}$$

We want to *gauge this symmetry* – i.e., make it local, $\alpha^a = \alpha^a(x)$ – by coupling to a gauge field, like in the Abelian case. We could, of course, just apply the Noether procedure described before: find the (global) Noether current j_μ^a, and couple it to a gauge field A_μ^a; then add terms to the action and transformation laws until we find invariance.

But, guided by the Abelian case, we know what we should do to short-cut the process. We can try replacing the normal derivatives ∂_μ, which now can also act on $\alpha^a(x)$ – not only on fields – with *covariant derivatives* D_μ, such that the fields are minimally coupled to the gauge field A_μ; that is, replace

$$\partial_\mu \phi_i(x) \to (D_\mu)_{ij}\phi_j(x). \tag{26.11}$$

Based on the Abelian case, we guess a similar formula, just that now multiplying A_μ^a by $(T_a)_{ij}$ in order to act correctly on the representation R, so

$$(D_\mu)_{ij} = \delta_{ij}\partial_\mu - ig(T_a^R)_{ij}A_\mu^a(x). \tag{26.12}$$

Note that for a field ϕ in the adjoint representation, where $(T_a)_{bc} = if_{abc}$, we have

$$(D_\mu)_{ab} = \delta_{ab}\partial_\mu + gf_{abc}A_\mu^c. \tag{26.13}$$

The point of introducing a covariant derivative is that it should transform *covariantly* – i.e., as ϕ itself –

$$(D_\mu)_{ij}\phi_j \to U_{ik}(D_\mu)_{kj}\phi_j. \tag{26.14}$$

We see that, by itself, $\partial_\mu \phi_i$ doesn't since

$$\partial_\mu \phi_i \to \partial_\mu(U\phi)_i = U_{ij}\partial_\mu\phi_j + (\partial_\mu U_{ij})\phi_j. \tag{26.15}$$

But now we must transform A_μ also:

$$A_\mu(x) \to A_\mu^U(x). \tag{26.16}$$

And we want to find this A_μ^U, such that

$$(D_\mu\phi)_i \to U_{ij}(D_\mu\phi)_j = U_{ij}(\partial_\mu\phi_j - ig(A_\mu)_{jk}\phi_k). \tag{26.17}$$

On the other hand, the extra term in the covariant derivative transforms as

$$-ig(T_a^R)_{ij}A_\mu^a\phi_j \to -ig(T_a^R)_{ij}A_\mu^{Ua}U_{jk}\phi_k. \tag{26.18}$$

In order for the sum of the terms to transform correctly, we see that we need to have

$$-igA^U(x) = U(\partial_\mu - igA_\mu)U^{-1}, \tag{26.19}$$

so

$$A^U(x) = U(x)A_\mu(x)U^{-1}(x) - \frac{i}{g}(\partial_\mu U)U^{-1}. \tag{26.20}$$

With this transformation law for A_μ, then, the $D_\mu\phi$ transforms covariantly into $U(D_\mu\phi)$. That means that the kinetic term for the scalar is the obvious generalization of the globally symmetric kinetic term,

$$-\frac{1}{2}\partial_\mu(\phi^i)^\dagger\partial^\mu\phi^i = -\frac{1}{2}\operatorname{Tr}[(\partial_\mu\phi)^\dagger(\partial^\mu\phi)], \qquad (26.21)$$

namely,

$$-\frac{1}{2}(D_\mu\phi^i)^\dagger D_\mu\phi^i = -\frac{1}{2}\operatorname{Tr}[(D_\mu\phi)^\dagger(D^\mu\phi)]. \qquad (26.22)$$

Indeed, in this case, the transformation of the kinetic term is (using the fact that for $SU(N)$ and $SO(N)$, $U^{-1} = U^\dagger$)

$$\operatorname{Tr}[(D_\mu\phi)(D^\mu\phi)^\dagger] \to \operatorname{Tr}[U(D_\mu\phi)(D^\mu\phi)^\dagger U^{-1}] = \operatorname{Tr}[(D_\mu\phi)^\dagger(D^\mu\phi)], \qquad (26.23)$$

so it is gauge invariant.

26.3 Pure Yang–Mills Theory

We want now to find a gauge invariant kinetic term for A_μ^a, since we have already found the gauge transformation.

Based on the Abelian case, we know that we need to find a field strength for A_μ^a first. The difference is that now we only want to find a field strength that *transforms covariantly*, which, in this case, means (as $F_{\mu\nu} \equiv F_{\mu\nu}^a T_a$)

$$F_{\mu\nu} \to U F_{\mu\nu} U^{-1}. \qquad (26.24)$$

Based on the covariant derivative in the adjoint representation case, $(D_\mu)_{ab}$, we guess

$$F_{\mu\nu}^a = \partial_\mu A_\nu^a - \partial_\nu A_\mu^a + g f^a{}_{bc} A_\mu^b A_\nu^c. \qquad (26.25)$$

Note, however, that $F_{\mu\nu}^a \neq D_\mu A_\nu^a - D_\nu A_\mu^a$, since this one would have a factor of 2 in the nonlinear term. We define as for the gauge field

$$F_{\mu\nu} \equiv F_{\mu\nu}^a T_a, \qquad (26.26)$$

and using the fact that $f_{abc} = -i(T_a)_{bc}$, we write the field strength as

$$F_{\mu\nu} = \partial_\mu A_\nu - \partial_\nu A_\mu - ig[A_\mu, A_\nu]. \qquad (26.27)$$

Then we find, for the first term, the transformation law

$$\begin{aligned}
\partial_\mu A_\nu - \partial_\nu A_\mu &\to \partial_\mu(UA_\nu U^{-1}) - \partial_\nu(UA_\mu U^{-1}) - \frac{i}{g}\partial_\mu(\partial_\nu U \cdot U^{-1}) + \frac{i}{g}\partial_\nu(\partial_\mu U \cdot U^{-1}) \\
&= U(\partial_\mu A_\nu - \partial_\nu A_\mu)U^{-1} + (\partial_\mu U A_\nu - \partial_\nu U A_\mu)U^{-1} \\
&\quad + U(A_\nu\partial_\mu U^{-1} - A_\mu\partial_\nu U^{-1}) - \frac{i}{g}(\partial_\nu U\partial_\mu U^{-1} - \partial_\mu U\partial_\nu U^{-1}),
\end{aligned} \qquad (26.28)$$

and, for the second term, using the fact that $\partial_\mu UU^{-1} + U\partial_\mu U^{-1} = 0$ (from $UU^{-1} = \mathbb{1}$ by derivation), we find

$$
\begin{aligned}
- ig[A_\mu, A_\nu] \to &- ig[UA_\mu U^{-1} - \frac{i}{g}\partial_\mu U \cdot U^{-1}, UA_\nu U^{-1} - \frac{i}{g}\partial_\nu U \cdot U^{-1}] \\
= &- igU[A_\mu, A_\nu]U^{-1} - (\partial_\mu UA_\nu - \partial_\nu UA_\mu)U^{-1} \\
&+ U(A_\mu \partial_\nu U^{-1} - A_\nu \partial_\mu U^{-1}) + \frac{i}{g}\partial_\mu U(-\partial_\nu U^{-1}) - \frac{i}{g}\partial_\nu U(-\partial_\mu U^{-1}).
\end{aligned}
$$
(26.29)

Summing them up, we find, indeed,

$$
F_{\mu\nu} \to UF_{\mu\nu}U^{-1} = U(\partial_\mu A_\nu - \partial_\nu A_\mu - ig[A_\mu, A_\nu])U^{-1}.
$$
(26.30)

The kinetic term for the gauge field is then constructed just like in the Abelian case, except with the trace that was there in the scalar field case also – i.e.,

$$
\mathcal{L}_{YM} = -\frac{1}{4}\operatorname{Tr}[F_{\mu\nu}F^{\mu\nu}].
$$
(26.31)

Indeed, since the trace acts only on the matrix indices in the adjoint representation, and $\operatorname{Tr}_{\text{adj}}[T_a T_b] = \delta_{ab}$, we find

$$
\mathcal{L}_{YM} = -\frac{1}{4}F^a_{\mu\nu}F^{b\mu\nu}\delta_{ab}.
$$
(26.32)

This Lagrangean is gauge invariant, since (using the cyclicity of the trace)

$$
\mathcal{L}_{YM} \to \operatorname{Tr}[UF_{\mu\nu}U^{-1}UF^{\mu\nu}U^{-1}] = \operatorname{Tr}[F_{\mu\nu}F^{\mu\nu}].
$$
(26.33)

It also is an obvious extension of the Abelian case – though, as we mentioned, we cannot say that it reduces to the Abelian case, since for $U(1)$ there is no adjoint representation.

Infinitesimal Transformation

For an infinitesimal transformation, the group element becomes (using $(T_a)^\dagger = T_a$)

$$
U(x) \simeq 1 + ig\alpha^a(x)T_a \Rightarrow U^{-1} \simeq 1 - ig\alpha^a(x)T_a.
$$
(26.34)

The gauge field then transforms as

$$
\begin{aligned}
\delta(A^a_\mu(x)T_a) &\simeq (1 + ig\alpha^b T_b)(A^a_\mu T_a)(1 - ig\alpha^a T_a) - A^a_\mu T_a - \frac{i}{g}\partial_\mu(1 + ig\alpha^a T_a)U^{-1} \\
&\simeq \partial_\mu \alpha^a T_a + ig[\alpha^b T_b, A^a_\mu T_a],
\end{aligned}
$$
(26.35)

and finally, defining $\alpha \equiv \alpha^a T_a$,

$$
\delta A_\mu = \partial_\mu \alpha - ig[A_\mu, \alpha] \equiv D_\mu \alpha,
$$
(26.36)

or, in components,

$$
\delta A^a_\mu(x) = \partial_\mu \alpha^a + gf^a{}_{bc}A^b_\mu \alpha^c \equiv D_\mu \alpha^a.
$$
(26.37)

In form language, defining first as in the Abelian case,

$$A \equiv A_\mu dx^\mu, \quad F \equiv \frac{1}{2}F_{\mu\nu}dx^\mu \wedge dx^\nu, \tag{26.38}$$

we find

$$F = dA - igA \wedge A. \tag{26.39}$$

Taking the exterior derivative of it, we find

$$dF = -i2gdA \wedge A \Rightarrow DF \equiv dF + 2igdA \wedge A = 0. \tag{26.40}$$

The factor of 2 in the nonlinear term in the covariant derivative is because there are two adjoint fields (gauge fields) on which the covariant derivative can act. Note that d is a one-form, so it gets a minus when it hops past another one-form, so $d(A \wedge A) = dA \wedge A - A \wedge dA$, however, $A = A^a T_a$ and $A \wedge A = A^{[a} \wedge A^{b]} T_{[a} T_{b]}$, meaning that $A \wedge dA = A^a \wedge dA^b T_{[a} \wedge T_{b]} = -dA^b \wedge A^a T_{[a} T_{b]} = +dA \wedge A$, hence the factor of 2.

The infinitesimal gauge transformation is

$$\delta A = D\alpha = d\alpha - ig[A_\mu, \alpha]. \tag{26.41}$$

The action in form language is

$$S_{YM} = -\frac{1}{4}\int d^4x \, \mathrm{Tr}[F \wedge *F]. \tag{26.42}$$

26.4 The Yang–Mills Equation

We now write the classical equations of motion of the pure Yang–Mills theory. The action is rewritten as

$$\begin{aligned} S_{YM} &= -\frac{1}{4}\int d^4x \, \mathrm{Tr}[F_{\mu\nu}F^{\mu\nu}] = -\frac{1}{4}\int d^4x F^a_{\mu\nu}F^{a\mu\nu} \\ &= -\frac{1}{4}\int d^4x (\partial_\mu A^a_\nu - \partial_\nu A^a_\mu + gf^a{}_{bc}A^b_\mu A^c_\nu)^2 \equiv -\frac{1}{4}\int d^4x [D_\mu A^a_\nu - \partial_\nu A^a_\mu]^2 \\ &= -\frac{1}{4}\int d^4x [(D_\mu A_\nu)(D^\mu A^\nu) + (\partial_\mu A_\nu)(\partial^\mu A^\nu) - 2(D_\mu A_\nu)(\partial^\nu A^\mu)], \end{aligned} \tag{26.43}$$

which is a covariant version of what we wrote in the Maxwell case. Varying with respect to A_μ, we find the equation of motion

$$D_\mu F^{a\mu\nu} = 0. \tag{26.44}$$

We can guess this form, as the equation of motion, since the equation of motion has to be a gauge covariant nonlinear generalization of the Maxwell equations of motion, is for observables. More explicitly, it is

$$\partial_\mu F^{a\mu\nu} + gf^a{}_{bc}A^b_\mu F^{c\mu\nu} = 0. \tag{26.45}$$

But like in the Maxwell case, the equation of motion is only half the Yang–Mills equations (the equivalent of the Maxwell equations at the non-Abelian case); the other half is the Bianchi identities, coming from the fact that we write the field strength F as a function of the gauge field A. As we saw, $F = dA + gA \wedge A$ implies $DF = dF + 2gA \wedge dA$, so the Bianchi identities are

$$DF \equiv dF - i2gA \wedge F = 0 \Rightarrow D_{[\mu}F^a_{\nu\rho]} = 0, \tag{26.46}$$

where we have used the fact that $A \wedge A \wedge A = 0$. In components, we get

$$\partial_{[\mu}F^a_{\nu\rho]} + gf^a{}_{bc}A^b_{[\mu}F^c_{\nu\rho]} = 0. \tag{26.47}$$

The commutator of two covariant derivatives gives the field strength, just like in the Maxwell case. Indeed,

$$[D_\mu, D_\nu] = [\partial_\mu - igT_aA^a_\mu, \partial_\nu - igT_bA^b_\nu]$$
$$= -ig(\partial_\mu A^a_\nu - \partial_\nu A^a_\mu)T_a - g^2 if_{ab}{}^c T_c A^a_\mu A^b_\nu = -igF^a_{\mu\nu}T_a. \tag{26.48}$$

The gauge field is subject to gauge invariance, so, like in the case of Maxwell theory, we must fix a gauge. We can use the same gauges like in the linear case – for instance, the *Lorenz gauge*, now $\partial^\mu A^a_\mu = 0$ – but in so doing, we now cannot reduce anymore the equations to the KG equation; the Yang–Mills equations are still nonlinear. We could, of course, consider some nonlinear gauge – for instance, adding some nonlinear terms to the Lorenz gauge – but we still will obtain a nonlinear system.

We can also add a source term to the Yang–Mills equations, like we did for the Maxwell case. The source will correspond to some matter system coupling with the Yang–Mills fields, like the electron source for the Maxwell equations. They will give a current j^a_μ, with interacting term $\int d^4x j^{a\mu}A^a_\mu$ in the action, and the equation of motion is then modified to

$$D_\mu F^{a\mu\nu} + j^{a\nu} = 0. \tag{26.49}$$

In the Maxwell case, $\partial_\mu F^{\mu\nu} + j^\nu = 0$ implies automatically current conservation by acting with ∂_ν and using that the derivatives commute. In the Yang–Mills case, this doesn't work. That means that $\partial_\mu j^{a\mu} \neq 0$, so the current is not conserved.

The point is that in a locally invariant non-Abelian theory, due to the fact that gauge fields can self-interact, usual charge is not conserved. The Noether theorem is only valid for global symmetries; in the local case, the local transformations are larger and include tranformations that don't conserve the current.

We can formally define a current $J^{a\mu}$ that includes a contribution from the gauge fields, $J^{a\mu} = j^{a\mu} + gf^a{}_{bc}A^b_\mu F^c_{\mu\nu}$, which amounts to adding the fields A^a_μ in the set of fields defining the Noether current. But then this conservation implies that the associated charge $Q^a = \int d^3x j^{a0}$ is not gauge invariant, or even covariant, so is not physical and, thus, cannot be measured (a change of gauge will change it).

We can also define a current $\tilde{j}^{a\mu}$ that is *covariantly conserved*, $D_\mu \tilde{j}^{a\mu} = 0$, but this is not conservation, $\partial_\mu \tilde{j}^{a\mu} \neq 0$, so, using Stokes's theorem (that uses d, not D), we find that $\tilde{Q}^a = \int d^3x \tilde{j}^{0a}$ is not conserved (time independent) since, with appropriate boundary conditions at infinity, $Q(t') - Q(t) = \int d^4x \partial^\mu \tilde{j}^a_\mu \neq 0$.

Important Concepts to Remember

- Non-Abelian gauge symmetry has a gauge field $A_\mu = A_\mu^a T_a$ in the Lie algebra – i.e., in the adjoint representation.
- The coupling of gauge fields to other fields is via the covariant derivative, $D_\mu \phi_i = (\delta_{ij}\partial_\mu - ig(A_\mu)_{ij})\phi_j$ – i.e., minimal coupling.
- The gauge field transforms as $A_\mu \to U A_\mu U^{-1} - (i/g)\partial_\mu U \cdot U^{-1}$, and the field strength $F_{\mu\nu}$ transforms covariantly, $F_{\mu\nu} \to U F_{\mu\nu} U^{-1}$.
- The field strength is $F_{\mu\nu} = \partial_\mu A_\mu - \partial_\nu A_\mu - ig[A_\mu, A_\nu]$, or $F = dA - igA \wedge A$.
- The kinetic term for matter is $-1/2\,\mathrm{Tr}[(D_\mu \phi)^\dagger D^\mu \phi]$ and for Yang–Mills gauge fields is $-(1/4)\,\mathrm{Tr}[F_{\mu\nu}F^{\mu\nu}]$.
- The infinitesimal gauge transformation is $\delta A = D_\mu \alpha$.
- The Yang–Mills equation of motion is $D_\mu F^{a\mu\nu} = 0$, and the Bianchi identity is $DF = dF - 2igA \wedge F = 0$.
- In the presence of a source, $D_\mu F^{a\mu\nu} + j^{a\nu} = 0$, and $[D_\mu, D_\nu] = -igF_{\mu\nu}^a T_a$.
- In the local non-Abelian case, we have no conserved and gauge invariant or covariant charge.

Further Reading

See chapter 15 in [4] and chapter 25 in [9].

Exercises

(1) Consider a scalar $\phi^{\bar{i}j}$ that transforms in the bi-fundamental representation of a symmetry group $U(N) \times U(N)$. Write its covariant derivative and kinetic terms if $U(N) \times U(N)$ is gauged.

(2) Calculate $D_\mu j^{a\mu}$ for a $j^{\mu a} A_\mu^a$ coupling.

(3) Show that

$$W = P e^{i\int_x^y A_\mu(x)dx} \equiv e^{iA_\mu^a(x)T_a dx^\mu} e^{iA_\mu^a(x+dx)T_a dx^\mu} \dots e^{iA_\mu^a(y)T_a dx^\mu} , \qquad (26.50)$$

where $A_\mu \equiv A_\mu^a T_a$, transforms under the gauge transformation with parameter $\alpha^a(x)$ as

$$W \to e^{i\alpha^a(y)T_a} W e^{-i\alpha^a(x)T_a}. \qquad (26.51)$$

(4) Consider the $SU(2)$ generators J_a normalized by $[J_a, J_b] = i\epsilon_{abc}J_c$. Calculate T_R and C_R for the fundamental and adjoint representations, and check that

$$T_R N_G = C_R N_R. \qquad (26.52)$$

27 The Dirac Monopole and Dirac Quantization

In this chapter, we will study an object with magnetic charge, the monopole, defined by Dirac, and the quantization relation for charges known as Dirac quantization. The monople defined by Dirac is in electromagnetism and is a singular structure, understood as a fundamental particle. In the next chapter, we will study the 't Hooft–Polyakov monopole, a soliton solution of non-Abelian theory that looks like a Dirac monopole at large distances but has a non-Abelian core.

27.1 Dirac Monopole from Maxwell Duality

We saw in Chapter 15 that there is a duality symmetry of the Maxwell's equations in the vacuum,

$$\partial_\mu F^{\mu\nu} = 0 \quad \text{and} \quad \partial_{[\mu} F_{\nu\rho]} = 0. \tag{27.1}$$

This *Maxwell duality* relates

$$F_{\mu\nu} \to (*F)_{\mu\nu} = \frac{1}{2}\epsilon_{\mu\nu\rho\sigma} F^{\rho\sigma}, \tag{27.2}$$

where $*^2 = -1$, meaning $(*F)_{\mu\nu} \to -F_{\mu\nu}$, so, in three-vector form, we have

$$\vec{E} \to \vec{B}; \quad \vec{B} \to -\vec{E}. \tag{27.3}$$

But now, considering the Maxwell's equations with sources,

$$\partial_\mu F^{\mu\nu} = -j^\nu, \quad \partial_{[\mu} F_{\nu\rho]} = 0, \tag{27.4}$$

or, in p-form language,

$$d * F = j, \quad dF = 0, \tag{27.5}$$

for a static source, $j^\mu = \rho/\epsilon_0 \delta^{\mu 0}$, we have

$$\vec{\nabla} \cdot \vec{E} = \frac{\rho_e}{\epsilon_0}, \quad \vec{\nabla} \cdot \vec{B} = 0. \tag{27.6}$$

Thus, clearly there is no Maxwell duality anymore. But that would be easily fixed if we would introduce a *magnetic current* k^ν, so that

$$\partial_\mu * F^{\mu\nu} = -k^\nu, \tag{27.7}$$

or, in form language,

$$dF = k. \tag{27.8}$$

For a static source, $k^\mu = \mu_0 \rho_m \delta^{\mu 0}$, we obtain

$$\vec{\nabla} \cdot \vec{B} = \mu_0 \rho_m. \tag{27.9}$$

That means that we can extend Maxwell duality by exchanging currents and charges,

$$j^\mu \leftrightarrow k^\mu : \quad \rho_e \leftrightarrow \rho_m : \quad q = \int d^3x \rho_e \leftrightarrow g = \int d^3x \rho_m. \tag{27.10}$$

Then under the duality, Gauss's law for electric charges,

$$\vec{\nabla} \cdot \vec{E} = \frac{q}{\epsilon_0} \delta^3(x) \tag{27.11}$$

becomes the same for magnetic charges,

$$\vec{\nabla} \cdot \vec{B} = \mu_0 g \delta^3(x). \tag{27.12}$$

We solve it by integrating over a sphere (ball) centered on the charge and then using the magnetic Gauss's law to turn it into an integral over the two-sphere of radius R, obtaining

$$\int_{B_R} d^3x \vec{\nabla} \cdot \vec{B} = \oint_{S_R^2} \vec{B} \cdot d\vec{S} = \mu_0 g. \tag{27.13}$$

This gives similarly to the electric case (when $\vec{E} = (e/4\pi\epsilon_0)\vec{r}/r^2$):

$$\vec{B} = \frac{\mu_0 g}{4\pi} \frac{\hat{r}}{r^2}. \tag{27.14}$$

This is called the *Dirac magnetic monopole*. We had said that in the multipole expansion of the electric fields, we have a monopole (point charge), dipole (separated, opposite, point charges), quadrupoles, etc., and for magnetic fields, we have only dipoles and higher. But that was in the absence of magnetic charge; otherwise, as we see here, we also have a monopole.

But Maxwell's equations are only half of the description of electromagnetism, finding the fields generated by charges. The other half is the action of those fields on charged particles – i.e., the Lorentz force law. For just electric charges, we had the four-force law,

$$\frac{dp^\mu}{d\tau} = qF^{\mu\nu}u_\nu. \tag{27.15}$$

But now we must construct the action on electric and magnetic charges. We do so by applying the duality in order to find a Maxwell duality invariant expression, $F_{\mu\nu} \to *F_{\mu\nu}$, and $q \to g$; therefore, the full full-force law is

$$\frac{dp^\mu}{d\tau} = (qF^{\mu\nu} + g * F^{\mu\nu})u_\nu. \tag{27.16}$$

In three-vectors, since the duality is $\vec{E} \to \vec{B}$ and $\vec{B} \to -\vec{E}$, we find

$$\vec{F} = q(\vec{E} + \vec{v} \times \vec{B}) + g(\vec{B} - \vec{v} \times \vec{E}). \tag{27.17}$$

Contradiction with Gauge Field

However, in writing the duality invariant Maxwell's equations with sources, and keeping the form in terms of F in terms of the gauge field $A_\mu = (-\phi, \vec{A})$, we find a contradiction: Since $\vec{B} = \vec{\nabla} \times \vec{A}$, we use Gauss's law (or rather, Stokes's law) twice and find

$$\mu_0 Q_m = \int_{M^3} d^3x \vec{\nabla} \cdot \vec{B} = \oint_{\partial M^3 = S_2} \vec{B} \cdot d\vec{S} = \oint_{S_2(\text{closed})} (\vec{\nabla} \times \vec{A}) d\vec{S} = \oint_{\partial S_2(\text{closed})} \vec{A} \cdot d\vec{l} = 0,$$

(27.18)

since S_2 is a closed surface (the boundary of the three-dimensional domain M_3), so it doesn't have a boundary, and the integral is zero. On the other hand, by our postulated Maxwell's equation with magnetic source, the left-hand side equals the magnetic charge, so we get a contradiction. We can see this easier in form language, since

$$F = dA, \quad \text{but} \quad dF = k \neq 0.$$

(27.19)

The solution is that $F = dA$ or, in the $A_0 = 0$ gauge, $\vec{B} = \vec{\nabla} \times \vec{A}$ is valid *only on patches* – i.e., only on *open surfaces that don't intersect the magnetic charges* – and not on the closed surface S surrounding a charge.

27.2 Gauge Fields on Patches

We must therefore consider a generalization of gauge fields – one that is defined only on patches (parts) of the physical domain. Before we get to that, however, we will formulate the problem in terms of p-forms, since it is easier to deal with them.

Formula for Charges in Terms of p-Forms

In form language, using the Maxwell equation with electric source integrated over a spatial volume $V = \Sigma_t^{(3)}$, we have

$$\int_V d * F = \int_{V = \Sigma_t^{(3)}} j \equiv \int_{\Sigma_t^{(3)}} j^\mu d\Sigma_\mu = \int d^3x j^0 = \frac{Q_e}{\epsilon_0}.$$

(27.20)

Using Stokes's law on the left-hand side, we find

$$\int_V d * F = \oint_{\partial V = \partial \Sigma_t^{(3)}} *F = \oint_{S_{\infty,t}^{(2)}} *F,$$

(27.21)

so that, finally ($Q_e = q$),

$$\oint_{S_{\infty,t}^{(2)}} *F = \frac{Q_e}{\epsilon_0}.$$

(27.22)

Analogously, using the Maxwell equation with magnetic source integrated over $V = \Sigma_t^{(3)}$,

$$\int_V dF = \int_{V = \Sigma_t^{(3)}} k \equiv \int_{\Sigma_t^{(3)}} k^\mu d\Sigma_\mu = \int d^3x k^0 = \mu_0 Q_m.$$

(27.23)

Using Stokes's law on the left-hand side, we find

$$\int_V dF = \oint_{\partial V = \partial \Sigma_t^{(3)} = S_{\infty,t}^{(2)}} F, \tag{27.24}$$

so that, finally ($Q_m = g$),

$$\oint_{S_{\infty,t}^{(2)}} F = \mu_0 Q_m. \tag{27.25}$$

For both Q_e and Q_m, $S_{\infty,t}^{(2)}$ is a closed surface at spatial infinity at fixed time t.

Patches and Magnetic Charge from Transition Functions

Since the magnetic charge is the integral of F over $S_{\infty,t}^{(2)}$, as we said, we need to divide the surface into two overlapping patches, $\mathcal{O}_\alpha$ and $\mathcal{O}_\beta$, with $\mathcal{O}_{\alpha\beta} = \mathcal{O}_\alpha \cap \mathcal{O}_\beta$, as in Figure 27.1a. And on each patch, we define a gauge field, so $A^{(\alpha)} = A_\mu^{(\alpha)} dx^\mu$ and $A^{(\beta)}$, so that

$$F = dA^{(\alpha)} = dA^{(\beta)}. \tag{27.26}$$

But that means that on the common patch $\mathcal{O}_{\alpha\beta}$, we have both $A^{(\alpha)}$ and $A^{(\beta)}$ as valid gauge fields, which means that they should be related by a gauge transformation, so

$$A^{(\alpha)} = A^{(\beta)} + d\lambda^{(\alpha\beta)}. \tag{27.27}$$

Here, $\lambda^{(\alpha\beta)}$ is called the *transition function*.

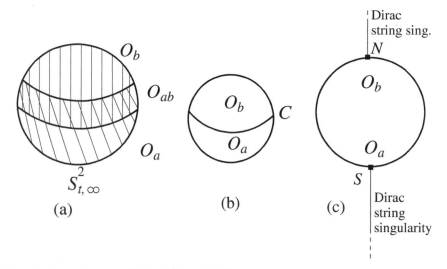

(a) Two overlapping patches 0_a and 0_b for the two-sphere S^2, with common patch 0_{ab}. (b) The case when the overlap 0_{ab} is an equator C. (c) The case when $0_a = S^2$ less the North Pole, and $0_b = S^2$ less the South Pole. The Dirac string singularity is a line going to infinity either at the North or the South Pole.

Explicitly, for the point-like magnetic charge (Dirac magnetic monopole), in the $A_0 = 0$ gauge, we can consider the patches

$$(\alpha) = S^2 - \text{North Pole}, \ \theta = \pi. \quad (\beta) = S^2 - \text{South Pole}, \ \theta = 0. \tag{27.28}$$

Then, on patch (α), we find the gauge field (vector potential)

$$\vec{A}^{(\alpha)} = \frac{\mu_0}{4\pi} \frac{g}{r} \frac{(-1 + \cos\theta)}{\sin\theta} \vec{e}_\phi, \tag{27.29}$$

where $\vec{e}_\phi$ is the unit vector in the direction of ϕ, in Cartesian coordinates

$$\vec{e}_\phi = (-\sin\phi, \cos\phi, 0). \tag{27.30}$$

We see that, indeed, the field is singular for $\sin\theta = 0$, but $\cos\theta \neq 1$ – i.e., for $\theta = \pi$ – which means that this field is defined on (α).

We can also verify that the formula is correct, since in spherical coordinates the curl is (note that we have $\theta \to \pi - \theta$ with respect to the usual definition)

$$\vec{\nabla} \times \vec{A} = \frac{1}{r\sin\theta}\left(-\frac{\partial}{\partial\theta}(A_\phi \sin\theta) + \frac{\partial A_\theta}{\partial\phi}\right)\hat{r} - \frac{1}{r}\left(\frac{1}{\sin\theta}\frac{\partial A_r}{\partial\phi} - \frac{\partial}{\partial r}(rA_\phi)\right)\vec{e}_\theta$$
$$+ \frac{1}{r}\left(-\frac{\partial}{\partial r}(rA_\theta) + \frac{\partial A_r}{\partial\theta}\right)\vec{e}_\phi, \tag{27.31}$$

and from it we derive the monopole magnetic field.

On the other hand, the gauge field in patch (β) is

$$\vec{A}^{(\beta)} = \frac{\mu_0}{4\pi} \frac{g}{r} \frac{(+1 + \cos\theta)}{\sin\theta} \vec{e}_\phi. \tag{27.32}$$

So now the field is singular for $\sin\theta = 0$, but $\cos\theta \neq -1$ – i.e., for $\theta = 0$ – which means the field is indeed defined on patch (β).

In form language and in Cartesian coordinates, we have

$$F = \frac{\mu_0 g}{4\pi r^3} \epsilon^{ijk} x^i dx^j \wedge dx^k, \tag{27.33}$$

which, in spherical coordinates, becomes

$$F = -\frac{\mu_0 g}{4\pi} \sin\theta d\theta \wedge d\phi = \mu_0 g \Omega_2, \tag{27.34}$$

where Ω_2 is the volume two-form on the unit S^2. Then, indeed,

$$\int_{S^2} F = \mu_0 g \int_{S^2} \Omega_2 = \mu_0 g. \tag{27.35}$$

Then the gauge field A ($F = dA$) on the two patches, in form language and in spherical coordinates, is

$$a^{(\alpha)} = \frac{\mu_0 g}{4\pi}(-1 + \cos\theta)d\phi$$
$$a^{(\beta)} = \frac{\mu_0 g}{4\pi}(+1 + \cos\theta)d\phi. \tag{27.36}$$

Going back to three-vectors, the difference of the two gauge fields, on different patches, is

$$\vec{A}^{(\alpha)} - \vec{A}^{(\beta)} = \frac{\mu_0}{2\pi}\frac{-g}{r\sin\theta}\vec{e}_\phi \equiv \vec{\nabla}\lambda^{(\alpha\beta)},$$

(27.37)

which means the transition function is

$$\lambda^{(\alpha\beta)} = -\frac{\mu_0}{2\pi}g\phi ,$$

(27.38)

This is correct, since the gradient in spherical coordinates is

$$\vec{\nabla} = \frac{\partial}{\partial r}\hat{r} + \frac{1}{r}\frac{\partial}{\partial\theta}\vec{e}_\theta + \frac{1}{r\sin\theta}\frac{\partial}{\partial\phi}\vec{e}_\phi.$$

(27.39)

We see that the difference (27.37), i.e., $\vec{\nabla}\lambda^{(\alpha\beta)}$ is single valued on the circle parametrized by $\phi \in [0, 2\pi]$, but $\lambda^{(\alpha\beta)}$ is not, since

$$\lambda^{(\alpha\beta)}(\phi = 2\pi) - \lambda^{(\alpha\beta)}(\phi = 0) = -\mu_0 g.$$

(27.40)

On the other hand, we can consider two patches $\mathcal{O}_C^\alpha, \mathcal{O}_C^b$ for S^2 that only have in common their boundary, a circle (equator) C – i.e., $\mathcal{O}_C^\alpha \cap \mathcal{O}_C^\beta = C$ (so $\mathcal{O}_C^\alpha \subset \mathcal{O}_\alpha$, $\mathcal{O}_C^\beta \subset \mathcal{O}_\beta$, $C \subset \mathcal{O}_{\alpha\beta}$, and $S^2 = \mathcal{O}_C^\alpha \cup \mathcal{O}_C^\beta$), as in Figure 27.1b.

Then the magnetic charge is given in terms of the integral of forms

$$\mu_0 Q_m = \int_{S^2 = \mathcal{O}_C^\alpha \cup \mathcal{O}_C^\beta} F = \int_{\mathcal{O}_C^\alpha} dA + \int_{\mathcal{O}_C^\beta} dA.$$

(27.41)

Using Stokes's theorem, and considering that the surface $\mathcal{O}_C^\alpha$ has an outward normal associated (via the right screw rule) with a direction on C, whereas the surface $\mathcal{O}_C^\beta$ has an outward normal associated (via the right screw rule) with the opposite direction on C, so the integral is on $-C$, we obtain

$$\mu_0 Q_m = \oint_{\partial\mathcal{O}_C^\alpha = +C} A + \oint_{\partial\mathcal{O}_C^\beta = -C} A = \oint_C (A^{(\alpha)} - A^{(\beta)}) = \oint_C d\lambda^{(\alpha\beta)},$$

(27.42)

which means that

$$\lambda^{(\alpha\beta)}(\phi = 2\pi) - \lambda^{(\alpha\beta)}(\phi = 0) = \oint_C d\lambda^{(\alpha\beta)} = \mu_0 g.$$

(27.43)

Therefore, *the magnetic charge (times μ_0) equals the nonsingle-valuedness of the transition function on the common boundary C.*

27.3 Topology and Dirac Quantization

We see that the magnetic charge is related to some topological quantity; as such, it should be quantized.

Quantized First Chern Number c_1

Mathematically, the quantity calculated in (27.43) is called the *first Chern number*, c_1 (associated with a gauge field, or "vector bundle," on a certain manifold), the integral of the *first Chern form* $F/(2\pi)$ on S^2,

$$c_1 = \int_{S^2} \frac{F}{2\pi}. \tag{27.44}$$

We now show that this quantity is an integer. Consider a scalar field that transforms under a $U(1)$ gauge transformation as

$$\phi \to e^{i\alpha}\phi. \tag{27.45}$$

Then, on the intersection of the patches $\mathcal{O}_{\alpha\beta}$, the scalar fields in the two patches must be related by a gauge transformation, just like in the case of the gauge fields,

$$\phi^{(\alpha)} = e^{i\lambda^{(\alpha\beta)}}\phi^{(\beta)}. \tag{27.46}$$

But then, since both $\phi^{(\alpha)}$ and $\phi^{(\beta)}$ must be well defined (single valued) under a rotation around C, it means that $e^{i\lambda^{(\alpha\beta)}}$ is single valued, i.e.,

$$e^{i\lambda^{(\alpha\beta)}(2\pi)} = e^{i\lambda^{(\alpha\beta)}(0)}, \tag{27.47}$$

so it means that

$$\int_{S^2} F = \lambda^{(\alpha\beta)}(2\pi) - \lambda^{(\alpha\beta)}(0) = 2\pi N, \tag{27.48}$$

where $N \in \mathbb{Z}$, which means that the first Chern number is quantized,

$$c_1 = \int_{S^2} \frac{F}{2\pi} = N \in \mathbb{Z}. \tag{27.49}$$

Dirac Quantization from $c_1 \in \mathbb{Z}$

We now derive an important relation, found by Dirac, quantizing together electric and magnetic charges, thus called *Dirac quantization*.

Consider that there are charged particles, with charge q, that are described by a field – for instance, a scalar field ϕ. Then the coupling of the field ϕ to electromagnetism is done through minimal coupling in the covariant derivative,

$$D\phi = (d - iqA)\phi. \tag{27.50}$$

Putting $\mu_0 = 1$ for simplicity, then we have the same calculation as before, so the magnetic charge is written as

$$g = \int_{S^2_\infty} F = \int_{S^2_\infty} \vec{B}\cdot d\vec{S} = \int_{\mathcal{O}^\alpha_C}(\vec{\nabla}\times\vec{A})\cdot d\vec{S} + \int_{\mathcal{O}^\beta_C}(\vec{\nabla}\times\vec{A})\cdot d\vec{S} = \oint_C (\vec{A}^{(\alpha)} - \vec{A}^{(\beta)})\cdot d\vec{l}. \tag{27.51}$$

But then on the common boundary $C \in \mathcal{O}_{\alpha\beta}$, the scalar ϕ changes by a gauge transformation on $C \in \mathcal{O}_{\alpha\beta}$, as

$$\phi^{(\alpha)} = e^{iq\lambda^{(\alpha\beta)}} \phi^{(\beta)}. \tag{27.52}$$

The same single-valuedness of $\phi^{(\alpha)}$ and $\phi^{(\beta)}$ around C, thus of $e^{iq\lambda^{(\alpha\beta)}}$, leads to

$$qg = q(\lambda^{(\alpha\beta)}(2\pi) - \lambda^{(\alpha\beta)}(0)) = 2\pi N, \tag{27.53}$$

or, reintroducing $\hbar$ by dimensional analysis,

$$qg = 2\pi \hbar N. \tag{27.54}$$

This is the *Dirac quantization condition*. Note that the only thing we have assumed is that there are charges q, defined by the field ϕ, which *only need be defined on C*, and there is *somewhere else*, namely at the origin (in the middle of S^2 and of C), a magnetic charge g.

This, then, has the far-reaching consequence that *if there is a single magnetic monopole somewhere in the Universe*, then the electric charge is quantized. This is, in fact, the only true theoretical argument for the quantization of electric charge, which is an experimentally observed fact. Therefore, this is a good argument for the existence of magnetic monopoles. Note, however, that the argument is about the existence of magnetic charge, not about the fact that there is a Dirac monopole (a point-like magnetic charge, analogous to an electron). We could very well have a soliton ('t Hooft–Polyakov monopole), like in the next chapter.

27.4 Dirac String Singularity and Dirac Quantization from It

We have seen that for the Dirac monopole, the magnetic vector potential $\vec{A}$ was defined everywhere except on a line: $\vec{A}^{(\alpha)}$ was defined everywhere except on the North Pole $\theta = \pi$ on the unit sphere, which translates into a "string" at $\theta = \pi$ extending from the monopole ($r = 0$) to infinity ($r = \infty$). Similarly, $\vec{A}^{(\beta)}$ was defined everywhere except on the string $\theta = 0$ extending from the monopole to infinity.

This singularity is called a *Dirac string singularity*, and, as we see, its position can be moved around by a gauge transformation (from the patch (α) to the patch (β), it goes from $\theta = \pi$ to $\theta = 0$), as in Figure 27.1c; therefore, it is not a physical singularity.

Dirac's interpretation of the Dirac string singularity is as follows. He considers a total *regular* (nonsingular, in terms of $F = dA$ and $dF = 0$) magnetic field $\vec{B}_{\text{reg}}$, formed by the monopole magnetic field $\vec{B}_{\text{mon}}$, plus the field of the Dirac string singularity,

$$\vec{B}_{\text{string}} = g\theta(-z)\delta(x)\delta(y)\hat{z}, \tag{27.55}$$

so that, in total,

$$\vec{B}_{\text{reg}} = \vec{B}_{\text{mon}} + \vec{B}_{\text{string}}. \tag{27.56}$$

Note that the Dirac string field is for the (β) patch (for the singulariy on the South Pole), which means it extends radially outward from the origin to infinity, along the negative z axis, via the Heaviside function $\theta(-z)$.

Then the total, regular magnetic field obeys the usual relations (in the absence of magnetic charges)

$$\vec{\nabla} \cdot \vec{B}_{\text{reg}} = 0 \Rightarrow \vec{B}_{\text{reg}} = \vec{\nabla} \times \vec{A}. \tag{27.57}$$

But then the Dirac quantization condition appears from the condition that the Aharonov–Bohm effect for a contour encircling the Dirac string is unobservable – i.e., the Dirac string has no physical meaning.

The Aharonov–Bohm effect is the fact that the vector potential $\vec{A}$ has some physical (observable) consequences. Indeed, if a magnetic field $\vec{B}$ is confined in some region, and we enclose it with a contour C along which $\vec{B}$ is negligible, it would seem that if only $\vec{B}$, and not $\vec{A}$, is physical, there should be no effect. But, in fact, this is not so, since by the Gauss (Stokes's) law,

$$\oint_{C=\partial S} \vec{A} \cdot d\vec{l} = \int_S \vec{B} \cdot d\vec{S} \neq 0, \tag{27.58}$$

and this object appears in the exponent acting on wave functions and (scalar) fields with charge q,

$$\psi \to \exp\left[\frac{iq}{\hbar} \oint_{C=\partial S} \vec{A} \cdot d\vec{l}\right] \psi. \tag{27.59}$$

The statement of Dirac is then that the Aharonov–Bohm effect for the added magnetic field, of the Dirac string, should be unobservable, i.e., that the *monodromy* is trivial,

$$\exp\left[\frac{iq}{\hbar} \oint_{C=\partial S} \vec{A}_{\text{string}} \cdot d\vec{l}\right] = \exp\left[\frac{iq}{\hbar} \int_S \vec{B} \cdot d\vec{S}\right] = e^{\frac{iqg}{\hbar}} = 1. \tag{27.60}$$

This, then, implies the same Dirac quantization condition,

$$qg = 2\pi\hbar N. \tag{27.61}$$

27.5 Dirac Quantization from Semiclassical Nonrelativistic Considerations

We can also obtain Dirac quantization from semiclassical considerations. Consider a nonrelativistic system of an electric charge q in the monopole field of a magnetic charge g. Then $\vec{E}_{\text{mon}} = 0$ and $\vec{B}_{\text{mon}} = (\mu_0 g/4\pi)\hat{r}/r^2$, and the Lorentz force exerted on the electric charge is

$$\dot{\vec{p}} = m\ddot{\vec{r}} = q\dot{\vec{r}} \times \vec{B} = \frac{\mu_0 qg}{4\pi} \frac{\dot{\vec{r}} \times \vec{r}}{r^3}. \tag{27.62}$$

The time derivative of the orbital angular momentum is then (for $\mu_0 = 1$)

$$\frac{d\vec{L}}{dt} = \frac{d}{dt}(m\vec{r} \times \dot{\vec{r}}) = m\vec{r} \times \ddot{\vec{r}} = \frac{qg}{4\pi r^3}\vec{r} \times (\dot{\vec{r}} \times \vec{r}) = \frac{d}{dt}\left(\frac{gq}{4\pi}\frac{\vec{r}}{r}\right), \tag{27.63}$$

where we have used the fact that

$$\frac{1}{r^3}[\vec{r} \times (\dot{\vec{r}} \times \vec{r})]_i = \frac{1}{r^3}\epsilon_{ijk}x^j(\epsilon^{klm}v_l x_m) = \frac{1}{r^3}(\delta_i^l\delta_j^m - \delta_i^m\delta_j^l)x^j v_l x_m$$

$$= \frac{v_i}{r} - \frac{x_i}{r^3}(\vec{r}\cdot\dot{\vec{r}}) = \left[\frac{\dot{\vec{r}}}{r} - \frac{\vec{r}(\vec{r}\cdot\dot{\vec{r}})}{r^3}\right]_i$$

$$= \frac{d}{dt}\frac{\vec{r}}{r}. \tag{27.64}$$

Finally, we obtain

$$\frac{d}{dt}\left(\vec{L} - \frac{qg}{4\pi}\frac{\vec{r}}{r}\right) \equiv \frac{d}{dt}\vec{J} = 0. \tag{27.65}$$

Since the first term is the orbital angular momentum, so in the absence of the second term (at $g = 0$), this relation is conservation of angular momentum, we must have that for $g \neq 0$, this is the conservation of the *total angular momentum* $\vec{J}$.

But then we can assume that at $\vec{L} = 0$, in quantum mechanics, $|\vec{J}|$ is quantized in half-integer units of $\hbar$, i.e.,

$$\frac{gq}{4\pi} = \frac{N}{2}\hbar \Rightarrow gq = 2\pi\hbar N, \tag{27.66}$$

which is the correct Dirac quantization condition.

The last observation of this chapter is that the energy density of the Dirac monopole diverges at $r \to 0$,

$$\mathcal{E} = \frac{\vec{B}^2}{2\mu_0} = \frac{g^2\mu_0}{32\pi^2 r^4} \propto \frac{1}{r^4} \to \infty, \tag{27.67}$$

but, moreover, that means that the total energy of the magnetic field of the Dirac monopole is infinite,

$$E = \int \mathcal{E}d^3x \sim \int_{r\sim 0} 4\pi r^2 dr \mathcal{E} \sim \frac{g^2\mu_0}{8\pi}\frac{1}{r} \to \infty, \tag{27.68}$$

just like the total energy of the electric field of an electron.

In the case of the electron, Born and Infeld proposed their nonlinear theory of electromagnetism in order for the electric field to have an upper bound so that the total energy of the electric field of the electron is finite. But the magnetic field is not bounded in it. The Born–Infeld action appears in string theory and has the same effect: the total energy of an "electron" is finite.

Instead, for magnetic field, we will replace in physical situations the Dirac monopole with a non-Abelian solitonic monopole, which at infinity looks like a Dirac monopole but has a finite energy (is a soliton) by being non-Abelian near its core.

Important Concepts to Remember

• The Dirac monopole appears by insisting on having duality invariant Maxwell equations in the presence of electric sources (currents) j^μ, leading to magnetic currents k^μ and charges Q_m.

- The duality $F_{\mu\nu} \to *F_{\mu\nu}$ is supplemented by $j^\mu \to k^\mu$ and $q \to g$. The Lorentz force law becomes $dp^\mu/d\tau = (qF^{\mu\nu} + g*F^{\mu\nu})u_\nu$.
- The presence of a magnetic charge g means that $dF \neq 0$, so $F \neq dA$ globally (on a closed surface S), but only on patches.
- We have $\int_{S^2_\infty} *F = q$, $\int_{S^2_\infty} F = g$.
- On patches, $F = dA^{(\alpha)}$ and $F = dA^{(\beta)}$, but $A^{(\alpha)}$ and $A^{(\beta)}$ are related by a gauge transformation on the common part of the patches, $A^{(\alpha)} = A^{(\beta)} + d\lambda^{(\alpha\beta)}$, where $\lambda^{(\alpha\beta)}$ are the transition functions.
- The magnetic charge is the nonsingle-valuedness of the transition functions around the common boundary = equator C of the patches, $\mu_0 g = \lambda(\phi = 2\pi) - \lambda(\phi = 0)$.
- The first Chern number is quantized, $c_1 = \int_{S^2} F/(2\pi) = N \in Z$.
- From single-valuedness of scalar fields with charge q under motion around C, we find the Dirac quantization condition, $gq = 2\pi\hbar N$.
- The Dirac quantization condition means that the existence of a single monopole in the Universe is enough to quantize charge, which is observed experimentally; hence, it is very likely that magnetic monopoles exist.
- The monopole has a Dirac string singularity, extending from the monopole to infinity, but the string can be moved around by a gauge transformation, so is not physical.
- The Dirac string also cannot be observed by an Aharonov–Bohm effect if Dirac quantization is satisfied.
- The Dirac quantization can also be obtained from semiclassical quantization of the total angular momentum $\vec{J}$ in the presence of both electric and magnetic charge.

Further Reading

See chapter 8 in [14] and section 1 in [18].

Exercises

(1) Consider an electron and a monopole at the same point at a distance R from a perfectly conducting infinite plane. What is $\vec{E}$ and $\vec{B}$ on the plane at the minimum distance point from the charges?

(2) Consider that the magnetic field for a particle at 0 is a delta function,

$$\vec{B}(x) = \frac{2\theta}{e}\delta(\vec{x}). \tag{27.69}$$

Take two such particles, and rotate one around the other. Calculate the Aharonov–Bohm phase

$$e^{i\oint_C \vec{A}\cdot d\vec{l}} \tag{27.70}$$

of the moving particle. Since the particles are identical, how would you interpret this result?

(3) In the presence of so-called *dyons*, particles that carry both electric and magnetic charges, the Dirac quantization condition for a particle with charges (q_1, g_1) and another one with (q_2, g_2) is generalized to the *Dirac-Schwinger-Zwanziger quantization condition*

$$q_1 g_2 - q_2 g_1 = 2\pi \hbar n , \quad n \in \mathbb{Z}. \tag{27.71}$$

Prove this relation using a generalization of the argument of the quantization of the total angular momentum of the system of the two particles.

(4) Consider two dyons satisfying the Dirac-Schwinger-Zwanziger quantization condition above, with the minimum value for the integer, $n = 1$, being that one of them has $q_1 = e, g_1 = h/e$. Calculate the total relative force between the dyons.

28

The 't Hooft–Polyakov Monopole Solution and Topology

In the last chapter, we have seen that there should be Dirac monopoles, due to Dirac's quantization condition. But the Dirac monopole had infinite energy $E \sim 1/r$ as $r \to 0$, which means it can only be a fundamental state, which makes sense at the quantum level, like the electron does. Yet we haven't seen one, so it is therefore much likelier that we have solitonic states that look like Dirac monopoles at large distances. Those can be very rare for the simple reason that they are heavy (as we will see later).

Therefore, in this chapter, we will study nonsingular solitonic monopoles in a non-Abelian theory. They have a nonsingular solitonic core, but at large distances they look Abelian. The solutions were found by Gerard 't Hooft and Alexander Polyakov in 1974.

The system in which they are found is a Yang–Mills–Higgs system for a group G – i.e., a non-Abelian gauge field A_μ^a coupled to a scalar field ϕ^a that, in its vacuum, breaks the gauge symmetry from the group G to a group H.

28.1 Setup for Georgi–Glashow Model

We will consider the special case of a field ϕ^a in the *adjoint* representation of G and, moreover, for the gauge group $G = SU(2) \simeq SO(3)$, which is called the *Georgi–Glashow model*. Then $a = 1, 2, 3$ and ϕ^a is real, and $f_{abc} = \epsilon_{abc}$. The Lagrangean is

$$\mathcal{L} = -\frac{1}{4}F_{\mu\nu}^a F^{a\mu\nu} - \frac{1}{2}(D_\mu\phi^a)(D^\mu\phi^a) - V(\phi), \tag{28.1}$$

where $A_\mu = A_\mu^a T_a$, $F_{\mu\nu} = \partial_\mu A_\nu - \partial_\nu A_\mu - ig[A_\mu, A_\nu]$, with the potential

$$V(\phi) = \frac{\lambda}{4}(|\phi|^2 - \phi_0^2), \tag{28.2}$$

and $|\phi|^2 \equiv \text{Tr}[\phi^2] = \phi^a\phi^a$. The covariant derivative on a scalar, transforming under the group G by $\phi_i \to (e^{i\alpha^a T_a})^i{}_j \phi^j$ in general is

$$(D_\mu\phi)^i = \partial_\mu\phi^j \delta_j^i - ig(T_a)^i{}_j A_\mu^a \phi^j, \tag{28.3}$$

which in our case becomes (for $(T_a)_{bc} = -if_{abc} = -i\epsilon_{abc}$),

$$(D_\mu\phi)^a = \partial_\mu\phi^a - g\epsilon^{abc}A_\mu^b\phi^c. \tag{28.4}$$

As we saw before, when talking about symmetry breaking, in the vacuum, the symmetry is broken from G to the subgroup H that leaves the vacuum solution $\langle\phi_i\rangle = \phi_{0i}$ invariant.

In the case of the $SU(2)$ model, for ϕ in the doublet representation, the vacuum could be put by a global $SU(2)$ transformation to the form $\langle \phi \rangle = \frac{1}{\sqrt{2}} \begin{pmatrix} 0 \\ v \end{pmatrix}$, which meant that it is invariant under the $U(1) \subset SU(2)$ that just rotates the first component by a phase. That means that the vacuum breaks the symmetry $G = SU(2) \to U(1) = H$.

In the case of the $SO(3)$ model with ϕ^a in the adjoint (Georgi–Glashow model), the vacuum solutions satisfy $|\phi| = \phi_0$, and by a gauge transformation $\phi \to U\phi U^{-1}$ (remember that the gauge field transforms by $A_\mu \to UA_\mu U^{-1} - \frac{i}{g}\partial_\mu U \cdot U^{-1}$, so the *global* symmetry transformation on an *adjoint* field is just the first term), we can put it to the form

$$\phi^a_{\mathrm{vac}} = \phi_0 \delta^{a3}, \tag{28.5}$$

which is left invariant by an $SO(2) = U(1)$ transformation that only acts on $a = 1, 2$. That means that again the symmetry breaking pattern in the vacuum is $G = SU(2) \to U(1) = H$.

The equations of motion of $\mathcal{L}$ are as follows. Since

$$\frac{\partial D_\mu \phi^a}{\partial A^b_\mu} = g\epsilon^{abc}\phi^c \delta^\mu_\nu , \tag{28.6}$$

the A_ν equation of motion is the YM equation with a scalar source,

$$D_\mu F^{a\mu\nu} = -g\epsilon^{abc}\phi^b(D^\nu \phi^c) = -g[\phi, D^\nu \phi]^a. \tag{28.7}$$

Since

$$\frac{\partial D_\mu \phi^a}{\partial \phi^b} = (D_\mu)^a{}_b, \tag{28.8}$$

and we can partially integrate it, the ϕ^a scalar equation of motion is a covariant version of the modified KG equation,

$$(D_\mu D^\mu \phi)^a = \lambda \phi^a(|\phi|^2 - \phi_0^2). \tag{28.9}$$

But, moreover, since the previous equations are written in a covariant manner, we must supplement them with the Bianchi identity for the field strength,

$$D_{[\mu}F^a_{\nu\rho]} = 0 \Rightarrow D_\mu * F^{a\mu\nu} = 0. \tag{28.10}$$

We define (color-) electric and (color-) magnetic fields in the same way as for electromagnetism, by

$$F^{a0i} = E^{ai}; \quad F^a_{ij} = \epsilon_{ijk}B^{ak}. \tag{28.11}$$

The energy density is then found in the usual way (like for the Abelian case) to be

$$\mathcal{E} = \frac{1}{2}\left[(E^a_i)^2 + (B^a_i)^2 + (D_0\phi^a)^2 + (D_i\phi^a)^2\right] + V(\phi). \tag{28.12}$$

28.2 Vacuum Manifold and 't Hooft–Polyakov Solution

The vacua of the theory are found by putting to zero all the squares in the energy density in order to obtain zero energy, i.e.,

$$F^a_{\mu\nu} = 0, \quad (D_\mu \phi)^a = 0, \quad V(\phi) = 0. \tag{28.13}$$

The first condition means that A^a_μ is a pure gauge (equivalent to zero by a gauge transformation), at least locally, while the last means that $|\phi_{\text{vac}}| = \phi_0$. It remains to satisfy $D_\mu \phi^a = 0$.

If $\partial_\mu \phi_{\text{vac}} = 0$ – namely, if we have a constant vacuum – then we can put $A^a_\mu = 0$, and, as we said earlier, we can use a global gauge transformation to fix $\phi_{\text{vac}} = \phi_0 \delta^{a3}$. The vacuum is invariant under $U(1)$, so we have an unbroken $H = U(1)$ gauge invariance. But the vacuum condition is $|\phi_{\text{vac}}| = \phi_0$ ($\phi^a_{\text{vac}} \phi^a_{\text{vac}} = \phi_0^2$), which defines a two-sphere S^2 in field space.

The *vacuum manifold* $\mathcal{M}$, or "moduli space of vacua," comprising only vacua not equivalent under the unbroken symmetry $H = U(1)$ is the "coset manifold" $\mathcal{M} = G/H$. Indeed, a transformation in G will take $\phi_0 \delta^{a3}$ into a general vacuum, whereas a transformation in H would just generate an equivalent vacuum.

That means that the vacuum manifold is the coset manifold G/H, the space of "equivalence classes" under the equivalence relation for $g \sim hg$, $\forall g \in G, h \in H$: we consider all equivalent elements to be a single element, the "equivalence class." The coset manifold G/H is the group manifold G, reduced to equivalence classes under H.

But an n−sphere of constant radius is the coset manifold $S^n = SO(n+1)/SO(n)$, as can be seen simply from the definition of the sphere in Euclidean $(n+1)$−dimensional space $\sum_{i=1}^{n+1} x_i^2 = R^2 =$constant. A point on this sphere is related to another by a rotation of the Euclidean coordinates, $x'^i = R^i_j x^j$, $R^i_j \in SO(n+1)$. But rotations around the point on the sphere (that leave it invariant) are rotations in the local n−dimensional space around the point – i.e., $SO(n)$ rotations – therefore,

$$S^n = SO(n+1)/SO(n). \tag{28.14}$$

In our case, we have, indeed,

$$S^2 = SO(3)/SO(2), \tag{28.15}$$

which matches against our observation that the space of vacua is an S^2 of constant $|\phi|$.

In a vacuum $\phi_{\text{vac}} = \phi_0 \delta^{a3}$, the $H = U(1)$ that is unbroken is the group of rotations around this direction, which is the generator

$$\frac{1}{\phi_0}(\phi^a_{\text{vac}} T_a) \equiv T_3. \tag{28.16}$$

Indeed, for $\phi^a_{\text{vac}} = \phi_0 \delta^{a3}$, the vacuum transforms under the $U(1)$ as $e^{i\alpha T_3} \phi^a_{\text{vac}} e^{-i\alpha T_3} = \phi^a_{\text{vac}}$, so it is invariant. This $U(1)$ will be identified with electromagnetism, so the charge associated with it is electric charge,

$$Q = \hbar g \frac{(\phi^a_{\text{vac}} T_a)}{\phi_0} \, , \tag{28.17}$$

where the basic (quantum of) charge is $\hbar g$, as usual. The electromagnetic gauge field is

$$A_\mu = \frac{\phi^a_{\text{vac}} A^a_\mu}{\phi_0}. \tag{28.18}$$

Finally, now we are ready to write the monopole ansatz for the fields. Since the solution is a *magnetic* monopole, we expect the electric field to be zero, so we work in the Coulomb gauge $A^a_0 = 0$ (so that for a static solution, with $\partial_t = 0$, the electric field is, indeed, zero).

At infinity, we consider a vacuum solution, but one that *depends on the direction*,

$$\phi^a_{\text{vac}} = \phi_0 \frac{x^a}{r}. \tag{28.19}$$

Note first that this means that we identify the gauge index a with the spatial index i, and second, that this was the only possibility that doesn't specify a preferred direction in field space, since the only three-vector index available, to identify with ϕ^a, is x^i. Then the ansatz for ϕ^a is

$$\phi^a = \frac{x^a}{r} \phi_0 h(\phi_0 g r). \tag{28.20}$$

Here $h(x)$ is a numerical function, and again, the only constant with mass dimension (of one) in our theory is ϕ_0, so the only possibility for a dimensionless variable is $\phi_0 r$ (and the multiplication by g is then really an ansatz).

Now, for the vacuum *at infinity*, we must impose the nontrivial constraint $D_i \phi^a = 0$ ($D_0 \phi^a = 0$ is trivial, since $\partial_t = 0$ and $A_0 = 0$), giving

$$\partial_i \phi^a_{\text{vac}} = \left(\frac{\delta^a_i}{r} - \frac{x^a x^i}{r^3} \right) \phi_0$$

$$= g \epsilon^{abc} A^b_i \phi^c_{\text{vac}} = g \phi_0 \epsilon^{abc} A^b_i \frac{x^c}{r} \, , \tag{28.21}$$

which is solved by

$$A^a_{i,\text{vac}} = -\epsilon_{ija} \frac{x^j}{g r^2}. \tag{28.22}$$

Moreover, the only possible ansatz for A^a_i that doesn't single out a special direction must involve $\epsilon_{ija} x^j$, as the only object with two three-dimensional indices available to us. Moreover, for a monopole solution, we need $A_i \sim 1/r$, so the full ansatz must be

$$A^a_i = -\epsilon_{ija} \frac{x^j}{g r^2} [1 - K(\phi_0 g r)]. \tag{28.23}$$

Asymptotically, at $r \to \infty$, since in order to have finite energy, we must have $A^a_i \to A^a_{i,\text{vac}}$, we must have $K \to 0$. Also, since we need $\phi^a \to \phi^a_{\text{vac}}$, again for finiteness of the energy, we must have $h \to 1$.

The same finiteness of energy can be used to find the solution at $r \to 0$. Finiteness of energy means avoiding the singularity in the energy of the Dirac monopole. The finiteness of the magnetic field energy means $\int d^3 x (\vec{B}^a)^2 \sim (\vec{B}^a)^2 r^2 dr < \infty$, leading to $B(r)$

diverging less than $1/r^{3/2}$. If $K(x)$ is Taylor-expandable, only integer powers are allowed for it, which means $A_i^a \sim x^j/r$ – i.e.,

$$1 - K(\phi_0 g r) \sim \phi_0 g r. \tag{28.24}$$

We also need the finiteness of the scalar kinetic term, $\int r^2 dr \, (D_i \phi^a)^2 < \infty$, which means that we need $D_i \phi^a \sim 1/r$. That leads to

$$h \sim 1. \tag{28.25}$$

In this way, the 't Hooft–Polyakov monopole solution avoids the singularity in the energy of the Dirac monopole.

This is the most we can do using analytical methods (of course, we can find subleading terms by solving the equations order by order near the $r \to \infty$ or $r \to 0$ asymptotics). For the full solution, however, we must resort to numerics, which again can be found by using the shooting method: impose the asymptotics at one end, with varying coefficients, and vary the coefficients until we fall on the other end on the expected asymptotics.

28.3 Topology of the Monopole

The vacuum condition, $D_\mu \phi^a = 0$, *for* $\phi^a = \phi_{\text{vac}}^a$ *(in a Higgs vacuum)*, leads in general to a relation between the full non-Abelian gauge field A_μ^a and the Maxwell part A_μ, together with ϕ_{vac}^a, as

$$A_\mu^a = -\frac{1}{g\phi_0^2} \epsilon^{abc} \phi_{\text{vac}}^b \partial_\mu \phi_{\text{vac}}^c + \frac{\phi_{\text{vac}}^a}{\phi_0} A_\mu. \tag{28.26}$$

Then we find for the Abelian part of the full field strength,

$$F_{\mu\nu} \equiv \frac{\phi_{\text{vac}}^a F_{\mu\nu}^a}{\phi_0} = \partial_\mu A_\nu - \partial_\nu A_\mu - \frac{1}{g\phi_0^3} \epsilon^{abc} \phi_{\text{vac}}^a \partial_\mu \phi_{\text{vac}}^b \partial_\nu \phi_{\text{vac}}^c, \tag{28.27}$$

which we leave as an exercise to prove.

Moreover, the full equations of motion for A_μ^a in the Higgs vacuum, $\phi^a = \phi_{\text{vac}}^a$, imply that the electromagnetic field strength satisfy the vacuum Maxwell's equations (equations of motion and Bianchi identity), satisfied by the Dirac monopole away from the source at $r = 0$,

$$\partial_\mu F^{\mu\nu} = 0, \quad \partial_{[\mu} F_{\nu\rho]} = 0, \tag{28.28}$$

just that now they are satisfied eveywhere we have $\phi^a = \phi_{\text{vac}}^a$. That is certainly true for our solution, sufficiently far away from the core (at $r \to \infty$). In that case, the first term in the Abelian field strength is zero, since we can calculate that $A_i = A_i^a \phi^a / \phi_0 \propto \epsilon_{ija} x^j x^a = 0$, and $A_0^a = 0$.

The magnetic charge g_m for the Abelian field of the static soliton solution is (according to the last chapter, and using $B_i = \frac{1}{2} \epsilon_{ijk} F_{jk}$)

$$g_m = \int_{\Sigma^2} B^i dS^i = -\frac{1}{2g} \int_{\Sigma^2_\infty} \epsilon^{ijk} \epsilon^{abc} \frac{\phi^a_{\text{vac}}}{\phi_0} \frac{(\partial_i \phi^b_{\text{vac}})}{\phi_0} \frac{(\partial_k \phi^c_{\text{vac}})}{\phi_0} dS^i. \tag{28.29}$$

But $\phi^a/\phi_0 \equiv n^a$ is a unit vector, defined on the unit sphere S^2 in field space, the moduli space of vacua $\mathcal{M} = S^2$.

Thus, the magnetic charge is defined by a map from the surface at spatial infinity, $\Sigma^2_\infty = S^2$, to the vacuum manifold,

$$n^a \equiv \frac{\phi^a_{\text{vac}}}{\phi_0} : \Sigma^2_\infty = S^2 \to \mathcal{M}. \tag{28.30}$$

But this is, by definition, the *homotopy group* $\pi_2(\mathcal{M} = S^2)$, which is characterized by an integer $\nu \in \mathbb{Z}$, or *winding number of $\mathcal{M}$ over S^2*, which is

$$\nu = -\frac{1}{8\pi} \int_{S^2} \epsilon^{ijk} \epsilon^{abc} n^a (\partial_j n^b)(\partial_k n^c) dS^i. \tag{28.31}$$

That means that we have the magnetic charge

$$g_m = \frac{4\pi \hbar \nu}{(\hbar g)}, \tag{28.32}$$

which is nothing but the Dirac quantization condition for an electric charge of

$$q_e = \frac{g\hbar}{2}. \tag{28.33}$$

Homotopy Groups

We have used the notion of homotopy groups, so we need to define it. Formally, the homotopy group $\pi_k(\mathcal{M})$ is the group of *homotopy classes*, i.e., equivalence classes of maps

$$a : S^k \to \mathcal{M}, \tag{28.34}$$

under the equivalence relation that the maps a and b are equivalent, if $a \cdot b^{-1} : S^k \to S^k$ is a trivial map – i.e., one that can be extended from the sphere S^k to its interior – the ball B^{k+1}.

Less formally, we see that the maps $a : S^k \to \mathcal{M}$ define possible "windings" of $\mathcal{M}$ around S^k, and equivalent maps a and b have the same "winding," so the map $a \cdot b^{-1}$ is one-to-one (and hence can be extended to the interior, the ball B^{k+1}). It is intuitively clear that we can wind an S^2 around another S^2 an integer number of times, which means that

$$\pi_2(S^2) = \mathbb{Z}, \tag{28.35}$$

and the integer that characterizes the equivalence classes is the ν above, a topological number. It is indeed clear that the form integrated in the definition of ν is the only one-form on S^2_∞ that we can construct that is invariant under $\mathcal{M}$ transformations.

28.4 Bogomolnyi Bound and BPS Monopole

We can rewrite the definition of the Abelian magnetic charge, using the fact that

$$\partial_i(\mathrm{Tr}(B_i\phi)) = \mathrm{Tr}(D_i(B_i\phi)) = \mathrm{Tr}(B_iD_i\phi),\qquad(28.36)$$

since the Bianchi identity is $D_iB_i = \frac{1}{2}\epsilon_{ijk}D_iF_{jk} = 0$, as

$$g_m = \int_{S^2_\infty}\vec{B}\cdot d\vec{S} = \frac{1}{\phi_0}\int_{S^2_\infty=\partial V}B_i^a\phi^a dS^i = \frac{1}{\phi_0}\int_V d^3x\, B_i^a(D_i\phi)^a,\qquad(28.37)$$

where in the second equality we have Stokes's theorem and (28.36).

In a similar way, we can write the definition of the Abelian electric charge, using the similar relation

$$\partial_i(\mathrm{Tr}(E_i\phi)) = \mathrm{Tr}(D_i(E_i\phi)) = \mathrm{Tr}(E_iD_i\phi),\qquad(28.38)$$

which now comes from the equation of motion in vacuum $D_iE_i = D_iF^{i0} = 0$, as

$$q_e = \int_{S^2_\infty}\vec{E}\cdot d\vec{S} = \frac{1}{\phi_0}\int_{S^2_\infty=\partial V}E_i^a\phi^a dS^i = \frac{1}{\phi_0}\int_V d^3x\, E_i^a(D_i\phi)^a.\qquad(28.39)$$

Using the formula for the energy density (28.12), and completing the squares in a similar manner to the 1+1 dimensional and the 2+1 dimensional cases (but more like in the complex case of 1+1 dimensions, in that we have a S^1-parametrized formula instead of a simple $\pm$ inside the squares), we find

$$M = \int d^3x\,\mathcal{E} = \int d^3x\left\{\frac{1}{2}\left[(E_i^a - D_i\phi^a\sin\theta)^2 + (B_i^a - D_i\phi^a\cos\theta)^2 + (D_0\phi^a)^2\right] + V(\phi)\right\}$$
$$+\phi_0(q_e\sin\theta + g_m\cos\theta),\qquad(28.40)$$

where θ is an arbitrary parameter. We leave the proof as an exercise, for which we must use (28.36) and (28.38).

Since the potential is positive definite, $V(\phi)\geq 0$, we obtain the bound

$$M\geq \phi_0(q_e\sin\theta + g_m\cos\theta).\qquad(28.41)$$

But this bound depends on the arbitrary parameter θ, so we must minimize further over it. Then

$$\frac{dM_{\min}}{d\theta} = \phi_0(q_e\cos\theta - g_m\sin\theta) = 0 \Rightarrow \tan\theta_{\min} = \frac{q_e}{g_m},\qquad(28.42)$$

which also means that $\cos\theta_{\min} = g_m/\sqrt{q_e^2+g_m^2}$. We thus obtain the most stringent bound,

$$M\geq \phi_0\frac{\cos\theta_{\min}}{g_m}(q_e^2+g_m^2) = \phi_0\sqrt{q_e^2+g_m^2}.\qquad(28.43)$$

This is the *BPS bound*, a bound found in this case by Bogomolnyi and used later by Prasad and Sommerfield.

The 't Hooft–Polyakov monopole has $q_e = 0$, so $\theta_{\min} = 0$, and, using the fact that $g_m = 4\pi/g$, we find

$$M \geq \phi_0 |g_m| = \phi_0 \frac{4\pi}{g}. \tag{28.44}$$

However, for the BPS bound to be saturated, we see that we need $V(\phi) = 0$, so we need to be in the *BPS limit*, $\lambda \to 0$. Only in this limit, $M = \phi_0 g_m = \phi_0 4\pi/g$.

Note that there are also solutions with both $g_m \neq 0$ and $q_e \neq 0$ (both electric and magnetic charges), called *dyons*, and found by Bernard Julia and Anthony Zee, but we will not study them here.

To saturate the bound for a magnetic monopole in the BPS limit, which has $q_e = 0$, thus $\theta_{\min} = 0$, we put all the square terms in the energy to zero, except the last (topological) term, for $\theta = \theta_{\min} = 0$, so

$$E_i^a = 0, \quad B_i^a = (D_i \phi)^a, \quad (D_0 \phi)^a = 0. \tag{28.45}$$

These equations are called *the BPS equations*, and their solutions are *BPS states*. Note that, like in the 1+1 and 2+1 dimensional cases, the BPS equations are first order in derivatives, so they are simpler to solve than the full equations of motion. Also note that the notion of BPS bound arises in supersymmetry in another way, as a bound coming from the structure of the supersymmetry algebra, for supersymmetry-preserving solutions. In that case, if we embed the non-supersymmetric theory into the supersymmetric one, we find that the BPS bound from the supersymmetry algebra reduces to the BPS bound for the fields, and the supersymmetric states are the BPS states.

For a static solution in the Coulomb gauge, $\partial_t = 0$ and $A_0 = 0$, we satisfy $E_i^a = 0$, and $(D_0 \phi)^a = 0$. Then substituting the ansatz (28.20), (28.22) into the remaining BPS equation, $B_i^a = D_i \phi^a$, we find two coupled first-order differential equations for h and K,

$$\frac{dh}{dr} = \frac{1}{2r^2}(1 - K^2)$$

$$\frac{dK}{dr} = -2hK. \tag{28.46}$$

In this BPS limit, for the BPS equations, we find the exact analytic solution

$$h(r) = \frac{1}{\tanh 2r} - \frac{1}{2r}$$

$$K(r) = \frac{2r}{\sinh 2r}. \tag{28.47}$$

28.5 Topology for the BPS and the Dirac Monopole

We have seen that the vacuum satisfies $V(\phi) = 0$ and $(D_\mu \phi)^a = 0$, leading also to

$$[D_\mu, D_\nu] \phi^a = -g(F_{\mu\nu} \phi)^a = 0, \tag{28.48}$$

leaving an unbroken group H (for $A_\mu^a \phi_{\text{vac}}^a$), and the vacuum manifold is G/H. We have also seen that the monopole solutions are classified by the maps $\phi : S_\infty^2 \to \mathcal{M}$, which means the homotopy group is $\pi_2(\mathcal{M}) = \pi_2(G/H)$. In our case, this is

$$\pi_2(SO(3)/SO(2)) = \pi_2(S^2) = \mathbb{Z}. \qquad (28.49)$$

The winding number at infinity is identified with the monopole number.

We see that we could figure out whether we have a monopole solution before writing any ansatz, simply by analyzing the homotopy group $\pi_2(\mathcal{M})$.

But there is another classification that is of relevance to our analysis. The magnetic charge that we have defined is defined at infinity, where there is a gauge field for H,

$$A : S^k \to H, \qquad (28.50)$$

where H is the unbroken gauge group. We have generalized a bit, to S^k, thinking ahead, though, in our case, $k = 2$. For gauge fields on a sphere, as we saw last time, we can define A locally on patches, and we can define two patches, $\mathcal{O}^+$ and $\mathcal{O}^-$, intersecting over $\mathcal{O}^{+-}$, as in (the generalization of) Figure 27.1a. If $\mathcal{O}^{+-}$ is the common boundary of $\mathcal{O}^+$ and $\mathcal{O}^-$, as in (the generalization of) Figure 27.1b, a sphere S^{k-1}, we have transition functions $\lambda^{+-} : \mathcal{O}^{+-} \to H$ relating the gauge fields A in the two patches.

The topological classification of this gauge field is then defined by the "winding" of the unbroken group H over S^{k-1} – i.e., by the homotopy group

$$\pi_{k-1}(H). \qquad (28.51)$$

In our physical case, $k = 2$, so that is

$$\pi_1(U(1)) = \mathbb{Z}, \qquad (28.52)$$

the winding number of the electromagnetic $U(1)$ over the circle (equator of the sphere), which corresponds, as we saw last time, to the Dirac monopole number.

The relation between the 't Hooft–Polyakov monopole number and the Dirac monopole number is provided by the theorem stating that

$$\pi_2(G/H) = \pi_1(H)_G. \qquad (28.53)$$

Here $\pi_1(H)_G$ are the equivalence classes of paths (circles) in H that are homotopic to trivial paths *in* G. In other words, it could be that some path is nontrivial in H, but is trivial in G (contractible to a point), or vice versa. But if G is simply connected, which is the case for $SU(2)$, then

$$\pi_1(H)_G = \pi_1(H). \qquad (28.54)$$

That means that, in our case,

$$\pi_2(S^2 = SU(2)/U(1)) = \pi_1(U(1))_{SU(2)} = \pi_1(U(1)) = \mathbb{Z}, \qquad (28.55)$$

and the classification of the 't Hooft monopole is the same as of the Dirac monopole. That also means that the Dirac monopole is embeddable into the 't Hooft–Polyakov monopole, as we have already seen.

Finally, there is one more bit of topological classification that will be useful later. If we have a gauge field for the group G, defined on S^4, then we have shown that its classification is done through the homotopy group $\pi_{k-1}(G) = \pi_3(G)$. But we have that for a non-Abelian group G (in particular for $G = SU(2)$),

$$\pi_3(G) = \mathbb{Z}. \tag{28.56}$$

The topological number is called instanton number, and the corresponding instanton solution will be studied in the next chapter. Of course, we are not interested in solutions on $\mathbb{R}^5$ or on an actual S^4, but all we need is that the fields are defined on a space that is *topologically* S^4.

Indeed, if we consider the total space to contain the point at infinity as well, then $\mathbb{R}^4 \cup \{\infty\} \simeq S^4$, by the equivalent of the stereographic projection (from the S^2 to the complex plane) in four dimensions. And the reason we could add the point at infinity to $\mathbb{R}^4$ is if all the fields are trivial (equal to zero) there.

Important Concepts to Remember

- In the vacuum, the Yang–Mills–Higgs system breaks $G \to H$, where H is the subgroup that leaves invariant the vacuum, and the manifold of vacua is $\mathcal{M} = G/H$.
- For the Georgi–Glashow model, this is $SU(2) \simeq SO(3) \to SO(2) \simeq U(1)$, and the manifold of vacua is $\mathcal{M} = G/H = S^2$.
- For the nontrivial vacuum of the monopole at infinity, $\phi_{vac}^a = \phi_0 x^a/r$ and $A_{i,vac}^a = -\epsilon_{ija}x^j/(gr^2)$.
- The 't Hooft–Polyakov monopole solution, $\phi^a = \phi_{vac}^a h(g\phi_0 r)$ and $A_i^a = A_{i,vac}^a[1 - K(g\phi_0 r)]$.
- For the monopole, the unbroken gauge field is $A_\mu = A_\mu^a \phi^a/\phi_0$, but the unbroken field strength, $F_{\mu\nu} = F_{\mu\nu}^a \phi^a/\phi_0$, differs from the field strength of A_μ by a term that carries the magnetic charge.
- The magnetic charge g_m of the monopole is proportional to the winding number ν characterizing the homotopy group $\pi_2(\mathcal{M}) = \mathbb{Z}$, so that $g_m = 4\pi\hbar\nu/(\hbar g)$.
- The BPS bound for the Yang–Mills–Higgs system is $M \geq \phi_0\sqrt{g_m^2 + q_e^2}$, and it can be saturated in the BPS limit $\lambda \to 0$.
- In the BPS limit, saturation of the Bogomolnyi (BPS) bound leads to the first-order BPS equations, solved by BPS states, which can be found analytically.
- Gauge fields for the group H, on S^k, are classified by $\pi_{k-1}(H)$.
- Dirac monopoles for $H = U(1)$, at infinity on S^2, are classified by $\pi_1(U(1)) = \mathbb{Z}$.
- Since $\pi_2(G/H) = \pi_1(H)_G$, and for simply connected groups, $\pi_1(H)_G = \pi_1(H)$, we embed the Dirac monopole into the 't Hooft monopole.
- For non-Abelian gauge fields for a group G on $S^4 \simeq \mathbb{R}^4 \cup \{\infty\}$, the classification is by $\pi_3(G) = \mathbb{Z}$.

Further Reading

See chapter 3 in [13], chapter 8 in [14], and section 1 in [18].

Exercises

(1)   Show that the unbroken gauge field strength is

$$F_{\mu\nu} \equiv \frac{\phi_{vac}^a}{\phi_0} F_{\mu\nu}^a = \partial_\mu A_\nu - \partial_\nu A_\mu + \frac{1}{\phi_0^3 g} \epsilon^{abc} \phi_{vac}^a \partial_\mu \phi_{vac}^b \partial_\nu \phi_{vac}^c. \qquad (28.57)$$

(2)   Prove the BPS bound formula

$$M = \int d^3x \, \mathcal{E} = \int d^3x \left\{ \frac{1}{2} \left[(E_i^a - D_i \phi^a \sin\theta)^2 + (B_i^a - D_i \phi^a \cos\theta)^2 + (D_0 \phi^a)^2 \right] + V(\phi) \right\}$$
$$+ \phi_0 (q \sin\theta + g \cos\theta). \qquad (28.58)$$

Use the fact that

$$\partial_i (\mathrm{Tr}(B_i \phi)) = \mathrm{Tr}(D_i(B_i \phi)) = \mathrm{Tr}(B_i D_i \phi), \qquad (28.59)$$

since the Bianchi identity is $D_i B_i = 0$; and a similar fact for E_i.

(3)   Check that the BPS equation $B_i^a = D_i \phi^a$ reduces to the set of equations (28.46) and that their solution is (28.47).

(4)   Calculate explicitly E_i^a, B_i^a, and then the mass M for the BPS monopole solution, to check explicitly the saturation of the BPS bound.

29

The BPST-'t Hooft Instanton Solution and Topology

In the last chapter, we saw that we can classify topology by homotopy groups, which lets us know whether we have some topological solution. The 't Hooft–Polyakov monopoles, solutions with (non-Abelian) magnetic charge, that have both a gauge field A_μ^a and scalar field ϕ^a, are classified by the homotopy group $\pi_2(\mathcal{M})$, maps of the S^2 at spatial infinity to $\mathcal{M}$ is the manifold of vacua, the "moduli space," parametrized by ϕ_{vac}^a. But pure gauge fields on the sphere S^k can be classified by the maps of the transition functions, from S^{k-1} to the gauge group G, $\pi_{k-1}(G)$.

The relevant case is $k = 4$, so gauge fields on S^4 being classified by $\pi_3(G)$, and for G non-Abelian, $\pi_3(S^4) = \mathbb{Z}$, which means there is a topological number that can be carried by some non-Abelian gauge field configuration. Of course, if we are not interested in higher dimensions, S^4 must seem a strange manifold to have. But it is only relevant that the manifold is *equivalent* to S^4, and S^4 can be obtained from $\mathbb{R}^4 \cup \{\infty\}$ (the compactification of $\mathbb{R}^4$) by a stereographic projection. The stereographic projection is the same as for the standard one (from S^2 to the compactified complex plane $\mathbb{C} \cup \{\infty\} = \mathbb{R}^2 \cup \{\infty\}$) – namely, we draw a line from the North Pole of S^4 through S^4 at some point x, and then it passes through the plane at the point y, the stereographic projection of x.

So how can the relevant manifold be $\mathbb{R}^4 \times \{\infty\}$? First, we need $\mathbb{R}^4$, so we need to consider gauge fields on an Euclidean version of our Minkowski space. Then, to consider the point at infinity as well (i.e., the compactification of $\mathbb{R}^4$), we need to make sure that *all* the fields go to zero at infinity, so the point at infinity is a single one, and trivial. For instance, that was *not* the case for the magnetic monopole, since there we had a scalar field that at infinity was in a direction-dependent vacuum, $\phi_{vac}^a \propto x^a/r$.

Therefore, in this chapter, we will consider a topological soliton solution of pure Euclidean Yang–Mills theory that will carry the topological charge associated with $\pi_3(G)$. But since, as we said, $\pi_3(G) = \mathbb{Z}$ independently of the non-Abelian group G, it is logical to assume (and that will, indeed, be the case) that the topological soliton solution carrying the charge of $\pi_3(G)$ for an arbitrary G will actually be the same one as for the minimal non-Abelian group, $SU(2)$, only embedded in the larger group G (with A_μ^a with more components, all the extra components put to zero).

That is why we consider pure Euclidean $SU(2)$ Yang–Mills. The "instanton" soliton solution was first found by Belavin, Polyakov, Schwartz, and Tyupkin (BPST), and then it was improved by 't Hooft, who also found the multi-soliton solutions and, more importantly, figured out the relevance of the instatons for quantum mechanics. There was also early work by Witten, which turned out that actually also gave multi-instantons, though it was not obvious.

29.1 Setup for Euclidean Yang–Mills and Self-Duality Condition

Euclidean space is only relevant because we construct the *Wick rotation* to Euclidean space of our own Minkowski space. That means we define

$$x_4 = it = ix^0. \tag{29.1}$$

Then the metric turns into the Euclidean metric,

$$ds^2 = dx_4^2 + dx_1^2 + dx_2^2 + dx_3^2 , \tag{29.2}$$

But together with x^0 all contravariant vectors transform in the same way, and since now $\partial_4 = -i\partial_t$, all covariant vectors must transform similarly under the Wick rotation, so, in particular, we also have

$$A_4 = -iA_0. \tag{29.3}$$

That means that the electric field of electromagnetism in Euclidean space also transforms as

$$E_i^{\text{Eucl}} = \partial_4 A_i - \partial_i A_4 = -iE_i^{\text{Mink}} , \tag{29.4}$$

whereas the magnetic field doesn't change, since it involves only spatial indices, $B_i = \frac{1}{2}\epsilon_{ijk}F_{jk}$.

The Wick rotation is used for calculation of quantum mechanical quantities like transition amplitudes, defined in Feynman's path integral formulation in terms of the phase e^{iS}, integrated over. But the phase oscillates wildly if the action S is real, like in Minkowski space. So Wick rotation to Euclidean space is defined such that we turn this $e^{iS^{\text{Mink}}}$ into $e^{-S^{\text{Eucl}}}$, such that the integral turns from a highly oscillatory one to a highly convergent one. That means that

$$iS^{\text{Mink}} = i\int dt\, L^M = -S^{\text{Eucl}} = -\int dx_4 L^E = -i\int dt L^E \Rightarrow L^E = -L^M. \tag{29.5}$$

Then, finally, the Euclidean Lagrangean for electromagnetism is

$$\mathcal{L}^E = -\mathcal{L}^M = -\frac{1}{2}(\vec{E}_M^2 - \vec{B}_M^2) = +\frac{1}{2}(\vec{E}_E^2 + \vec{B}_E^2) = +\frac{1}{4}F_{\mu\nu}^E F_E^{\mu\nu} \tag{29.6}$$

where $F_{4i}^E = E_i^E$ and $F_{ij}^E = \epsilon_{ijk}B_k^E$.

For a Yang–Mills field, we only need to add a trace. But, moreover, we will use a convention for Yang–Mills fields that is very useful. Until now, we have used the standard notation, with $F_{\mu\nu} = \partial_\mu A_\nu - \partial_\nu A_\mu - ig[A_\mu, A_\nu]$, but it is sometimes useful to factor out the dependence on g in the action by redefining $A_\mu = \tilde{A}_\mu/g$ and $F_{\mu\nu} = \tilde{F}_{\mu\nu}/g$, in which case

$$\tilde{F}_{\mu\nu} = \partial_\mu\tilde{A}_\nu - \partial_\nu\tilde{A}_\mu - i[\tilde{A}_\mu, \tilde{A}_\nu], \tag{29.7}$$

and here we will use this definition, dropping the tildes. Then with this convention, the Euclidean Yang–Mills Lagrangean is

$$\mathcal{L}_{YM}^E = +\frac{1}{4g^2}\text{Tr}[F_{\mu\nu}F^{\mu\nu}]. \tag{29.8}$$

Moreover, we define the (Poincaré) dual field strength in Euclidean space as (note that in Euclidean space the position of the indices, up or down, doesn't matter, so I will put all of them down)

$$* F_{\mu\nu} = \frac{1}{2} \epsilon_{\mu\nu\rho\sigma} F_{\rho\sigma},\tag{29.9}$$

where $\epsilon_{1234} = +1$. But since we are in Euclidean space, it means that now $*^2 = 1$, since

$$\frac{1}{2} \epsilon_{\mu\nu\rho\sigma} \frac{1}{2} \epsilon_{\rho\sigma\lambda\tau} = \delta_{\mu\nu}^{\lambda\tau}.\tag{29.10}$$

This is in contrast with the Minkowski space, where, as we saw, $*^2 = -1$, since (29.10) has a minus on the right-hand side.

But the positive sign means that now we can rewrite the Yang–Mills action by completing squares and leaving over a topological quantity, just like in the BPS bounds in lower dimensions,

$$S_{YM}^E = +\frac{1}{4g^2} \int d^4x (F_{\mu\nu}^a)^2$$
$$= \int d^4x \left[\pm \frac{1}{4g^2} F_{\mu\nu}^a * F^{a\mu\nu} + \frac{1}{8g^2} (F_{\mu\nu}^a \mp *F_{\mu\nu}^a)^2 \right].\tag{29.11}$$

Since the last term is positive, we have a kind of BPS bound, but for the action, not the energy,

$$S_{YM}^E \geq \frac{1}{4g^2} \left| \int d^4x F_{\mu\nu}^a * F^{a\mu\nu} \right|.\tag{29.12}$$

The bound is saturated only for (anti-)self-dual configurations,

$$F_{\mu\nu}^a = \pm F_{\mu\nu}^a ,\tag{29.13}$$

which are the equations of motion for the instanton/anti-instanton solutions. These solutions then will be topological, with topological charge associated with the remaining expression for the action.

The plus sign in the equations of motion is called self-dual, and the minus sign is called anti-self-dual. We note again that these self-dual configurations are only possible because of the Euclidean signature, where $*^2 = +1$. In Minkowski space, where $*^2 = -1$, the same kind of BPS bound would be saturated only for $F_{\mu\nu}^a = \pm i F_{\mu\nu}^a$, which doesn't have real solutions. We only have real solutions in Euclidean space.

Moreover, the bound on the action leaves a topological invariant,

$$\int d^4x \operatorname{Tr}[F_{\mu\nu} * F^{\mu\nu}] = \int d^4x \frac{1}{2} \epsilon^{\mu\nu\rho\sigma} \operatorname{Tr}[F_{\mu\nu} F_{\rho\sigma}] = \int 2 \operatorname{Tr}[F \wedge F] ,\tag{29.14}$$

where the last expression is in form language. We can check that the result is a topological invariant (i.e., independent on the metric of the space) since it is a total derivative (an exterior derivative d in form language), so it is given by a homotopy group.

Indeed, we have

$$\operatorname{Tr}[F \wedge F] = \operatorname{Tr} \left\{ d \left[dA \wedge A + \frac{2}{3} A \wedge A \wedge A \right] \right\} \equiv d\mathcal{L}_{\text{CS}},\tag{29.15}$$

since we can check that

$$d\mathcal{L}_{\mathrm{CS}} = dA \wedge dA + 2dA \wedge A \wedge A = \mathrm{Tr}(F \wedge F), \tag{29.16}$$

since $F = dA + A \wedge A$. Using the Stokes theorem, we find

$$\int_{M_4} d^4x \, \mathrm{Tr}[F \wedge F] = \int_{\partial M_4 = \Sigma^3_\infty} \mathcal{L}_{\mathrm{CS}}, \tag{29.17}$$

where the three-form integrated at infinity is called the *Chern–Simons form*. The fact that the four-form is a total derivative (is an *exact* form) can be obtained also from the fact that its total derivative is zero (it is a *closed* form),

$$d \, \mathrm{Tr}[F \wedge F] = 2 \, \mathrm{Tr}[dF \wedge F] = 4 \, \mathrm{Tr}[dA \wedge A \wedge F] = 4 \, \mathrm{Tr}[dA \wedge A \wedge (dA + A \wedge A)] = 0, \tag{29.18}$$

where we have used the Bianchi identity $DF = dF - 2dA \wedge A = 0$, but more importantly, that we have five antisymmetrized indices, when there are only four coordinates.

29.2 Chern Forms

We now make a mathematical interlude, to define some topological invariants for non-Abelian gauge fields. We have seen that the *first Chern form* for an Abelian field is $\frac{F}{2\pi}$, its integral over an S^2, the *first Chern number*, being the topological invariant associated with magnetic monopole charge.

In a non-Abelian theory, we have the simplest possible generalization of the first Chern form, namely,

$$C_1 = \frac{1}{2\pi} \, \mathrm{Tr}[F] \,, \tag{29.19}$$

which picks out the Abelian part. Indeed, for the gauge group $U(N)$, which we will mostly be interested in, the trace picks out the $U(1)$ part, since $SU(N)$ means matrices of determinant one, $1 = \det M = \exp[\mathrm{Tr}[\ln M]]$, and since $M = \exp[i\alpha^a T_a]$, it means that $\mathrm{Tr}[T_a] = 0$. This leaves only the $U(1)$ associated with the determinant of matrices (the generator T proportional to the identity) as the only one with nonzero trace. The first Chern number is its integral,

$$c_1 = \int_{M_2} C_1. \tag{29.20}$$

For $SU(N)$, the case we are mostly interested in, this is zero.

The *second Chern form* is, however,

$$C_2 = \frac{1}{8\pi^2}[\mathrm{Tr}[F \wedge F] - \mathrm{Tr} \, F \wedge \mathrm{Tr} \, F], \tag{29.21}$$

and in our relevant case of $SU(N)$, when there is no $U(1)$ part, we have

$$C_2(SU(N)) = \frac{1}{8\pi^2} \, \mathrm{Tr}[F \wedge F]. \tag{29.22}$$

This is our four-form from the Lagrangean, with the normalization chosen so that its integral, the *second Chern number*, is an integer,

$$c_2 = \int_{M_4} C_2 = N \in \mathbb{Z}. \tag{29.23}$$

This topological number is also called the *instanton number* in the physics literature, or the *Pontryagin index* in the mathematics literature.

The case we are interested in is for gauge group $G = SU(2)$, defined on $\mathbb{R}^4$, with boundary condition at infinity that the field strength vanishes at infinity,

$$F_{\mu\nu} \to 0 \quad \text{for} \quad r \to \infty. \tag{29.24}$$

This is a necessary condition for finite energy (it would be sufficient if we would say that $F_{\mu\nu}$ drops faster than $1/r^2$ at infinity).

In this case, it means that on the Σ^3_∞, the three-surface at infinity, the gauge field is a pure gauge,

$$A_\infty = -dU_\infty \cdot U^{-1}. \tag{29.25}$$

[Remember that a gauge transformation is (in the convention we use here, with g factored out, and in form language) $A^U = UAU^{-1} - idU \cdot U^{-1}$, so by putting $A = 0$, the pure gauge field, gauge transformed from 0, is (29.25).] In this case, the gauge group element U_∞, for $\Sigma^3_\infty = S^3_\infty$, is a map $S^3_\infty \to SU(2)$, which is an element in an equivalence class of the homotopy group $\pi_3(SU(2)) = \mathbb{Z}$, characterized by the instanton number.

In this case, since $F(\Sigma^3_\infty) = 0$, we obtain for the second Chern number

$$c_2 = \int_{\mathbb{R}^4} C_2 = \int_{\partial\mathbb{R}^4 = \Sigma^3_\infty} \mathcal{L}_{CS} = \frac{1}{8\pi^2} \int_{\Sigma^3_\infty} \text{Tr}\left[F \wedge A - \frac{1}{3} A \wedge A \wedge A \right]$$

$$= -\frac{1}{24\pi^2} \int_{\Sigma^3_\infty} \text{Tr}[A \wedge A \wedge A]$$

$$= \frac{i}{24\pi^2} \int_{\Sigma^3_\infty} \text{Tr}[dU_\infty \cdot U_\infty^{-1} \wedge dU_\infty \cdot U_\infty^{-1} \wedge dU_\infty \cdot U_\infty^{-1}]. \tag{29.26}$$

But we can consider a Σ^3_∞ that is made up of two $\mathbb{R}^3$ surfaces, one at $x_4 = -\infty$, the other at $x_4 = +\infty$, together with the surface at infinity in $\mathbb{R}^3$ times the x_4 line, as in Figure 29.1. Then on the $\mathbb{R}^3$ we have a topological number = "winding number," equal to the integral in (29.26), integrated over $\mathbb{R}^3$.

This *winding number* is associated with the same $\pi_3(G)$ (since it is the same form, integrated over a different surface), just that now the S^3 is really the compactified $\mathbb{R}^3$ at $x_4 = \pm\infty$, i.e., $\mathbb{R}^3 \cup \{\infty\}$, where again we can only add the point at infinity in $\mathbb{R}^3$ since we must have the boundary condition $U \to \mathbb{1}$ on the boundary of the space $\partial\mathbb{R}^3$. The U_∞ on $\mathbb{R}^3$ of winding number n is called a "large gauge transformation." It, in fact, turns out that the same winding number on $\mathbb{R}^3$ can be associated with the 't Hooft–Polyakov (non-Abelian) monopole number (for $\pi_2(G/H)$, where S^2 is the sphere at infinity), and U_∞ is a gauge transformation relating it to the vacuum.

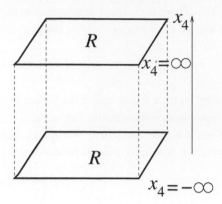

Figure 29.1 The domain of the instanton, bounded by two domains, at $x_4 = -\infty$ and $x_4 = +\infty$ (and a surface at spatial infinity times time), such that the instanton number is the difference of the winding numbers on the two surfaces.

Then, as we see from the formula (29.26) for c_2, the instanton number equals the *difference* (since in Σ_3, the two $\mathbb{R}^3$ surfaces come with opposite directions, relative to x_4) of the winding numbers, at $x_4 = +\infty$ and $x_4 = -\infty$.

29.3 The Instanton Solution

We turn now to the explicity instanton solution, in the formulation of 't Hooft.

We first define the quantities $\Sigma_{\mu\nu}$ and $\tilde{\Sigma}_{\mu\nu}$, antisymmetric in $\mu\nu$, whose components are all 2×2 matrices, by

$$\Sigma_{i4}(\tilde{\Sigma}_{i4}) = \pm\sigma_i, \quad \Sigma_{ij}(\tilde{\Sigma}_{ij}) = \epsilon_{ijk}\sigma_k, \tag{29.27}$$

where σ_i are the Pauli matrices. Equivalently, we define the *'t Hooft symbols*, also antisymmetric in $\mu\nu$, $\eta_{\mu\nu}^a$, and $\tilde{\eta}_{\mu\nu}^a$ by

$$\eta_{ij}^a(\tilde{\eta}_{ij}) = \epsilon^{aij}, \quad \eta_{i4}^a(\tilde{\eta}_{i4}^a) = \pm\delta_i^a. \tag{29.28}$$

The reason for these definitions is that $\Sigma_{\mu\nu}$ is self-dual, whereas $\tilde{\Sigma}_{\mu\nu}$ is anti-self-dual.

$$\frac{1}{2}\epsilon_{\mu\nu\alpha\beta}\Sigma_{\alpha\beta} = +\Sigma_{\mu\nu}, \quad \frac{1}{2}\epsilon_{\mu\nu\alpha\beta}\tilde{\Sigma}_{\alpha\beta} = -\tilde{\Sigma}_{\mu\nu}, \tag{29.29}$$

and the 't Hooft symbols are their decomposition in terms of $T_a = \sigma_a/2$, the generators of $SU(2)$,

$$\Sigma_{\mu\nu} = \eta_{\mu\nu}^a \frac{\sigma_a}{2}, \quad \tilde{\Sigma}_{\mu\nu} = \tilde{\eta}_{\mu\nu}^a \frac{\sigma_a}{2}. \tag{29.30}$$

Then the **ansatz for the instanton solution** is

$$A_\mu(x) = \tilde{\Sigma}_{\mu\nu}\partial_\nu \ln \phi(x) \Rightarrow$$
$$A_\mu^a(x) = \tilde{\eta}_{\mu\nu}^a \partial_\nu \ln \phi(x). \tag{29.31}$$

Substituting the ansatz into the equation of motion for the instanton – i.e., $F_{\mu\nu} = *F_{\mu\nu}$ – gives the equation

$$\frac{1}{\phi}\Delta_4\phi = 0, \tag{29.32}$$

where Δ_4 refers to the Laplacean in four Euclidean dimensions. The proof of this statement is left as an exercise.

If ϕ is nonsingular, then the equation reduces to $\Delta_4\phi = 0$. But this equation, on four-dimensional Euclidean space, has as the only nonsingular solution the constant, which results in $A_\mu = 0$. This is so, since we can decompose $\mathbb{R}^4$ as $\mathbb{C}^2$, and on $\mathbb{C}$ we have that $\Delta\phi = \partial\bar{\partial}\phi = 0$ has the general solution $\phi = f(z) + g(\bar{z})$, but the only holomorphic (with no singularity) function on the whole complex plane is the constant. Similarly, on $\mathbb{C}^2$, $\Delta_4\phi = \partial_i\bar{\partial}_i\phi = 0$ with the same result.

But if ϕ is singular, we can have nontrivial solutions – for instance, the basic solution

$$\phi = \frac{1}{|x|^2}. \tag{29.33}$$

Then, for $x \neq 0$, we have

$$\frac{1}{\phi}\Delta_4\phi = |x|^2 \partial_\mu \left(\frac{-2x_\mu}{x^4}\right) = 0. \tag{29.34}$$

That means the result can only be nonzero for $x = 0$ – i.e., proportional to a delta function. But, in fact, $1/|x|^2$ is a solution of the Poisson equation on $\mathbb{R}^4$ – i.e., a Green's function for the four-dimesional Laplacean. More precisely,

$$\Delta_4 \frac{1}{|x|^2} = -4\pi^2\delta^4(x). \tag{29.35}$$

To prove that, we integrate over a ball B^4 of radius R centered around 0, and after using the Stokes's theorem, we obtain

$$-4\pi^2 = \int_{B^4} d^4x \frac{1}{|x|^2} = \oint_{\partial B^4 = S_R^3} d\Sigma^\mu \partial_\mu \frac{1}{|x|^2} = \left(\int d\Omega\right) R^3 \frac{\partial}{\partial R}\frac{1}{R^2}, \tag{29.36}$$

and since $\int_{S^3} d\Omega = 2\pi^2$, we obtain an identity ($-4\pi^2 = -4\pi^2$).

Finally, that means that

$$\frac{1}{\phi}\Delta_4\phi = -4\pi^2|x|^2\delta^4(x) = 0, \tag{29.37}$$

which is zero even at $x = 0$, because of the prefactor $|x|^2$. That means that $\phi = 1/|x|^2$ is a singular solution. But, except for the $1/\phi$ factor, the equation is linear, which means we can add various solutions to obtain a new one.

We can therefore add several $1/|x|^2$ solutions centered at different points, and multiplied by different coefficients, together with a constant one, 1, to find the *multi-instanton solution* (or N–instanton solution),

$$\phi(x) = 1 + \sum_{i=1}^{N} \frac{\rho_i^2}{|x_\mu - a_\mu^i|^2}. \tag{29.38}$$

Note that for $x_\mu \to a_\mu^i$, the corresponding term is the only one that diverges, so $1/\phi \simeq |x_\mu - a_\mu^i|^2$, which is why (29.38) is a solution of $(\Delta_4 \phi)/\phi = 0$.

For $N = 1$ – i.e., for the single instanton – substituting ϕ into A_μ^a, we find

$$A_\mu^a = \tilde\eta_{\mu\nu}^a |x_\mu - a_\mu|^2 \frac{\partial_\nu [\rho^2/(x_\mu - a_\mu)^2]}{(x_\mu - a_\mu)^2 + \rho^2} = -2\tilde\eta_{\mu\nu}^a \frac{\rho^2}{(x_\mu - a_\mu)^2} \frac{(x - a)_\nu}{(x_\mu - a_\mu)^2 + \rho^2}. \quad (29.39)$$

But this gauge field is singular at $x_\mu = a_\mu$. Of course, what is relevant is whether the field strength is singular or not, but, in fact, we find that by a gauge transformation, we can transform the gauge field to a nonsingular form.

We use the gauge transformation group element

$$U(y) = \frac{y_4 + iy_i\sigma_i}{|y|}, \quad y_\mu \equiv x_\mu - a_\mu, \quad (29.40)$$

which is, indeed, an element of $SU(2)$, since, as we have shown in Chapter 3, it is of the type

$$U = e^{i\vec\alpha \cdot \vec\sigma} = \cos|\vec\alpha| + i\frac{\vec\alpha \cdot \vec\sigma}{|\vec\alpha|} \sin|\vec\alpha|, \quad (29.41)$$

and the previous qualifies.

Then we calculate

$$[U(y)]^{-1}\partial_{y^\mu} U(y) = -2i\tilde\Sigma_{\mu\nu}\frac{y_\nu}{y^2} \Rightarrow U\partial_\mu U^{-1} = -2i\Sigma_{\mu\nu}\frac{y_\nu}{y^2}, \quad (29.42)$$

and substitute in the gauge transformation of A_μ to obtain

$$^U A_\mu(x) = U(y)\left[A_\mu + \frac{i}{g}\partial_\mu\right] U^{-1}(y) = +\frac{2}{g}\Sigma_{\mu\nu}\frac{y_\nu}{y^2 + \rho^2}, \quad (29.43)$$

which we leave as an exercise to prove. Note that in the formula, $\tilde\Sigma_{\mu\nu}$ is replaced by $\Sigma_{\mu\nu}$, and now the gauge field is nonsingular,

$$^U A_\mu^a = +\frac{2}{g}\eta_{\mu\nu}^a \frac{(x - a)_\nu}{(x - a)^2 + \rho^2}. \quad (29.44)$$

Moreover, we can check that at infinity it becomes a pure gauge, which we can see since $A_\mu \sim 1/r^3$, but $^U A_\mu \sim 1/r$ at $r \to \infty$. That means that, indeed, at $r \to \infty$,

$$^U A_\mu^a \to \frac{i}{g} U(x)\partial_\mu U^{-1}. \quad (29.45)$$

The action for the instanton can be calculated. First, (after some algebra, left as an exercise) we obtain that the Euclidean Lagrangean is

$$\mathcal{L}^E = \frac{1}{4}\,\mathrm{Tr}[F_{\mu\nu}F^{\mu\nu}] = \frac{48}{g^2}\frac{\rho^4}{[(x - a)^2 + \rho^2]^4}. \quad (29.46)$$

Integrating it to obtain the Euclidean action, we find

$$S^E = \int d^4 y \mathcal{L}^E = \frac{48\rho^4}{g^2} \int d^4 y \frac{1}{[y^2 + \rho^4]^4} = \frac{48 \cdot 2\pi^2 \rho^4}{g^2} \int_0^\infty y^3 dy \frac{1}{[y^2 + \rho^4]^4}$$

$$= \frac{48\pi^2}{g^2} \rho^4 \int_{\rho^2}^\infty (w - \rho^2) \frac{dw}{w^4} = \frac{8\pi^2}{g^2}. \tag{29.47}$$

Moreover, we already saw that the action *on the self-dual configurations* is proportional to the instanton number $c_2 = N$, so the action for the multi-instanton solution is

$$S_N = N S_1 = \frac{8\pi^2 N}{g^2}. \tag{29.48}$$

Note that the various instanton numbers divide the gauge field configurations into topological classes.

29.4 Quantum Interpretation of the Instanton

As we said, an important contribution of 't Hooft was to find the quantum interpretation of the instanton, which is why it has become an important component of non-perturbative studies of gauge theories.

Indeed, in quantum mechanics, transition amplitudes T (whose probabilities are $|T|^2$) are given by "path integrals." When Wick rotating to Euclidean space in order to calculate and expanding around the minimum configuration, we obtain

$$T \propto \int \mathcal{D}(\ldots) e^{-S_{min} - \delta S} = e^{-S_{min}} \int \mathcal{D}\phi e^{-\delta S} \sim e^{-S_{min}}. \tag{29.49}$$

Therefore, the transition amplitude is given in the first (semiclassical) approximation by the action of the minimum configuration in the allowed topological class.

But, as we said, we could consider Minkowski space with $\mathbb{R}^3$ boundaries at $t = \pm\infty$, and then after the Wick rotation to Euclidean space, we can complete this to the surface Σ_∞^3 relevant for the instanton (see Figure 29.1), which will be bounded by gauge field configurations with winding number (related to monopole number) N at $x_4 = -\infty$ and $N + 1$ at $x_4 = +\infty$.

But the path integral for the transition amplitude T must have as boundary conditions for the field configurations to be integrated over the states at $x_4 = -\infty$ and $x_4 = +\infty$, which by the above definition have winding numbers N and $N + 1$. Then, finally,

$$T(\{A\}_1(N) \to \{A\}_2(N+1)) \propto e^{-S_{min}(A_1,A_2)} = e^{-S_{inst}} = e^{-\frac{8\pi^2}{g^2}}. \tag{29.50}$$

For a small g, this gives an extremely small transition probability, which is why we have not seen monopoles popping out of the vacuum: the transition probability for $N = 0$ to turn into $N = 1$ is ridiculously small.

Moreover, we can extend, at a more formal level, this reasoning to topological solitons in lower dimensions. Namely, we could cook up models such that the monopole (solution in three spatial dimensions) gives the transition amplitude to go between 2+1 dimensional theories with different vortex numbers, and vortices (solutions in two spatial dimesions) give the transition amplitudes to go between 1+1 dimensional theories with different kink numbers.

Important Concepts to Remember

- The instanton solution of BPST and 't Hooft is associated with the integer of the homotopy group $\pi_3(G)$, for gauge fields on $S^4 \simeq \mathbb{R}^4 \cup \{\infty\}$.
- The instanton is a solution of pure $SU(2)$ Yang–Mills theory in Euclidean space (Wick rotated Minkowski space), and any non-Abelian instanton is this instanton embedded in the larger group G.
- Instantons are self-dual solutions saturating a BPS bound, with topological charge $\propto \int \text{Tr}[F \wedge F] = \int d\mathcal{L}_{CS}$.
- For $SU(N)$, the instanton number (or Pontryagin number) equals the second Chern number, the integral of the second Chern class. The first Chern class is $\text{Tr}\, F/(2\pi)$, so it picks up only the $U(1)$ part of an $U(N)$ group.
- The instanton number equals the difference in winding numbers, given by the maps $U: S^3 \rightarrow G$ on compactified $\mathbb{R}^3$, and related to monopole number.
- The instanton ansatz is $A_\mu = \tilde{\Sigma}_{\mu\nu}\partial_\nu \ln \phi(x)$, with $\tilde{\Sigma}_{\mu\nu}$ constant self-dual matrices and $\phi(x)$ a harmonic function $1 + \sum_i \rho_i^2/(x - d^i)^2$.
- After a gauge transformation, the gauge field becomes nonsingular and is written in terms of $\Sigma_{\mu\nu}$.
- The action for the instanton is $S_1 = 8\pi^2/g^2$ and for N instantons is $S_N = NS_1$.
- The instanton action S_1 governs the transition amplitudes between states on spatial $\mathbb{R}^3$ with different winding numbers, matching to states of different monopole numbers, by $T \propto e^{-S_1}$.
- Monopoles govern transitions between vortices, and vortices transitions between kinks.

Further Reading

See chapter 4 in [13] and chapter 10 in [14].

Exercises

(1)   Show that for the ansatz

$$A_\mu^a(x) = \tilde{\eta}_{\mu\nu}^a \partial_\nu \ln \phi ,$$ (29.51)

imposing the self-duality condition $F_{\mu\nu} = *F_{\mu\nu}$ gives

$$\frac{1}{\phi}\Delta_4\phi = 0.$$ (29.52)

(2)   Prove that the instanton solution $A_\mu^a(x)$, transformed by

$$U(y) = \frac{y_4 + iy_iT_i}{|y|}, \quad y_\mu = x_\mu - a_\mu,$$

(29.53)

becomes

$$^UA_\mu^a(x) = +\frac{2}{g}\frac{\eta_{\mu\nu}^a(x-a)_\nu}{(x-a)^2 + \rho^2}.$$

(29.54)

(3)   Prove that the Euclidean Yang–Mills Lagrangean on the instanton solution gives (29.46).

(4)   Check explicitly that the action of the two-instanton solution, with ϕ in (29.38), gives twice the action of the single instanton solution.

(5)   Consider a gauge group parameter

$$U = e^{if(r)\vec{n}\cdot\vec{\sigma}}, \quad \vec{n} = \frac{\vec{r}}{r}.$$

(29.55)

Find the condition on $f(r)$ such that at infinity, we have

$$A_i^a = \epsilon_{ija}\frac{x^j}{r^2},$$

(29.56)

and show that in this case we have a winding number of 1.

30 General Topology and Reduction on an Ansatz

In this chapter, we try to put together the various informations we had on topology into a coherent picture and to organize the way we construct ansätze.

30.1 Topological Classification of Scalars and Homotopy Groups

We have seen that the homotopy groups are groups of maps from spheres to manifolds, $\pi_n(\mathcal{M})$ is defined by the map $S^n \to \mathcal{M}$. We have also seen that generally the topology of solitons is described by these homotopy groups.

Nonlinear interactions are necessary in general to stabilize the theory and impose finiteness of the energy, though we have seen that in a linear, purely gauge theory (Maxwell's electromagnetism without sources), we could have a notion of topology as well.

For a purely scalar theory, with scalars taking value in some (group, or coset) manifold $\mathcal{M}$ – i.e., with *global* symmetry group G – the energy of a nontrivial field configuration will be in general infinite, as seen from Derrick's theorem. We say that we have *global solitons*.

In order to have finite energy, either for global or local (coupled to gauge fields) solitons, the field at infinity must be in the vacuum manifold $\mathcal{V} \subset \mathcal{M}$ that minimizes (makes zero) the potential.

In that case, we can classify the scalars by the map from the spatial sphere at infinity S^{d-1}_∞, to the vacuum manifold,

$$\phi(\infty) : S^{d-1}_\infty \to \mathcal{V}, \tag{30.1}$$

which defines the homotopy group $\pi_{d-1}(\mathcal{V})$. But, moreover, if the scalar field takes different values in different directions (different points in S^{d-1}_∞), then the global soliton has infinite energy.

To have a finite energy, the scalar field must go to a constant (the same value) in all directions S^{d-1}_∞. In that case, we can identify all the points at infinity, and we have effectively compactified space, $\mathbb{R}^d \cup \{\infty\} \simeq S^d$ by the stereographic projection. Then the scalar field map over the whole of space, $\phi : \mathbb{R}^d \to \mathcal{M}$, is extended to the map

$$\phi : S^d \to \mathcal{M}, \tag{30.2}$$

characterized by the homotopy group $\pi_d(\mathcal{M})$. The classic examples of the this behavior are models with energy depending only on $\partial_\mu \phi$, like the nonlinear sigma models

($\mathcal{L} = g_{IJ}(\phi)\partial_\mu \phi^I \partial^\mu \phi^J$) and the Skyrme models (models with Skyrme term or general-izations, which have higher powers of $\partial_\mu \phi$).

d = 1

In one spatial dimension, we could have purely scalar solitons of finite energy according to Derrick's theorem. The topological classification is according to the values of the scalars at the two points at infinity $x = \pm\infty$. Now we have two possible types, as we saw:

- In the case of the kink, the points at infinity are distinct, $\phi(+\infty) \neq \phi(-\infty)$, and the classification (according to the general principle previously noted) is by $\pi_0(V)$. For the kink, this group has the elements $\{-1, 0, +1\}$.
- In the case of sine-Gordon, we can identify the field modulo 2π because of the periodicity of the potential. Then the scalar manifold is effectively $\mathcal{M} = S^1$. Moreover, because of the identification, now $\phi(+\infty) - \phi(-\infty)$, which means we can identify $x = +\infty$ with $x = -\infty$, and topologically, the spatial line becomes also S^1. That means that the relevant homotopy group classifying the solutions is $\pi_1(S^1) = \mathbb{Z}$, which is indeed correct.

d = 2

In $d = 2$ spatial dimensions, we can only have global solitons of infinite energy, per Derrick's theorem. Indeed, we have seen these global solitons, with periodic field $\theta = N\alpha$ (α is the polar angle in the two spatial dimensions), for the rotor model in Chapter 22*. There, energy was logarithmically divergent, $E \propto \ln R$, where R was the size of the system. According to the general principle, the classification of the solutions is according to the map $\phi(\infty) \colon S^1_\infty \to V$ – i.e., by $\pi_1(V)$. Then the only nontrivial case is for $V = U(1) = S^1$, when $\pi_1(S^1) = \mathbb{Z}$, like in the case of the rotor model soliton.

For a model with $SO(n)$ symmetry for the scalars, and if $V = S^{n-1} = SO(n)/SO(n-1)$, for $n > 2$, $\pi_1(S^{n-1}) = \{0\}$, so there are no solitons.

For a model coupled to gauge fields, we now can have finite energy local solitons, since the energy contains covariant derivatives $D_i\phi$, which can go to zero, even when the field becomes nontrivial at infinity.

For a nonlinear sigma model or Skyrme model, if the symmetry group is $SO(3)$ and $\mathcal{M} = S^2$, then we can consider a constant field at infinity compactifying space, and the fields are classified by $\phi : S^2 \to \mathcal{M}$, with homotopy group $\pi_2(\mathcal{M}) = \mathbb{Z}$. This is the case of the $O(3)$ sigma model and Skyrme model in 2 dimensions.

d = 3

In $d = 3$ spatial dimensions (the general physical case), we can have scalars pointing in different directions at infinity parametrized by the homotopy group $\pi_2(V)$. For $V = S^2 = SO(3)/SO(2)$, we have $\pi_2(V) = \mathbb{Z}$, and this is the monopole number.

We can have global monopoles of infinite energy (due to the gradient term at infinity), $\phi^a = x^a/r \equiv \theta^a$, or local monopoles. The latter case is the 't Hooft–Polyakov monopole for gauge group $G = SO(3)$, with vacuum manifold $G/H = SO(3)/SO(2)$.

For a space compactified by a constant boundary condition for the field at infinity, $\phi : S^3 \to \mathcal{M}$, so the relevant homotopy group for topological classification is $\pi_3(\mathcal{M})$. There are now 2 relevant cases:

- The simplest case is the Skyrme model with $\mathcal{M} = S^3$, when $\pi_3(S^3) = \mathbb{Z}$. But since as we said before, we can embed $S^3 \simeq SU(2)$ into any $SU(N)$ or any compact, non-Abelian simple group G, then $\pi_3(G) = \mathbb{Z}$, and we can find Skyrmion solutions.
- But we can also have $\mathcal{M} = S^2$, and then $\pi_3(S^2) = \mathbb{Z}$ defines the Hopf index. The Skyrme-Faddeev model is an example, and it then has topological solitons.

30.2 Gauge Field Classifications

We have seen that the solutions for the scalar fields can be classified by their topology from the homotopy groups. But in a theory with gauge fields, we can (also) classify the gauge fields by their topology.

For a gauge field in some group H defined over a sphere S^k, like the sphere at infinity S^k_∞, the topological classification is defined as follows. The gauge field is subject to gauge transformations, but that results in a consistency condition. We can consider two patches – for instance, the patch "north of the equator" O^+ and the patch "south of the equator" O^-, and their intersection, the equator $O^{+-} = O^+ \cap O^- = S^{k-1}$. The gauge field cannot be defined globally, only on patches, and then the gauge fields on the two patches differ by a gauge transformation, or transition function, with values in the same group H. The transition function must "wind" an integer number of times around S^{k-1}. Thus, the topological classification for the gauge fields on S^k is given in terms of maps

$$f : S^{k-1} \to H, \tag{30.3}$$

defining the homotopy group $\pi_{k-1}(H)$.

The simplest case is of a linear (noninteracting) gauge field – i.e., Maxwell electromagnetism – for the gauge group $U(1)$:

- We can consider time-dependent ansätze, in which case we can have a topological number associated with the whole time-dependent solution and related to the existence of a particular ansatz. This is the case of the Hopfion solution, which was related to solving the Maxwell equations in terms of two complex scalars: either α and β in Bateman's ansatz $\vec{F} = \vec{E} + i\vec{B} = \vec{\nabla}\alpha \times \vec{\nabla}\beta$, or θ and ϕ in writing $F_{\mu\nu}$ and its dual $*F_{\mu\nu}$ as the two-forms on $S^3 \simeq \mathbb{R}^3 \cup \{\infty\}$ pulled back from an S^2. Then the space is compactified by the fact that, at infinity, the complex scalars are constant (in all directions), and as a result, we have Hopf maps

$$\phi, \theta : S^3 \simeq \mathbb{R}^3 \cup \{\infty\} \to S^2 \simeq \mathbb{C} \cup \{\infty\}, \tag{30.4}$$

characterized by a Hopf index. Even in this case, the existence of the topological index (Hopf index) is related to the conservation (constancy in time) of the helicities H_{ab}, $a, b = e$ or m.

- We can also consider static gauge field solutions. Consider the $H = U(1) = S^1$ valued fields on the physical space $\mathbb{R}^3$. In the absence of a constant boundary condition at infinity, the relevant topology is the one on the sphere at infinity S^2_∞, thus $k = 2$ in the general case. Then the topology is defined by $\pi_1(H) = \pi_1(S^1) = \mathbb{Z}$, defining the *Dirac monopole number*. The corresponding solutions are the singular Dirac monopoles.

The next case is of non-Abelian gauge fields, for a group G, in physical space $\mathbb{R}^3$, with nonconstant boundary condition at infinity. Considered by themselves, the same logic as for the monopole number would apply, and the relevant homotopy group would be $\pi_1(H)$, where H is the subgroup relevant at infinity. But for this to give a monopole number, it would assume that we must have a subgroup $H = U(1) \subset G$, and then, moreover, we would need that

$$\pi_1(H)_G = \pi_1(H), \tag{30.5}$$

where $\pi_1(H)_G$ are the equivalence classes of paths ($=$ circles, S^1) in H that are homotopically trivial *in G*, and the equality is true if G is simply connected, like for $G = SU(2) \simeq S^3$. Moreover, we have a theorem that

$$\pi_2(G/H) = \pi_1(H)_G, \tag{30.6}$$

which now relates the monopole number of the 't Hooft–Polyakov (scalar + gauge field) monopole, defined via the scalars by $\pi_2(G/H)$, to the Dirac monopole number: For $H = U(1)$ and $G = SU(2)$, we obtain

$$\pi_2(S^2 = SU(2)/U(1)) = \pi_1(U(1))_{SU(2)} = \pi_1(U(1)). \tag{30.7}$$

Finally, the last relevant case is of the non-Abelian gauge field on $S^4 \simeq \mathbb{R}^4 \cup \{\infty\}$, corresponding to an instanton configuration on the four-dimensional Euclidean space, with a constant boundary condition at infinity, more precisely zero (vanishing gauge field), so that we can compactify space. Then the instanton number is defined by the gauge field

$$A^a_\mu : S^4 \to H, \tag{30.8}$$

classified by $\pi_3(H)$. Then for $H = SU(2) \simeq S^3$ and any G that includes this H, we have $\pi_3(G) = \mathbb{Z}$, so we can embed the BPST instanton, for $SU(2)$ in the theory.

30.3 Derrick's Argument with Gauge Fields

Like in Chapter 24, we consider rescaled fields (now including the gauge fields)

$$\tilde{\phi}(x) = \phi(\lambda x), \quad \tilde{A}^a_\mu(x) = \lambda A^a_\mu(\lambda x), \tag{30.9}$$

where we have included an extra λ for gauge fields, since they are one-forms, just like $\partial/\partial x^\mu$. We consider the energy of the tilde configuration, in the case there is also a kinetic term for the gauge fields,

$$E(\lambda) = \int d^d x \left[\frac{1}{4}\tilde{F}_{ij}^2 + \frac{1}{2}(\tilde{F}_{0i})^2 + \frac{1}{2}(\vec{D}\tilde{\phi})^2 + V(\tilde{\phi})\right] \equiv E_F(\lambda) + E_{\text{kin}}(\lambda) + E_{\text{pot}}(\lambda).$$

(30.10)

Now by the same steps as in Derrick's case, we find (since F^2 now comes with λ^4)

$$E(\lambda) = \lambda^{4-d} E_F + \lambda^{2-d} E_{\text{kin}} + \lambda^{-d} E_{\text{pot}}.$$

(30.11)

Unlike the purely scalar case, now there is a minimum over λ of this function both for $d = 2$ and for $d = 3$, since in both cases there is an increasing power and a decreasing power. These are the vortex and monopole solutions. Moreover, for the purely gauge fields case ($E_{\text{kin}} = E_{\text{pot}} = 0$) in $d = 4$, the energy is independent of λ. Indeed, we saw that we can have purely gauge field instantons in $d = 4$, and they have an arbitrary scale (unlike the other solitons, which have a fixed size).

30.4 Reduction on an Ansatz

In finding a soliton solution, it is important to use the symmetries of the problem, and write *the most general ansatz consistent with the symmetries* (in order to obtain the most general soliton solution). We can either replace the ansatz in the equations of motion, and thus directly verify that the ansatz is correct – i.e., it doesn't contradict the equations of motion, replace it in the action, or (for static solutions) in the energy functional. In most cases, specifically *if we have a consistent truncation* – i.e., if the ansatz solves the equation of motion – the two procedures are equivalent, and it is easier to replace the ansatz in the action or energy functional, and obtain simpler equations of motion from the thus reduced action. If we wrote the most general solution consistent with the symmetries, the ansatz will automatically give a consistent truncation (since we haven't dropped any needed functions in a would-be solution).

But how do we write an ansatz? We analyze some examples.

1. For a rotationally invariant solution (which is often the case), we write generically (for generic fields ϕ^I)

$$\phi^I(R \cdot \vec{x}) = \phi^I(x),$$

(30.12)

where R is a rotation matrix in coordinate space, which leads to

$$\phi^I(\vec{x}) = f^I(r),$$

(30.13)

where r is the radial coordinate, and, moreover, differentiability at the origin gives the condition

$$f'^I(r = 0) = 0.$$

(30.14)

Without it, we would obtain a discontinuous Cartesian derivative, when going in different directions (different angles). This is the case for the vortex solutions.

2. For a solution that combines rotational invariance under $SO(d)$ with invariance under an internal group $SO(n)$, such that the solution is only invariant under a common subgroup of $SO(d) \times SO(n)$, we can impose

$$\phi^I(R \cdot \vec{x}) = [D(R)]^I{}_J \phi^J(\vec{x}), \qquad (30.15)$$

where R is a spatial, $SO(d)$, rotation and $D(R)$ a representation of the spatial rotation group $SO(d)$, for the matrix R, which is also a representation of the internal group $SO(n)$, on a vector space with index I.

An important special case is when $D(R) = R$,

$$\Phi^I(R \cdot \vec{x}) = R^I{}_J \phi^J(\vec{x}). \qquad (30.16)$$

This is only possible if $SO(n)$ is larger or equal to $SO(d)$, and we say that we "embed the rotations in the internal group." Consider the isotropy group of $SO(d)$, the group of transformations that leaves a point invariant, specifically $SO(d-1)$ (rotations around the vector $\vec{x}$). For it, $R \cdot \vec{x} = \vec{x}$, so we must have $\phi^I(x) = R^I{}_J \phi^J(\vec{x})$ – i.e., that ϕ is invariant under the isotropy group of spatial rotations. The only possibility is that ϕ is proportional to $\vec{x}$, with the constant of proportionality a function of the radius r,

$$\vec{\phi}(\vec{x}) = g(r)\vec{x} = f(r)\frac{\vec{x}}{r}. \qquad (30.17)$$

This is the "hedgehog" ansatz, and we have seen it for the Skyrmion. Moreover, since ϕ must be continuous at $r = 0$, $g(r)$ must be finite – i.e., $f(r) \sim \mathcal{O}(r)$ near $r = 0$.

For the case of gauge fields, the analysis is more complicated, but a similar logic applies.

We see from the previous examples that, generically, we reduce the ansatz to one or several functions of one coordinate (usually r), so we can replace it in the energy functional and by variation obtain a single (or several) differential equation for them. Of course, in this way, the equation is second order in derivative. If we use instead a saturation of a Bogomolnyi-type bound, we obtain more than one first-order equation (BPS equation), which is easier to solve.

Important Concepts to Remember

- Scalar fields are generically classified by the homotopy group $\pi_{d-1}(\mathcal{V})$, for maps between the sphere at infinity S_∞^{d-1} and the vacuum manifold $\mathcal{V}$.
- If the field at infinity takes different values in different directions, the global soliton has infinite energy (for $d \neq 1$).
- If the field at infinity takes a constant value, the solution is classified by $\pi_d(\mathcal{M})$.
- Gauge fields on S^k with values in H are classified by $\pi_{k-1}(H)$.
- For fields that are nontrivial at infinity (even if only other fields, like scalars are), we consider S^k to be the sphere at infinity. Then for $k = 2$ we have $\pi_1(H)$, and this is the Dirac monopole number, nontrivial for $H = U(1)$ or $G \supset U(1)$.

- The scalar monopole number $\pi_2(G/H) = \pi_1(H)_G$ is identified with the Dirac monopole number $\pi_1(H)$ for $H = U(1)$ and $G = SU(2)$.
- For $k = 4$, we have the instantons, which are pure gauge fields vanishing at infinity, thus effectively defined on S^4, with values in $SU(2) \simeq S^3$. Then $\pi_3(S^3) = \mathbb{Z}$.
- If we have the most general ansatz compatible with the symmetries, we can substitute it in the action or energy functional and obtain simpler equations.
- For a rotationally invariant ansatz, $\phi^I(\vec{x}) = f^I(r)$ and $f'(0) = 0$.
- For an ansatz that embeds rotational invariance into the internal symmetry, we have the hedgehog ansatz $\vec{\phi}(\vec{x}) = f(r)\vec{x}/r$.

Further Reading

See chapter 4 in [14].

Exercises

(1) Redo Derrick's argument for the nonlinear sigma model and the nonlinear sigma model minimally coupled to an Abelian gauge field.

(2) Consider a set of scalar fields in four-dimensional Euclidean space, transforming under $SU(2) \simeq SO(3)$. Write a nontrivial ansatz combining rotational invariance with $SO(3)$ invariance (invariant under a common subgroup).

(3) Write an ansatz for a Skyrmion solution for the gauge group $SU(3)$.

(4) Write all the possible different ansätze for the instanton solution for the gauge group $SU(3)$.

Other Soliton Types. Nontopological Solitons: Q-Balls; Unstable Solitons: Sphalerons

Until now, we have considered only solitons defined by topology – i.e., (stable) topological solitons. But in this chapter, we will see that they are not the only examples of solitons. There are also non-topological solitons, the most famous example of which is the Q-ball, which are at least classically stable. And there are also unstable solitons, specifically the sphalerons, which, nevertheless, can play a role.

31.1 Q-Balls

An interesting classically stable object that is not defined by topology but rather by charge conservation is the *Q-ball* (the name involves a pun: a cue ball, a ball used to break the grouping of the other balls in billiards), found by Coleman.

It is a soliton defined in a theory of complex scalar fields with an unbroken conserved (global or local) charge Q.

The simplest model is a single complex scalar field in 3+1 dimensions, with Lagrangean

$$\mathcal{L} = -\frac{1}{2}(\partial_\mu \phi)^2 - U(\phi^* \phi), \tag{31.1}$$

and, thus, invariant under a global $U(1)$ symmetry $\phi \to e^{i\alpha}\phi$.

The Q-ball solution is a solution of given charge Q, and this condition is enforced in the energy with a Lagrange multiplier ω, thus writing the energy functional

$$
\begin{aligned}
E_\omega &= E + \omega \left[Q - \frac{i}{2} \int d^3x (\phi^* \overset{\leftrightarrow}{\partial_t} \phi) \right] \\
&= \int d^3x \left[\frac{1}{2}|\dot{\phi}|^2 + |\vec{\nabla}\phi|^2 + U(\phi^* \phi) \right] + \omega \left[Q - \frac{i}{2} \int d^3x (\phi^* \dot{\phi} - \phi \dot{\phi}^*) \right].
\end{aligned} \tag{31.2}
$$

We complete squares and find

$$E_\omega = \omega Q + \int d^3x \left[\frac{1}{2}|\dot{\phi} + i\omega\phi|^2 + U(\phi^* \phi) - \omega^2 \phi^2 \phi + \frac{1}{2}|\vec{\nabla}\phi|^2 \right]. \tag{31.3}$$

We see that the energy is minimized if we put $\dot{\phi} + i\omega\phi = 0$, or

$$\phi(\vec{x}, t) = e^{-i\omega t}\phi(\vec{x}), \tag{31.4}$$

so the Lagrange multiplier ω is actually a frequency of rotation of the complex scalar in the $U(1)$ symmetry direction. Substituting back into the energy functional, we obtain

$$E_\omega = \omega Q + \int d^3x \left[\frac{1}{2}|\vec{\nabla}\phi(\vec{x})|^2 + U_\omega(\phi^*(\vec{x})\phi(\vec{x}))\right], \tag{31.5}$$

where the modified potential is

$$U_\omega(\phi^*\phi) = U(\phi^*\phi) - \omega^2\phi^*\phi, \tag{31.6}$$

and, by minimizing with respect to ω, the total charge of the Q-ball is

$$Q = \omega^2 \int d^3x \phi^*(\vec{x})\phi(\vec{x}). \tag{31.7}$$

To find the solution, we need to minimize the energy functional with respect to $\phi(\vec{x})$ and with ω, for fixed Q. We assume spherical symmetry, $\phi(\vec{x}) = \phi(r)$, and substituting in the energy functional, we get

$$E_\omega = \omega Q + 4\pi \int_0^\infty r^2 dr \left[\frac{1}{2}|\phi'|^2 + U_\omega(\phi^*\phi)\right], \tag{31.8}$$

which when minimized with respect to ϕ gives the equation of motion

$$\frac{d^2\phi}{dr^2} + \frac{2}{r}\frac{d\phi}{dr} = U'_\omega(\phi^*\phi). \tag{31.9}$$

From now on, we can assume that $\phi(r)$ is real, namely that all the phase of ϕ was the time-dependent phase $e^{-i\omega t}$ and that any constant initial phase can be redefined by a global $U(1)$ symmetry transformation.

The resulting equation of motion is the same as in the case of the kink solutions, with two modifications: since we are in three spatial dimensions with a radial ansatz instead of just in one dimension, we have the second term on the left-hand side, and the potential U is changed to U_ω. But otherwise the same interpretation applies: the equation is the same as for classical motion $\phi(r) \to x(t)$ in the inverted potential $V = -U_\omega$ (see Figure 31.1). The second term on the left-hand side now is simply understood as a friction term (force proportional to the velocity): indeed, if we put it on the right-hand side, it would be an extra resisting (negative) force proportional to the velocity $d\phi/dr \to dx/dt$.

The boundary conditions on the solution are the following: At $r = 0$, by the general analysis of the last chapter, we must have $d\phi/dr = 0$ for continuity of the derivative of the field. At $r = \infty$, we want $d\phi/dr = 0$ in order to have a finite energy, and moreover $\phi = 0$ in order to have a localized solution. Since then $\omega^2|\phi|^2 \to 0$, if we have $U(\phi^*\phi)$ have a minimum at $\phi = 0$, then so does $U_\omega(\phi^*\phi^*)$, and defining it to be $U_{\min} = 0$, the potential term in the energy is, therefore, finite.

So the potential U must have an absolute minimum at $\phi = 0$, which means that in a Taylor expansion near $\phi = 0$,

$$U \simeq M^2|\phi|^2 + \cdots, \tag{31.10}$$

where M is the (classical) mass of the field quanta (particles). In order for the classical Q-ball solution to be stable against decay, it must have an energy smaller than the mass of

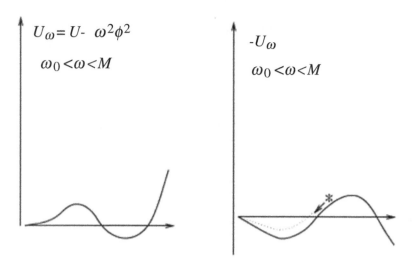

Figure 31.1 The potential for the bounce, and the motion in the inverse potential.

Q field quanta, $E < MQ$, or otherwise it will break apart. Since $U(\phi = 0) = 0$, it means that, also, $U_\omega(\phi = 0) = 0$.

For the inverted potential $V = -U_\omega$, the boundary conditions mean that we must start at some nonzero $\phi = \phi_0$ with zero velocity on what must be an absolute maximum of V and fall towards $\phi = 0$, where (due to the friction) we also must reach with zero velocity, meaning that $\phi = 0$ must be a local minimum of V. In terms of U_ω, it means that it must have an absolute minimum at $\phi_0 \neq 0$ and a local maximum at $\phi = 0$.

The condition on the potential is that U/ϕ^2 should have a minimum at $\phi_0 \neq 0$. Then,

$$\frac{U(\phi)}{\phi^2} \geq \frac{U(\phi_0)}{\phi_0^2} \equiv \omega_0^2 \Rightarrow U - \omega_0^2 \phi^2 \geq 0 \Rightarrow U_{\omega_0} \geq 0. \tag{31.11}$$

Moreover, $U_{\omega_0} = 0$ only at $\phi = 0$ and $\phi = \phi_0$, so for $\omega > \omega_0$, U_ω has an absolute minimum at ϕ_0.

The range of ω for which this is possible is

$$\omega_0 \leq \omega \leq M. \tag{31.12}$$

Indeed, for $\omega < \omega_0$, there is no absolute minimum at ϕ_0 for U_ω, so there is no absolute maximum at ϕ_0 for $V = -U_\omega$, and there is no "particle rolling down the potential" trajectory. And for $\omega > M$, the potential $U_\omega \simeq M\phi^2 - \omega\phi^2$ starts off negative – which means that $V = -U_\omega$ starts off positive, and, therefore, there is no local maximum at $\phi = 0$, so no solution reaching it with zero velocity. Indeed, then, there is just a downhill slope in V from $\phi = \phi_0$ to $\phi = 0$.

The last step, after finding as solution of the equations of motion for $\phi(r)$ at given ω, is to minimize over ω. We have seen in the original form for the energy functional that this will just enforce the total charge condition,

$$Q = 4\pi\omega^2 \int_0^r r^2 dr \phi^2(r), \tag{31.13}$$

given a fixed Q. Coleman goes on to prove theorems stating that there is always (for a potential as in [31.13]) a minimum Q, Q_{min}, such that for $Q \geq Q_{min}$ there is a Q-ball solution, and that solution is classically stable (i.e., small fluctuations don't destroy the solution).

31.2 Sphalerons

There is one more interesting category of solitons – this time solitons defined by topology, yet even classically unstable – the "sphalerons." The term comes from the ancient Greek, where "sphaleros" means "unstable," or "about to fall," a reminder that the solution is unstable.

The physical situation for which they are relevant is of different vacua separated by energy barriers, in which case we can have tunneling between the vacua driven either by quantum processes or by thermal fluctuations. Just like the classical instanton solution is relevant for tunneling between topologically distinct sectors, the sphaleron solution is relevant for these transitions. We will not describe the relevant transitions here, only the classical sphaleron solutions.

To understand sphalerons, let's consider a set of scalar fields ϕ^i, $i = 1,\dots,n$ with potential $U(\phi)$ and a static, spherically symmetric ansatz, $\phi^i = \phi^i(r)$. Then the equations of motion will be, as in previous examples (including the Q-balls),

$$\frac{d^2\phi^i}{dr^2} + \frac{d-1}{r}\frac{d\phi^i}{dr} = \frac{\partial U(\phi)}{\partial \phi^i}. \tag{31.14}$$

As we have described before, this looks like the motion of an n-dimensional particle in a potential $V(x^i) = -U$, with friction if $d > 1$.

Consider, then, a potential U that depends on two real scalars, or a complex one, and consider two vacua in the complex plane, ϕ_1 and ϕ_2, separated by a valley, or *saddle point*: there is a path of minimum energy between them that passes by a maximum point ϕ^*, which is, however, a minimum in the transverse direction (so is a "saddle point"), as in Figure 31.2.

Formally then, we consider paths $C(w)$ between the vacua $\phi_C(w)$, where w is a parameter, $0 \leq w \leq 1$, with $\phi_C(0) = \phi_1$ and $\phi_C(1) = \phi_2$. The maximum along the path is

$$V_C = \max_{w\in[0,1]} V(\phi_C(w)), \tag{31.15}$$

and then we consider the minimum among them, to be the saddle point,

$$V_* = \min_C V_C = V(\phi*). \tag{31.16}$$

Consider further a solution $\phi(r)$ that passes between ϕ_1 and ϕ_2, through ϕ_*. To understand it, consider the particle motion in $V(x) = -U(\phi)$. It starts at the maximum x_1 of V and goes through the "minimum" at x_* until climbing back to the maximum at x_2. But x_* is now a saddle point and is only a minimum along the path C_*, but in the

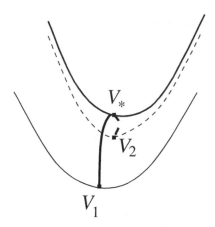

Figure 31.2   A sphaleron, as a path in field configuration space in between two vacua V_1 and V_2, that goes through a saddle point V_*.

transverse direction it is a local maximum, which means that, indeed, the path of the particle is unstable: if there is any deviation from it, it will "fall sideways" – hence the name sphaleron: the solution is classically unstable to small perturbations.

However, the sphaleron solution is more complicated than that: To define it, we consider the energy E of a field configuration and consider it as a *functional of the field configurations* $\phi(x)$. The constant field configurations = vacua (or topological solitons configurations!) are minima of the energy functional. The sphaleron is actually a saddle point *in field configuration space* between two such vacua of the energy functional.

Thus to find it, we consider paths $C(w)$ in field configuration space, $\phi_C(x, w)$, $w \in [0, 1]$, interpolating between the constant, vacuum field configurations $\phi(x, 0) = \phi_1$ and $\phi(x, 1) = \phi_2$. The maximum energy along the path is

$$E_C = \max_{w \in [0,1]} E(w) , \quad E(w) \equiv E[\phi_C(x, w)], \tag{31.17}$$

and the saddle point is the minimum of that among paths,

$$E_* = \min_C E_C. \tag{31.18}$$

31.3 Sphaleron on a Circle

A simple example of sphaleron is for a real scalar in one spatial dimension $\phi(x)$ with a Higgs potential, the model that has the kink solution, with action

$$S = \int d^2x \left[-\frac{1}{2}(\partial_\mu \phi)^2 - V(\phi) \right] , \quad V(\phi) = \frac{\lambda}{4}\left(\phi^2 - \frac{\mu^2}{\lambda} \right)^2 , \tag{31.19}$$

but where the scalar is periodic on a circle of radius R,

$$\phi(x + 2\pi R, t) = \phi(x, t). \tag{31.20}$$

Then consider the energy functional of a static field configuration,

$$E = \int_0^{2\pi R} dx \left[\frac{1}{2} \partial_x \phi^2 + \frac{\lambda}{4} \left(\phi^2 - \frac{\mu^2}{\lambda} \right)^2 \right], \qquad (31.21)$$

leading to the equation of motion

$$\frac{d^2\phi}{dx^2} - \lambda\phi \left(\phi^2 - \frac{\mu^2}{\lambda} \right) = 0. \qquad (31.22)$$

There are three constant solutions:

$$\phi_0 = 0 \quad \phi_1 = -\sqrt{\frac{\mu^2}{\lambda}}, \quad \phi_2 = +\sqrt{\frac{\mu^2}{\lambda}}, \qquad (31.23)$$

the first being an unstable one (at the local maximum of the potential) and the other two being vacua (minima of the potential).

Then, on the space of constant fields $\phi(x) = \phi$, we can consider the path

$$\phi_C(x; w) = \phi_C(w) = (1 - 2w)\sqrt{\frac{\mu^2}{\lambda}} \qquad (31.24)$$

that interpolates between the vacua ϕ_1 and ϕ_2, passing through ϕ_0. That is, a saddle point: it is a maximum along the path, though *in the full field configuration space*, there are directions along which it is a minimum. Then,

$$E(w) = E[\phi_C(x; w)] = 2\pi R \frac{\lambda}{4} \frac{\mu^4}{\lambda^2} (1 - (1 - 2w)^2)^2 = \frac{8\pi R\mu^4}{\lambda} w^2 (1 - w)^2. \qquad (31.25)$$

The maximum over w is at $w_0 = 1/2$, and we have the maximal energy

$$E_C = E_{C0} = \frac{\pi R\mu^4}{2\lambda}. \qquad (31.26)$$

This field configuration,

$$\phi_{C0}(x, w_0) = 0 = \phi_0; \qquad (31.27)$$

in other words, the maximum of the potential solution for small R is the sphaleron.

But for large R, there is a nontrivial sphaleron. It is a more general solution of the equations of motion (31.22). The equation of motion can be reduced by simple rescalings to the equation satisfied by the Jacobi elliptic function of order k, $sn_k(u)$ or $sn(k, u)$,

$$\frac{d^2y}{du^2} + y[1 + k^2 - 2k^2 y^2] = 0. \qquad (31.28)$$

The Jacobi elliptic function has the properties

$$sn_k(0) = 0, \quad sn_k(u + 2K_k) = -sn_k(u), \quad sn_k(K_k) = 1, \qquad (31.29)$$

where K_k (or $K(k)$) is the complete elliptic integral of the first kind. Then, finally, the solutions to the equations of motion (31.22) are

$$\phi_{n,k}(x) = \sqrt{\frac{2}{1 + k^2}} k \sqrt{\frac{\mu^2}{\lambda}} sn_k \left(\sqrt{\frac{2}{1 + k^2}} \mu x \right). \qquad (31.30)$$

From the conditions on $sn_k(u)$, we see that there is a periodicity of $4K_k$ – which, if equated to $2\pi R$, leads to

$$R = \frac{\sqrt{2(1+k^2)}K_k n}{\pi\mu}.$$

(31.31)

The relation fixes k in terms of R, for various n values. We see that, then,

$$\phi(0) = 0 = \phi_0 , \quad \phi_{n,1}(\pi R/2) = \sqrt{\frac{\mu^2}{\lambda}} = \phi_2 \quad \phi_{n,1}(-\pi R/2) = -\sqrt{\frac{\mu^2}{\lambda}} = \phi_1,$$

(31.32)

so, again, the sphaleron interpolates between the vacuum solutions, passing through the maximum (the former saddle point and sphaleron).

Thus, in this case, unlike a general one, the sphaleron really is a solution that passes through a saddle point in going between two vacua in real space (not in field configuration space).

31.4 Other Sphalerons

We look now at sphaleron solutions more relevant to real physics but, therefore, more complicated to analyze. We will only sketch the main points in these cases.

Complex Kink As a Sphaleron

Consider now the complex Higgs model, again in one spatial dimension, with a complex version of the action (31.19). The kink solution, saturating a BPS bound, was defined by topology in the real scalar case. But by embedding it in the complex scalar case,

$$\phi(x) = \sqrt{\frac{\mu^2}{\lambda}} \tanh\left(\frac{\mu x}{\sqrt{2}}\right)$$

(31.33)

still saturates a BPS bound but cannot truly be said to be defined by topology. Indeed, in the real case, there are two separate vacua, $\phi = \pm\sqrt{\mu^2\lambda}$; in the complex case, there is a connected continuum of vacua,

$$\phi_v = \sqrt{\frac{\mu^2}{\lambda}} e^{i\theta},$$

(31.34)

so we can continuously deform the two vacua at the end points of the kink into each other.

But we can then think of the kink as a sphaleron since it is a saddle point (mountain pass) solution going from a vacuum to another. In fact, we must go to the gauged model to have this interpretation since otherwise there is an unstable direction, but the resulting field configuration is not the vacuum field configuration but one that goes from a vacuum to another via a series of vacua (the solution follows the vacuum circle).

The Electroweak Sphaleron

The relevance of sphalerons to the real world comes from the fact that there is a sphaleron for electroweak theory, which is an $SU(2)$ gauge theory with scalars and fermions. One can write an ansatz (that we will not explain here) for the path in field configuration space and find its maximum. Then one minimizes this or, rather, one substitutes the ansatz in the energy and minimizes the energy, finding some field equations. The resulting saddle point, a solution of the electroweak field equations, is usually called the sphaleron and connects two vacua.

Important Concepts to Remember

- A Q-ball is a non-topological soliton in theories with scalar fields and a global or local symmetry, made stable by the having a conserved charge Q.
- The solution is found by imposing the charge conservation with a Lagrange multiplier ω, which gives $\phi(\vec{x}, t) = e^{j\omega t}\phi(\vec{x})$, and with a modified potential $U_\omega = U - \omega^2 |\phi|^2$.
- The existence of Q-balls depends on U having an absolute minimum at $\phi = 0$, and U_ω at $\phi_0 \neq 0$, and a Q-ball solution exists for a $Q \geq Q_{\min}$.
- A sphaleron is an unstable solution associated with a saddle point (mountain pass) connecting two vacua.
- The sphaleron is a path going through a mountain pass between two vacua in field configuration space, or otherwise it is the saddle point itself, and as a solution, it connects two vacua of V through a saddle point in V.
- For a real scalar on a circle with radius R, with a Higgs potential, at large R, the sphaleron is a solution that interpolates between two vacua, passing through a saddle point.
- For the complex version of the kink model, the kink is a kind of sphaleron.
- In the electroweak theory, there is a sphaleron solution, associated with a saddle point in field configuration space that connects two vacua through a saddle point.

Further Reading

See chapter 11 in [14] and chapter 6 in [19] for sphalerons. The original paper on Q-balls by Coleman is [20], and a review of nontopological solitons is [21].

Exercises

(1)   Check explicitly that the solutions (31.30) solve the equations of motion for the scalar sphaleron (31.22).
(2)   Give two examples (with different functional dependence) of potentials that admit Q-ball solutions.

(3) Solve the equation of motion for $\phi(r)$ in the case of one of the potentials at exercise 31.4.

(4) Consider the real scalar Higgs model in 2+1 dimensions instead of 1+1 dimensions like for the sphaleron on a circle. Write the equation of motion for $\phi(r)$ and transform it to the (nonlinear Schrödinger) form $\frac{d^2\psi}{dx^2} = V(\psi)$. Then find the vacua of $V(\psi)$ and investigate whether there is a corresponding sphaleron for it.

32 Moduli Space; Soliton Scattering in Moduli Space Approximation; Collective Coordinates

In this chapter, we will learn about the moduli space of soliton solutions, the space of parameters that don't change the energy, and how to extract physics from it: how to calculate the low velocity scattering of solitons, and how to quantize the solitons by writing a certain "quantum mechanics for the parameters of multi-soliton solutions, the collective coordinates."

This is the third and final example of quantum mechanics in this book, and as in the previous two cases, it only deals with quantum *mechanics* aspects (since the collective coordinate quantization method is related to the notion of solitons that features predominantly in the book), but not with quantum *field theory* aspects.

32.1 Moduli Space and Soliton Scattering in Moduli Space Approximation

A soliton solution always comes with some parameters that can be varied without varying the energy: if nothing else, the position of the soliton (since soliton means that they can move and scatter), but also, in general, more complicated things like internal rotations, etc. In the case of the simplest soliton, the 1+1 dimensional kink, the position x_0 is such a parameter as we have the static solution

$$\phi_s(x) = \sqrt{\frac{\mu^2}{\lambda}} \tanh\left(\frac{\mu}{\sqrt{2}}(x - x_0)\right).$$
(32.1)

But what happens if we make the parameter a function of time? If we replace $x_0 = x_0(t)$ in the (32.1),

$$\phi_{(0)} = \phi_s(x - x_0(t))$$
(32.2)

and replace the ansatz in the action, the gradient plus potential part will give the same classical energy of the kink E_{cl}, and only the kinetic energy will be nontrivial, giving

$$E = \frac{1}{2}\int dx [\partial_t \phi_s(x - x_0(t))]^2 + E_{\text{cl}} = \frac{1}{2}\int dx\, \phi_s'^2(x - x_0(t))\, \dot{x}_0(t)^2 + E_{\text{cl}}.$$
(32.3)

We note that the energy is of the type

$$E = E_{\text{cl}} + \frac{1}{2}\int dx\, g_{00}(x_0(t))\dot{x}_0(t)^2,$$
(32.4)

i.e., of the type of a *nonlinear sigma model* in the one-dimensional space of functions $x_0(t)$, with a "metric" $g_{00}(x_0(t))$ in front of $\dot{x}_0^2(t)$.

The equation of motion for $x_0(t)$, obtained by varying the energy functional with respect to it, is

$$\phi_s'^2(x - x_0(t))\ddot{x}_0(t) + \mathcal{O}(\dot{x}_0(t)^2) = 0. \tag{32.5}$$

That means that for $\dot{x}_0(t)^2 \equiv v^2 \ll 1$, we can write $\ddot{x}_0(t) = 0$, with the solution

$$x_0(t) = x_0 + vt, \tag{32.6}$$

and replace it in the above $\phi(x,t) = \phi_s(x - x_0(t))$. Of course, as we see, this is now in general not a solution of the equations of motion. We need to modify the solution by the addition of a first-order correction,

$$\phi(x,t) = \phi_{(0)}(x,t) + \phi_{(1)}(x,t) = \phi_s(x - x_0(t)) + \phi_{(1)}(x,t), \tag{32.7}$$

and replacing this in the full action, we can find perturbatively in $\dot{x}_0(t) = v$ the full solution: first $\phi_{(1)}$, then replacing back in the action we find a $\phi_{(2)}$, etc.

Of course, in this simple case, we know the full answer: the theory is relativistic, so the full solution is the Lorentz boosted one,

$$\phi(x,t) = \phi(\gamma(x - x_0 - vt)), \quad \gamma = 1/\sqrt{1 - v^2}, \tag{32.8}$$

but the procedure is more general.

In particular, we can consider solutions that correspond to more than one soliton. The most important example is the case when there is a set of BPS equations (a Bogomolnyi limit), which are first order in derivatives. As we saw, in this case, there is a topological charge, and the energy of solutions that saturated the bound, thus satisfy the BPS equations, is proportional to it.

That in general means that there is a static multi-soliton solution: Consider a soliton solution with (integer) charge N localized at the same point, thus with energy $E = NE_{\text{cl}}$ (E_{cl} is the energy of a single soliton), which means that there is no interaction potential between the solitons. Then we can separate the solution into a *static* solution with solitons localized at various positions $x_i(t)$, $i = 1, \ldots, N$, $\phi(x,t) = \phi(x, \{x_i(t)\})$, and the energy will be the same one.

But even more generally (and unlike the kink case), we can have other parameters than the soliton positions whose variation doesn't change the energy, called the *moduli*. The full set of moduli, including positions, will be denoted by a^I, $I = 1, \ldots, M$ (and $M \geq N$), and the space of them is called *moduli space*, $\mathcal{M}_M$. In fact, usually $M = kN$, with k a positive integer (for instance, $k = 2$ for vortices and $k = 4$ for monopoles), and then the interpretation is that we have N solitons, each with k degrees of freedom. If the solitons are close together, it might not be possible to separate the moduli into positions and others, so it makes no sense to consider the positions separately. Then we can consider small time variations of the moduli a^I – i.e., $a^I(t) \equiv v^I$. We then consider the first-order ansatz

$$\phi(x,t) = \phi_{(0)}(x,t) + \phi_{(1)}(x,t) = \phi_s(x, \{a^I(t)\}) + \phi_{(1)}(x,t), \tag{32.9}$$

and by substituting in the equations of motion we find (perturbatively) the form of $\phi_{(1)}(x, t)$. The process could be continued in principle indefinitely. If we now replace the solution in the energy functional, we find a nonlinear sigma model with metric $g_{IJ}(a^K(t))$,

$$E = NE_{\text{cl}} + \frac{1}{2} \int d^d x \, g_{IJ}(a^K(t)) \dot{a}^I \dot{a}^J + \mathcal{O}(v^4). \tag{32.10}$$

We call g_{IJ} the *metric on moduli space*. Indeed, the motion in the moduli space of a^I is a geodesic motion on $\mathcal{M}_M$ (we will see in part III what that means more precisely: basically, shortest path in the space, which extremizes the above energy functional, as it should).

Dropping the higher order terms in v^I (the $\mathcal{O}(v^4)$ terms mentioned in the above) is called the *moduli space approximation*. In this approximation, the motion in a^I is geodesic on the moduli space and, since some of the a^I are soliton positions, it corresponds to *scattering of solitons*: the solitons come close together, interact, and then go further apart.

But the procedure described is difficult in general: we usually don't know the exact multi-soliton solution, and even then substituting the solution in the energy functional is a difficult calculation. Sometimes we don't have to, and one can directly calculate the metric on moduli space using indirect methods. For instance, that was done by Atiyah and Hitchin (famous mathematicians) for the two-monopole moduli space: they were able first to show that the metric must be Kähler and then use the symmetries of the problem to completely constrain it.

32.2 Example: ANO Vortices in the Abelian-Higgs Model

To see more concretely the methods described earlier, we examine a concrete example that has been much studied and is of interest to real systems, like superconductivity: the Abelian Higgs model in 2+1 dimensions and its ANO vortices.

The BPS equations for the Abelian-Higgs model in 2+1 dimensions, at $\lambda = e^2$ (in the Bogomolnyi limit), were given in (25.25), which are $D_0\phi = 0, F_{0i} = 0$ (satisfied on the static solution by $\dot{\phi} = 0, \dot{\vec{A}} = 0$ and the gauge $A_0 = 0$) and

$$\begin{aligned} (D_1 + iD_2)\phi &= 0 \\ B &\equiv F_{12} = -e(|\phi|^2 - \phi_0^2). \end{aligned} \tag{32.11}$$

Using complex coordinates, $z = x_1 + ix_2$, and defining $\partial \equiv \partial_z \equiv (\partial_1 - i\partial_2)/2$ (since $\partial x_1/\partial z = 1/2, \partial x_2/\partial z = -i/2$), and so, similarly,

$$A \equiv A_z \equiv \frac{A_1 - iA_2}{2}, \tag{32.12}$$

we find that the first BPS equation becomes

$$A = A_z = \frac{i}{e} \bar{\partial} \ln \bar{\phi}, \tag{32.13}$$

and substituting in the second BPS equation, we obtain the equivalent of (25.38) for the general case of N solitons,

$$\Delta \ln |\phi| = 2\pi \sum_{i=1}^{N} \delta^2(\vec{x} - \vec{x}_i) + e^2(|\phi|^2 - \phi_0^2). \tag{32.14}$$

As observed by Taubes, who first rigorously proved the existence of multi-soliton solutions, and then by Samols, who wrote the recipe for calculating the solution, the presence of the delta functions on the right-hand side means that near each soliton, $z \simeq z_i$, the solution must behave as $\phi \propto z - z_i$, so $|\phi| \propto |z - z_i|$. Thus, we can separate a polynomial from ϕ,

$$H_0(z) = \prod_{i=1}^{N} (z - z_i), \tag{32.15}$$

and write

$$\phi(z, \bar{z}) = \phi_0 e^{f(z,\bar{z})/2} \equiv \phi_0 e^{-\psi(z,\bar{z})/2} H_0(z). \tag{32.16}$$

But, in fact, having an independent function of z means a vortex exists somewhere, since then $z = re^{i\theta}$ appears without $\bar{z}$, thus $e^{i\theta}$ in ϕ gives a topology. One can, therefore, prove rigorously that, once H_0 is separated and there are no more vortices, $\psi(z, \bar{z})$ must be a *real function*.

Moreover, substituting the ansatz in A in (32.13) we find $(H_0 = H_0(z), \bar{H}_0 = \bar{H}_0(\bar{z}))$

$$A \equiv A_z = -\frac{i}{2e} \partial \psi + \frac{i}{e} \frac{\partial \bar{H}_0}{\bar{H}_0} = -\frac{i}{2e} \partial \psi = A_{(0)}$$

$$\bar{A} \equiv A_{\bar{z}} = +\frac{i}{2e} \bar{\partial} \psi - \frac{i}{e} \frac{\partial H_0}{H_0} = +\frac{i}{2e} \bar{\partial} \psi = \bar{A}_{(0)}. \tag{32.17}$$

Substituting it also in the equation for $|\phi|$, (32.14), and using the fact that in complex coordinates $4\partial\bar{\partial} = \Delta$, we find

$$-2\partial\bar{\partial}\psi + 4\bar{\partial}\partial H_0 = 2\pi \sum_{i=1}^{N} \delta^2(\vec{x} - \vec{x}_i) + e\phi_0^2(e^{-\psi}|H_0|^2 - 1). \tag{32.18}$$

But, just like we proved that $\epsilon^{ij}\partial_i\partial_j\theta = 2\pi\delta^2(r)$, we can prove that

$$4\bar{\partial}\partial \ln(z - z_i) = 2\pi\delta^2(z - z_i), \tag{32.19}$$

by using the Green–Riemann theorem (Stokes in two dimensions), which we leave as an exercise. Then we obtain

$$2\bar{\partial}\partial\psi = e\phi_0^2(1 - e^{-\psi}|H_0|^2). \tag{32.20}$$

Moreover, the boundary condition at infinity is that $|\phi|^2 = \phi_0 e^{-\psi}|H_0|^2 \rightarrow \phi_0$ (the field is in the vacuum at infinity), which amounts to the boundary condition for ψ,

$$\psi \rightarrow \ln|H_0|^2 \quad \text{as} \quad |z| \rightarrow \infty. \tag{32.21}$$

The boundary condition at $z = z_i$ is $\phi \simeq A(z - z_i)$, so it amounts to $e^{-\psi/2(z=z_i)}$ finite.

Now note that the gauge condition is not satisfied anymore, since $D_0\phi$ cannot be put to zero if $A_0 = 0$. Instead, insisting on the BPS condition, so $D_0\phi = 0$ and $F_{0i} = 0$, we find

$$D_0\phi = 0 \Rightarrow A_0^{(1)} = \frac{1}{ie}\frac{\dot\phi}{\phi} = \frac{1}{ie}\left(\sum_{i=1}^{N}\frac{\dot z_i}{z - z_i} - \frac{\dot z_i}{2}\partial_{z_i}\psi - \frac{\dot{\bar z}_i}{2}\partial_{\bar z_i}\psi\right)$$

$$\partial_0 A_z^{(0)+(1)} = \partial_z A_0 \Rightarrow \partial_0 A_z = \frac{1}{ie}\left(\sum_{i=1}^{N}\frac{\dot z_i}{(z - z_i)^2} - \frac{\dot z_i}{2}\partial_{z_i}\partial\psi - \frac{\dot{\bar z}_i}{2}\partial_{\bar z_i}\partial\psi\right). \qquad (32.22)$$

We see that, as advertised, at least A_0 must change at this order, though A_z doesn't need to change ($A_0^{(0)}$ already satisfies it).

Next, we want to construct the effective action, which means to construct the metric on the moduli space. Following Samols, we consider the kinetic energy in the $A_0 = 0$ gauge,

$$T = \frac{1}{2}\int d^2x(\dot{\vec{A}}^2 + |\dot\phi|^2) = \frac{1}{2}\int d^2x(4\dot A\dot{\bar A} + \dot\phi\dot{\bar\phi}). \qquad (32.23)$$

We can now rewrite it as (using the formula for A_z, from the BPS equations)

$$T = \frac{1}{2}\int d^2z[4\partial_z(\dot\phi/\phi)\bar\partial_{\bar z}(\dot\phi/\phi) + e^{-\psi}|H_0|^2(\dot\phi/\phi)^2], \qquad (32.24)$$

and consider the integration of a domain that excludes a small circle around each z_i, since this gives a vanishing total contribution, as we see from the original formula for T (the integrand is smooth). Then we partially integrate ∂_z to obtain

$$T = 2\int_{\mathbb{R}^2 - S_\epsilon} d^2z\partial_z[(\dot\phi/\bar\phi)\bar\partial_{\bar z}(\dot\phi/\phi)] + \frac{1}{2}\int_{\mathbb{R}^2 - S_\epsilon} d^2z(\dot\phi/\bar\phi)[-\Delta(\dot\phi/\phi) + e^{-\psi}|H_0|^2(\dot\phi/\phi)].$$
$$(32.25)$$

But by the equation of motion for ϕ, the integrand in the second term gives (derivatives of) delta functions at z_i, which were excluded from the domain of integration. Using Gauss's law for the first integral, it is given by integrals along circles of radius ϵ around each of the z_i,

$$T = \sum_{i=1}^{N}\int_0^{2\pi} d\theta_i\dot{\bar z}_i\bar\partial_{\bar z}(\dot\phi/\phi) = \sum_{i,j=1}^{N}\int_0^{2\pi} d\theta_i\bar\partial_{\bar z}\partial_{z_j}\ln|\phi|^2\dot z_i\dot{\bar z}_j. \qquad (32.26)$$

We Taylor expand ψ near z_i to find

$$\psi = a_i + b_i(z - z_i) + \bar b_i(\bar z - \bar z_i) + c_i(z - z_i)^2 + d_i(z - z_i)(\bar z - \bar z_i) + \bar c_i(\bar z - \bar z_i)^2 + \cdots \quad (32.27)$$

Then, using the equation of motion for ψ near z_i to find that

$$4\partial_z\bar\partial_{\bar z}\psi = -1, \qquad (32.28)$$

we finally find that

$$\bar\partial_{\bar z}\partial_{z_j}\ln|\phi|^2(z = z_i) = \frac{1}{4}\delta_{ij} + \partial_{z_j}\bar b_i, \qquad (32.29)$$

so the kinetic energy is

$$T = \frac{\pi}{2}\sum_{i,j=1}^{N}\left(\delta_{ij} + 4\partial_{z_j}\bar b_i\right)\dot z_j\dot{\bar z}_i, \qquad (32.30)$$

and the metric on moduli space is

$$ds^2 = \pi \sum_{i,j=1}^{N} \left(\delta_{ij} + 4\partial_{z_j}\bar{b}_i \right) dz_j d\bar{z}_i. \qquad (32.31)$$

To calculate the metric on the moduli space when the vortices are well separated, we can consider the fact that small scalar field fluctuations far from the source will obey a linear KG equation,

$$\Delta\delta\phi - e\phi_0^2\delta\phi = 0, \qquad (32.32)$$

which has as a solution in two dimensions (when the Laplacean is $\Delta = \partial_r^2 + 1/r\partial_r$) the Bessel function

$$\delta\phi \propto K_0(mr), \quad m \equiv \phi_0\sqrt{e}. \qquad (32.33)$$

In the linearized regime, we see that $\delta\phi \simeq -\delta\psi/2$. One can perform an exact analysis and find that, in fact,

$$\partial_{z_j}b_i \simeq \frac{q^2}{8\pi^2} \sum_{i \neq j} K_0(m|z_i - z_j|), \qquad (32.34)$$

where q is a numerical constant determined from the nonlinear solution ($q \simeq -10.6$).

Then the field theory kinetic energy for well separated vortices can be rewritten as ($z_{ij} \equiv z_i - z_j$)

$$L_{\text{eff}} = T = \frac{\pi}{2} \sum_i \dot{z}_i\dot{\bar{z}}_i - \frac{q^2}{8\pi} \sum_{i \neq j} K_0(m|z_{ij}|)\dot{z}_{ij}\dot{\bar{z}}_{ij}. \qquad (32.35)$$

This takes the form of an effective Lagrangean for particles, with the first term being kinetic and the second potential. That allows us to calculate the interaction potential of two vortices separated by a distance y and with a relative velocity v:

$$V_{\text{int}} \simeq -\frac{q^2}{6\pi}K_0(my)v^2. \qquad (32.36)$$

32.3 Collective Coordinates and Their Quantization

Since, as we saw, solitons generically keep their shape and mass and thus act like some extended particles, the moduli corresponding to the positions of the solitons, $x_i(t)$, $i = 1, \ldots, N$, at least when they are well separated, can be considered in some sense like the positions of some particles. But if the mass of these particles is not extremely large, we must apply quantum mechanics to them as well, and we must view the classical field theory adopted until now as a zeroth-order approximation. Using the full quantum field theory formalism that will be developed in another book would be exagerated, and we can instead use a certain quantization procedure that reminds of the quantization of regular particles. We will thus endeavor to describe this quantization. Note, however, that the procedure can be derived also from the path integral formalism of quantum field theory, as shown by Gervais, Jevicki, and Sakita.

For every position coordinate x_i there is a canonically conjugate momentum p_i. But we can extend the canonically conjugate pair to include all parameters = moduli a^I, or q^A from now on to use the standard notation from classical mechanics for a canonically conjugate p_A. In this context, we call the moduli *collective coordinates*, since they act like (generalized) coordinates for the collective motion of the field (soliton). The quantization method will then be called *collective coordinate quantization*.

We expand a general time-dependent perturbation of the soliton $\phi(x, t)$ (x can be in more than one dimension) into a background = the soliton, and a perturbation. Further, we expand the perturbation into a complete set $\eta_n(x)$ of eigenmodes of the linearized kinetic operator around the soliton, times time-dependent coefficients. That is, we replace $\phi = \phi_{\text{soliton}}(x) + \delta\phi$ into the equation of motion for ϕ, and obtain an equation $Kin(\phi_{\text{soliton}})\delta\phi = 0$. We now factor out the time dependence $\delta\phi = e^{i\omega t}\eta(x)$ and write an eigenvalue equation (similar to the Schrödinger case)

$$Kin'(\phi_{\text{sol}})\delta\eta_n(x) = -\omega_n^2\delta\eta_n(x). \tag{32.37}$$

Here $\{\eta_n(x)\}_n$ are a complete, orthonormal set of eigenfunctions of Kin' (the kinetic operator less the time derivatives). The decomposition is then

$$\phi(x, t) = \phi_{\text{soliton}}(x) + \sum_{n\in\mathbb{N}} q_n(t)\eta_n(x). \tag{32.38}$$

In this decomposition, $q_0(t) = X(t)$ (soliton position) or "q^A" (in a more general setting) are the *zero modes*, or moduli, which don't modify the energy – i.e., have eigenvalue $\omega_0 = 0$.

To be more concrete, consider the case of the simplest model, a scalar in 1+1 dimensions with canonical kinetic term and a potential,

$$S = \int dx\, dt\, \left[\frac{1}{2}\dot{\phi}^2 - \frac{1}{2}\phi'^2 - U(\phi)\right]. \tag{32.39}$$

Then the equation of motion for ϕ is

$$-\frac{d^2}{dt^2}\phi + \frac{d^2}{dx^2}\phi - \frac{dU}{d\phi} = 0. \tag{32.40}$$

Writing $\phi = \phi_{\text{soliton}} + \delta\phi$, we obtain the linearized equation of motion (given that the soliton itself satisfies equation [32.40])

$$-\frac{d^2}{dt^2}\delta\phi + \frac{d^2}{dx^2}\delta\phi - \frac{d^2U}{d\phi^2}\bigg|_{\phi=\phi_{\text{soliton}}}\delta\phi = 0. \tag{32.41}$$

Replacing $\delta\phi = e^{i\omega t}\eta$, we find the eigenvalue equation

$$\left[\frac{d^2}{dx^2} - \frac{d^2U}{d\phi^2}\bigg|_{\phi=\phi_{\text{soliton}}}\right]\eta_n(x) = \omega_n^2\eta_n(x). \tag{32.42}$$

Then η_0, corresponding to $\omega_0 = 0$, are the derivatives with respect to the moduli of the classical solution, and the corresponding $q_0(t)$ are the moduli $q^A = (X(t), \dots)$.

Indeed, in the simplest case, when there is just one modulus, the position $X(t)$, we know $\phi_{\text{soliton}}(x - X(t))$ is a solution (to lowest order), and by Taylor expanding it we get

$$\phi_{\text{soliton}}(x - X(t)) \simeq \phi_{\text{soliton}}(x) - X(t)\phi'_{\text{soliton}}(x) + \cdots \qquad (32.43)$$

That means that $\phi'_{\text{soliton}}(x)$ is a zero mode (a mode of zero energy).

Then one makes a change of basis and erases the zero mode from the sum over n but changes x with $x - X(t)$ on the right-hand side of the expansion, to obtain

$$\phi(x, t) = \phi_{\text{soliton}}(x - X(t)) + \sum_{m=1}^{\infty} q_m(t)\eta_m(x - X(t)). \qquad (32.44)$$

In this way, in the "vacuum" on the space of field configurations defined by the soliton, the relation defines a change of variables from the continuum $\phi(x, t) = \phi_x(t)$ to the discrete (yet still infinite) set $q^I = \{X(t), q_m(t)\}$.

In this form, remembering that we started with *orthonormal* modes $q_n(t)$, the mode $q_0 = N\phi'_{\text{soliton}}$ (N is the normalization constant) must be orthogonal to the other ones, so

$$\int dx \, \eta_0(x - X(t))\eta_m(x - X(t)) = 0, \qquad (32.45)$$

where

$$\eta_0(x) = \frac{\phi'_{\text{soliton}}(x)}{\sqrt{\int dx \, (\phi'_{\text{soliton}}(x))^2}}. \qquad (32.46)$$

32.4 Change of Basis in a Hamiltonian and Quantization

For a general quantum system with discrete degrees of freedom, consider a change of coordinates between a basis $\{q^I\}$ and a basis $\{\tilde{q}^a\}$, by

$$\tilde{q}^a = \tilde{q}^a(q_I). \qquad (32.47)$$

The original coordinates, $\tilde{q}^a(t)$, have canonically conjugate momenta $\tilde{p}_a(t)$, and the $q^I(t)$ have canonically conjugate momenta $p_I(t)$. The Poisson brackets are canonical,

$$\{\tilde{q}^a, \tilde{p}_b\}_{P.B.} = \delta^a_b; \quad \{q^I, p_J\}_{P.B.} = \delta^I_J. \qquad (32.48)$$

The change of basis is especially useful in the case that in one basis, say $\{\tilde{q}^a\}$, the Hamiltonian has a standard kinetic term, with Hamiltonian

$$H = \sum_a \frac{\tilde{p}_a^2}{2} + V(\tilde{q}). \qquad (32.49)$$

For instance, it could be that we have a collection of harmonic oscillators with frequency ω, and then

$$V(\tilde{q}) = \sum_a \frac{\omega(\tilde{q}^a)^2}{2}. \qquad (32.50)$$

The change in basis on the coordinates implies a change in basis on the momenta, more precisely,

$$\frac{\partial \tilde{q}^a}{\partial q_K} \tilde{p}_a(t) = p_K(t). \tag{32.51}$$

The relation can be inverted, to give

$$\tilde{p}_a(t) = \frac{\partial \tilde{q}^a}{\partial q^I} g^{IJ}(q) p_J(t), \tag{32.52}$$

where g^{IJ} is the matrix inverse of the metric

$$g_{IJ}(q) = \sum_a \frac{\partial \tilde{q}^b}{\partial q^I} \frac{\partial \tilde{q}^b}{\partial q^J}. \tag{32.53}$$

Indeed, multiplying by $\partial \tilde{q}^a / \partial q^K$ the inverted relation, we obtain the original one.

At the classical level, going from $\{\tilde{q}^a, \tilde{p}_a\}$ to $\{q^I, P_I\}$ in the Hamiltonian results in

$$H(p_I, q^I) = \sum_{I,J} \frac{1}{2} g^{IJ} p_I p_J + \frac{\omega^2}{2} \sum_a \tilde{q}^a(q) \tilde{q}^a(q). \tag{32.54}$$

Quantization

When quantizing, we replace $\{,\}_{P.B.}$ with $i\hbar[,]$, and we can then represent

$$\tilde{p}_a \rightarrow \hat{p}_a = i\hbar \frac{\partial}{\partial \tilde{q}^a}, \tag{32.55}$$

so that the (naive) Hamiltonian in the harmonic oscillator basis is

$$H = -\frac{\hbar^2}{2} \sum_a \frac{\partial^2}{\partial (\tilde{q}^a)^2} + V(\tilde{q}). \tag{32.56}$$

Doing the transformation $\tilde{q}^a = \tilde{q}^a(q^I)$ on it, we obtain

$$H = -\frac{\hbar^2}{2} \sum_{I,J} \frac{1}{\sqrt{g(q)}} \frac{\partial}{\partial q^I} g^{IJ}(q) \sqrt{g(q)} \frac{\partial}{\partial q^J} + V(\tilde{q}(q)). \tag{32.57}$$

Here

$$g(q) = \det g_{IJ}, \tag{32.58}$$

and is the Jacobian of the transformation from $\tilde{q}^a$ to q^I.

But then the scalar product between two functions of position (in the Schrödinger representation), which is standard in the $\tilde{q}^a$ picture, becomes in the q^I picture

$$(\psi_1, \psi_2)_{\tilde{q}} \equiv \int \left(\prod_a d\tilde{q}^a \right) \psi_1^*(\tilde{q}) \psi_2(\tilde{q})$$

$$= \int \left(\prod_I dq^I \right) \sqrt{g(q)} \psi_1^*(\tilde{q}(q)) \psi_2(\tilde{q}(q)) \equiv (\psi_1, \psi_2)_q. \tag{32.59}$$

But in order to identify p with $\partial/\partial q$, we must have a trivial scalar product. To obtain this, we redefine our Hilbert space, by writing $\psi_i = g^{-1/4}\tilde{\psi}$ so as to get rid of the $\sqrt{g}$ in the product. Then on $\tilde{\psi}$ we have the usual action of momenta, with $\partial/\partial q^I$, and going back to ψ we obtain, therefore,

$$p_I\psi = g^{-1/4}\frac{\partial}{\partial q^I}g^{1/4}\psi\,, \tag{32.60}$$

or

$$\frac{\partial}{\partial q^I} = g^{1/4}p_I g^{-1/4}. \tag{32.61}$$

Replacing in the Hamiltonian, we obtain the quantum Hamiltonian

$$\hat{H} = \frac{1}{2}g(\hat{q})^{-1/4}\sum_{I,J}\hat{p}_I g^{IJ}(\hat{q})\sqrt{g(\hat{q})}\hat{p}_J g(\hat{q})^{-1/4} + V(\tilde{q}(\hat{q})). \tag{32.62}$$

Note that now any possible ordering ambiguities in the quantum Hamiltonian have been resolved by the coordinate transformation: in (32.62), $\hat{p}$ and $\hat{q}$ don't commute, but the order is well defined.

It is useful to consider also the *Weyl ordering*, since it is known that the Hamiltonian in the path integral formalism maps to the Weyl ordered one in the operator formalism of quantum mechanics. Weyl (or symmetric) ordering of $\hat{p}^k$ and $\hat{q}^l$, $\{\hat{p}^k, \hat{q}^l\}_W$, is defined by

$$(\alpha\hat{p} + \beta\hat{q})^N = \sum_{k,l}\alpha^k\beta^l\{\hat{p}^k, \hat{q}^l\}_W\frac{N!}{l!\,k!}. \tag{32.63}$$

For the canonical operators $\hat{q}, \hat{p}$, we can prove the relation

$$\hat{p}_I g^{IJ}(\hat{q})\hat{p}_J = \{\hat{p}_I g^{IJ}(\hat{q})\hat{p}_J\}_W + \frac{\hbar}{4}\partial_J\partial_I g^{IJ}(\hat{q}). \tag{32.64}$$

We leave the proof as an exercise.

One can then show that the quantum Hamiltonian can be put into the form

$$\hat{H} = \frac{1}{2}\{\hat{p}_I\hat{p}_J g^{IJ}(\hat{q})\}_W + V(\tilde{q}(q)) + \Delta V(q), \tag{32.65}$$

where

$$\begin{aligned}
\Delta V(q) &= \frac{\hbar^2}{8}\left[\partial_J\partial_I g^{IJ}(\hat{q}) - 4[g(\hat{q})]^{-1/4}\partial_J(g^{IJ}(\hat{q})[g(\hat{q})]^{1/2}\partial_I(g(\hat{q})^{-1/4}))\right]\\
&= \frac{\hbar^2}{8}\Gamma^I_{JL}(q)\Gamma^J_{IM}(q)g^{lm}(q)\\
\Gamma^I_{JK} &\equiv \frac{1}{2}g^{IM}(\partial_J g_{MK} + \partial_K g_{MJ} - \partial_M g_{JK}).
\end{aligned} \tag{32.66}$$

32.5 Application to Collective Coordinate Quantization

We now apply the previously described method to the quantization of collective coordinates. The difference is that a now is not a discrete index but is continuous: $\tilde{q}^a(t)$ is replaced by $\phi^x(t)$, so $\sum_a$ becomes $\int dx$ now.

One could now substitute $\phi^x(t) = \phi(x,t)$ from (32.44) into the action and obtain the Hamiltonian in the form

$$H = \frac{1}{2} g^{IJ}(q) p_I p_J + V(q),$$
(32.67)

at the classical level, where $p_I = \{P(t), p_m\}$ and $q^I = \{X(t), q_m\}$. The metric is now

$$g_{IJ}(q) = \int dx \partial_I \phi^x(t) \partial_J \phi^x(t) = \int dx \frac{\partial \phi(x,t)}{\partial q^I} \frac{\partial \phi(x,t)}{\partial q^J}.$$
(32.68)

Because of the translational invariance of the integral, and the fact that $X(t)$ appears in $\phi(x,t)$ only in the combination $x - X(t)$, the metric will be independent of $X(t)$ – i.e.,

$$g_{IJ} = g_{IJ}(q_m)$$
(32.69)

only.

At the quantum level, we need to be careful and use the correct quantization procedure. We formally start by identifying, from the Poisson brackets

$$\{\phi^x(t), \pi^x(t)\}_{P.B.} = \delta^{xy} \Rightarrow \{\phi(x,t), \pi(y,t)\}_{P.B.} = \delta(x-y),$$
(32.70)

instead of the classical $\pi(x,t) = \dot{\phi}(x,t)$ (from the canonical kinetic term for $\phi(x,t)$), the quantum mechanical

$$\pi^x(t) = i\hbar \frac{\partial}{\partial \phi^x(t)}.$$
(32.71)

The matrix g_{IJ} becomes, then, explicitly (remember that 0 stands for $X(t)$)

$$g_{00}(q) = \int dx \, \phi'(x,t)^2 = (\phi', \phi')$$

$$g_{0n}(q) = g_{n0}(q) = -\int dx \, \eta_n(x) \phi'(x,t) = -(\eta_n, \phi')$$

$$g_{nm}(q) = \delta_{nm}.$$
(32.72)

We can then apply the general formalism and find the full quantum Hamiltonian, but the expressions are quite complicated, and we will not show them. We just want to note that the collective coordinates q^A, here exemplified by $X(t)$, will appear in general in it, not just the q_m (nonzero modes).

Then we can use the quantum Hamiltonian to calculate quantum corrections to solitons and their scattering, and we only need to use regular quantum mechanics of an infinite, though discrete, set of degrees of freedom.

Important Concepts to Remember

- The moduli space of (multi-)soliton solutions is the space of parameters, including soliton positions, whose modification leaves the energy unchanged.
- Varying soliton positions (and in general moduli a^I) with respect to time, $x_0 \to x_0(t)$, we get a nonlinear sigma model in time, with metric g_{IJ} = metric on moduli space.
- The equation of motion for $x_0(t)$ gives free motion, $x_0(t) = x_0 + vt$, but only in the first approximation. Otherwise, for a single soliton, we obtain Lorentz boosting with $\gamma(x - x_0 - vt)$.
- For multi-solitons, we need to add corrections in v to the solutions, $\phi(x,t) = \phi_s(x, \{a^I(t)\}) + \phi_{(1)}(x,t)$, since $x_0 \to x_0(t)$ doesn't solve the equations of motion anymore.
- Motion in the moduli space $\mathcal{M}_M$ ($M > N$, usually $M = kN$) is geodesic, giving for $N > 1$ solitons scattering, with an interaction potential depending on velocity, and is valid in the moduli space approximation.
- For ANO vortices, the multi-soliton ansatz is $\phi(z,\bar{z}) = \phi_0 e^{-\psi(z\bar{z})/2} H_0(z)$, with $H_0(z) = \prod_{i=1}^{N}(z - z_i)$ a holomorphic polynomial depending on the soliton positions z_i, and ψ a real function.
- The metric on the moduli space of vortices is $ds^2 = \sum_{ij}[\delta_{ij} + 4\partial_{z_i}\bar{b}_i]dz^i d\bar{z}^j$, with b_j the linear term in the Taylor expansion of ψ in z.
- At large vortex separation z_{ij}, the effective Lagrangean is proportional to $K_0(mz_{ij})\dot{z}_{ij}\dot{\bar{z}}_{ij}$, giving an interaction potential proportional to $K_0(mz)v^2$.
- Solitons act as particles, keeping their shape and mass, so they can be quantized in a similar manner, using collective coordinate quantization.
- Collective coordinates are the moduli, generalizing positions, which are parameters for the collective motion of the fields, keeping the shape of the solitons.
- We expand the field in soliton plus small fluctuations, $\phi = \phi_{sol} + \delta\phi$, and the variation in eigenmodes of the kinetic operator, $Kin'(\phi_{sol})\eta_n(x) = -\omega_n^2\eta_n(x)$, $\delta\phi = \sum_{n\geq 0} q_n(t)\eta_n(x)$.
- For $\omega_0 = 0$, we have the zero modes = moduli, $\eta_0(x)$ corresponding to $q_0(t) = X(t)$ or q^A more generally.
- The collective coordinate ansatz is to erase the zero mode from the sum and replace x with $x - X(t)$, so $\phi(x,t) = \phi_{sol}(x - X(t)) + \sum_{n\geq 1} q_n(t)\eta_n(x)$.
- When changing basis in a quantum Hamiltonian, $\tilde{q}^a = \tilde{q}^a(q^I)$, we obtain a sigma model for $H(p_I, q^I)$.
- When quantizing, if the Hamiltonian is canonical in some basis, $H = 1/2 \sum_a \tilde{p}_a^2 + \ldots$, we can replace $\tilde{p}_a$ with $i\hbar\partial/\partial\tilde{q}^a$ and then change variables to obtain a correctly ordered Hamiltonian in terms of $\hat{q}^I$ and $\hat{p}_I$.
- Weyl (symmetric) order corresponds to the Hamiltonian in the path integral for quantum mechanics.
- For collective coordinate quantization, we start with $\phi(x,t) = \phi^x(t)$ and $\pi^x(t) = i\hbar\partial/\partial\phi^x(t)$ and change variables to the discrete set $q_m(t)$.

Further Reading

For moduli space approximation and soliton scattering, see chapter 4 in [14]. The work to prove the existence of multi-vortex solitons was by Taubes [22]. Then Samols, in [23], found a formula for the first-order metric on the moduli space, and, in [24], an explicit one was found. For a formalism for the effective Lagrangean and resulting interaction potential of vortices, see [25] and [26]. For collective coordinate quantization, see chapter 8 in [13]; the review [27]; and the original articles [28, 29] about the method, and [30] about the application to soliton scattering, [31], and the refinement of quantization in [32].

Exercises

(1)   Prove the relation (32.19), by using the Green–Riemann theorem.

(2)   Prove the relation (32.64).

(3)   Show that one can put the quantum mechanics Hamiltonian for q^I in the form (32.65).

(4)   Solve the eigenfunction-eigenvalue equation (32.42) for the real Higgs potential in 1+1 dimensions, which admits the kink solution, to find (ω_n, η_n).

(5)   For the case at exercise 32.5, calculate the metric g^{IJ}, and thus the Hamiltonian in collective coordinate quantization (32.67).

PART III

OTHER SPINS OR STATISTICS; GENERAL RELATIVITY

33 Chern–Simons Terms: Emergent Gauge Fields, the Quantum Hall Effect (Integer and Fractional), Anyonic statistics

Most of the book has been dealing with physics in 3+1 dimensions, which is the standard case for the real world. But there are specific phenomena that appear in 2+1 dimensions, when we can consider that one dimension of a physical system is negligible, such as in the case of a graphene sheet, which is a couple of atoms thick.

In this chapter, we consider, therefore, field theories specific to 2+1 dimensions and their physical implications. There is basically only one such example, the Chern–Simons gauge field, but its applications are diverse. We will see its applications to the Quantum Hall Effect, how to obtain emergent statistical CS fields in solids, how anyons (particles with fractional statistics) and particle with non-Abelian statistics are obtained from CS fields in general, and, in particular, in the Quantum Hall Effect.

33.1 Chern–Simons Gauge Field

Consider a $U(1)$ gauge field A_μ, subject to the usual gauge invariance, $\delta A_\mu = \partial_\mu \alpha$, and having the same field strength, $F_{\mu\nu} = \partial_\mu A_\nu - \partial_\nu A_\mu$. In 3+1 dimensions, that meant that the kinetic term for the gauge field must be only a function of $F_{\mu\nu}$, which, together with the need to have a two-derivative kinetic term, restricted the possibilities to the unique term $-1/4 \int d^4x F_{\mu\nu}^2$.

But in 2+1 dimensions, there is still another possibility: the *Chern–Simons (CS) kinetic term*, which appears because now we have a constant matrix $\epsilon^{\mu\nu\rho}$:

$$S_{CS} = \frac{k}{4\pi} \int d^{2+1}x \, \epsilon^{\mu\nu\rho} A_\mu \partial_\nu A_\rho , \qquad (33.1)$$

where k is a number that will soon be shown to be necessarily an integer. In form language, we would write

$$S_{CS} = \frac{k}{4\pi} \int d^{2+1}x A \wedge dA. \qquad (33.2)$$

If the action for A_μ is restricted to the Chern–Simons kinetic term, the equation of motion is, varying with respect to A_μ in both positions (and partially integrating the derivative away when doing so) and forming the combination $F_{\mu\nu}$,

$$\frac{k}{4\pi} \epsilon^{\mu\nu\rho} F_{\nu\rho} = 0. \qquad (33.3)$$

Thus, we need to have $F_{\mu\nu} = 0$, or A_μ to be *locally* (though not necessarily globally) a pure gauge.

That means that the theory of Chern–Simons gauge fields is a *topological theory* that only depends on the global properties of the space and of the gauge field. The fact that the action is topological can also be seen by the fact that action can be written without reference to the details of any curved space it lives in (its "metric," see later in the book), and the form of the action in flat space is the same as in curved space. In fact, Chern–Simons gauge theory was used by Witten in his mathematical works on topology that gave him the Fields Medal, though we will not describe that, since it relates to *quantum* field theory.

Instead, we will consider the effect of the Chern–Simons gauge field on other fields. Generically, we can say that the effect of all other fields coupling to A_μ can be described by a current J^μ, giving a term

$$S_{\text{source}} = \int d^{2+1}x \, J^\mu A_\mu \,, \tag{33.4}$$

to be added to S_{CS}. Then the total action is, therefore,

$$S = S_{CS} + S_{\text{source}} = \int d^{2+1}x \left[\frac{k}{4\pi} \epsilon^{\mu\nu\rho} A_\mu \partial_\nu A_\rho + J^\mu A_\mu \right] \,, \tag{33.5}$$

with the equation of motion

$$\frac{k}{4\pi} \epsilon^{\mu\nu\rho} F_{\nu\rho} = J^\mu \Rightarrow F_{\mu\nu} = \frac{2\pi}{k} \epsilon_{\mu\nu\rho} J^\rho \,, \tag{33.6}$$

where we have multiplied by $\epsilon_{\mu\sigma\tau}$ and used $\epsilon_{\mu\sigma\tau}\epsilon^{\mu\nu\rho} = -2\delta_{\sigma\tau}^{\nu\rho}$.

So the equations of motion derived from the Chern–Simons action are gauge invariant, which is not the case of the Maxwell action, but the action seems to not be gauge invariant. We can put it in a somewhat gauge-invariant form by considering that the instanton action can be written as an exterior derivative of the Chern–Simons form, as we saw in (29.15). In form language, and restricting to our current case of a $U(1)$ gauge field, we have $F \wedge F = d(A \wedge dA) = d\mathcal{L}_{CS}$. Then we obtain

$$S_{CS} = \frac{k}{4\pi} \int_{S=\partial M} d^3x \, \epsilon^{\mu\nu\rho} A_\mu \partial_\nu A_\rho = \frac{k}{16\pi} \int_M d^4x \, \epsilon^{\mu\nu\rho\sigma} F_{\mu\nu} F_{\rho\sigma}. \tag{33.7}$$

So, more precisely, a four-dimensional gauge-invariant action, the $U(1)$ equivalent of the instanton action (or Pontryagin index), reduces on the 2+1 dimensional boundary (via Stokes's law, $\int_{\partial M} \mathcal{L} = \int_M d\mathcal{L}$) to the Chern–Simons action.

But the gauge variation of the Chern–Simons action is not zero, and, in fact, it leads to the quantization of the *Chern–Simons level*, the coefficient k. Indeed, under a gauge transformation $\delta A_\mu = \partial_\mu \lambda$, the CS action changes by

$$\delta S_{CS} = \frac{k}{2\pi} \int_M d^{2+1}x \, \epsilon^{\mu\nu\rho} (\partial_\mu \lambda) \partial_\nu A_\rho = \frac{k}{2\pi} \int_M d^{2+1}x \, \partial_\mu (\lambda \epsilon^{\mu\nu\rho} \partial_\nu A_\rho)$$

$$= \frac{k}{2\pi} \int_{S=\partial M} dS^\mu \, \lambda \epsilon_{\mu\nu\rho} \partial^\nu A^\rho. \tag{33.8}$$

Note here an important subtle point: we have partially integrated the derivative on the second A *before* varying it so as to obtain another contribution from it as well, for a total factor of 2. Were it not for this, we would have obtained half the result in (33.8). If S is a flat two-dimensional spatial domain, $dS^\mu = dS\delta^{\mu 0}$, and $\epsilon^{0\mu\nu}\partial_\mu A_\nu = F_{12}$, therefore,

$$\delta S_{CS} = \frac{k}{2\pi}\int_S dS\, F_{12}\lambda. \tag{33.9}$$

Choosing λ to be space independent and only $\mu = 0$, i.e., time, dependent, so $\delta A_0 = \partial_0\lambda$ only, and considering a surface S composed of a surface S_1 at time t_1 and a surface S_2 at time t_2, so $S = S_1 - S_2$, we obtain (in 2+1 dimensions, $F_{12} = B$ is the magnetic field)

$$\delta S_{CS} = \left[\frac{\int B\, dS}{2\pi}\right] k(\lambda(t_2) - \lambda(t_1)). \tag{33.10}$$

Consider now a periodic time direction, as needed formally for putting a quantum mechanical system at finite temperature (we will not explain that further, as it is a quantum field theory issue), and that $t_2 - t_1$ is that periodicity. Then, since λ is a gauge parameter, the action of the transformation on a field like an electron is by $\psi \to e^{ie\lambda/\hbar}\psi$, which means that λ need to be periodic only up to $2\pi\hbar/e$. Considering, moreover, that, as seen in the chapter on Dirac quantization, $\int B\, dS/(2\pi)$ must be an integer n times $\hbar/e$, we find

$$\delta S_{CS} = kn\frac{2\pi\hbar}{e}\frac{\hbar}{e}. \tag{33.11}$$

Yet the action needs to be gauge invariant also only up to $2\pi\hbar$, since in quantum mechanics it always appears only in the combination $e^{iS/\hbar}$. That means that $\hbar k/e^2$ is quantized, $k = k'e^2/\hbar$. Moreover, it is standard to consider the usual rescaling $A_\mu \to A_\mu/e$ (that puts $1/e^2$ in front of the Maxwell action), in which case the Chern–Simons action has no e^2 in front, only an integer $k = k'$ (and $1/\hbar$, usually put to 1).

33.2 Topological Response Theory and the Quantum Hall Effect

Until now, we have studied the Chern–Simons term abstractly, without asking where it comes from. The simplest possibility is that A_μ is the electromagnetic field, but then we know there shouldn't be any CS term in the fundamental theory, only the Maxwell kinetic term.

But in that case, we can consider the action as not a *fundamental action* but, rather, as a (topological) response theory: this *effective action* encodes the response of the full system to an external electromagnetic field A_μ. By varying the action with respect to A_μ, we obtain the current generated in the system by A_μ. Then, if

$$S_{\text{eff}} = \frac{e^2}{\hbar}\frac{k}{4\pi}\int d^{2+1}x\, \epsilon^{\mu\nu\rho} A_\mu \partial_\nu A_\rho, \tag{33.12}$$

for an external electromagnetic field A_μ, we obtain the response of a current

$$j^\mu \equiv \frac{\delta S}{\delta A_\mu} = \frac{e^2}{\hbar}\frac{k}{2\pi}\epsilon^{\mu\nu\rho}\partial_\nu A_\rho. \tag{33.13}$$

We write this in components, noting that $j^0 = \rho$ is the charge density, j^i is the current, and in 2+1 dimensions $F_{0i} = E_i$ and $F_{ij} = \epsilon_{ij}B$,

$$\rho(B) = \frac{e^2 k}{2\pi\hbar} B$$

$$j^i(E) = \frac{e^2 k}{2\pi\hbar} \epsilon^{ij} E_j. \tag{33.14}$$

The second relation describes for us what is the relevant physics. The current generated by an electric field is $\sigma \cdot \vec{E}$, and now this is a tensorial relation. This is the characteristic of the *Hall effect*: Applying a magnetic field B perpendicular to a two-dimensional material (in the transverse z direction), for an applied electric field in the x direction, we have a *Hall conductance* σ_H in the y direction (and reversely for $x \leftrightarrow y$). That means that

$$\sigma_H = \frac{e^2 k}{2\pi\hbar} = \sigma_0 k, \tag{33.15}$$

which is exactly the quantized relation for the conductivity observed experimentally, for the *(Integer) Quantum Hall Effect*.

Then the first equation of motion can be derived from the second, using current conservation, $\partial_0 \rho + \partial_i j^i = 0$, acting with ∂_i on the second equation and using Maxwell's equation in two dimensions, $\vec{\nabla} \times \vec{E} = -\partial_t \vec{B}$, which becomes in 2+1 dimensions $\epsilon^{ij}\partial_i E_j = -\partial_0 B$:

$$\partial_0 \rho = -\partial_i j^i = -\sigma_H \epsilon^{ij}\partial_i E_j = \sigma_H \partial_0 B \Rightarrow \rho(B) - \rho_0 = \sigma_H B. \tag{33.16}$$

In conclusion, the CS action, as a topological response effective action, contains in it the physics of the Integer Quantum Hall Effect.

33.3 Emergent (Statistical) Chern–Simons Gauge Field in Solids

But there is another possibility for the relevance of CS terms in real 2+1 dimensional materials. That is that A_μ is not a fundamental field present in the material (after all, the only such field is electromagnetism, which we just discussed) but, rather, that A_μ is an *emergent* gauge field, a composite field that appears under certain circumstances.

Consider a generic system of electrons interacting via a potential v, in an external electromagnetic field A_μ, with microscopic Hamiltonian

$$H_e = \sum_{j=1}^{N} \frac{|\vec{p}_j - e\vec{A}(\vec{r}_j)|^2}{2m_b} + \sum_{i<j} v(\vec{r}_i - \vec{r}_j) + \sum_{i=1}^{N} eA_0(\vec{r}_i). \tag{33.17}$$

The quantum state of the electrons is described by a multi-electron wave function $\Psi_e(\vec{r}_1, \ldots, \vec{r}_N)$, satisfying a Schrödinger equation

$$H_e \Psi_e(\vec{r}_1, \ldots, \vec{r}_N) = E\Psi_e(\vec{r}_1, \ldots, \vec{r}_N). \tag{33.18}$$

One can make a canonical transformation $\Psi_e \to \Phi_e = U\Psi_e$ on the wave function, with a certain unitary matrix U,

$$\Phi_e(\vec{r}_1,\ldots,\vec{r}_N) = U\Psi_e(\vec{r}_1,\ldots,\vec{r}_N) \equiv \left[\prod_{i<j} e^{-i\frac{\theta}{\pi}\alpha(\vec{r}_i - \vec{r}_j)}\right] \Psi_e(\vec{r}_1,\ldots,\vec{r}_k), \tag{33.19}$$

where $\alpha(\vec{r}_i - \vec{r}_j)$ is the angle made by the relative distance between the electrons, $\vec{r}_{ij} = \vec{r}_i - \vec{r}_j$, with a fixed axis, and θ is a constant.

Under the canonical transformation, the Hamiltonian is changed by $H_e \to H'_e = UH_eU^{-1}$. To calculate it, first observe that

$$U^{-1}(\vec{p}_i - e\vec{A}(\vec{r}_i))U = \vec{p}_i - e\vec{A}(\vec{r}_i) - e\vec{a}(\vec{r}), \tag{33.20}$$

where we have defined the emergent (or statistical) gauge field $\vec{a}$ by

$$e\vec{a}(\vec{r}_i) = \frac{\theta}{\pi}\sum_{j\neq i}\vec{\nabla}_i\alpha(\vec{r}_i - \vec{r}_j) = \frac{\theta}{\pi}\sum_{j\neq i}\frac{\hat{z}\times(\vec{r}_i - \vec{r}_j)}{|\vec{r}_i - \vec{r}_j|^2}, \tag{33.21}$$

$\hat{z}$ being the direction perpendicular to the material (third spatial direction). Since $\alpha_{ij} = \alpha(\vec{r}_i - \vec{r}_j)$ is the angle made by the distance between two electrons with a fixed direction, then

$$\alpha_{ij} = \alpha_{ji} + \pi. \tag{33.22}$$

Then the transformed Hamiltonian is

$$H'_e = \sum_{j=1}^{N}\frac{|\vec{p}_j - e\vec{A}(\vec{r}_j) - e\vec{a}(\vec{r}_j)|^2}{2m_b} + \sum_{i<j}v(\vec{r}_i - \vec{r}_j) + \sum_{i=1}^{N}eA_0(\vec{r}_i). \tag{33.23}$$

But now the wave function has different properties under interchange of particles, since $\alpha_{ij} = \alpha_{ji} + \pi$, so when interchanging two electrons with positions $\vec{r}_i$ and $\vec{r}_j$, besides the original minus (for fermions), we get an extra phase factor of $e^{i\theta}$, for a total

$$\Phi_e(\ldots,\vec{r}_j,\ldots,\vec{r}_i,\ldots) = -e^{i\theta}\Phi_e(\ldots,\vec{r}_i,\ldots,\vec{r}_j,\ldots), \tag{33.24}$$

That means that if $\theta = (2k+1)\pi$, the statistics is changed from Fermi to Bose and, more generally, for arbitrary θ, we obtain *fractional, or anyonic, statistics*. Thus, a_μ is called a statistical gauge field since it changes the statistics of the particles.

Before we explain that better, we observe that there is a magnetic field associated with the emergent (statistical) gauge field a_μ, proportional to the charge density:

$$f_{12}(\vec{r}_i) \equiv b(\vec{r}_i) = \vec{\nabla}\times\vec{a}(\vec{r}_i) = \frac{2\theta}{e}\sum_{j\neq i}\delta(\vec{r}_i - \vec{r}_j) = \frac{2\theta}{e^2}\rho_{\text{charge}}. \tag{33.25}$$

This equation of motion comes, indeed, from the Chern–Simons Lagrangean with a source, more precisely,

$$S = \int d^{2+1}x\left[\frac{e^2}{4\theta}\epsilon^{\mu\nu\rho}a_\mu\partial_\nu a_\rho + J^\mu a_\mu\right]. \tag{33.26}$$

Comparing with the Chern–Simons action with the correctly quantized level k, we see that we have

$$\theta = \frac{\pi}{k}. \tag{33.27}$$

33.4 Anyons and Fractional Statistics

By considering a Chern–Simons gauge field, and in particular a statistical gauge field, we have obtained particles with different statistical properties, under the interchange of two of them, namely with an arbitrary phase $e^{i\theta}$, called *anyons*. This implies a strange property: if we interchange two bosons or two fermions twice, we get the same wave function, since $(\pm 1)^2 = +1$, but if we interchange two anyons twice, we don't come back to the same state, but one with phase $e^{2i\theta}$.

The simplest way to obtain anyonic statistics is the one we already described: to attach a delta function magnetic flux at the position of the particles, as in (33.25). Indeed, integrating over a small disk, we get a magnetic flux of

$$\Phi = \int_S f_{12} dx^1 \wedge dx^2 = \frac{2\theta}{e}. \tag{33.28}$$

If we consider the boundary of this disk, $\partial S = C$, via Stokes's law, we obtain

$$\Phi[S] = \int_{C=\partial S} \vec{A} \cdot d\vec{x} = \frac{2\pi}{e}. \tag{33.29}$$

Aharonov–Bohm Effect

But this situation leads to an Aharonov–Bohm effect. The effect stems from the question: can the gauge field A_μ have physical consequences, or is it just a convenient mathematical construct, and only $F_{\mu\nu}$ contains physics? The answer is that there are physical situations where A_μ contains physics – more precisely if we go in a loop around a region with magnetic field, like for the previously discussed contour C, around the delta function magnetic field.

In the absence of A_μ, since the momentum operator acts on the wave function of a particle as $\vec{p}\psi = -i\hbar \frac{\partial}{\partial x}\psi$, for a particle going on a path P, the wave function must have a phase factor

$$e^{\frac{i}{\hbar}\int_P \vec{p}\cdot d\vec{x}}. \tag{33.30}$$

Then introducing a vector potential $\vec{A}$ (in the gauge $A_0 = 0$) amounts to the shift $\vec{p} \to \vec{p} - q\vec{A}$, for a total phase factor

$$e^{\frac{i}{\hbar}\int_P (\vec{p}-q\vec{A})\cdot d\vec{x}} \equiv e^{i\delta}. \tag{33.31}$$

If we consider the path P to be a closed loop C, then we obtain a phase

$$\delta = \frac{q}{\hbar}\oint_{C=\partial S} \vec{A}\cdot d\vec{x} = \frac{q}{\hbar}\int_S \vec{B}\cdot d\vec{S} = \frac{q\Phi}{\hbar}. \tag{33.32}$$

This means that the phase factor of a wave function for a particle going around a region of magnetic field is nontrivial, even if the path itself never intersects the magnetic field, thus proving the physicality of $\vec{A}$. This is the Aharonov–Bohm effect.

Anyonic Exchange

Now consider two anyons, and one (of charge e) going in a closed loop C around the other. This can be interpreted as a double interchange of the particles. Instead of the wave function to come back to its original state (like for bosons and fermions), it will aquire a phase factor of

$$\exp\left(ie \oint_C \vec{A} \cdot d\vec{x}\right) = \exp(2i\theta), \qquad (33.33)$$

which means that under a single interchange, we have a factor $e^{i\theta}$. Therefore, θ, introduced in the equation of motion (33.25) and the action (33.26), is indeed the anyonic parameter, and the Chern–Simons term is the kinetic term for the statistical gauge field.

33.5 Anyons and the Fractional Quantum Hall Effect (FQHE)

If one goes to higher magnetic fields, it is found experimentally that the conductivity plateaus become not integer, but fractional, multiples of $\sigma_0 = e^2/(2\pi\hbar)$. There are many relevant fractions, but not all have simple theoretical explanations. One case that does is the case of a "filling fraction,"

$$\nu \equiv \frac{\sigma_H}{\sigma_0} = \frac{1}{r}, \qquad (33.34)$$

where r is an integer, which we explain now.

Some of the physics of the Fractional Quantum Hall is described well by an effective action for an emergent (statistical) gauge field a_μ, coupled to the electromagnetic gauge field A_μ. The emergent gauge field must have a Chern–Simons kinetic term, with correctly quantized coefficient, $k/(4\pi)$, and conventionally we write $k = -r$. Its coupling to the electromagnetic field A_μ must, because of symmetry issues[§], be described also by a Chern–Simons type action, the "BF term," $C\epsilon^{\mu\nu\rho}A_\mu\partial_\nu a_\rho$. Its coefficient C can be guessed as follows. The BF term can be decomposed into two CS terms of opposite level by writing a decomposition that preserves the norm of the fields, $|A_\mu|^2 + |a_\mu|^2 = |a_\mu^1|^2 + |a_\mu^2|^2$, more precisely,

$$A_\mu = \frac{a_\mu^1 + a_\mu^2}{\sqrt{2}}; \quad a_\mu = \frac{a_\mu^1 - a_\mu^2}{\sqrt{2}}, \qquad (33.35)$$

[§] The Chern–Simons action is not invariant under parity, the change of spatial directions by a minus, $\vec{x} \to -\vec{x}$; rather, it is multiplied by a minus sign. We can show that this needs to happen to the coupling of A_μ and a_μ.

which leads to

$$C\epsilon^{\mu\nu\rho}A_\nu\partial_\nu a_\rho = \frac{C}{2}\epsilon^{\mu\nu\rho}a^1_\mu\partial_\nu a^1_\rho - \frac{C}{2}\epsilon^{\mu\nu\rho}a^2_\mu\partial_\nu a^2_\rho. \tag{33.36}$$

Since the CS terms must have coefficients quantized in units of $1/(4\pi)$, it follows that the BF term must have coefficients quantized in units of $1/(2\pi)$. We are then led to the effective action for the Fractional Quantum Hall Effect,

$$S_{\text{eff}} = \int_{M_3} d^3x \left[\frac{1}{2\pi}\epsilon^{\mu\nu\rho}A_\mu\partial_\nu a_\rho - \frac{r}{4\pi}\epsilon^{\mu\nu\rho}a_\mu\partial_\nu a_\rho \right], \tag{33.37}$$

to which we must, of course, add the action of electromagnetism.

Classically, we can write the equation of motion for a_μ,

$$f_{\mu\nu} = \frac{1}{r}F_{\mu\nu}, \tag{33.38}$$

and solve it by identifying the gauge fields (up to a gauge transformation),

$$a_\mu = \frac{A_\mu}{r}. \tag{33.39}$$

Replacing it back in the effective action, we would get

$$S'_{\text{eff}} = \frac{1}{r}\frac{1}{4\pi}\int_{M_3} d^3x \epsilon^{\mu\nu\rho}A_\mu\partial_\nu A_\rho. \tag{33.40}$$

This should be valid at the classical level only, since, as we saw, at the quantum level, we need to have an integer multiple of $1/(4\pi)$, whereas here we have a fractional one, $1/r$.

By the calculation that we already did, we know that the coefficient of the CS action for the electromagnetic field (when the effective action is viewed as a response action), multiplied by 2, is the Hall conductivity, so now we have (reintroducing e^2 and $\hbar$)

$$\sigma_H = \frac{1}{r}\frac{e^2}{2\pi\hbar} = \frac{1}{r}\sigma_0. \tag{33.41}$$

The presence of anyons could be found simply by the existence of the original statistical gauge field, since we saw that we only needed the correctly quantized Chern–Simons term for that. However, we want to be more precise: we needed also a coupling of a_μ to some particle current, and there is already a coupling of A_μ to a_μ. Then, adding a source term to the effective action, corresponding to some "quasiparticles" of charge q under the statistical gauge field a_μ,

$$\int_{M_3} J^\mu a_\mu = q \int dt\, a_0(x_0, t), \tag{33.42}$$

the equation of motion for a_0 is changed to

$$\frac{F_{12}}{2\pi} - \frac{rf_{12}}{2\pi} + q\delta(x - x_0) = 0. \tag{33.43}$$

This means that either the real magnetic field F_{12} is a delta function or the statistical gauge field is. But it is conceptually more difficult to accept that the real magnetic field becomes a delta function, not the least because there are no magnetic monopoles so the delta function would be the idealization of a small flux tube, but by the vanishing of the magnetic charge, the flux tube must return (in the opposite direction) somewhere nearby.

So the reasonable identification is to say that we have

$$\frac{f_{12}}{2\pi} = \frac{q}{r}\delta(x - x_0),$$

(33.44)

leading (as we saw before) to an anyonic parameter

$$\theta = \frac{\pi}{r}\frac{q}{e}.$$

(33.45)

Replacing (33.44) into the effective action, the BF coupling of A_μ to a_μ becomes just a source for A_0, leading to an electromagnetic charge density (by definition, the variation of the effective action with respect to A_0)

$$J_0 = \frac{\delta S_{eff}}{\delta A_0} = \frac{f_{12}}{2\pi} = \frac{q}{r}\delta(x - x_0).$$

(33.46)

That means that we have a fractional unit of ordinary electric charge, $Q = q/r$ (carried by the "quasiparticles").

We see that the physics of the Fractional Quantum Hall Effect leads to anyonic quasiparticles with fractional electric charge.

Nonabelian Statistics

We should mention that the anyonic statistics, with a general phase $e^{i\theta}$ multiplying the wave function under the exchange of two identical particles, is not the most general possibility. In fact, as Moore and Read showed, we can introduce, already in the context of the Fractional Quantum Hall Effect, a statistics that is non-Abelian. The general possibilities under interchange are not characterized just by the permutation group $\mathbb{Z}_N$, but by the *braid group* $\mathcal{B}_N$. Anyons are in Abelian representations of $\mathcal{B}_N$, but there are also non-Abelian representations, leading to non-Abelians anyons, or *nonabelions*.

The nonabelions transform under the interchange of two excitations r and s as

$$\psi_{p:\{i_1,...,i_r,...,i_s,...,i_N\}}(z_1, \ldots, z_{i_r}, \ldots, z_{i_s}, \ldots, z_N) = \sum_q B_{pq}[i_1, \ldots, i_N]\psi_{q:\{i_1,...,i_N\}}(z_1, \ldots, z_N).$$

(33.47)

Here p is a shorthand for the set of indices $\{i_1, \ldots, i_N\}$.

We will not describe the construction of Moore and Read, since it will require the introduction of several other difficult concepts.

Important Concepts to Remember

- In 2+1 dimensions, for a gauge field, we can have a Chern–Simons (CS) kinetic term, $(k/4\pi)\epsilon^{\mu\nu\rho}A_\mu\partial_\nu A_\rho$.
- The CS term is topological, the boundary term for a four-dimensional $\int F \wedge F$ term, and, in general, not gauge invariant.
- The requirement of gauge invariance leads, at the quantum level, to the quantization of k in units of $e^2/\hbar$.
- The CS action for the electromagnetic field A_μ, viewed as a (topological) response effective action, leads to a quantized Hall conductivity – i.e., an Integer Quantum Hall Effect.

- In solids, we can introduce an emergent, statistical gauge field $e\vec{a}(\vec{r}_i) = (\theta/\pi) \sum_{j \neq i} \vec{\nabla}_i \alpha(\vec{r}_i - \vec{r}_j)$ if we do a canonical transformation on the wave function by the phases $\prod_{i<j} e^{-i\frac{\theta}{\pi}\alpha(\vec{r}_i - \vec{r}_j)}$.
- The effect is to change the statistics of the particles by adding the phase $e^{i\theta}$ when we interchange two of them in the wave function.
- The statistical gauge field leds to a magnetic field f_{12} localized at the positions of the particles, $f_{12}(\vec{r}_i) = \frac{2\theta}{e} \sum_{j \neq i} \delta(\vec{r}_i - \vec{r}_j)$, which can be obtained from the Chern–Simons kinetic term, coupled to a current made up of the delta functions, for $\theta = \pi/k$.
- The Aharonov–Bohm effect is due to the phase $\exp\left(i\frac{e}{\hbar} \oint_C \vec{A} \cdot d\vec{x}\right) = \exp(ie\Phi/\hbar)$ on wave functions, where Φ is the magnetic flux.
- Anyons are particles with a factor of $e^{i\theta}$ in the multiparticle wave function under the interchange of two identical particles.
- They can be obtained from the magnetic field flux localized at the position of the particles (delta function) and leading to an Aharonov–Bohm phase, the flux coming from the Chern–Simons action.
- The Fractional Quantum Hall Effect (FQHE) for filling fraction, $\nu = 1/r$, can be obtained from the Chern–Simons action for the statistical gauge field, coupled to the electromagnetic field through a BF term.
- The FQHE leads to anyonic particles with fractional electric charge.
- One can also define non-Abelian statistics, for non-Abelian factors under the interchange of identical particles, and they can be obtained in the FQHE.

Further Reading

For an introduction to Chern–Simons theory, see the review [33]. For an introduction to anyons, see the review [34]. For a modern treatment of the Quantum Hall Effect and its relation to Chern–Simons fields, see Ed Witten's lectures [35]. The nonabelions were introduced in [36].

Exercises

(1)   Consider a system with a Hall conductivity σ_H, but with zero direct conductivity $\sigma_{xx} = \sigma_{yy} = 0$. Calculate the resistivity ρ_H. What can you deduce about the material?

(2)   Consider a system with statistical gauge field action (33.26), and only electronic states and no fractional quasiparticles. Show that we can only have bosonic or fermionic wavefunction descriptions.

(3)   Consider a superconducting ring threaded by n units of magnetic flux $\Phi = nh/e$. What is the Aharonov–Bohm phase around the ring? What if the superconductor exhibits a Fractional Quantum Hall Effect with quasiparticles of electric charge $1/r$?

(4)   Consider a system described by a Fractional Quantum Hall Effect action (33.37). Calculate its topological electromagnetic response.

34

Chern–Simons and Self-Duality in Odd Dimensions, Its Duality to Topologically Massive Theory and Dualities in General

In the previous chapter, we saw that in 2+1 dimensions we can have a different kinetic term for a gauge field, the Chern–Simons term. It has a single derivative and is topological in nature (as it can be obtained on the boundary of a 3+1 dimensional space with a topological action).

In this chapter, we see if we can extend this notion of Chern–Simons to other dimensions, specifically odd dimensions, $d = 2k + 1$. In the process, we will find that these actions are "dual" to more common actions, and derive a notion of "duality" that applies to more general field theories. We will see the duality relations between different theories in various dimensions and, finally, see that we have a basic duality relation in 1+1 dimensions that can act as a template for the duality transformations.

34.1 Vector Theories in 2+1 Dimensions

We can ask: what are the possible theories that can be constructed in 2+1 dimensions from a vector field A_μ?

Maxwell Theory

One answer is the usual Maxwell theory, a gauge theory with kinetic term

$$S_{\text{Maxwell}} = -\frac{1}{4} \int d^{2+1}x F_{\mu\nu}^2. \tag{34.1}$$

It propagates a single mode. In four dimensions there are two, the two possible polarizations for A_μ, along the two spatial directions transverse to the propagation direction $\hat{z} = \vec{k}/k$, of the momentum of the wave. Since in two spatial dimensions, there is a single direction transverse to the propagation (momentum) direction, we have a single mode. More formally, the gauge invariance $\delta A_\mu = \partial_\mu \alpha$ can be used to fix one component to zero – for instance, $A_0 = 0$ (the Coulomb gauge). Further, on top of that we can also put $\partial^\mu A_\mu = 0$, which is the Lorenz gauge condition (obtaining, thus, the radiation gauge), after which the equation of motion for A_μ is just KG: $\Box A_\mu = 0$, so it contains no more conditions. Then in momentum space we have $k^\mu A_\mu = 0$, which means that $\vec{k} \cdot \vec{A} = 0$, or the polarization is transverse to the momentum.

A note on (mass) dimensions. Since the action should be dimensionless, and the volume element d^3x has dimension -3, the Lagrangean should have dimension 3, and the two derivatives already give dimension 2, leaving for A_μ of dimension 1/2. On the other hand, in the last chapter, where we considered only a CS term, the coefficient in most of the chapter was a number, so A_μ had the standard dimension of 1, such that $A\partial A$ has dimension 3. Though when quantizing, we noticed that naturally, we obtained also an e^2 coefficient, which could be rescaled away. Therefore, the relation between the two definitions of A_μ is given by a rescaling with the coupling e, which in three dimensions has dimensions, specifically dimension 1/2, $[e] = 1/2$. That means that adding a Maxwell term in the case with just a numerical coefficient $k/(4\pi)$ must be done by dividing it by e^2, which has dimension 1.

Proca Field

But we can further add a mass term to the action and obtain a Proca field,

$$S_{\text{Proca}} = \int d^{2+1}x \left[-\frac{1}{4}F_{\mu\nu}^2 - \frac{m^2}{2}A_\mu A^\mu \right], \tag{34.2}$$

whose equation of motion is

$$\partial^\mu F_{\mu\nu} - m^2 A_\nu = 0, \tag{34.3}$$

as we saw before. In this case, there is no gauge invariance, but by taking ∂^ν on the equation of motion, we see that classically, we have

$$\partial^\nu A_\nu \propto \partial^\mu \partial^\nu F_{\mu\nu} = 0, \tag{34.4}$$

so the same "Lorenz condition" from before. Then, in 2+1 dimensions, we eliminate one component of A_μ by this transversality condition (polarization perpendicular to the momentum), being left with two propagating modes that, because of the "Lorenz condition," satisfy the KG equation

$$(\Box - m^2)A_\mu = 0. \tag{34.5}$$

Topological Mass Term

Instead of adding a single mass term, we can add a "topological mass term" – i.e., a Chern–Simons term – to the Maxwell action. The action in this case is, if A_μ has dimension 1,

$$S = \int d^{2+1}x \left[-\frac{1}{4e^2}F_{\mu\nu}F^{\mu\nu} + \frac{k}{4\pi}\epsilon^{\mu\nu\rho}A_\mu \partial_\nu A_\rho \right], \tag{34.6}$$

or, by rescaling $A_\mu = eA'_\mu$, write

$$S_{\text{topmass}} = \int d^{2+1}x \left[-\frac{1}{4}F_{\mu\nu}F^{\mu\nu} - \frac{m}{2}\epsilon^{\mu\nu\rho}A_\mu \partial_\nu A_\rho \right], \tag{34.7}$$

where $m = -ke^2/(2\pi)$ has dimension 1, as it should, and A'_μ has dimension 1/2 (and we have removed the primes for simplicity). The equation of motion is

$$-\partial_\nu F^{\mu\nu} + \frac{m}{2}\epsilon^{\mu\nu\rho}F_{\nu\rho} = 0, \tag{34.8}$$

and it is explicitly gauge invariant. With the quantization condition, the action itself is gauge invariant.

In either case, we note that at low energies, which is the same as at small derivatives on fields, $[\partial_\mu] \ll m \propto e^2$, the Maxwell term can be neglected with respect to the CS term, so we are back to the CS theory. Moreover, now we can use the gauge invariance in the same way as for the pure Maxwell theory to show that we have the same single mode (spatial and transverse to the momentum) propagated by the equation of motion.

"Self-Dual" Action

Finally, we can consider the Chern–Simons term and a mass term, with action

$$S_{\text{SD}} = \int d^{2+1}x \left[-\frac{m^2}{2} A_\mu A^\mu + \frac{m}{2} \epsilon^{\mu\nu\rho} A_\mu \partial_\nu A_\rho \right], \tag{34.9}$$

and equation of motion

$$m A^\mu = \frac{1}{2} \epsilon^{\mu\nu\rho} F'_{\nu\rho}. \tag{34.10}$$

This equation of motion can be called a "self-duality condition" in 2+1 dimensions. Indeed, in 3+1 dimensions, a "self-duality condition" would be to have

$$F_{\mu\nu} = \tilde{F}_{\mu\nu} \equiv \frac{1}{2} \epsilon_{\mu\nu\rho\sigma} F^{\rho\sigma}. \tag{34.11}$$

But, if we do a "dimensional reduction" to 2+1 dimensions, by putting $A_3 = 0$ and $\partial_3 A_i(x^k, x^3) = m A_i(x^k) f(x^3)$, $i = 0, 1, 2$, from the self-duality relation for F_{3i} we would obtain exactly the equation of motion,

$$m A_i = \frac{1}{2} \epsilon^{3ijk} F_{jk}. \tag{34.12}$$

We note that the "self-duality" condition in three dimensions (34.10) implies both $\partial^\mu A_\mu = 0$ (using the Bianchi identity for $F_{\mu\nu}$, $\partial_{[\mu} F_{\nu\rho]} = 0$) and the Proca equation, since by multiplying it with $\epsilon^{\mu\lambda\tau}$ (and using $\epsilon_{\mu\nu\rho} \epsilon^{\mu\lambda\tau} = -2\delta^{\lambda\tau}_{\nu\rho}$) and then with $m\partial^\lambda$, we obtain

$$- \partial_\lambda F^{\lambda\tau} = m \epsilon^{\mu\lambda\tau} \partial_\lambda A_\mu = -\frac{m}{2} \epsilon^{\mu\lambda\tau} F_{\mu\lambda} = -m^2 A^\tau. \tag{34.13}$$

In fact, note that the sign of m was not important in this relation.

Thus, in a sense, the self-duality condition is a "square root" of the Proca equation. We can be more precise about it and note that the product of two kinetic operators for the self-dual condition, with opposite signs for m, gives the kinetic operator for the Proca equation.

We first note that we can write the self-dual condition as

$$(\epsilon^{\mu\nu\rho} \partial_\nu - m\delta^{\mu\rho}) A_\rho = 0. \tag{34.14}$$

Next, taking the product of two such factors, we obtain

$$\begin{aligned}(\epsilon^{\mu\nu\rho} \partial_\nu - m\delta^{\mu\rho})(\epsilon^{\rho\lambda\sigma} \partial_\lambda + m\delta^{\rho\sigma}) A_\sigma &= \left[(-\delta^{\mu\lambda}\delta^{\nu\sigma} + \delta^{\mu\sigma}\delta^{\nu\lambda}) \partial_\nu \partial_\lambda - m^2 \delta^{\mu\sigma} \right] A_\sigma \\ &= [-\partial_\mu \partial_\sigma + \delta^{\mu\sigma} \partial^2 - m^2 \delta^{\mu\sigma}] A_\sigma \\ &= -\partial^\sigma \partial_\mu A_\sigma + \partial^\sigma \partial_\sigma A_\mu - m^2 A_\mu \\ &= \partial^\sigma F_{\sigma\mu} - m^2 A_\mu \,, \end{aligned} \tag{34.15}$$

which proves our statement. This makes it clear why the self-duality condition propagates only one mode, whereas the Proca equation propagates two: half of the modes of the Proca equation are propagated by the self-dual condition with one sign for m, and the other half by the self-dual condition with the other sign for m.

34.2 Self-Duality in Odd Dimensions

We can now generalize the previous analysis to the case of a general odd dimension. We must, however, distinguish between $d = 4k + 1$ and $d = 4k - 1$ dimensions because of reality properties.

$d = 4k - 1$ Dimensions

The relevant cases for physics are the three-dimensional case already studied, and seven and 11 dimensions (in higher dimensional theories like supergravity and M theory).

We consider a self-duality condition similar to the one for A_μ, just that now necessarily for p-forms, specifically for $p = (2k - 1)$-forms,

$$A_{\mu_1...\mu_{2k-1}} = \frac{1}{(2k)!\, m} \epsilon_{\mu_1...\mu_{2k-1}\nu_1...\nu_{2k}} F^{\nu_1...\nu_{2k}}, \tag{34.16}$$

where $F_{(2k)}$ is the field strength of $A_{(2k-1)}$ – i.e., $F_{(2k)} = dA_{(2k-1)}$ – or

$$F_{\mu_1...\mu_{2k}} = (2k)\partial_{[\mu_1} A_{\mu_2...\mu_{2k}]}. \tag{34.17}$$

We notice that the self-duality condition can be written as

$$\left[\frac{1}{(2k-1)!} \epsilon^{\mu_1...\mu_{2k-1}\rho\nu_1...\nu_{2k-1}} \partial_\rho - \delta^{\mu_1...\mu_{2k-1}}_{\nu_1...\nu_{2k-1}} \right] A_{\nu_1...\nu_{2k-1}} = 0 \tag{34.18}$$

Then, we observe the important fact that in $d = 4k - 1$ Minkowski space we have the equality (there is a minus sign from the η_{00} metric, canceled by another sign for moving ρ past $2k - 1$ indices)

$$\frac{1}{(2k-1)!} \epsilon^{\mu_1...\mu_{2k-1}\rho\nu_1...\nu_{2k-1}} \frac{1}{(2k-1)!} \epsilon_{\nu_1...\nu_{2k-1}\lambda\sigma_1...\sigma_{2k-1}} = 2k\delta^{\rho\mu_1...\mu_{2k-1}}_{\lambda\sigma_1...\sigma_{2k-1}}. \tag{34.19}$$

This allows one to again write the product of two kinetic operators for self-duality as an Proca kinetic operator,

$$\left[\frac{1}{(2k-1)!} \epsilon^{\mu_1...\mu_{2k-1}\rho\nu_1...\nu_{2k-1}} \partial_\rho - m\delta^{\mu_1...\mu_{2k-1}}_{\nu_1...\nu_{2k-1}} \right]$$
$$\times \left[\frac{1}{(2k-1)!} \epsilon^{\nu_1...\nu_{2k-1}\lambda\sigma_1...\sigma_{2k-1}} \partial_\lambda + m\delta^{\nu_1...\nu_{2k-1}}_{\sigma_1...\sigma_{2k-1}} \right] A_{\sigma_1...\sigma_{2k-1}}$$

$$= (2k)\partial^\rho \partial_{[\rho} A_{\mu_1...\mu_{2k-1}]} - m^2 A_{\mu_1...\mu_{2k-1}}$$

$$= \partial^\rho F_{\rho\mu_1...\mu_{2k-1}} - m^2 A_{\mu_1...\mu_{2k-1}}. \tag{34.20}$$

So again we can say that half of the modes of the Proca equation are propagated by the self-dual operator with one sign for m, and the other by the self-dual operator with the other sign for m.

The self-dual action giving the self-dual condition as its equation of motion is

$$S_{\text{SD}} = \int d^{4k-1}x \left[\frac{m}{2(2k-1)!} \epsilon^{\mu_1...\mu_{2k-1}\rho\nu_1...\nu_{2k-1}} A_{\mu_1...\mu_{2k-1}} \partial_\rho A_{\nu_1...\nu_{2k-1}} - \frac{m^2}{2} A^2_{\mu_1...\mu_{2k-1}} \right], \tag{34.21}$$

whereas the Proca action is

$$S_{\text{Proca}} = \int d^{4k-1}x \left[-\frac{1}{4k} F^2_{\mu_1...\mu_{2k}} - \frac{m^2}{2} A^2_{\mu_1...\mu_{2k-1}} \right]. \tag{34.22}$$

Note that one could also consider a "topologically massive" action by considering only the Maxwell type and CS type terms,

$$S_{\text{topmass}} = \int d^{2k-1}x \left[-\frac{1}{4k} F^2_{\mu_1...\mu_{2k}} - \frac{m}{2(2k-1)!} \epsilon^{\mu_1...\mu_{2k-1}\rho\nu_1...\nu_{2k-1}} A_{\mu_1...\mu_{2k-1}} \partial_\rho A_{\nu_1...\nu_{2k-1}} \right], \tag{34.23}$$

which, unlike the Proca one, is gauge invariant and, in fact, propagates only half the degrees of freedom of the Proca equation, like the self-dual action.

$d = 4k + 1$ Dimensions

The relevant cases for physics are five and nine dimensions (in higher dimensional theories like supergravity and string theory).

Now we have a problem because we obtain an extra minus sign (for ρ going past an extra index) in the analog of (34.19),

$$\frac{1}{(2k)!} \epsilon^{\mu_1...\mu_{2k}\rho\nu_1...\nu_{2k}} \frac{1}{(2k)!} \epsilon_{\nu_1...\nu_{2k}\lambda\sigma_1...\sigma_{2k}} = -(2k+1)\delta^{\rho\mu_1...\mu_{2k}}_{\lambda\sigma_1...\sigma_{2k}}. \tag{34.24}$$

That means that if we write the self-dual condition like in $d = 4k-1$ dimensions, namely as

$$\left[\frac{1}{(2k)!} \epsilon^{\mu_1...\mu_{2k}\rho\nu_1...\nu_{2k}} \partial_\rho - \delta^{\mu_1...\mu_{2k}}_{\nu_1...\nu_{2k}} \right] A_{\nu_1...\nu_{2k}} = 0, \tag{34.25}$$

the product of two such kinetic operators gives a tachyonic Proca operator,

$$\left[\frac{1}{(2k)!} \epsilon^{\mu_1...\mu_{2k}\rho\nu_1...\nu_{2k}} \partial_\rho - m\delta^{\mu_1...\mu_{2k}}_{\nu_1...\nu_{2k}} \right] \times$$

$$\times \left[\frac{1}{(2k)!} \epsilon^{\nu_1...\nu_{2k}\lambda\sigma_1...\sigma_{2k}} \partial_\lambda + m\delta^{\nu_1...\nu_{2k}}_{\sigma_1...\sigma_{2k}} \right] A_{\sigma_1...\sigma_{2k-1}}$$

$$= -(2k+1)\partial^\rho \partial_{[\rho} A_{\mu_1...\mu_{2k}]} - m^2 A_{\mu_1...\mu_{2k}}$$

$$= -[\partial^\rho F_{\rho\mu_1...\mu_{2k-1}} + m^2 A_{\mu_1...\mu_{2k-1}}]. \tag{34.26}$$

The solution to the conundrum seems simple: add an extra i to the self-duality condition to write it as

$$A_{\mu_1...\mu_{2k}} = \frac{i}{(2k+1)!\,m}\epsilon_{\mu_1...\mu_{2k}\nu_1...\nu_{2k+1}}F^{\nu_1...\nu_{2k+1}} \tag{34.27}$$

or, equivalently,

$$\left[\frac{i}{(2k)!}\epsilon^{\mu_1...\mu_{2k}\rho\nu_1...\nu_{2k}}\partial_\rho - \delta^{\mu_1...\mu_{2k}}_{\nu_1...\nu_{2k}}\right]A_{\nu_1...\nu_{2k}} = 0, \tag{34.28}$$

which will lead to the correct Proca equation. But that implies that the field $A_{\mu_1...\mu_{2k}}$ must necessarily be complex.

The action that gives the self-duality condition must then be the action for the complex field $A_{\mu_1...\mu_{2k}}$,

$$S = \int d^{4k+1}x \left[\frac{im}{(2k)!}\epsilon^{\mu_1...\mu_{2k}\rho\nu_1...\nu_{2k}}A^*_{\mu_1...\mu_{2k}}\partial_\rho A_{\nu_1...\nu_{2k}} - m^2 A^*_{\mu_1...\mu_{2k}}A^{\mu_1...\mu_{2k}}\right]. \tag{34.29}$$

We can convince ourselves that the action is real by taking its complex conjugate, and partially integrating the derivative.

However, then, decomposing the complex field into real components,

$$A_{\mu_1...\mu_{2k}} = \frac{A^{(1)}_{\mu_1...\mu_{2k}} + iA^{(2)}_{\mu_1...\mu_{2k}}}{\sqrt{2}}, \tag{34.30}$$

we first note that the terms $A^{(1)}dA^{(1)}$ and $A^{(2)}dA^{(2)}$ in the action vanish (by partially integrating and showing that the term is equal to minus itself), and we obtain the action

$$S = \int d^{4k+1}x \left[\frac{m}{(2k)!}\epsilon^{\mu_1...\mu_{2k}\rho\nu_1...\nu_{2k}}A^{(2)}_{\mu_1...\mu_{2k}\rho\nu_1...\nu_{2k}}\partial_\rho A^{(1)}_{\nu_1...\nu_{2k}} - \frac{m^2}{2}(A^{(1)}_{\mu_1...\mu_{2k}})^2 - \frac{m^2}{2}(A^{(2)}_{\mu_1...\mu_{2k}})^2\right]. \tag{34.31}$$

This is, however, not the action of a new field but, rather, the "first-order form" of the Proca action, with $A^{(2)}$ being an independent field for the field strength of $A^{(1)}$.

Indeed, writing the equation of motion of $A^{(2)}_{\mu_1...\mu_{2k}}$, we obtain

$$mA^{(2)}_{\mu_1...\mu_{2k}} = \frac{1}{(2k)!}\epsilon_{\mu_1...\mu_{2k}\rho\nu_1...\nu_{2k}}\partial^\rho A^{(1)\nu_1...\nu_{2k}} = \frac{1}{(2k+1)!}\epsilon_{\mu_1...\nu_{2k}\rho\nu_1...\nu_{2k}}F^{(1)\rho\nu_1...\nu_{2k}}, \tag{34.32}$$

and replacing it in the action, the $\epsilon A^{(2)}\partial A^{(1)}$ and $m^2(A^{(2)})^2/2$ terms give the same kinetic term contribution with factors of $+2$ and -1, respectively, and the resulting action is just the Proca action for $A^{(1)}$,

$$S_{\text{Proca}} = \int d^{4k+1}x \left[-\frac{1}{2(2k)}(F^{(1)}_{\rho\mu_1...\mu_{2k}})^2 - \frac{m^2}{2}(A^{(1)}_{\mu_1...\mu_{2k}})^2\right]. \tag{34.33}$$

34.3 Duality of Self-Dual Action and Topologically Massive Action

We now want to show that, despite appearing to be different, the "topological mass term" action (34.7) and the self-dual action (34.9) in 2+1 dimensions are, in fact, equivalent physically. The transformation between one and the other is a transformation called *duality* that inverts the coupling and exchanges the fields of one with the fields of the other in a given way.

The duality transformation is achieved by writing a "first-order form" for the action or a "master action for the duality" with two independent fields instead of one, and we find that if we eliminate one via its equation of motion, we get one model, and if we eliminate the other, we get the other model.

As a first step, we rewrite the "topological mass term" action (34.7) by defining

$$F^\mu \equiv \epsilon^{\mu\nu\rho}\partial_\nu A_\rho,$$ (34.34)

and considering that $\epsilon^{\mu\nu\rho}\epsilon_{\sigma\nu\rho} = -2\delta^\mu_\sigma$, we can write

$$S_{\text{topmass}} = \int d^{2+1}x \left[\frac{1}{2}F_\mu F^\mu - \frac{m}{2}F^\mu A_\mu\right].$$ (34.35)

Now it remains to write a "first-order action" for (34.35), which will also be a "master action for the duality," by introducing an auxiliary field f^μ, besides the previously defined F^μ. The master action is

$$S_{\text{master}}^{(2+1)} = \int d^{2+1}x \left[-\frac{1}{2}f_\mu f^\mu + f_\mu F^\mu - \frac{m}{2}F^\mu A_\mu\right].$$ (34.36)

The equation of motion for the auxiliary field f_μ is then $f^\mu = F^\mu$, and replacing this back, we obtain (34.35). That means that, indeed, the action is a "first-order form" for (34.35).

But it is also a master action for duality. Indeed, if we now keep f_μ but solve for A_μ, we get the equation of motion

$$\epsilon^{\mu\nu\rho}\partial_\nu(f_\rho - mA_\mu) = 0 \Rightarrow A_\mu = \frac{f_\mu}{m} \equiv \tilde{A}_\mu.$$ (34.37)

Here we have denoted the gauge field $\tilde{A}_\mu$ to emphasize we get a different, dual, action.

Replacing in the master action, we find

$$S_{\text{master}}^{(2+1)} = \int d^{2+1}x \left[-\frac{1}{2}f_\mu f^\mu + \frac{1}{2m}\epsilon^{\mu\nu\rho}f_\mu\partial_\nu f_\rho\right]$$

$$= \int d^{2+1}x \left[-\frac{m^2}{2}\tilde{A}_\mu\tilde{A}^\mu + \frac{m}{2}\epsilon^{\mu\nu\rho}\tilde{A}_\mu\partial_\nu\tilde{A}_\rho\right] = S_{\text{SD}}.$$ (34.38)

This is then the "self-dual action," proving our statement.

The relation between the original variable A_μ, the one in the topologically massive action, and the dual variable $\tilde{A}_\mu$, the one in the "self-dual action," is

$$\epsilon^{\mu\nu\rho}\partial_\nu A_\rho = F^\mu = f^\mu = m\tilde{A}_\mu.$$ (34.39)

We see that if we would put $A_\mu = \tilde{A}_\mu$, we would get the "self-duality" relation. Then this relation is an example of "duality" relation, the "dimensional reduction" of the four-dimensional duality relation

$$\tilde{F}^{\mu\nu} = \frac{1}{2}\epsilon^{\mu\nu\rho\sigma}F_{\rho\sigma}.$$ (34.40)

We can easily extend the construction to $d = 4k - 1$ dimensions.

We first define

$$F^{\mu_1\cdots\mu_{2k-1}} \equiv \epsilon^{\mu_1\cdots\mu_{2k-1}\rho\nu_1\cdots\nu_{2k-1}}\partial_\rho A_{\nu_1\cdots\nu_{2k-1}}, \tag{34.41}$$

and then, using (34.19), we rewrite the topological mass term action (34.23) in terms of it, as

$$S_{\text{topmass}}^{(4k-1)} = \int d^{4k-1}x\left[\frac{1}{2}\left(\frac{1}{(2k-1)!}F^{\mu_1\cdots\mu_{2k-1}}\right)^2 - \frac{m}{2(2k)!}A_{\mu_1\cdots\mu_{2k-1}}F^{\mu_1\cdots\mu_{2k-1}}\right]. \tag{34.42}$$

Then, introducing an auxiliary field $f^{\mu_1\cdots\mu_{2k-1}}$, we can write the master action

$$S_{\text{master}}^{(4k-1)} = \int d^{4k-1}x\left[-\frac{1}{2}\left(\frac{1}{(2k-1)!}f^{\mu_1\cdots\mu_{2k-1}}\right)^2\right.$$
$$\left. +\frac{1}{(2k-1)!}f_{\mu_1\cdots\mu_{2k-1}}\frac{1}{(2k-1)!}F^{\mu_1\cdots\mu_{2k-1}} - \frac{m}{2(2k)!}A_{\mu_1\cdots\mu_{2k-1}}F^{\mu_1\cdots\mu_{2k-1}}\right]. \tag{34.43}$$

Again, the equation of motion for $f^{\mu_1\cdots\mu_{2k-1}}$ is $f^{\mu_1\cdots\mu_{2k-1}} = F^{\mu_1\cdots\mu_{2k-1}}$, so by solving for it, we go back to the topological mass term action; thus, we have a first order action for it.

But if instead we solve for $A_{\mu_1\cdots\mu_{2k-1}}$, we obtain a dual action. The equation of motion is

$$\epsilon^{\mu_1\cdots\mu_{2k-1}\rho\nu_1\cdots\nu_{2k-1}}\partial_\rho\left(\frac{1}{(2k-1)!}f_{\nu_1\cdots\nu_{2k-1}} - mA_{\nu_1\cdots\nu_{2k-1}}\right) = 0$$
$$\Rightarrow A_{\mu_1\cdots\mu_{2k-1}} = \frac{1}{m(2k-1)!}f_{\mu_1\cdots\mu_{2k-1}} \equiv \tilde{A}_{\mu_1\cdots\mu_{2k-1}}. \tag{34.44}$$

Replacing back in the master action for the duality, we obtain

$$S_{\text{master}}^{(4k-1)} = \int d^{4k-1}\left[-\frac{1}{2}\left(\frac{1}{(2k-1)!}f_{\mu_1\cdots\mu_{2k-1}}\right)^2 + \frac{1}{2m(2k-1)!}\epsilon^{\mu_1\cdots\mu_{2k-1}\rho\nu_1\cdots\nu_{2k-1}}\times\right.$$
$$\left.\times\frac{1}{(2k-1)!}f_{\mu_1\cdots\mu_{2k-1}}\partial_\rho\frac{1}{(2k-1)!}f_{\nu_1\cdots\nu_{2k-1}}\right]$$
$$= \int d^{4k-1}x\left[-\frac{m^2}{2}\tilde{A}_{\mu_1\cdots\mu_{2k-1}}^2 + \frac{m}{2(2k-1)!}\epsilon^{\mu_1\cdots\mu_{2k-1}\rho\nu_1\cdots\nu_{2k-1}}\tilde{A}_{\mu_1\cdots\mu_{2k-1}}\partial_\rho\tilde{A}_{\nu_1\cdots\nu_{2k-1}}\right]$$
$$= S_{\text{SD}}^{(4k-1)}. \tag{34.45}$$

34.4 Duality in 1+1 Dimensions: "T-Duality"

The dualization procedure that we have already followed is more general. Here we consider the simplest case of a theory of scalars in 1+1 dimensions, a "sigma model," and the duality is called "T-duality." It arises in the context of viewing the field theory as the theory on the "worldsheet" of a string, in string theory. In the next chapter, we will use some four-dimensional examples.

Consider the Euclidean space version of a 1+1 dimensional "sigma model" theory for D bosons X^I, $I = 1, \ldots, D$, with the action

$$S = \int d^2\sigma \, \delta^{\mu\nu} g_{IJ}(X) \partial_\mu X^I \partial_\nu X^J. \tag{34.46}$$

We split the sum over I, J by separating one of them, called X^0, which has the property that translation in it is a symmetry of the "metric" g_{IJ} – i.e., g_{IJ} doesn't depend explicitly on it. Then,

$$S = \int d^2\sigma \, \delta^{\mu\nu} \left[g_{00} \partial_\mu X^0 \partial_\nu X^0 + 2g_{0i} \partial_\mu X^0 \partial_\nu X^i + g_{ij} \partial_\mu X^i \partial_\nu X^j \right]. \tag{34.47}$$

We could have added to this action a a term with $\epsilon^{\mu\nu}$ and an antisymmetric B-field $B_{IJ}(X)$,

$$S = \int d^2\sigma \, \epsilon^{\mu\nu} \partial_\mu X^I \partial_\nu X^J B_{IJ}(X), \tag{34.48}$$

but for simplicity of calculations, we will consider $B_{IJ} = 0$.

We now write the master action as a first-order form of the original action. Moreover, besides the auxiliary field for the original field, we need another auxiliary field that can become the dual field. In the example we had before, the dual field and the auxiliary field were one and the same, but now they are not. We consider an auxiliary field V_μ, which should become equal to $\partial_\mu X^0$ when the equations of motion are satisfied, and another auxiliary field $\hat{X}^0$. The latter is taken to be the Lagrange multiplier for the constraint $\epsilon^{\mu\nu} \partial_\mu V_\nu = 0$, whose solution is, locally, $V_\mu = \partial_\mu X^0$ (for some locally defined X^0 field).

The master action is, then,

$$S_{\text{master}} = \int d^2\sigma \left\{ \delta^{\mu\nu} [g_{00} V_\mu V_\nu + 2g_{0i} V_\mu \partial_\nu X^i + g_{ij} \partial_\mu X^i \partial_\nu X^j] + 2\epsilon^{\mu\nu} \hat{X}^0 \partial_\mu V_\nu \right\}. \tag{34.49}$$

Indeed, by varying with respect to the Lagrange multiplier $\hat{X}^0$, we get $\epsilon^{\mu\nu} \partial_\mu V_\nu = 0$, which means $V_\mu = \partial_\mu X^0$. But if instead we vary with respect to V_μ, we obtain the equation of motion

$$V_\mu = \hat{g}_{00} \epsilon^{\nu\rho} \delta_{\rho\mu} \partial_\nu \hat{X}^0 - \frac{g_{0i}}{g_{00}} \partial_\mu \hat{X}^i, \tag{34.50}$$

where $\hat{X}^i = X^i$, $\hat{g}_{00} = 1/g_{00}$.

Then, replacing this back into the master action, we obtain a dual action, in terms of new variables $(\hat{X}^0, \hat{X}^i)$, with new "metric" $\hat{g}_{IJ}$. Note that the Lagrange multiplier has become the variable dual to X^0. Moreover, now a B-field has been generated, and the dual action is

$$\tilde{S} = \int d^2\sigma \left[\delta^{\mu\nu} \hat{g}_{IJ} \partial_\mu \hat{X}^I \partial_\nu \hat{X}^J + \epsilon^{\mu\nu} \hat{B}_{IJ} \partial_\mu \hat{X}^I \partial_\nu \hat{X}^J \right], \tag{34.51}$$

where the "T-dual metric and B-field components" are

$$\hat{g}_{00} = \frac{1}{g_{00}}; \quad \hat{g}_{ij} = g_{ij} - \frac{g_{0i} g_{0j}}{g_{00}}; \quad \hat{B}_{0i} = \frac{g_{0i}}{g_{00}}. \tag{34.52}$$

We can equate the V_μ in the dual formulation, (34.50), with the original one, $V_\mu = \partial_\mu X^0$, to find the relation between the original and dual variables:

$$\partial_\mu X^0 = \hat{g}_{00}\epsilon^{\nu\rho}\delta_{\rho\mu}\partial_\nu \hat{X}^0 - \frac{g_{0i}}{g_{00}}\partial_\mu \hat{X}^i = \hat{g}_{00}\epsilon_{\rho\mu}\partial^\rho \hat{X}^0 - \frac{g_{0i}}{g_{00}}\partial_\mu \hat{X}^i. \quad (34.53)$$

Since the "metric" g_{00} appears in front of the kinetic term for X^0, $\partial^\mu X^0 \partial_\mu X^0$, which is dualized, and then $\hat{g}_{00} = 1/g_{00}$, it means that g_{00} has the interpretation of coupling of the scalar theory and is inverted under duality: $g' = 1/g$. That means that the duality exchanges weak coupling with strong coupling, which is a characteristic of most dualities.

34.5 Comments on Chern–Simons Theories in Higher Dimensions

In this chapter, we have generalized the theories with Chern–Simons terms in higher dimensions, replacing the gauge field A_μ with a p-form field $A_{\mu_1...\mu_p}$ and otherwise keeping the structure of the Chern–Simons term. Since p-form fields are also spin 1, and the number of derivatives in the action remains the same, this was a natural generalization.

But there also exists in fact a generalization of the notion of Chern–Simons gauge field in all odd dimensions $d = 2k + 1$. Namely, we extend the requirement that the CS term is obtained on the boundary of space from a topological term for a $U(1)$ gauge field A_μ. More precisely, we require that

$$S = \int_{S=\partial M} d^{2k+1}x\mathcal{L}_{CS} = \int_M d^{2k+2}x dL_{CS} \equiv \int_M d^{2k+2}x F \wedge \cdots \wedge F. \quad (34.54)$$

In components, that gives

$$\epsilon^{\mu_1...\mu_{2k+2}}\partial_{\mu_1}\tilde{\mathcal{L}}^{CS}_{\mu_2...\mu_{2k+2}} = \epsilon^{\mu_1...\mu_{2k+2}}F_{\mu_1\mu_2}\cdots F_{\mu_{2k+1}\mu_{2k+2}}, \quad (34.55)$$

which leads to the Chern–Simons Lagrangean in odd dimensions $d = 2k + 1$,

$$\mathcal{L}_{CS} = k\epsilon^{\mu_1...\mu_{2k+1}}A_{\mu_1}F_{\mu_2\mu_3}\cdots F_{\mu_{2k}\mu_{2k+1}}. \quad (34.56)$$

Of course, that means that in dimensions higher than three, this kinetic term has more than one derivative. In $d = 5$, it will have two derivatives, just like the Maxwell term, so it will contribute equally at all energies. But in $d \geq 7$, it will have more than two derivatives, so it will be dominant *in the UV* (at high energies), which is somewhat strange.

Important Concepts to Remember

- In 2+1 dimensions, for a vector field A_μ we can construct: Maxwell theory with one polarization, Proca theory with two, topological mass term theory (which is Maxwell with CS) with one, and self-dual theory with one.
- The self-dual kinetic operator (imposing $*F = A$) can be squared, with different signs for m, to give the Proca kinetic operator. Roughly, $(\epsilon\partial + m)(\epsilon\partial - m) = \partial^2 - m^2$.

- We can extend self-dual theory to $d = 4k - 1$ dimensions, for a "gauge field" $A_{\mu_1 \ldots \mu_{2k-1}}$, and still have $(\epsilon \partial + m)(\epsilon \partial - m) = \partial^2 - m^2$.
- In $d = 4k + 1$ dimensions, we can put an i in the self-duality condition, which means we need a complex $A_{\mu_1 \ldots \mu_{2k}}$, but the resulting theory is not different from Proca, but, rather, it is a first-order form of the Proca theory.
- The topologically massive and self-dual actions are dual to each other, which means we can write a first-order form for both of them that acts as a "master action" for the duality: eliminating one field leads to one, eliminating the other leads to the other.
- We can extend the duality of topologically massive and self-dual cases to $d = 4k - 1$ dimensions.
- An Euclidean sigma model for scalars X^I has a "T-duality" that acts on a scalar X^0 whose translation is a symmetry of the metric g_{IJ}. The master action involves $\hat{X}^0 \epsilon^{\mu\nu} \partial_\mu V_\nu$, forcing $V_\mu = \partial_\mu X^0$.
- T-duality inverts the coupling, since $\hat{g}_{00} = 1/g_{00}$, hence is a strong/weak duality.
- We can extend CS theory to any odd dimension $d = 2k + 1$, by $d\mathcal{L}_{CS} = F \wedge \cdots \wedge F$, so $\mathcal{L}_{CS} = A \wedge F \cdots \wedge F$.

Further Reading

Self-duality in odd dimensions was found in [37].

Exercises

(1)   Calculate the number of on-shell degrees of freedom for a "self-dual" action in $d = 4k - 1$ dimensions, generalizing the three-dimensional case.

(2)   Calculate the T-duality transformation rules, transforming the fields under the duality transformation, in the case that the field B_{IJ} is nonzero before the duality.

(3)   Consider the Chern–Simons Lagrangian in five dimensions.

   (a) Calculate the topological response corresponding to it.
   (b) Dimensionally reduce it to four dimensions for the ansatz $A_5 = c$ (constant).

(4)   Consider a $U(1)$ gauge field A_μ in 1+1 dimensions, and for it the Maxwell term, and a mass term. Which combination of these terms describes a nontrivial, dynamical, theory, and why?

35

Particle–Vortex Duality in Three Dimensions, Particle–String Duality in Four Dimensions, and p-Form Fields in Four Dimensions

In this chapter, we will apply the notions of dualities and p-forms to our 3+1 dimensions. After reviewing the usual Maxwell duality in 3+1 dimensions, and the "particle–vortex" duality in 2+1 dimensions, we find out how to extend the latter to a "particle–string" duality in 3+1 dimensions, where an antisymmetric form field $B_{\mu\nu}$ will play a role. Finally, we consider the properties that general antisymmetric fields $B_{\mu\nu}$ and $C_{\mu\nu\rho}$ have and how their Poincaré duality to a field and a constant can be understood.

35.1 Maxwell Duality in 3+1 Dimensions

As we have already described when discussing the Dirac monopole, Maxwell's equations (in vacuum: $\partial_\mu F^{\mu\nu} = 0$ and $\partial_{[\mu} F_{\nu\rho]} = 0$) have a symmetry under the interchange of the electromagnetic field with its dual:

$$F_{\mu\nu} \rightarrow *F_{\mu\nu} \equiv \frac{1}{2} \epsilon_{\mu\nu\rho\sigma} F^{\rho\sigma}. \tag{35.1}$$

Since $*^2 = -1$, the duality also takes $*F_{\mu\nu} \rightarrow -F_{\mu\nu}$. In terms of $\vec{E}$ and $\vec{B}$, the duality exchanges

$$\vec{E} \rightarrow \vec{B}, \quad \vec{B} \rightarrow -\vec{E}. \tag{35.2}$$

In the presence of both electric and magnetic charges, the equations

$$\partial_\mu F^{\mu\nu} = -j^\nu, \quad \partial_\mu * F^{\mu\nu} = -k^\nu \tag{35.3}$$

have the same symmetry, supplanted with the action on the electric and magnetic currents and charges,

$$j^\mu \leftrightarrow k^\mu, \quad \rho_e \leftrightarrow \rho_m, \quad q = \int d^3x \rho_e \leftrightarrow g = \int d^3x \rho_m. \tag{35.4}$$

In this way, we see that the duality exchanges fields of the same sort (electromagnetic), so can be thought of as a *self-duality* since it maps Maxwell theory to itself. It also exchanges fundamental particles (electrons of charge e) with fundamental particles (Dirac monopoles of charge g), though in the latter case they can also be solitonic (the 't Hooft–Polyakov monopoles, behaving at large distances as Dirac monopoles).

Moreover, the duality can be performed at the level of an action, by introducing a master action for the duality. It can be done, in fact, in several ways, but we will describe the simplest (though it is not necessarily the best for physics). For the Maxwell action

$$S^{(3+1)}_{\text{original}} = -\frac{1}{4}\int d^{3+1}x F_{\mu\nu}F^{\mu\nu}, \tag{35.5}$$

where as usual $F_{\mu\nu} = \partial_\mu A_\nu - \partial_\nu A_\mu$, consider the first-order form, acting as a master action for the duality,

$$S^{(3+1)}_{\text{master}} = \int d^{3+1}x \left[\frac{1}{4}(2\epsilon^{\mu\nu\rho\sigma}F_{\mu\nu}\partial_\rho \tilde{A}_\sigma - F_{\mu\nu}F^{\mu\nu})\right], \tag{35.6}$$

where now $F_{\mu\nu}$ is an independent field. Then, varying with respect to $\tilde{A}_\mu$, thought of as a Lagrange multiplier, we obtain

$$\epsilon^{\mu\nu\rho\sigma}\partial_\nu F_{\rho\sigma} = 0 \Rightarrow F_{\rho\sigma} = \partial_\rho A_\sigma - \partial_\sigma A_\rho. \tag{35.7}$$

Substituting it back in the master action, we get the original action.

If instead we vary over $F_{\mu\nu}$, we obtain the equation of motion

$$F^{\mu\nu} = \frac{1}{2}\epsilon^{\mu\nu\rho\sigma}2\partial_\rho \tilde{A}_\sigma. \tag{35.8}$$

Replacing it back in the master action, we obtain the dual action

$$S^{(3+1)}_{\text{dual}} = -\frac{1}{4}\int d^{3+1}x(\partial_\mu \tilde{A}_\nu - \partial_\nu \tilde{A}_\mu)^2, \tag{35.9}$$

which is formally the same Maxwell action, except in dual variables, since equating the two formulas for $F_{\mu\nu}$, we obtain

$$F_{\mu\nu} = \partial_\mu A_\nu - \partial_\nu A_\mu = \frac{1}{2}\epsilon_{\mu\nu\rho\sigma}(\partial^\rho \tilde{A}^\sigma - \partial^\sigma \tilde{A}^\rho). \tag{35.10}$$

35.2 Particle–Vortex Duality in 2+1 Dimensions

In 2+1 dimensions, we can do a similar transformation for the fields that will exchange particles for vortices.

The starting point action is the Abelian-Higgs action with some general potential, assumed to have a vortex solution. Therefore,

$$S_{\text{original}} = \int d^3x \left[-\frac{1}{2}|D_\mu\Phi|^2 - V(|\Phi|) - \frac{1}{4}F^2_{\mu\nu}\right]. \tag{35.11}$$

An N−vortex solution will have $\Phi = \Phi(r)e^{i\alpha(\theta)}$, with $\alpha(\theta) = N\theta$ and, therefore, $\epsilon^{ij}\partial_i\partial_j\alpha \propto \delta(r)$.

In the presence of vortices, we write the field as

$$\Phi = \Phi(r)e^{i\alpha}, \quad \alpha = \alpha_{\text{vortex}} + \alpha_{\text{smooth}}, \tag{35.12}$$

where α_{smooth} satisfies $\epsilon^{ij}\partial_i\partial_j\alpha_{\text{smooth}} = 0$.

Then the action becomes

$$S_{\text{original}} = -\frac{1}{2} \int d^3x \left[(\partial_\mu \Phi_0)^2 + (\partial_\mu \alpha_{\text{smooth}} + \partial_\mu \alpha_{\text{vortex}} + ea_\mu)^2 \Phi_0^2 \right]$$

$$- \int d^3x \left[V(\Phi_0) + \frac{1}{4} F_{\mu\nu}^2 \right]. \tag{35.13}$$

As before, we write a first-order action for it, which serves as a master action for the duality, by introducing a field λ_μ that becomes $\lambda_{\mu,\,\text{smooth}} = \partial_\mu \alpha_{\text{smooth}}$ on-shell. The latter condition is imposed with a Lagrange multiplier b_μ as $\epsilon^{\mu\nu\rho} \partial_\nu \lambda_{\rho,\,\text{smooth}} = 0$. Thus, the master action is

$$S_{\text{master}} = \int d^3x \left[-\frac{1}{2}(\partial_\mu \Phi_0)^2 - \frac{1}{2}(\lambda_{\mu,\text{smooth}} + \lambda_{\mu,\text{vortex}} + ea_\mu)^2 \Phi_0^2 \right.$$

$$\left. + \epsilon^{\mu\nu\rho} b_\mu \partial_\nu \lambda_{\rho,\,\text{smooth}} - V(\Phi_0) - \frac{1}{4} F_{\mu\nu}^2 \right], \tag{35.14}$$

Then, indeed, varying with respect to b_μ leads to $\lambda_{\mu\,\text{smooth}} = \partial_\mu \alpha_{\text{smooth}}$, and substituting back in the master action, we reobtain the original action. But if instead we vary with respect to $\lambda_{\mu,\,\text{smooth}}$, we obtain

$$(\lambda_\mu + ea_\mu)\Phi_0^2 = \epsilon^{\mu\nu\rho} \partial_\nu b_\rho. \tag{35.15}$$

If we susbtitute this back in the master action, we obtain the dual action, written in terms of the dual variable b_μ (the Lagrange multipliers),

$$S_{\text{dual}} = \int d^3x \left[-\frac{(f_{\mu\nu}^b)^2}{4\Phi_0^2} - \frac{1}{2}(\partial_\mu \Phi_0)^2 - e\epsilon^{\mu\nu\rho} b_\mu \partial_\nu a_\rho - \frac{2\pi}{e} b_\mu j_{\text{vortex}}^\mu - V(\Phi_0) - \frac{1}{4} F_{\mu\nu}^2 \right].$$

$$\tag{35.16}$$

Here $f_{\mu\nu}^b = \partial_\mu b_\nu - \partial_\nu b_\mu$ and

$$j_{\text{vortex}}^\mu = \frac{e}{2\pi} \epsilon^{\mu\nu\rho} \partial_\nu \lambda_{\rho,\text{vortex}}. \tag{35.17}$$

The original theory has a $U(1)$ invariance, which is global if the field a_μ is "frozen," or external. The $U(1)$ Noether current associated with it is, as usual,

$$j_\mu = -ie(\Phi^\dagger \partial_\mu \Phi - \Phi \partial_\mu \Phi^\dagger) = e\Phi_0^2 \partial_\mu \alpha, \tag{35.18}$$

whereas in the dual theory we have a topological (vortex) current. If Φ_0 is kept fixed, the vortex current is

$$j_{\text{vortex}}^\mu = \frac{e}{2\pi} \epsilon^{\mu\nu\rho} \partial_\nu \partial_\rho \alpha = \frac{1}{2\pi\Phi_0^2} \epsilon^{\mu\nu\rho} \partial_\nu j_\rho. \tag{35.19}$$

Thus the particle and vortex currents are dual to each other, justifying the name particle-vortex duality for the transformation. Moreover, we can relate the fields by equating the values for λ_μ in the two reductions

$$\partial_\mu \alpha + ea_\mu = \lambda_\mu + ea_\mu = \frac{1}{\Phi_0^2} \epsilon^{\mu\nu\rho} \partial_\nu b_\rho. \tag{35.20}$$

The fields are, therefore, indeed also dual to each other, as expected.

Finally, note that the duality is strong/weak coupling, since in the original action Φ_0^2 appears in front of the kinetic term for α, as a coupling factor $1/g^2$, whereas in the dual action we have $1/\Phi_0^2$ playing the role of $1/g'^2$, thus $g' = 1/g$.

35.3 Particle–String Duality in 3+1 Dimensions

We can now extend the particle–vortex duality to 3+1 dimensions, where it should relate now particles and vortex strings. Indeed, a vortex in two spatial dimensions lifts up to a vortex string in three spatial dimensions (like a flux tube in a superconductor or a cosmic string). Thus, for a theory that admits these vortex strings, we should be able to write a particle–string duality that exchanges them for particles.

An obvious starting point is the Abelian-Higgs theory in 3+1 dimensions,

$$S = \int d^4x \left[-\frac{1}{2}|(\partial_\mu - ieu_\mu)\psi|^2 - \frac{1}{4}\Gamma_{\mu\nu}^2 - V(|\phi|) \right], \tag{35.21}$$

As before, we write $\phi = \phi(r)e^{i\theta}$, and $\theta = \theta_{\text{vortex}} + \theta_{\text{smooth}}$, where $\epsilon_{\mu\nu\rho}\partial_\mu\partial_\nu\theta_{\text{smooth}} = 0$, and θ_{vortex} satisfies

$$\epsilon^{zij}\partial_i\partial_j\theta \propto \delta^2(r), \tag{35.22}$$

where z is a direction parallel to some string and i,j are transverse to it, so that the action becomes

$$S_{\text{original}} = -\frac{1}{2}\int d^4x \left[(\partial_\mu\phi_0)^2 + (\partial_\mu\theta_{\text{smooth}} + \partial_\mu\theta_{\text{vortex}} + ea_\mu)^2\phi_0^2 + \frac{1}{2}F_{\mu\nu}^2 + 2V(\phi_0) \right]. \tag{35.23}$$

We next write a first-order formulation for it, in terms of a field λ_μ that on-shell equals $\partial_\mu\theta$, and that acts as a master action for the duality. The relation $\partial_{[\mu}\lambda_{\nu]} = 0$ is imposed with a Lagrange multiplier $b_{\mu\nu}$ so that the master action is

$$S_{\text{master}} = -\frac{1}{2}\int d^4x[(\partial_\mu\phi_0)^2 + (\lambda_{\mu,\text{smooth}} + \lambda_{\mu,\text{vortex}} + ea_\mu)^2\phi_0^2]$$

$$+ \int d^4x \left[\epsilon^{\mu\nu\rho\sigma}b_{\mu\nu}\partial_\rho\lambda_{\sigma,\text{smooth}} - V(\phi_0) - \frac{1}{4}F_{\mu\nu}^2 \right]. \tag{35.24}$$

If we vary with respect to $b_{\mu\nu}$, we go back to the original action. But if we vary with respect to $\lambda_{\mu,\text{ smooth}}$ instead, we obtain the equation of motion

$$(\lambda_\mu + ea_\mu)\phi_0^2 = \epsilon_{\mu\nu\rho\sigma}\partial^\nu b^{\rho\sigma}. \tag{35.25}$$

Replacing this back in the master action, we obtain the dual action,

$$S_{\text{dual}} = \int d^4x \left[-\frac{(2\partial_{[\mu}b_{\nu\rho]})^2}{2\phi_0^2} - e\epsilon^{\mu\nu\rho\sigma}b_{\mu\nu}\partial_\rho a_\sigma \right.$$

$$\left. -\frac{2\pi}{e}J_{\text{vortex}}^{\mu\nu}b_{\mu\nu} - V(\phi_0) - \frac{1}{4}F_{\mu\nu}^2 - \frac{1}{2}(\partial_\mu\phi_0)^2 \right], \tag{35.26}$$

where now, however, the topological "vortex current" is not quite a current, but an antisymmetric tensor,

$$J^{\mu\nu}_{\text{vortex}} = \frac{e}{2\pi}\epsilon^{\mu\nu\rho\sigma}\partial_\rho\partial_\sigma\theta_{\text{vortex}} = \frac{e}{2\pi}\epsilon^{\mu\nu\rho\sigma}\partial_\rho\partial_\sigma\theta. \quad (35.27)$$

It satisfies the "current conservation" relations $\partial_\mu J^{\mu\nu}_{\text{vortex}} = \partial_\nu J^{\mu\nu}_{\text{vortex}} = 0$, satisfied automatically because of its topological nature.

Note that we can now define

$$H_{\mu\nu\rho} = 3\partial_{[\mu}b_{\nu\rho]} \quad (35.28)$$

as the field strength of the dual field $b_{\mu\nu}$, therefore the dual field theory is a gauge theory in terms of antisymmetric tensors $b_{\mu\nu}$.

The original Abelian-Higgs model still has a $U(1)$ symmetry, global for "frozen" a_μ, with Noether current

$$j_\mu = -\frac{ie}{2}(\phi^*\partial_\mu\phi - \phi\partial_\mu\phi^*) = e\phi_0^2\partial_\mu\theta, \quad (35.29)$$

That means that we have the natural extension to 3+1 dimensions of the particle–vortex current relation,

$$J^{\mu\nu}_{\text{vortex}} = \frac{1}{2\pi\phi_0^2}\epsilon^{\mu\nu\rho\sigma}\partial_\rho j_\sigma. \quad (35.30)$$

Moreover, again equating the two formulas for λ_μ, we have the duality relation for fields,

$$\partial_\mu\theta + ea_\mu = \lambda_\mu = \frac{1}{\phi_0^2}\epsilon_{\mu\nu\rho\sigma}\partial^\nu b^{\rho\sigma}. \quad (35.31)$$

We still need to understand better what does the "vortex current" $J^{\mu\nu}_{\text{vortex}}$ means. In fact, it has an equivalent representation that is more illuminating. Consider a vortex string source, $X^\mu = X^\mu(\sigma^a)$, where $\sigma^a - (\sigma^0,\sigma^1)$ are intrinsic coordinates on the surface spanned by the string as it moves in spacetime, called *worldsheet*. Since it is a source for $b_{\mu\nu}$, it should couple to it linearly. It is easy to realize that the only way this coupling can happen is via

$$S_{\text{source}} = \int d^2\sigma b_{\mu\nu}(X^\mu(\sigma^a))\epsilon^{ab}\partial_a X^\mu\partial_b X^\nu. \quad (35.32)$$

Indeed then, $b_{ab} \equiv b_{\mu\nu}(X(\sigma^a))\partial_a X^\mu\partial_b X^\nu$ is the field "induced" on the worldsheet by $b_{\mu\nu}$, and because of antisymmetry this can only couple to ϵ^{ab}.

But then, by varying with respect to $b_{\mu\nu}$, we should obtain the "vortex current," generated by the vortex source,

$$J^{\mu\nu}_{\text{vortex}}(x^\mu) = \int d^2\sigma\delta^4(x^\mu - X^\mu(\sigma^a))\epsilon^{ab}\partial_a X^\mu\partial_b X^\nu. \quad (35.33)$$

Now, however, it is not completely clear what physics does the dual action (35.26) describe.

What we can say is that again we have a strong/weak type of duality, since again ϕ_0^2 appears as a coupling in front of the kinetic term for the original variable θ, whereas $1/\phi_0^2$ appears in front of the kinetic term for the dual variable $b_{\mu\nu}$, so the coupling is inverted, $g' = 1/g$.

35.4 Poincaré Duality and Applications to 3+1 Dimensions

We can extend the Maxwell duality (of the Maxwell's equations), which relates field strengths of gauge fields through the epsilon tensor, to *Poincaré duality*, which similarly relates field strengths of p-form gauge fields through the epsilon tensor.

For a n–form $A_{\mu_1...\mu_n}$, the field strength is

$$F_{\mu_1...\mu_{n+1}} = (n+1)\partial_{[\mu_1}A_{\mu_2...\mu_{n+1}]}, \tag{35.34}$$

and the action is

$$S = -\int \frac{F^2_{\mu_1...\mu_{n+1}}}{2(n+1)!} = -\int \frac{F_{(n+1)} \wedge *F_{(n+1)}}{2(n+1)!}. \tag{35.35}$$

As in the Maxwell case, the equations of motion and Bianchi identities in vacuum (no sources) are

$$d * F = 0, \quad dF = 0 \tag{35.36}$$

respectively. One can also extend them by considering sources for both the equations of motion and Bianchi identities, as in the case of Maxwell. The duality symmetry of these equations is now

$$F_{(n+1)} \to *F_{(D-n-1)} \Rightarrow$$
$$(n+1)\partial_{[\mu_1}A_{\mu_2...\mu_{n+1}]} = F_{\mu_1...\mu_{n+1}} \to$$
$$*F_{\mu_1...\mu_{D-n-1}} = \frac{1}{(n+1)!}\epsilon_{\mu_1...\mu_D}F^{\mu_{D-n}...\mu_D} = (D-n-1)\partial_{[\mu_1}\tilde{A}_{\mu_2...\mu_{D-n-1}]}. \tag{35.37}$$

Note that this duality is different than the particle–vortex duality considered before, which related a gauge field with a field strength.

We can also extend the equations of motion and duality to the case of sources. Now, however, the singular source must be extended in $n - 1$ spatial directions, so the delta function must be in $D - n$ transverse directions.

In form language, the equations of motion with source become

$$d(*F)_{D-n-1} = (*J)_{D-n}, \tag{35.38}$$

where the current source is

$$(*J)_{1...D-n} = e_n\delta^{D-n}(y), \tag{35.39}$$

where e_n is the analog of the electric charge, in this case density of "electric" charge per unit volume of the source.

We can also extend the Bianchi indentity, to introduce the analog of magnetic charge,

$$dF_{n+1} = X_{n+2} \equiv d\omega_{n+1} \to F_{n+1} = dA_n + \omega_{n+1}. \tag{35.40}$$

The magnetically charged current source is

$$X_{1...n+2} = g_{D-n-2}\delta^{n+2}(y), \tag{35.41}$$

where g_{D-n-2} is the "magnetic" charge density per unit volume of the source.

The electric-type charges are defined by the Gauss law

$$e_n = \int_{S^{D-n-1}} (*F)_{D-n-1} = \int_{M^{D-n}} *J_{D-n}, \tag{35.42}$$

and the magnetic-type charges are defined by the magnetic Gauss's law

$$g_{D-n-2} = \int_{S^{n+1}} F_{n+1} = \int_{M^{n+2}} X_{n+2}, \tag{35.43}$$

and they satisfy a generalized Dirac quantization,

$$e_n g_{D-n-2} = 4\pi \frac{k}{2}. \tag{35.44}$$

The Poincaré duality can be extended to exchange the electric and magnetic charges and currents,

$$(*J)_{(D-n)} \leftrightarrow X_{(n+2)} \Rightarrow e_n \leftrightarrow g_{D-n-2}. \tag{35.45}$$

To apply this formalism to 3+1 dimensions, we note that the only two possibilities other than Maxwell duality are of a duality between an antisymmetric field $B_{\mu\nu}$ and a scalar,

$$H_{\mu\nu\rho} = 3\partial_{[\mu}B_{\nu\rho]} = \frac{1}{3!}\epsilon_{\mu\nu\rho\sigma}\partial^\sigma\phi, \tag{35.46}$$

and the "duality" between an antisymmetric field $C_{\mu\nu\rho}$ and a constant m,

$$F_{\mu\nu\rho\sigma} = 4\partial_{[\mu}C_{\nu\rho\sigma]} = m\epsilon_{\mu\nu\rho\sigma}. \tag{35.47}$$

In the first case, as we have already noted, the field $B_{\mu\nu}$ is sourced by a string source – i.e., a source extended in one spatial dimension. The Poincaré dual source would be dual to an object localized not only in the spatial directions but in the time direction as well – i.e., an instanton. Thus, Poincaré duality would relate a string to an instanton.

In the second case, the field $C_{\mu\nu\rho}$ is sourced by a surface (wall) source – i.e., a source extended in two spatial directions. In this case, strictly speaking, there is no Poincaré dual, since the field strength itself is just a scalar function.

Important Concepts to Remember

- Maxwell duality is a self-duality that interchanges particles with particles, and gauge fields with gauge fields, via a master action.
- Particle–vortex duality is a duality in 2+1 dimensions that exchanges particles and vortices, relates the phase α of the scalar in the Abelian-Higgs model to a gauge field b_μ through a master action, and exchanges a Noether current with a topological current.
- Particle–string duality is a duality in 3+1 dimensions that generalizes particle–vortex duality, exchanges particles and strings – relating the phase α of the scalar in the Abelian-Higgs model to an antisymmetric field $b_{\mu\nu}$ via a master action – and exchanges a Noether current with a topological vortex string current $J^{\mu\nu}_{\text{vortex}}$.

- Poincaré duality relates a field strength of a p-form field with the dual field strength $*F$ (via the epsilon tensor), and extended sources to extended sources.
- In 3+1 dimensions, Poincaré duality relates a $B_{\mu\nu}$ field to a scalar ϕ and a string to an instanton, and it also relates $C_{\mu\nu\rho}$ to a constant.

Further Reading

See the original papers [38, 39].

Exercises

(1)   Write a self-dual action (that goes over into itself under duality) for particle–vortex duality in 2+1 dimensions, involving a scalar and a gauge field.

(2)   Calculate the number of on-shell degrees of freedom for a p-form field with Maxwell-type kinetic term in $d + 1$ spacetime dimensions, and show that it is invariant under Poincaré duality.

(3)   Derive the Poincaré duality in four dimensions between a $B_{\mu\nu}$ gauge field and a scalar ϕ as a transformation of the action, like we did in the case of two dual gauge fields A_μ.

(4)   Compare the transformation in exercise 3 with the particle–string duality in the text. Are they equivalent?

36 Fermions and Dirac Spinors

After dealing with spin 0 and spin 1 fields until now, we finally move on to fields of other spin. In this chapter, we will describe spinor fields, which are associated with fermionic statistics and have half-integer spin.

36.1 Spinors, Dirac and Weyl

We have already seen in Chapter 11 that the Lorentz group has some spinor representations. The Lorentz algebra was shown to be equivalent to the one of $SU(2) \times SU(2)$, and representations of the Lorentz group were determined by the "spins" for the two, as (j, j'). But, moreover, the correct classification is in terms of representations of the little group, the diagonal $SU(2)$ characterized by a "spin" J. Then we saw that the basic spinor representation of the diagonal $SU(2)$, $J = 1/2$, corresponds to the Lorentz group representations $(1/2, 0)$ and $(0, 1/2)$. More generally, half-integer spin J representations, by the spin-statistics theorem from chapter 12 are associated with Fermi–Dirac statistics – i.e., fermions – and are called spinors. This is one way to define spinors.

Another way to define them is related to the fact that, as proven for spin 1/2 in Chapter 11 (and 12), a spinor changes by a factor of -1 by a 2π rotation.

Note that we will only deal with spin 1/2 fermions (spinors) in this book, since spin 3/2 is only relevant for a supersymmetric theory of gravity, supergravity, and even higher spins (5/2, 7/2, etc.) are only relevant for higher dimensional (Kaluza–Klein) theories – like, for instance string theory. But both of these definitions apply to any spinor.

We will define now a more fundamental and abstract one. We start with the observation that the spin $J = 1/2$ representation of $SU(2)$ is defined by the Pauli matrices

$$\sigma_1 = \begin{pmatrix} 0 & 1 \\ 1 & 0 \end{pmatrix}, \quad \sigma_2 = \begin{pmatrix} 0 & -i \\ i & 0 \end{pmatrix}, \quad \sigma_3 = \begin{pmatrix} 1 & 0 \\ 0 & -1 \end{pmatrix}, \tag{36.1}$$

with the generators $J_i = \sigma_i/2$. Since the Pauli matrices satisfy the product formula

$$\sigma_i \sigma_j = \delta_{ij} + i\epsilon_{ijk}\sigma_k, \tag{36.2}$$

it follows that, indeed, the J_i values satisfy the $SU(2)$ Lie algebra,

$$\left[\frac{\sigma_i}{2}, \frac{\sigma_j}{2}\right] = i\epsilon_{ijk}\frac{\sigma_k}{2}. \tag{36.3}$$

But it also means that we have a different kind of algebra, namely

$$\{\sigma_i, \sigma_j\} = 2\delta_{ij}. \tag{36.4}$$

Unlike a Lie algebra, now we have an anticommutator, and on the right-hand side is a metric times the identity.

In fact, this is a particular case of a more general algebra, defined on any manifold with a metric $g_{\mu\nu}$, known as a *Clifford algebra*,

$$\{\gamma_\mu, \gamma_\nu\} = 2g_{\mu\nu}. \tag{36.5}$$

Its representations are, in all generality, called spinors (bosons – i.e., integer spin representations of the Lorentz group – are not representations of the Clifford algebra). This is our third, and most general, definition of spinors. Note that by representations of the Clifford algebra, we mean the same things as for representations of a Lie algebra: just like for a Lie algebra with generators T_a, we mean that T_a is now a matrix $T_a(R)^i{}_j$, acting on a vector space of element ϕ^j (giving the representation), for the Clifford algebra with "generators" γ^μ, we mean that γ^μ is now $(\gamma^\mu)^\alpha{}_\beta$ acting on ψ^β.

Weyl Spinors

- The (1/2 0) representation of the Lorentz group is called a right-handed Weyl spinor, denoted ψ_a. It is acted upon by $\sigma_i/2$ as

$$\psi_a \to (e^{i\theta^i \frac{\sigma_i}{2}})_a{}^b \psi_b \tag{36.6}$$

- The (0 1/2) representation of the Lorentz group is called a left-handed Weyl spinor, denoted $\eta^{\dot{a}}$, and is acted upon by $\sigma_i/2$ as

$$\eta^{\dot{a}} \to (e^{i\theta^i \frac{\sigma_i}{2}})^{\dot{a}}{}_{\dot{b}} \eta^{\dot{b}}. \tag{36.7}$$

Under a Lorentz transformation, we will embed these transformations via the diagonal reduction. But then we obtain not a usual representation but a *projective representation* of the Lorentz group – i.e., modulo a phase –

$$R[g_1]R[g_2] = e^{i\alpha(g_1, g_2)} R[g_1 g_2]. \tag{36.8}$$

In the case of the Lorentz group $SO(3, 1)$, the phase is only ± 1, the -1 arising for a 2π rotation. But since, as we saw in Chapter 11, $SO(3, 1) = Sl(2, \mathbb{C})/\mathbb{Z}_2$, we actually have a representation for $Sl(2, \mathbb{C})$, the universal cover of the Lorentz group.

For a Lorentz group element, defined in terms of the generators $J_i = \frac{1}{2}\epsilon_{ijk}J_{jk}$ (angular momentum for spatial rotations) and $K_i = J_{0i}$ (boosts),

$$\Lambda^\mu{}_\nu = (e^{i\omega^i K_i} e^{i\theta^i J_i})^\mu{}_\nu, \tag{36.9}$$

we have defined $M_i = (J_i + iK_i)/2$ and $N_i = (J_i - iK_i)/2$, so the group element acting on the right-handed Weyl spinors, defined by $N_i = 0$, so $J_i = iK_i = \sigma_i/2$, is

$$R(\Lambda) = e^{\frac{\omega^i \sigma_i}{2}} e^{i\frac{\theta^i \sigma_i}{2}}. \tag{36.10}$$

Similarly, for the left-handed Weyl spinors, defined by $M_i = 0$, so $J_i = -iK_i = \sigma_i/2$, we have the group element

$$L(\Lambda) = e^{-\frac{\omega^i \sigma_i}{2}} e^{i\frac{\theta^i \sigma_i}{2}}. \tag{36.11}$$

But these are matrices acting on the two-dimensional Weyl spinor representations of the *Lorentz group*, in terms of $\sigma_i/2$. We want now to construct the representations of the *full Clifford algebra* in four Minkowski dimensions,

$$\{\gamma^\mu, \gamma^\nu\} = 2\eta^{\mu\nu}, \quad \mu, \nu = 0, 1, 2, 3. \tag{36.12}$$

Dirac Spinor

The basic representation is four-dimensional – in terms of spinors ψ^α, $\alpha = 1, 2, 3, 4$ – meaning that, as we said, γ^μ transforms the spinors as

$$(\gamma^\mu)^\alpha{}_\beta \psi^\beta = (\psi')^\alpha. \tag{36.13}$$

This is called a *Dirac spinor*, and it is composed in the obvious way of the two types of Weyl spinors as

$$\psi^\alpha = \begin{pmatrix} \psi_a \\ \eta^{\dot{a}} \end{pmatrix} = \begin{pmatrix} \psi_R \\ \psi_L \end{pmatrix}. \tag{36.14}$$

36.2 Gamma Matrices

For the "generators" γ^μ (for four-dimensional Minkowski space) of the Dirac spinor representation themselves, called *gamma matrices*, we have several possibilities – i.e., representations.

We can, in fact, construct them as tensor products of the σ_i, viewed as Clifford gamma matrices for $\mathbb{R}^2$. We note that the $\sigma_i, i = 1, 2, 3$ actually even form a Clifford algebra for three-dimensional space ($\mathbb{R}^3$), $\{\gamma_i, \gamma_j\} = 2\delta_{ij}$, though one thinks of them as two-dimensional ones.

Then, one possible representation for γ^μ in four Minkowski dimensions is constructed as

$$\gamma^i = \sigma_2 \otimes \sigma^i = -i \begin{pmatrix} \mathbf{0} & \sigma^i \\ -\sigma^i & \mathbf{0} \end{pmatrix}$$

$$\gamma^0 = -i\sigma_1 \otimes \mathbb{1} = -i \begin{pmatrix} \mathbf{0} & \mathbb{1} \\ \mathbb{1} & \mathbf{0} \end{pmatrix}. \tag{36.15}$$

Since $(a \otimes b) \cdot (c \otimes d) = a \cdot c \otimes b \cdot d$, from the Pauli matrix relations we obtain

$$\{\gamma^i, \gamma^j\} = \sigma_2^2 \otimes \{\sigma^i, \sigma^j\} = \mathbb{1} \otimes 2\delta_{ij} = 2\delta_{ij}\,\mathbb{1}_{4\times 4}$$
$$\{\gamma^i, \gamma^0\} = -i\{\sigma_2, \sigma_1\} \otimes \sigma^i = 0$$
$$\{\gamma^0, \gamma^0\} = 2(\gamma^0)^2 = -2\sigma_1^2 \otimes \mathbb{1} = -2\,\mathbb{1}_{4\times 4}. \tag{36.16}$$

That means that we have defined indeed a representation of the Clifford algebra, called the *Weyl, or chiral, representation.*

The γ^μ defined above satisfy

$$(\gamma^0)^\dagger = (-i\sigma_1)^\dagger \otimes \mathbb{1} = -\gamma^0, \quad (\gamma^i)^\dagger = (\sigma_2)^\dagger \otimes (\sigma^i)^\dagger = \sigma_2 \otimes \sigma^i = \gamma^i, \tag{36.17}$$

which together are written as (note that $\gamma^\mu \gamma^\nu = -\gamma^\nu \gamma^\mu$ for $\mu \neq \nu$)

$$\gamma^0 (\gamma^\mu)^\dagger \gamma^0 = +\gamma^\mu \Rightarrow (\gamma^\mu)^\dagger = +\gamma^0 \gamma^\mu \gamma^0. \tag{36.18}$$

Defining the matrices

$$\sigma^\mu = (\,\mathbb{1}, \sigma^i), \quad \bar\sigma^\mu = (\,\mathbb{1}, -\sigma^i), \tag{36.19}$$

we can write the gamma matrices in the Weyl representation together as

$$\gamma^\mu_{\text{Weyl}} = -i \begin{pmatrix} \mathbf{0} & \sigma^\mu \\ \bar\sigma^\mu & \mathbf{0} \end{pmatrix}. \tag{36.20}$$

The generators of the Lorentz group in the spinor representation are

$$J_{\mu\nu} = \frac{1}{4}[\gamma_\mu, \gamma_\nu]. \tag{36.21}$$

Indeed, one can check that they satisfy the Lorentz Lie algebra, which we leave as an exercise.

Moreover, we also find that if

$$\Lambda_S = e^{i\theta^{\mu\nu}(J_{\mu\nu})_{\text{spinor}}}, \quad \Lambda^\mu{}_\nu = \left(e^{i\theta^{\rho\sigma}(J_{\rho\sigma})_{\text{vector}}}\right)^\mu{}_\nu, \tag{36.22}$$

we have

$$\Lambda_S \gamma^\mu \Lambda_S^{-1} = \Lambda^\mu{}_\nu \gamma^\nu. \tag{36.23}$$

The interpretation is that, indeed, γ^μ is a Lorentz vector (as the notation suggests), namely when acting with a Lorentz transformation on the matrix indices, the transformation is equivalently acting on the vector index.

The Hermiticity properties of the Lorentz generators are given by

$$(J_{\mu\nu})^\dagger = +\frac{i}{4}[\gamma_\nu^\dagger, \gamma_\mu^\dagger] = +\frac{i}{4}\gamma^0[\gamma_\nu, \gamma_\mu]\gamma^0 = -\gamma^0 J_{\mu\nu}\gamma^0, \tag{36.24}$$

or (since $(\gamma^0)^{-1} = -\gamma^0$)

$$\gamma^0 (J_{\mu\nu})^\dagger \gamma^0 = -J_{\mu\nu} \Rightarrow \gamma^0 (J_{\mu\nu})^\dagger (\gamma^0)^{-1} = J_{\mu\nu}. \tag{36.25}$$

Since $[\gamma^0, \gamma^0] = 0$, but $\{\gamma^0, \gamma^i\} = 0$, we find in components

$$(J_{ij})^\dagger = -J_{ij}, \quad (J_{0i})^\dagger = +J_{0i}. \tag{36.26}$$

We can now find the relation of $\Lambda_S^\dagger$ to Λ_S^{-1}, since

$$\gamma^0\Lambda_S^\dagger\gamma^0 = -\gamma^0\Lambda_S^\dagger(\gamma^0)^{-1} = -\gamma^0\left[e^{i\theta^{\mu\nu}J_{\mu\nu}}\right]^\dagger(\gamma^0)^{-1} = -\exp\left[-i\theta^{\mu\nu}\gamma^0(J_{\mu\nu})^\dagger(\gamma^0)^{-1}\right]$$

$$= -e^{-i\theta^{\mu\nu}J_{\mu\nu}} = -\Lambda_S^{-1}. \tag{36.27}$$

Another way of writing it is as a commutation relation,

$$\Lambda_S^\dagger\gamma^0 = +\gamma^0\Lambda_S^{-1}, \tag{36.28}$$

which implies we can construct a Lorentz invariant out of a Dirac spinor ψ by $\psi^\dagger i\gamma^0\psi$, since it transforms as

$$\psi^\dagger i\gamma^0\psi \to i\psi^\dagger\Lambda_S^\dagger\gamma^0\Lambda_S\psi = i\psi^\dagger\gamma^0\Lambda_S^{-1}\Lambda_S\psi = i\psi^\dagger\gamma^0\psi. \tag{36.29}$$

We can then define the *Dirac conjugate spinor* as

$$\bar\psi \equiv \psi^\dagger i\gamma^0, \tag{36.30}$$

such that $\bar\psi\psi$ is Lorentz invariant.

The fact that the two-dimensional gamma matrices, the σ_i, $i=1,2$ actually form a three-dimensional Clifford algebra, where $\sigma_3 = -i\sigma_1\sigma_2$, suggests that this is a more general property in higher *even* dimensions (in odd dimensions, the products of gamma matrices gives something proportional to $\mathbb{1}$, for instance, $\sigma_1\sigma_2\sigma_3 = 1$): the products of the gamma matrices acts as another gamma matrix, a $d+1$th one. We thus define

$$\gamma_5 = -i\gamma^0\gamma^1\gamma^2\gamma^3, \tag{36.31}$$

where the overall sign is conventional, but the i is such that $\gamma_5^2 = 1$. Indeed, we find (by anticommuting the different gamma matrices, $\gamma^i\gamma^j = -\gamma^j\gamma^i$ for $i \neq j$, and $(\gamma^0)^2 = -1$, $(\gamma^i)^2 = +1$)

$$\gamma_5^2 = (-i)^2\gamma^0\gamma^1\gamma^2\gamma^3\gamma^0\gamma^1\gamma^2\gamma^3 = -\gamma^1\gamma^2\gamma^3\gamma^1\gamma^2\gamma^3 = -\gamma^2\gamma^3\gamma^2\gamma^3 = (\gamma^3)^2 = 1. \tag{36.32}$$

For the Hermiticity property we find

$$\gamma_5^\dagger = +i\gamma^{3\dagger}\gamma^{2\dagger}\gamma^{1\dagger}\gamma^{0\dagger} = -i\gamma^3\gamma^2\gamma^1\gamma^0 = -i\gamma^0\gamma^1\gamma^2\gamma^3 = \gamma_5. \tag{36.33}$$

We can write γ_5 in an invariant way using the Levi-Civitta symbol, using the fact that different γ values anticommute (and $\gamma^0 = -\gamma_0$, $\gamma^i = \gamma_i$),

$$\gamma_5 = +\frac{i}{4!}\epsilon^{\mu\nu\rho\sigma}\gamma_\mu\gamma_\nu\gamma_\rho\gamma_\sigma. \tag{36.34}$$

Moreover, since different gammas anticommute and the same commute, we find that

$$\gamma_5\gamma^\mu = -i\gamma^0\gamma^1\gamma^2\gamma^3\gamma^\mu = +i\gamma^\mu\gamma^0\gamma^1\gamma^2\gamma^3 = -\gamma^\mu\gamma_5, \tag{36.35}$$

which completes the properties of the gamma matrices, showing that γ_5 completes a five-dimensional Clifford algebra.

In the Weyl representation, we can check that

$$\gamma_5 = \begin{pmatrix} \mathbb{1} & 0 \\ 0 & -\mathbb{1} \end{pmatrix}, \tag{36.36}$$

and this simple form is why the Weyl representation has the name of the Weyl spinors: the Weyl spinors are eigenvectors of γ_5 with eigenvalues ± 1,

$$\gamma_5 \psi_R = +\psi_R, \quad \gamma_5 \psi_L = -\psi_L. \tag{36.37}$$

Then we can construct the *projectors onto the R/L subspaces* as

$$P_R = \frac{1 + \gamma_5}{2}, \quad P_L = \frac{1 - \gamma_5}{2}. \tag{36.38}$$

Then from (36.38) we find

$$\frac{1 + \gamma_5}{2}\psi = \psi_R, \quad \frac{1 - \gamma_5}{2}\psi = \psi_L, \tag{36.39}$$

and, moreover,

$$\frac{1 - \gamma_5}{2}\psi_R = 0, \quad \frac{1 + \gamma_5}{2}\psi_L = 0. \tag{36.40}$$

In terms of the projectors, we find from $\gamma_5^2 = 1$ that

$$P_R^2 = P_R, \quad P_L^2 = P_L, \quad P_R P_L = \frac{1 - \gamma_5^2}{2} = 0 = P_L P_R, \tag{36.41}$$

as appropriate for projectors. We also find that

$$P_L^\dagger = \left(\frac{1 - \gamma_5}{2}\right)^\dagger = P_L, \quad P_R^\dagger = P_R, \tag{36.42}$$

but $\gamma^0 P_L = P_R \gamma^0$, so that

$$\overline{P_L \psi} = \psi^\dagger P_L^\dagger i\gamma^0 = \psi^\dagger i\gamma^0 P_R. \tag{36.43}$$

36.3 Majorana Spinors

We can define another representation for spinors in four dimensions, which is equivalent, from a representation theory viewpoint, with the Weyl one, the Majorana representation. While the Weyl spinors are acted upon by $SU(2)$, so are intrinsically complex, we can define a representation by a *reality condition*,

$$\bar{\psi} = \psi^C \equiv \psi^T C, \tag{36.44}$$

or "Dirac conjugate = Majorana conjugate," where the Majorana conjugate ψ^C is defined in terms of a *charge conjugation matrix* C. There is a general theory of the charge conjugation matrix in all dimensions, which will not be described here, but for us it suffices to say that in four Minkowski dimensions it is defined by the relations (in other dimensions the signs on the right hand side on the two relations below are dimension dependent)

$$C^T = -C, \quad C\gamma^\mu C^{-1} = -(\gamma^\mu)^T. \tag{36.45}$$

In the Weyl representation for the gamma matrices, we can prove that

$$C_{\alpha\beta} = \begin{pmatrix} -\epsilon^{ab} & \mathbf{0} \\ \mathbf{0} & \epsilon^{\dot{a}\dot{b}} \end{pmatrix}. \tag{36.46}$$

We leave this as an exercise. In the Weyl representation, we also find

$$\epsilon^{ab} = \begin{pmatrix} 0 & 1 \\ -1 & 0 \end{pmatrix} = i\sigma^2 \Rightarrow C = \begin{pmatrix} -i\sigma_2 & \mathbf{0} \\ \mathbf{0} & i\sigma_2 \end{pmatrix} = -i\gamma^0\gamma^2, \ C^{-1} = -C. \tag{36.47}$$

But there is another useful representation for the gamma matrices, the *Majorana representation*, where the gamma matrices are purely real,

$$\gamma^0 = \begin{pmatrix} \mathbf{0} & i\sigma_2 \\ i\sigma_2 & \mathbf{0} \end{pmatrix}, \quad \gamma^1 = \begin{pmatrix} \sigma_3 & \mathbf{0} \\ \mathbf{0} & \sigma_3 \end{pmatrix}, \quad \gamma^2 = \begin{pmatrix} \mathbf{0} & i\sigma_2 \\ -i\sigma_2 & \mathbf{0} \end{pmatrix}, \quad \gamma^3 = \begin{pmatrix} -\sigma_1 & \mathbf{0} \\ \mathbf{0} & -\sigma_1 \end{pmatrix}. \tag{36.48}$$

For the Weyl spinor components of a Dirac spinor $\psi = \begin{pmatrix} \psi_a \\ \tilde{\eta}^{\dot{a}} \end{pmatrix}$, we define raising and lowering of indices with the Levi-Civitta tensor,

$$\psi^b = \psi_a \epsilon^{ab}, \tag{36.49}$$

but care must be taken about the order of them, and we have the product

$$\psi\chi \equiv \psi^a \chi_a = \psi_b \epsilon^{ba} \chi_a = -\psi_b \chi^b. \tag{36.50}$$

Then, for a Majorana spinor, for which $\bar{\psi} = \psi^T C$, we have

$$\bar{\psi}^\alpha \chi_\alpha = \psi^\alpha C_{\alpha\beta} \chi_b = \psi^a \chi_a + \tilde{\psi}_{\dot{a}} \tilde{\chi}^{\dot{a}} = \psi\chi + \tilde{\psi}\tilde{\chi}, \tag{36.51}$$

where we saw that we can also define $\tilde{\psi}\tilde{\chi} \equiv \tilde{\psi}_{\dot{a}} \tilde{\chi}^{\dot{a}}$.

But for a general Dirac spinor, and for gamma matrices in the Majorana representation, we can write

$$\bar{\psi} = \psi^\dagger i\gamma^0 = (\psi^*)^T i \begin{pmatrix} \mathbf{0} & \epsilon^{ab} \\ \epsilon^{\dot{a}\dot{b}} & \mathbf{0} \end{pmatrix}, \tag{36.52}$$

which means that

$$\bar{\psi}^{\dot{a}} = i\epsilon^{\dot{a}a}(\psi_a)^* \tag{36.53}$$

so that ψ_R and ψ_L transform as complex conjugate representations and interchanged.

Important Concepts to Remember

- We can define spinors either as half-integer spin representations of the Lorentz group, or as objects transforming with a -1 under a 2π rotation, or as representations of the Clifford algebra.
- Spinors are fermions, by the spin-statistics theorem.
- The Clifford algebra, on a generic manifold, is $\{\gamma_\mu, \gamma_\nu\} = 2g_{\mu\nu}$, and γ_μ is called a gamma matrix, similar to the generators of a Lie algebra.
- Spinors are projective representations of the Lorentz group, or representations of its universal cover, $Sl(2, \mathbb{C})$.

- Weyl spinors are the (1/2 0) and (0 1/2) representations (right-handed and left-handed) of the Lorentz group, and we can construct the Dirac representation of the Clifford algebra by putting them together in
$$\psi = \begin{pmatrix} \psi_a \\ \tilde{\eta}^{\dot{a}} \end{pmatrix} = \begin{pmatrix} \psi_R \\ \psi_L \end{pmatrix}.$$
- The matrix $\gamma_5 = -i\gamma^0\gamma^1\gamma^2\gamma^3$ acts as a fifth gamma matrix.
- In the Weyl representation, $\gamma^\mu = \begin{pmatrix} 0 & \sigma^\mu \\ \bar{\sigma}^\mu & 0 \end{pmatrix}$ and $\gamma_5 = \begin{pmatrix} \mathbb{1} & 0 \\ 0 & \mathbb{1} \end{pmatrix}$.
- The Lorentz generators in the spinor representation are $J_{\mu\nu} = \frac{1}{4}[\gamma_\mu, \gamma_\nu]$.
- The Dirac conjugate spinor is $\bar{\psi} = \psi^\dagger i\gamma^0$ and then $\bar{\psi}\psi$ is Lorentz invariant.
- $P_R = (1+\gamma_5)/2$ and $P_L = (1-\gamma_5)/2$ are projectors onto Weyl spinors.
- Majorana spinors are defined by the reality condition $\bar{\psi} = \psi^c = \psi^T C$, where C is the charge conjugation matrix.
- C satisfies $C^T = -C$ and $C\gamma^\mu C^{-1} = -(\gamma_\mu)^T$.
- In the Majorana representation, the gamma matrices are all real.

Further Reading

See chapter 10 in [9], chapter 5.2 in [10] and chapter 3 in [4].

Exercises

(1)   Show that
$$J_{\mu\nu} = \frac{1}{2}[\gamma_\mu, \gamma_\nu] \tag{36.54}$$
satisfies the Lorentz algebra, using only the Clifford algebra.

(2)   Prove that
$$\Lambda_S^{-1}\gamma^\mu\Lambda_S = \Lambda^\mu{}_\nu\gamma^\nu, \tag{36.55}$$
where
$$\Lambda_S = e^{i\theta^{\mu\nu}(J_{\mu\nu})_{\text{spinor}}}; \quad \Lambda^\mu{}_\nu = (e^{i\theta^{\rho\sigma}(J_{\rho\sigma})_{\text{vector}}})^\mu{}_\nu. \tag{36.56}$$

(3)   Prove that in the Weyl representation for γ^μ,
$$C_{\alpha\beta} = \begin{pmatrix} -\epsilon^{ab} & 0 \\ 0 & \epsilon^{\dot{a}\dot{b}} \end{pmatrix} \tag{36.57}$$
satisfies
$$C\gamma^\mu C^{-1} = -(\gamma^\mu)^T. \tag{36.58}$$

(4)   Find a representations for gamma matrices in 5+1 dimensions, based on the Weyl representation in 3+1 dimensions, in a similar manner with how we extended the 1+1 dimensional gamma matrices to 3+1 dimensional ones.

(5)   Show that the gamma matrices in 3+1 dimensions $\mathbb{1}, \gamma^\mu, \gamma^5, \gamma^{\mu\nu} \equiv \frac{1}{2}[\gamma^\mu, \gamma^\nu]$ form a complete set for 4×4 matrices.

     Then show that $\gamma_5 \gamma^{\mu\nu}$, $\gamma^{\mu\nu\rho} \equiv \gamma^{[\mu}\gamma^\nu\gamma^{\rho]}$ (totally antisymmetric product) and $\gamma^{\mu\nu\rho\sigma} \equiv \gamma^{[\mu}\gamma^\nu\gamma^\rho\gamma^{\sigma]}$ are obtained in terms of them.

37 The Dirac Equation and Its Solutions

In this chapter, we continue the study of spinors (i.e., spin 1/2 fermions), by describing their classical field theory: the Lagrangean, equations of motion, and solutions.

There is one caveat. Spinors are fermions, which means they obey Fermi–Dirac statistics, which also means the Pauli exclusion principle: there is at most one spinor in a given state. That means that we cannot have "many particles in a state," which would be necessary for a quantum field to become classical. Nevertheless, we have a useful fiction called classical field theory for spinors, which amounts to using just the classical equation of motion. It corresponds to the formal classical limit $\hbar \to 0$ of quantum mechanics, but as we said, the corresponding field strictly speaking still has no classical limit and always contains quantum fluctuations.

37.1 Lorentz Invariant Lagrangeans

We start by writing Lagrangeans that are invariant under Lorentz transformations and can describe spinors.

We have seen in the last chapter that $\bar{\psi}\psi$ is Lorentz invariant by construction (where we defined $\bar{\psi} = \psi^\dagger i \gamma^0$). Moreover, we saw that γ^μ transforms under a Lorentz transformations on μ in the spinor representation, so $\gamma^\mu \partial_\mu$ transforms under Lorentz transformations into $\Lambda_S \gamma^\mu \partial_\mu \Lambda_S^{-1}$. That means that another invariant is $\bar{\psi}\gamma^\mu \partial_\mu \psi$. Since ∂_μ has the same dimension as a mass m, a good Lagrangean for ψ is

$$S_\psi = -\int d^4x\, \bar{\psi}(\gamma^\mu \partial_\mu + m)\psi. \tag{37.1}$$

Note that for any vector, its contraction with γ^μ is denoted by a slash, so $\gamma^\mu \partial_\mu = \slashed{\partial}$, $\gamma^\mu p_\mu = \slashed{p}$. Also note that in the previous *Dirac action* for the $\psi \in \mathbb{C}$, the fields ψ and $\bar{\psi}$ are considered to be independent, just like we considered for the complex scalar ϕ and ϕ^* to be independent.

Then, varying S_ψ with respect to $\bar{\psi}$, we obtain the equation of motion

$$(\gamma^\mu \partial_\mu + m)\psi = (\slashed{\partial} + m)\psi = 0, \tag{37.2}$$

called the *Dirac equation*. Dirac, of course, found first the equation from considerations of it being a sort of "square root of the KG equation," as we will see later, and from symmetry considerations – and only then wrote the action.

The previous action was written for the Dirac spinor ψ, the fundamental representation of the Clifford algebra. But we would like to decompose it in terms of the irreducible representations (irreps) of the Lorentz group, the Weyl spinors ψ_R and ψ_L. Since $\psi = \begin{pmatrix} \psi_R \\ \psi_L \end{pmatrix}$, and in the Weyl representation $i\gamma^0 = \begin{pmatrix} 0 & \mathbb{1} \\ \mathbb{1} & 0 \end{pmatrix}$, the first Lorentz invariant is

$$m\bar{\psi}\psi = m\psi^\dagger i\gamma^0 \psi = m \begin{pmatrix} \psi_R^\dagger & \psi_L^\dagger \end{pmatrix} \begin{pmatrix} 0 & \mathbb{1} \\ \mathbb{1} & 0 \end{pmatrix} \begin{pmatrix} \psi_R \\ \psi_L \end{pmatrix} = m(\psi_R^\dagger \psi_L + \psi_L^\dagger \psi_R). \tag{37.3}$$

This term is then called a *Dirac mass term*, and, as we see, it mixes the irreps of the Lorentz group, ψ_R and ψ_L.

But that cannot be the only possibility. Since ψ_R and ψ_L are irreps, there must be Lagrangeans only for each of them.

Indeed, we saw in the last chapter that for two right-handed spinors ψ_a and χ_a, we have a Lorentz invariant,

$$\psi\chi = \psi^a \chi_a = \psi_b \epsilon^{ba} \chi_a = \psi_R^T i\sigma_2 \chi_R. \tag{37.4}$$

Then, for a single right-handed spinor ψ_R, we can write a new mass term,

$$\mathcal{L}_{Mm} = m\psi_R^T i\sigma_2 \psi_R = m\psi_b \epsilon^{ba} \psi_a, \tag{37.5}$$

which is called a *Majorana mass term*. And we don't need to have a ψ_L to construct a Lagrangean; we can have one for just the irrep ψ_R. Similarly, we can write the same Majorana mass term for a ψ_L, $\psi_L^T i\sigma_2 \psi_L$.

Note that the Majorana mass term only makes sense if the spinor ψ_a is actually anticommuting,

$$\{\psi_a, \psi_b\} = 0, \tag{37.6}$$

since it is contracted with the antisymmetric symbol $\epsilon^{ab} = i\sigma_2$. But that is as it should be, since by the spin-statistics theorem we know that spinors should satisfy Fermi–Dirac statistics. In the classical limit, formally $\hbar \to 0$, all fermions A, B anticommute (the quantization conditions involve anticommutators, and on the right-hand side, we have $i\hbar \to 0$ multiplying everything),

$$\{A, B\} = 0. \tag{37.7}$$

Thus, we effectively have found a manifestation of the spin-statistics theorem. Mathematically, for anticommuting objects, we say we have *Grassmann numbers* (instead of usual, commuting, numbers).

We now aim to separate the kinetic terms in terms determined by Weyl spinors. In terms of the projectors P_R, P_L, the Weyl spinors are

$$\psi_R/\psi_L = P_L/P_R\psi = \frac{1 \pm \gamma_5}{2}\psi. \tag{37.8}$$

Since $\gamma_5^\dagger = \gamma_5$, and $\gamma_5\gamma_\mu = -\gamma_\mu\gamma_5$, we obtain $P_L^\dagger = P_L$, $P_R^\dagger = P_R$, and from them,

$$P_L^\dagger i\gamma^0 = i\gamma^0 P_R. \tag{37.9}$$

Moreover, then, ($P_R^2 = P_R, P_R P_L = 0$)

$$P_R = P_R^2 = \bar{P}_L P_R, \tag{37.10}$$

so we can write a mass term by inserting P_R as

$$\bar{\psi} P_R \psi = (\overline{P_L \psi})(P_R \psi) = \bar{\psi}_L \psi_R, \tag{37.11}$$

so it obtains half the Dirac mass term. Similarly, the other half is

$$\bar{\psi} P_L \psi = (\overline{P_R \psi})(P_L \psi) = \bar{\psi}_R \psi_L. \tag{37.12}$$

For the kinetic term, we note that

$$\slashed{\partial} P_R = \partial_\mu \gamma^\mu P_R = P_L \slashed{\partial}. \tag{37.13}$$

Then we can insert a P_R in the kinetic term and obtain

$$\bar{\psi} \slashed{\partial} P_R \psi = \bar{\psi} \slashed{\partial} P_R^2 \psi = \bar{\psi} P_L \slashed{\partial} P_R \psi = (\overline{P_R \psi}) \slashed{\partial} (P_R \psi) = \bar{\psi}_R \slashed{\partial} \psi_R. \tag{37.14}$$

This is the kinetic term just for the irrep (Weyl spinor) ψ_R, unlike the mass term. So it can be used together with a Majorana mass to give an action for ψ_R only.

Similarly, inserting P_L, we find

$$\bar{\psi} \slashed{\partial} P_L \psi = (\overline{P_L \psi}) \slashed{\partial} (P_L \psi) = \bar{\psi}_L \slashed{\partial} \psi_L, \tag{37.15}$$

a kinetic term for ψ_L only. The kinetic term for the Dirac spinor is thus the sum of the two terms,

$$\bar{\psi} \slashed{\partial} \psi = \bar{\psi} \slashed{\partial} (P_R + P_L) \psi = \bar{\psi}_R \slashed{\partial} \psi_R + \bar{\psi}_L \slashed{\partial} \psi_L. \tag{37.16}$$

37.2 The Dirac Equation

A solution of the Dirac equation,

$$(\slashed{\partial} + m)\psi = 0, \tag{37.17}$$

is also a solution of the KG equation, since

$$\slashed{\partial}\slashed{\partial} = \partial_\mu \partial_\nu \gamma^\mu \gamma^\nu = \partial_\mu \partial_\nu \frac{1}{2}\{\gamma^\mu, \gamma^\nu\} = \partial_\mu \partial_\nu \eta^{\mu\nu} = \Box, \tag{37.18}$$

so by multiplying (from the left) the Dirac equation with $(\slashed{\partial} - m)$, we obtain the KG equation,

$$0 = (\slashed{\partial} - m)(\slashed{\partial} + m)\psi = (\Box - m^2)\psi. \tag{37.19}$$

This is the way Dirac found his equation, as a "square root of the KG equation."

The result means that any solution of the Dirac equation is also a solution of KG – i.e., it must be of the type

$$e^{\pm i p \cdot x} u(p), \tag{37.20}$$

where $p^2 = -m^2$, as for any solution of the KG equation – and $u(p)$ is a spinor (matrix, or rather, four-component spinor).

37.3 Solutions of the Dirac Equation

1. The first set of solutions has positive frequencies – i.e., they have $e^{ip \cdot x} = e^{-i\omega t + i\vec{p} \cdot \vec{x}}$,

$$\psi(x) = \int \frac{d^3 p}{(2\pi)^3} u(p) \, e^{ip \cdot x} \Big|_{p^2 = -m^2, p^0 > 0}. \tag{37.21}$$

Except for the restrictions $p^2 = -m^2$ (from the KG equation) and $p^0 > 0$ (for positive frequency), this is just a Fourier transform, so the coefficient $u(p)$ (Fourier transformed field) satisfies

$$(i\not{p} + m)u(p) = 0, \tag{37.22}$$

where

$$\not{p} \equiv p_\mu \gamma^\mu = p^\nu \gamma^\mu \eta_{\mu\nu}. \tag{37.23}$$

Using the gamma matrices in the Weyl basis, $\gamma^\mu_{\text{Weyl}} = -i \begin{pmatrix} 0 & \sigma^\mu \\ \bar\sigma^\mu & 0 \end{pmatrix}$, we obtain the matrix equations

$$\begin{pmatrix} m\,\mathbb{1} & p \cdot \sigma \\ p \cdot \bar\sigma & m\,\mathbb{1} \end{pmatrix} u_s(p) = 0. \tag{37.24}$$

We have anticipated the fact that there are two independent solutions by writing an index $s = 1, 2$. The two solutions appear because the ψ^α have four components, and the Dirac equation relates half of the components to the other half – i.e., it reduces the number of degrees of freedom of the spinor by half. Moreover, the ψ will be associated with a spin 1/2 fermion like the electron, which has two helicities (spin projections onto the momentum), $\pm 1/2$.

To continue the analysis, we first go into the rest frame, where $p^\mu = (m, \vec{0})$, so $p \cdot \sigma = p \cdot \bar\sigma = -m\,\mathbb{1}$, leading to the equation

$$m \begin{pmatrix} \mathbb{1} & -\mathbb{1} \\ -\mathbb{1} & \mathbb{1} \end{pmatrix} u_s(p) = 0. \tag{37.25}$$

The general solution is of the type $\psi = \begin{pmatrix} \xi \\ \xi \end{pmatrix}$ for an arbitrary two-component spinor ξ. But we will define the basis vectors

$$\xi_s: \quad \xi_1 = \begin{pmatrix} 1 \\ 0 \end{pmatrix}, \quad \xi_2 = \begin{pmatrix} 0 \\ 1 \end{pmatrix}, \tag{37.26}$$

and use the normalization of $\sqrt{m}$, so we have

$$u_s = \sqrt{m} \begin{pmatrix} \xi_s \\ \xi_s \end{pmatrix}: \quad u_1 = \sqrt{m} \begin{pmatrix} 1 \\ 0 \\ 1 \\ 0 \end{pmatrix}, \quad u_2 = \sqrt{m} \begin{pmatrix} 0 \\ 1 \\ 0 \\ 1 \end{pmatrix}. \tag{37.27}$$

2. The second set of solutions has negative frequencies, so it is

$$\psi(x) = \int \frac{d^3 p}{(2\pi)^3} v_s(p) \, e^{-ip \cdot x}\Big|_{p^2 = -m^2, p^0 > 0}. \tag{37.28}$$

The Fourier transform coefficients $v_s(p)$ now satisfy the equation

$$(-i\not{p} + m)v_s(p) = 0, \tag{37.29}$$

or explicitly in the Weyl representation for gamma matrices

$$\begin{pmatrix} m\,\mathbb{1} & -p \cdot \sigma \\ -p \cdot \bar{\sigma} & m\,\mathbb{1} \end{pmatrix} v_s(p) = 0. \tag{37.30}$$

In the rest frame, we obtain

$$m \begin{pmatrix} \mathbb{1} & \mathbb{1} \\ \mathbb{1} & \mathbb{1} \end{pmatrix} v_s(p) = 0, \tag{37.31}$$

with the two solutions (for $s = 1, 2$),

$$v_s(p) = \sqrt{m} \begin{pmatrix} \xi_s \\ -\xi_s \end{pmatrix}, \tag{37.32}$$

with the same normalization.

Next, we consider a general momentum, but we orient the direction z along it, so

$$p \cdot \sigma = -p^0 \sigma^0 + p_z \sigma_3 = \begin{pmatrix} -E + p_z & 0 \\ 0 & -E - p_z \end{pmatrix}$$
$$p \cdot \bar{\sigma} = -p^0 \sigma^0 - p_z \sigma_3 = \begin{pmatrix} -E - p_z & 0 \\ 0 & -E + p_z \end{pmatrix}. \tag{37.33}$$

Since the matrices are diagonal, we can take the matrix square root, so we obtain

$$\sqrt{-p \cdot \sigma} = \begin{pmatrix} \sqrt{E - p_z} & 0 \\ 0 & \sqrt{E + p_z} \end{pmatrix} \equiv M$$
$$\sqrt{-p \cdot \bar{\sigma}} = \begin{pmatrix} \sqrt{E + p_z} & 0 \\ 0 & \sqrt{E - p_z} \end{pmatrix} \equiv N, \tag{37.34}$$

and defining $A \equiv \sqrt{E - p_z}$, $B = \sqrt{E + p_z}$, we obtain

$$(AB)^2 = (E - p_z)(E + p_z) = m^2 \Rightarrow MN = NM = m\,\mathbb{1}. \tag{37.35}$$

We then guess that in the general situation, the solution for $u_s(p)$ is

$$u_s(p) = \begin{pmatrix} K\xi_s \\ L\xi_s \end{pmatrix} \tag{37.36}$$

for some 2×2 matrices K and L. Substituting in the Dirac equation, we find

$$\begin{pmatrix} m\,\mathbb{1} & -M^2 \\ -N^2 & m\,\mathbb{1} \end{pmatrix} \begin{pmatrix} K\xi_s \\ L\xi_s \end{pmatrix} = 0, \quad \forall s. \tag{37.37}$$

Since the equation is valid for both ξ_s – that is, for any ξ – we can erase it and obtain two matrix equations,

$$
\begin{aligned}
mK - M^2 L &= M(NK - ML) = 0 \\
-N^2 K + mL &= N(-NK + ML) = 0,
\end{aligned}
\tag{37.38}
$$

where we have used $m = MN = NM$. We see that both equations mean

$$
NK = ML \,,
\tag{37.39}
$$

solved by $K = M, L = N$, which means that the solutions are

$$
u_s(p) = \begin{pmatrix} \sqrt{-p \cdot \sigma} \xi_s \\ \sqrt{-p \cdot \bar{\sigma}} \xi_s \end{pmatrix}.
\tag{37.40}
$$

We have found this result in one reference frame for spatial rotations, but the relation is written in a Lorentz covariant way (using $p \cdot \sigma = p^\mu \sigma^\mu \eta_{\mu\nu}$ and $p \cdot \bar{\sigma} = p^\mu \bar{\sigma}^\nu \eta_{\mu\nu}$, which act only on the two-dimensional spinor indices), so the result is valid in any reference frame. We also note that the normalization matches the one of the rest frames, since, there, $\sqrt{-p \cdot \sigma} = \sqrt{-p \cdot \bar{\sigma}} = m$.

For the negative frequency solution, we similarly write the ansatz

$$
v_s(p) = \begin{pmatrix} P \xi_s \\ -Q \xi_s \end{pmatrix}.
\tag{37.41}
$$

Substituting in the Dirac equation, we find

$$
\begin{pmatrix} m \, \mathbb{1} & +M^2 \\ +N^2 & m \, \mathbb{1} \end{pmatrix} \begin{pmatrix} P \xi_s \\ -Q \xi_s \end{pmatrix} = 0, \quad \forall s.
\tag{37.42}
$$

Again, since this is true for any ξ, we can erase it and obtain two matrix equations,

$$
\begin{aligned}
mP - M^2 Q &= M(NP - MQ) = 0 \\
N^2 P - mQ &= N(NP - MQ) = 0,
\end{aligned}
\tag{37.43}
$$

both equations meaning

$$
NP = MQ,
\tag{37.44}
$$

solved by $P = M, Q = N$, meaning that $v_s(p)$ is

$$
v_s(p) = \begin{pmatrix} \sqrt{-p \cdot \sigma} \xi_s \\ -\sqrt{-p \cdot \bar{\sigma}} \xi_s \end{pmatrix}.
\tag{37.45}
$$

Again, this solution is valid in any reference frame, and the normalization matches the one in the rest frame.

37.4 Normalizations

We now show that the solutions discussed in the previous section are orthonormal, in a well-defined sense.

We first note that, since $(\sigma^\mu)^\dagger = \sigma^\mu$ and $(\bar\sigma^\mu)^\dagger = \bar\sigma^\mu$, we have also

$$(\sqrt{-p\cdot\sigma})^\dagger = \sqrt{-p\cdot\sigma}, \quad (\sqrt{-p\cdot\bar\sigma})^\dagger = \sqrt{-p\cdot\bar\sigma}. \tag{37.46}$$

Then, in the Weyl representation for the gamma matrices,

$$\bar{u}_s(p)u_{s'}(p) = u_s^\dagger(p)(i\gamma^0)u_{s'}(p) = \begin{pmatrix} \sqrt{-p\cdot\sigma}\xi_s \\ \sqrt{-p\cdot\bar\sigma}\xi_s \end{pmatrix}^\dagger \begin{pmatrix} 0 & \mathbb{1} \\ \mathbb{1} & 0 \end{pmatrix} \begin{pmatrix} \sqrt{-p\cdot\sigma}\xi_{s'} \\ \sqrt{-p\cdot\bar\sigma}\xi_{s'} \end{pmatrix}$$

$$= \begin{pmatrix} \xi_s \\ \xi_s \end{pmatrix}^\dagger \begin{pmatrix} \sqrt{(p\cdot\sigma)(p\cdot\bar\sigma)} & 0 \\ 0 & \sqrt{(p\cdot\bar\sigma)(p\cdot\sigma)} \end{pmatrix} \begin{pmatrix} \xi_{s'} \\ \xi_{s'} \end{pmatrix}, \tag{37.47}$$

where in the last equality we have *rearranged* the result (we have acted with the "switching matrix" $\begin{pmatrix} 0 & \mathbb{1} \\ \mathbb{1} & 0 \end{pmatrix}$ which switches the up and down parts of the spinor, and then we have rewritten the result using a new matrix).

But since, as we can explicitly check, $\sigma_{(\mu}\bar\sigma_{\nu)} = \bar\sigma_{(\mu}\sigma_{\nu)}$, we have

$$(p\cdot\sigma)(p\cdot\bar\sigma) = (p\cdot\bar\sigma)(p\cdot\sigma) = (-E\,\mathbb{1}+\vec{p}\cdot\vec\sigma)(-E\,\mathbb{1}-\vec{p}\cdot\vec\sigma) = E^2\,\mathbb{1}-(\vec{p}\cdot\vec\sigma)^2 = m^2\,\mathbb{1}, \tag{37.48}$$

where we have used $p^i p^j \sigma_i \sigma_j = p^i p^j \frac{1}{2}\{\sigma_i, \sigma_j\} = p^i p^j \delta_{ij}\,\mathbb{1}_2 = \vec{p}^2\,\mathbb{1}_2$. Then we finally obtain

$$\bar{u}_s(p)u_{s'}(p) = \begin{pmatrix} \xi_s \\ \xi_s \end{pmatrix}^\dagger m\,\mathbb{1}_4 \begin{pmatrix} \xi_{s'} \\ \xi_{s'} \end{pmatrix} = 2m\xi_s^T\xi_{s'} = 2m\delta_{ss'}. \tag{37.49}$$

Similarly, we obtain

$$\bar{v}_s(p)v_{s'}(p) = v_s^\dagger(p)(i\gamma^0)v_{s'}(p) = \begin{pmatrix} \sqrt{-p\cdot\sigma}\xi_s \\ -\sqrt{-p\cdot\bar\sigma}\xi_s \end{pmatrix}^\dagger \begin{pmatrix} 0 & \mathbb{1} \\ \mathbb{1} & 0 \end{pmatrix} \begin{pmatrix} \sqrt{-p\cdot\sigma}\xi_{s'} \\ -\sqrt{-p\cdot\bar\sigma}\xi_{s'} \end{pmatrix}$$

$$= \begin{pmatrix} \xi_s \\ -\xi_s \end{pmatrix}^\dagger \begin{pmatrix} \sqrt{(p\cdot\sigma)(p\cdot\bar\sigma)} & 0 \\ 0 & \sqrt{(p\cdot\bar\sigma)(p\cdot\sigma)} \end{pmatrix} \begin{pmatrix} -\xi_{s'} \\ \xi_{s'} \end{pmatrix}, \tag{37.50}$$

and, using the same relations as before, we find

$$\bar{v}_s(p)v_{s'}(p) = \begin{pmatrix} \xi_s \\ -\xi_s \end{pmatrix}^\dagger m\,\mathbb{1}_4 \begin{pmatrix} -\xi_{s'} \\ \xi_{s'} \end{pmatrix} = -2m\xi_s^T\xi_{s'} = -2m\delta_{ss'}. \tag{37.51}$$

Moreover, the u_s and v_s solutions are orthogonal, since we find

$$\bar{u}_s(p)v_{s'}(p) = \begin{pmatrix} \xi_s \\ \xi_s \end{pmatrix}^\dagger \begin{pmatrix} \sqrt{(p\cdot\sigma)(p\cdot\bar\sigma)} & 0 \\ 0 & \sqrt{(p\cdot\bar\sigma)(p\cdot\sigma)} \end{pmatrix} \begin{pmatrix} -\xi_{s'} \\ \xi_{s'} \end{pmatrix} = \begin{pmatrix} \xi_s \\ \xi_s \end{pmatrix}^\dagger m\,\mathbb{1}_4 \begin{pmatrix} -\xi_{s'} \\ \xi_{s'} \end{pmatrix} = 0, \tag{37.52}$$

as well as

$$\bar{v}_s(p)u_{s'}(p) = \begin{pmatrix} \xi_s \\ -\xi_s \end{pmatrix}^\dagger \begin{pmatrix} \sqrt{(p\cdot\sigma)(p\cdot\bar\sigma)} & 0 \\ 0 & \sqrt{(p\cdot\bar\sigma)(p\cdot\sigma)} \end{pmatrix} \begin{pmatrix} \xi_{s'} \\ \xi_{s'} \end{pmatrix} = \begin{pmatrix} \xi_s \\ -\xi_s \end{pmatrix}^\dagger m\,\mathbb{1}_4 \begin{pmatrix} \xi_{s'} \\ \xi_{s'} \end{pmatrix} = 0. \tag{37.53}$$

On the other hand, we can consider normalization conditions using the Hermitean conjugate instead of the Dirac conjugate. Then we find

$$u_s^\dagger(p)u_{s'}(p) = \begin{pmatrix} \sqrt{-p\cdot\sigma}\,\xi_s \\ \sqrt{-p\cdot\bar\sigma}\,\xi_s \end{pmatrix}^\dagger \begin{pmatrix} \sqrt{-p\cdot\sigma}\,\xi_{s'} \\ \sqrt{-p\cdot\bar\sigma}\,\xi_{s'} \end{pmatrix} = \begin{pmatrix} \xi_s \\ \xi_s \end{pmatrix}^\dagger \begin{pmatrix} -p\cdot\sigma & 0 \\ 0 & -p\cdot\bar\sigma \end{pmatrix} \begin{pmatrix} \xi_{s'} \\ \xi_{s'} \end{pmatrix},$$
(37.54)

and since $-p\cdot\sigma = E\,\mathbb{1} - \vec{p}\cdot\vec\sigma$ and $-p\cdot\bar\sigma = E\,\mathbb{1} + \vec{p}\cdot\vec\sigma$, we obtain

$$u_s^\dagger(p)u_{s'}(p) = 2E\xi_s^\dagger\xi_{s'} - \xi_s^\dagger\vec{p}\cdot\vec\sigma\xi_{s'} + \xi_s^\dagger\vec{p}\cdot\vec\sigma\xi_{s'} = 2E\delta_{ss'}.$$
(37.55)

Similarly, for v_s,

$$v_s^\dagger(p)v_{s'}(p) = \begin{pmatrix} \sqrt{-p\cdot\sigma}\,\xi_s \\ -\sqrt{-p\cdot\bar\sigma}\,\xi_s \end{pmatrix}^\dagger \begin{pmatrix} \sqrt{-p\cdot\sigma}\,\xi_{s'} \\ -\sqrt{-p\cdot\bar\sigma}\,\xi_{s'} \end{pmatrix} = \begin{pmatrix} \xi_s \\ -\xi_s \end{pmatrix}^\dagger \begin{pmatrix} -p\cdot\sigma & 0 \\ 0 & -p\cdot\bar\sigma \end{pmatrix} \begin{pmatrix} \xi_{s'} \\ -\xi_{s'} \end{pmatrix},$$
(37.56)

and we finally obtain

$$v_s^\dagger(p)v_{s'}(p) = 2E\xi_s^\dagger\xi_{s'} - \xi_s^\dagger\vec{p}\cdot\vec\sigma\xi_{s'} + \xi_s^\dagger\vec{p}\cdot\vec\sigma\xi_{s'} = 2E\delta_{ss'}.$$
(37.57)

Now, however, u_s and v_s are not orthonormal anymore,

$$u_s^\dagger(p)v_{s'}(p) = \begin{pmatrix} \xi_s \\ \xi_s \end{pmatrix}^\dagger \begin{pmatrix} -p\cdot\sigma & 0 \\ 0 & -p\cdot\bar\sigma \end{pmatrix} \begin{pmatrix} \xi_{s'} \\ -\xi_{s'} \end{pmatrix} = -2\xi_s^\dagger\vec{p}\cdot\vec\sigma\xi_{s'} \neq 0$$

$$v_s^\dagger(p)u_{s'}(p) = \begin{pmatrix} \xi_s \\ -\xi_s \end{pmatrix}^\dagger \begin{pmatrix} -p\cdot\sigma & 0 \\ 0 & -p\cdot\bar\sigma \end{pmatrix} \begin{pmatrix} \xi_{s'} \\ \xi_{s'} \end{pmatrix} = -2\xi_s^\dagger\vec{p}\cdot\vec\sigma\xi_{s'} \neq 0.$$
(37.58)

Yet we can fix this, if we just replace p^μ with $\tilde{p}^\mu = (p^0, -\vec{p})$ in the second spinor,

$$u_s^\dagger(p)v_{s'}(\tilde{p}) = \begin{pmatrix} \xi_s \\ \xi_s \end{pmatrix}^\dagger \begin{pmatrix} \sqrt{(p\cdot\sigma)(\tilde{p}\cdot\sigma)} & 0 \\ 0 & -\sqrt{(p\cdot\bar\sigma)(\tilde{p}\cdot\bar\sigma)} \end{pmatrix} \begin{pmatrix} \xi_{s'} \\ \xi_{s'} \end{pmatrix} = 0$$

$$v_s^\dagger(p)u_{s'}(\tilde{p}) = \begin{pmatrix} \xi_s \\ \xi_s \end{pmatrix}^\dagger \begin{pmatrix} \sqrt{(p\cdot\sigma)(\tilde{p}\cdot\sigma)} & 0 \\ 0 & -\sqrt{(p\cdot\bar\sigma)(\tilde{p}\cdot\bar\sigma)} \end{pmatrix} \begin{pmatrix} \xi_{s'} \\ \xi_{s'} \end{pmatrix} = 0,$$
(37.59)

where we have used the fact that $\tilde{p}\cdot\bar\sigma = p\cdot\sigma, \tilde{p}\cdot\sigma = p\cdot\bar\sigma$, so

$$(p\cdot\sigma)(\tilde{p}\cdot\sigma) = (p\cdot\sigma)(p\cdot\bar\sigma) = m^2\,\mathbb{1}$$
$$(p\cdot\bar\sigma)(\tilde{p}\cdot\bar\sigma) = (p\cdot\bar\sigma)(p\cdot\sigma) = m^2\,\mathbb{1}.$$
(37.60)

To conclude, let us remark that Dirac found that the negative frequency (energy) solutions $v_s(p)$ are associated with "holes" in a completely filled Dirac sea of negative energies, which are the positrons (antiparticle for the electrons). The further explanation of this leads to quantum field theory, so we will not do it here. The two $u_s(p)$ solutions are related to the electrons of two possible helicities.

Important Concepts to Remember

- The classical field theory of spinor fields is a useful abstraction, though there is no really classical spinor field.
- The Lagrangean for a Dirac spinor is $-\int \bar\psi(\partial\!\!\!/ + m)\psi$, and the corresponding Dirac equation is $(\partial\!\!\!/ + m)\psi = 0$.

- A Dirac mass term $m\bar{\psi}\psi = m(\bar{\psi}_R\psi_L + \bar{\psi}_L\psi_R)$ mixes the irreps ψ_R and ψ_L, whereas a Majorana mass term, $m\psi_R^T i\sigma_2\psi_R$, doesn't.
- (Classical) spinors are anticommuting – i.e., Grassmann numbers – and in quantum mechanics, they obey Fermi–Dirac statistics.
- The kinetic term $\bar{\psi}\partial\!\!\!/\psi$ doesn't mix ψ_R and ψ_L.
- We can split both the mass and kinetic terms into two parts by inserting P_R and P_L in the middle: $\bar{\psi}P_R\psi = \bar{\psi}_L\psi_R$ and $\bar{\psi}\partial\!\!\!/P_R\psi = \bar{\psi}_R\partial\!\!\!/\psi_R$.
- The Dirac equation solves the KG equation, of which it is a sort of "square root," in the sense that $(\partial\!\!\!/ - m)(\partial\!\!\!/ + m)\psi = (\Box - m^2)\psi = 0$.
- The Dirac equation halves the number of degrees of freedom (unlike KG, which just resctricts the possible momenta).
- The solutions of the Dirac equation are $e^{ip\cdot x}u_s(p)$ and $e^{-ip\cdot x}v_s(p)$, for electrons and positrons, respectively.
- The $u_s(p)$, $v_s(p)$ solutions are orthonormal in the Dirac conjugate, $\bar{u}_s(p)u_{s'}(p) = 2m\delta_{ss'} = \bar{v}_s(p)v_{s'}(p)$, and $\bar{u}_s(p)v_{s'}(p) = 0 = \bar{v}_s(p)u_{s'}(p)$.
- The solutions are orthonormal in the Hermitean conjugate in a different sense, $u_s^\dagger(p)u_{s'}(p) = 2E\delta_{ss'} = v_s^\dagger(p)v_{s'}(p)$, but only $u_s^\dagger(p)v_{s'}(\tilde{p}) = 0 = v_s^\dagger u_{s'}(\tilde{p})$, where $\tilde{p} = (p^0, -\vec{p})$.

Further Reading

See chapters 10 and 11 in [9], chapter 5.2 in [10], and chapter 3 in [4].

Exercises

(1) Calculate the helicity $\frac{\vec{p}\cdot\vec{S}}{|\vec{p}|s}$ for $u_s(p)$ and $v_s(p)$.

(2) For four-dimensional gamma matrices γ^μ, prove that

$$(a) \quad \mathrm{Tr}[\gamma^\mu\gamma^\nu\gamma^\rho\gamma^\sigma] = 4(\eta^{\mu\nu}\eta^{\rho\sigma} - \eta^{\mu\rho}\eta^{\nu\sigma} + \eta^{\mu\sigma}\eta^{\nu\rho})$$
$$(b) \quad \gamma_\mu\partial\!\!\!/\gamma^\mu = -2\partial\!\!\!/. \tag{37.61}$$

(3) Prove the *Gordon identity* for *on-shell* spinors (satisfying the equation of motion – i.e., the Dirac equation),

$$\bar{u}(q)\gamma^\mu u(p) = \bar{u}(q)\left[\frac{q^\mu + p^\mu}{2m} + \frac{i\sigma^{\mu\nu}(q_\nu - p_\nu)}{2m}\right]u(p), \tag{37.62}$$

where

$$\sigma_{\mu\nu} \equiv \frac{i}{2}[\gamma_\mu, \gamma_\nu]. \tag{37.63}$$

(4) For four-dimensional gamma matrices γ^μ, γ_5, show that $\mathrm{Tr}[\gamma^\mu] = \mathrm{Tr}[\gamma_5] = 0$, and then calculate $\mathrm{Tr}[\gamma_5\gamma^\mu\gamma^\nu\gamma^\rho\gamma^\sigma]$.

38

General Relativity: Metric and General Coordinate Invariance

In this chapter, we will describe the basics of general relativity, foundation, and kinematics, and the dynamics will be left for the next chapter. As the name suggests, general relativity is a generalization of special relativity. It is a classical field theory for gravity, but one that is special, since it turns out to be related to the geometry of spacetime. The result of this fact is that it is very hard to write a quantum theory corresponding to it, and it is not known how to do that yet. String theory is a way to embed the classical field theory of gravity into a larger quantizable theory, but even there, we don't know how to quantize it except perturbatively.

38.1 Intrinsically Curved Spaces and General Relativity

Thus, in order to describe general relativity, we start with a quick review of *special relativity*. In short, we have seen that special relativity is the statement of invariance – or, more generally, covariance – of the physics under the Lorentz group $SO(3, 1)$. The *line element* of Minkowski space,

$$ds^2 = -dt^2 + d\vec{x}^2 = \eta_{\mu\nu} dx^\mu dx^\nu, \tag{38.1}$$

is invariant under linear transformations of coordinates, $x'^\mu = \Lambda^\mu{}_\nu x^\nu$, where $\Lambda^\mu{}_\nu \in SO(3, 1)$ is a Lorentz group element. More generally, any physical law is written in a Lorentz covariant form, and the action is Lorentz invariant.

In *general relativity*, then, we seek invariance under *general* coordinate transformation, not just linear. But for that, we need to consider the most general spacetime possible, defined by the *line element* – i.e., the differential distance between two points – which is, in the most general case,

$$ds^2 = g_{\mu\nu}(x^\rho) dx^\mu dx^\nu, \tag{38.2}$$

where $g_{\mu\nu}(x^\rho)$ is a matrix of arbitrary functions of spacetime called "the metric" (see Figure 38.1a). The x^μ are (arbitrary) parametrizations of the spacetime manifold – i.e., coordinates on the manifold. Since $dx^\mu dx^\nu$ is symmetric in $(\mu\nu)$, the matrix $g_{\mu\nu}$ is a symmetric matrix. Note that sometimes, by an abuse the notation, the line element ds^2 is also called "the metric."

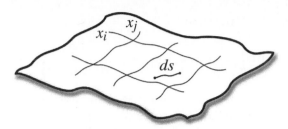

Figure 38.1   A generic curved space, in which the functional form of the distance between two points depends on some local coordinates.

Example 1

The notion of metric is a familiar one, but perhaps one didn't think in these terms. To understand this better, we consider the familiar example of the usual two-sphere in angular coordinates θ and ϕ, with line element

$$ds^2 = d\theta^2 + \sin^2\theta d\theta^2 \equiv g_{\mu\nu}dx^\mu dx^\nu, \tag{38.3}$$

where $x^\mu = (\theta, \phi)$, leading to

$$g_{\theta\theta} = 1, \quad g_{\phi\phi} = \sin^2\theta, \quad g_{\theta\phi} = g_{\phi\theta} = 0. \tag{38.4}$$

Example 2

But we can also consider the two-sphere embedded in three-dimensional Euclidean space – that is, consider the three-dimensional Euclidean line element

$$ds^2 = dx_1^2 + dx_2^2 + dx_3^2 , \tag{38.5}$$

and in this space, we consider the sphere constraint,

$$x_1^2 + x_2^2 + x_3^2 = R^2. \tag{38.6}$$

Then we can solve for x_3 as a function of independent coordinates (x_1, x_2) on the two-sphere,

$$x_3 = \sqrt{R^2 - x_1^2 - x_2^2}. \tag{38.7}$$

We also differentiate the sphere constraint and substitute the x_3 to obtain

$$x_1 dx_1 + x_2 dx_2 + x_3 dx_3 = 0 \Rightarrow dx_3 = -\frac{x_1 dx_1 + x_2 dx_2}{\sqrt{R^2 - x_1^2 - x_2^2}}. \tag{38.8}$$

Now, substituting this into the Euclidean metric, we obtain the *induced metric* on the two-sphere,

$$ds^2_{\text{induced}} = dx_1^2 + dx_2^2 + \frac{(x_1 dx_1 + x_2 dx_2)^2}{R^2 - x_1^2 - x_2^2} \equiv g_{\mu\nu}(x^\rho)dx^\mu dx^\nu, \tag{38.9}$$

where $x^\mu = (x_1, x_2)$, so

$$g_{11} = 1 + \frac{x_1^2}{R^2 - x_1^2 - x_2^2}, \quad g_{22} = 1 + \frac{x_2^2}{R^2 - x_1^2 - x_2^2}, \quad g_{12} = g_{21} = \frac{x_1 x_2}{R^2 - x_1^2 - x_2^2}. \quad (38.10)$$

In this case, the metric on the two-sphere is an *induced metric*, due to the constraint imposed on the Euclidean space.

The natural question is: can we describe in this way any curved space? For a general d–dimensional metric $ds^2 = g_{\mu\nu} dx^\mu dx^\nu$, $\mu = 1, \ldots, d$, the matrix $g_{\mu\nu}$ is symmetric, so it has $\frac{d(d+1)}{2}$ independent components (independent functions). That means that it cannot be embedded into $d + 1$ dimensional spacetime, since in that case, we would have only one independent function, $x^{d+1} = x^{d+1}(x^\mu)$. Since we can do arbitrary coordinate transformations $x'^\mu = x'^\mu(x^\nu)$, corresponding to d arbitrary functions of coordinates, we have a remaining number of $\frac{d(d-1)}{2}$ independent metric components $g_{ab}(x^\rho)$ that cannot be changed by coordinate transformations. This would be the *minimum* number of extra dimensions of an embedding space, since the functions $x^i = x^i(x^\mu)$, $i = d + 1, \ldots d(d + 1)/2$, can be used to fix $g_{ab}(x^\rho)$. In the case $d = 2$, it is a coincidence that $d(d + 1)/2 = d + 1$, but this is not true for $d > 2$.

Note that the general coordinate transformation $x'^\mu = x'^\mu(x^\nu)$ acts on the *fields* $g_{\mu\nu}(x^\rho)$ (as we know already, a function of spacetime that transforms in a well-defined way is a field; we saw that the symmetric traceless tensor, which we will see is the physical propagating mode, is a spin 2 field) and changes them, allowing to fix d of them. This is a redundancy in the description of the same type as gauge transformations, so general coordinate invariance is a type of gauge invariance.

So we can embed spacetimes into higher dimensional spaces, but we need several extra dimensions. But, in fact, the situation is worse than that, since we can also have a different *signature* of the embedding space, depending on what space we want to embed. For instance, there is a famous example found by Lobachevsky, hence called (two-dimensional) Lobachevsky space. It is the space defined by the three-dimensional constraint

$$x^2 + y^2 - z^2 = -R^2. \quad (38.11)$$

It can be shown that it cannot be embedded into three-dimensional Euclidean space, but rather in three-dimensional Minkowski space, with metric

$$ds^2 = dx^2 + dy^2 - dz^2. \quad (38.12)$$

We could ask: why is it important? Because we live in Minkowski space, it is not surprising that we can embed the space into a Minkowski space. But the surprising thing is that the signature *on the two-dimensional space* is Euclidean, not Minkowski, like for the sphere. In the two-dimensional case, possible signatures are $(++)$ and $(--)$, which are the same under the replacement $ds^2 \to -ds^2$ and are called the Euclidean case, with $\det g_{\mu\nu} > 0$, and $(-+)$ and $(+-)$, also equivalent, called the Minkowski case, with $\det g_{\mu\nu} < 0$. In the case of the Lobachevsky space, we calculate the induced metric completely similarly to the two-sphere case,

$$dz = \frac{xdx + ydy}{z} = \frac{xdx + ydy}{\sqrt{R^2 + x^2 + y^2}} \Rightarrow ds^2_{induced} = dx^2 + dy^2 + \frac{(xdx + ydy)^2}{R^2 + x^2 + y^2} \equiv g_{\mu\nu} dx^\mu dx^\nu.$$

(38.13)

As we see, we have $\det g_{\mu\nu} > 0$, so, indeed, the signature of Lobachevsky space is Euclidean, despite it being embeddable only into Minkowski space.

Thus all two-dimensional surfaces of Euclidean signature can be embedded into three-dimensional space (as the counting of number of independent functions suggests), just that the signature is not definite: it can be either Euclidean or Minkowski, depending on the two-dimensional surface. The same is true in higher dimensions where, however, we even have more choices for signature than just two.

That means that the notion of describing spaces by their embedding into a higher dimensional flat one is not very useful. We must instead consider spaces as *intrinsically curved*, a conceptual leap that took some time until it was realized, more precisely until the early nineteenth century, when the notion of *non-Euclidean geometry* was introduced. Lobachevsky (based on the previously discussed example that bears his name) and Janos Bolyai published it independently, and the giant of mathematics Gauss is believed to have known about it for some time beforehand but never published. A bit later, Riemann defined the notion of (Riemannian) manifold as we understand it today.

Summarizing, in order to define a general theory of relativity, we need to consider intrinsically curved spaces defined by a general metric field $g_{\mu\nu}(x^\rho)$, under the action of the "gauge invariance" of general coordinate transformations. But what would that correspond to physically?

38.2 Einstein's Theory of General Relativity

Einstein thus started to define his general relativity theory from two physical assumptions, intended to answer this question.

(1) **Gravity Is Geometry**

That means that matter follows *geodesics* in a curved spacetime – i.e., the lines of shortest distance in the spacetimes. But to us, this appears as the effect of gravity on the matter. A pictorial way to understand this is to consider a planar rubber sheet, on which we put a massive object, making it curve the sheet. Then, when we throw a small test ball towards the big one, it gets deflected by following the geodesic on the curved "dip" around the big ball, as in Figure 38.2. Similarly, we can think of putting a golf ball towards the hole and just missing it: it also gets deflected for the same reason.

The second assumption is, in a sense, the opposite of the first.

(2) **Matter Sources Gravity**

That means that matter generates the gravitational field, which by the first assumption is just the geometry, the curvature of spacetime. In the picture with the rubber sheet, we understand that the big ball curves it, so this assumption is also pictorially described by it. Though, of course, we have to remember that this is just a useful image, since

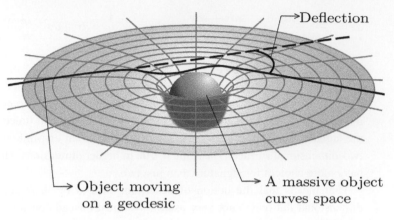

Figure 38.2   Matter curves space (a massive object creates a curvature of spacetime) and then matter (light objects) move on a geodesic, being deflected and creating the effect of gravity.

in reality the result of the curvature is the gravitational field of the Earth acting on the masses we put on the sheet.

The two physical assumptions can be translated into two physical principles with a mathematical formulation, defining the kinematics, and one equation for the dynamics (Einstein's equation), which we will describe in the next chapter.

(A) **Physics Is Invariant (or, More Generally, Covariant) under General Coordinate Transformations**

This principle is the exact generalization of the Lorentz invariance principle of special relativity, with which we started the chapter. For a general transformation of coordinates, $x'^{\mu} = x'^{\mu}(x^{\nu})$, the invariance of the line element implies

$$ds^2 = g_{\mu\nu}(x)dx^{\mu}dx^{\nu} = ds'^2 = g'_{\mu\nu}(x)dx'^{\mu}dx'^{\nu}, \qquad (38.14)$$

which gives the transformation law for the field $g_{\mu\nu}(x)$,

$$g'_{\mu\nu}(x') = g_{\rho\sigma}(x)\frac{\partial x^{\rho}}{\partial x'^{\mu}}\frac{\partial x^{\sigma}}{\partial x'^{\nu}}. \qquad (38.15)$$

As we already described $g_{\mu\nu}(x)$ is viewed as a field in spacetime, acted upon by the gauge invariance of general coordinate transformations. Thus, invariance under $x'^{\mu} = x'^{\mu}(x^{\nu})$ is realized as a gauge invariance of the theory we need to construct.

(B) **The Equivalence Principle**

This principle starts with a startling observation about Newtonian theory, whose importance lay unappreciated until Einstein's genius exposed it. When formulating physics, Newton gave a force law, $\vec{F} = m\vec{a}$ ($\vec{a}$ is acceleration), and a law for gravity, $\vec{F}_G = m\vec{g}$ ($\vec{g}$ is the local gravitational field). But it was not realized that there is no a priori need to have the same constant m in both relations, since they are physically distinct. The force law describes inertia, so the proportionality constant is an *inertial mass* m_i, whereas in the gravitational law we have a *gravitational mass* m_g,

$$\vec{F} = m_i\vec{a}, \quad \vec{F}_G = m_g\vec{g}. \tag{38.16}$$

Then their equality is actually an important physical principle, the *equivalence principle*,

$$m_i = m_g, \tag{38.17}$$

which must be added (it doesn't follow logically from the previous principle). If it is true, it means that *there is no difference between gravity and acceleration* (the two are completely equivalent). This important concept was realized by Einstein, and is the first formulation of the equivalence principle.

Another formulation was obtained by Einstein making a "thought experiment" (Gedanken experiment in German, the tool Einstein developed to analyze theories). Consider a person inside an elevator that is freely falling in the Earth's gravitational field from a very large height. Then, if $m_i = m_g$, the person inside cannot distinguish this situation from the situation of him being weightless (in the absence of gravity), by any *local* experiment. Therefore, the second formulation of the equivalence principle is that *we cannot distinguish the free-falling elevator from a weightless one by any local experiments*.

Of course, if we consider a very large elevator, so that we can perform somewhat non-local experiments, we can discover that we are in a gravitational field of an object: gravity in slightly different points is slightly different in direction and magnitude (it points radially to the center of the Earth and drops as $1/r^2$ from the center). We say that we have *tidal forces* that tend to break apart or crush the person inside the elevator. Also, if we wait for a long enough time, we will eventually crash the elevator on the surface of the Earth, so locality refers also to time, not just space.

The infinitesimal form of the general coordinate transformation = gauge transformation in (38.15) is obtained by substituting $x^\mu = x'^\mu + \xi^\mu$, with ξ^μ small, into it, and obtaining

$$g'_{\mu\nu}(x^\lambda - \xi^\lambda) = (\delta^\rho_\mu + \partial_\mu\xi^\rho)(\delta^\sigma_\nu + \partial_\nu\xi^\sigma)g_{\rho\sigma}(x). \tag{38.18}$$

But, using Taylor expansion around x^λ, we obtain also

$$g'_{\mu\nu}(x^\lambda - \xi^\lambda) = g'_{\mu\nu}(x^\lambda) - \partial_\lambda g'_{\mu\nu}\xi^\lambda. \tag{38.19}$$

Combining the two equations and dropping the terms quadratic in the small variation ξ^μ, we obtain

$$\delta g_{\mu\nu}(x) \equiv g'_{\mu\nu}(x) - g_{\mu\nu}(x) \simeq \xi^\lambda\partial_\lambda g'_{\mu\nu}(x) + (\partial_\mu\xi^\rho)g_{\rho\nu} + (\partial_\nu\xi^\sigma)g_{\mu\sigma}$$
$$\simeq \xi^\lambda\partial_\lambda g_{\mu\nu}(x) + (\partial_\mu\xi^\rho)g_{\rho\nu} + (\partial_\nu\xi^\rho)g_{\mu\rho}. \tag{38.20}$$

The first term is a translation, coming from the Taylor expansion (we have replaced $g'_{\mu\nu}$ by $g_{\mu\nu}$ since their difference is higher order in the small parameter ξ^μ). The next two terms are of the type of a generalized gauge transformation, $\delta A_\mu = \partial_\mu\alpha$, just appropriate to the action on a tensor instead of a vector, somewhat similar to the action of a gauge invariance on the $p-$form, which we already described.

In the global case, $\partial_\lambda \xi^\mu = 0$, we would have only the translation, so we can describe general coordinate transformations as a local version of translations. The theory we are describing is, therefore, a kind of a *gauge theory of translations*.

38.3 Kinematics

We now need to describe the kinematics of general relativity. We start with a simple generalization of the kinematics of special relativity. We define a general relativity generalization of tensors.

As before, a *contravariant tensor* transforms like dx^μ – namely,

$$A'^\mu = \frac{\partial x'^\mu}{\partial x^\nu} A^\nu , \quad dx'^\mu = \frac{\partial x'^\mu}{\partial x^\nu} dx^\nu . \tag{38.21}$$

A *covariant tensor* transforms like ∂_μ acting on a scalar ϕ – namely,

$$B'_\mu = \frac{\partial x^\nu}{\partial x'^\mu} B_\nu, \quad \partial'_\mu \phi' = \frac{\partial x^\nu}{\partial x'^\mu} \partial_\nu \phi. \tag{38.22}$$

A general tensor, with mixed indices, transforms by the products of individual factors – for instance,

$$T'^\mu{}_\nu(x') = \frac{\partial x'^\mu}{\partial x^\rho} \frac{\partial x^\sigma}{\partial x'^\nu} T^\rho{}_\sigma(x), \tag{38.23}$$

with an obvious generalization to $T^{\mu_1 \ldots \mu_p}_{\nu_1 \ldots \nu_q}$.

We will not describe how to deal with spinors, since it is more complicated, and it would take too long time at this point. We will do that in the last chapter of the book in order to deal with a more complicated example of gravitational solution. For now, we will just say that we need to consider a new formulation of general relativity, not in terms of the metric $g_{\mu\nu}$, but other objects called the *spin connection and the vielbein*.

Now we can go back to $g_{\mu\nu}$ and say that it must be a general relativity symmetric tensor (with two covariant indices), since $ds^2 = g_{\mu\nu} dx^\mu dx^\nu$ must be invariant, so $g_{\mu\nu}$ must transform as $\partial_\mu \partial_\nu$.

For any spacetime (any manifold), in a small neighborhood of a point x, the space looks approximately flat, which means that we can always make a coordinate transformation that fixes the metric and its first derivative to the flat metric – i.e.,

$$g_{\mu\nu}(x) = \eta_{\mu\nu} + \mathcal{O}(\delta x^2). \tag{38.24}$$

More precisely, we define the *tangent space* at each point x and regular coordinates in it; on it, we have $SO(3,1)$ (Lorentz) invariance, but the tangent space and its Lorentz invariance is *local* – i.e., it depends on x^ρ.

Thus, we can define a local Lorentz invariance on the spacetime manifold, and a related concept is the gauge invariance of the general coordinate transformations. But then it follows that, like in any gauge theory, the normal derivative ∂_μ acting on a tensor is not a tensor. We must construct a *covariant derivative* that transforms covariantly under the gauge transformations (general coordinate transformations, in this case). For the covariant derivative acting on a field in the fundamental, ϕ^a, of a $SO(m, n)$ group like the Lorentz group $SO(3, 1)$, the gauge field is in the adjoint, (ab) (antisymmetric product), so we have

$$D_\mu \phi^a = \partial_\mu \phi^a + (A^a{}_b)_\mu \phi^b. \tag{38.25}$$

To construct something similar in our case, for our local invariance, we note that coordinate indices μ and gauge indices a are the same, so we can define

$$D_\mu T^\nu = \partial_\mu T^\nu + (\Gamma^\nu{}_\sigma)_\mu T^\sigma, \tag{38.26}$$

where we put the brackets just to emphasize the similarity with the $SO(m, n)$ gauge field case. In fact, in the this way, we defined $\Gamma^\mu{}_{\nu\rho}$, called the *Christoffel symbol*. By its definition, it is a kind of "gauge field of gravity."

We can define also a covariant derivative on a general tensor – for instance,

$$D_\mu T^\rho_\nu = \partial_\mu T^\rho_\nu + \Gamma^\rho{}_{\sigma\mu} T^\sigma_\nu - \Gamma^\sigma{}_{\mu\nu} T^\rho_\sigma, \tag{38.27}$$

with the obvious generalization for an arbitrary tensor.

The Christoffel symbol being like a gauge field, it must be possible to put it *locally* to zero by a gauge transformation – i.e., by a general coordinate transformation. On the other hand, we can always put $g_{\mu\nu}$ to $\eta_{\mu\nu}$ by a gauge transformation, as we already explained. If the two were different coordinate transformations, it would mean that there is another special space at each point besides the tangent space, and that is clearly not possible. That means that we can put at the same time $\Gamma^\mu{}_{\nu\rho}$ to 0, and $g_{\mu\nu}$ to $\eta_{\mu\nu}$, which means that in this system of coordinates

$$D_\mu g_{\nu\rho} = \partial_\mu g_{\nu\rho} - \Gamma^\sigma{}_{\nu\mu} g_{\sigma\rho} - \Gamma^\sigma{}_{\rho\mu} g_{\sigma\nu} = 0. \tag{38.28}$$

But $g_{\mu\nu}$ is a tensor, as we said, and then so must be $D_\mu g_{\nu\rho}$, and a tensor transforms multiplicatively, so if it is zero in one coordinate system, it is zero always. The solution of $D_\mu g_{\nu\rho} = 0$ is

$$\Gamma^\mu{}_{\nu\rho} = \frac{1}{2} g^{\mu\sigma} (\partial_\rho g_{\nu\sigma} + \partial_\nu g_{\sigma\rho} - \partial_\sigma g_{\nu\rho}), \tag{38.29}$$

a fact that we leave as an exercise to prove. The solution is symmetric in $(\nu\rho)$, as we can see, and in order to find it, we defined the *inverse metric* $g^{\mu\nu}$, which is just the matrix inverse of the metric – i.e.,

$$g_{\mu\rho} g^{\rho\sigma} = \delta^\sigma_\mu. \tag{38.30}$$

38.4 Motion of Free Particles

We next define the motion of the particles in a curved spacetime, which is known as a "geodesic motion," or motion that minimizes the total length of the path, $\int ds$. The fact that motion of free particles in curved spacetime follows geodesics is a simple generalization of the special relativity case, where the Lagrangean was (see Chapter 4)

$$S = -mc^2 \int ds = -mc^2 \int d\tau \sqrt{-\frac{dx^\mu}{d\tau}\frac{dx^\nu}{d\tau}\eta_{\mu\nu}}, \qquad (38.31)$$

the two forms being equivalent only "on-shell" – i.e., when satisfying the equation of motion, $x^\mu = x^\mu(\tau)$, since $-d\tau^2 \equiv ds^2 = g_{\mu\nu}dx^\mu dx^\nu$.

The general relativity generalization is obtained by the replacement of the Minkowski metric $\eta_{\mu\nu}$ with the general metric $g_{\mu\nu}$, leading to the action

$$S = -mc^2 \int ds = -mc^2 \int d\tau \sqrt{-\frac{dx^\mu}{d\tau}\frac{dx^\nu}{d\tau}g_{\mu\nu}(x)}. \qquad (38.32)$$

We find the equation of motion as in the special relativity case, by varying the action with respect to $x^\mu(\tau)$, to obtain

$$\delta S = -mc^2 \int d\tau \left[-\frac{2\dot{x}^\mu g_{\mu\nu}(x)\delta\dot{x}^\nu + \dot{x}^\mu\dot{x}^\nu\partial_\rho g_{\mu\nu}\delta x^\rho}{2\sqrt{-\dot{x}^\mu\dot{x}^\nu g_{\mu\nu}}} \right]$$

$$= -mc^2 \int d\tau \delta x^\mu \left\{ \frac{d}{d\tau}\left[\frac{g_{\mu\nu}\dot{x}^\nu}{\sqrt{-\dot{x}^\mu\dot{x}^\nu g_{\mu\nu}}} \right] - \frac{\dot{x}^\rho\dot{x}^\sigma\partial_\mu g_{\rho\sigma}}{2\sqrt{-\dot{x}^\mu\dot{x}^\nu g_{\mu\nu}}} \right\} = 0, \quad (38.33)$$

where we have denoted $\frac{d}{d\tau}x^\mu$ by $\dot{x}^\mu$ and in the second equality we have used a partial integration for $\delta\dot{x}^\nu$.

Calculating the conjugate momentum to $\dot{x}^\mu$, which is just normal momentum,

$$p^\mu = \frac{\delta S}{\delta\dot{x}^\mu} = mc^2 \frac{g_{\mu\nu}\dot{x}^\nu}{\sqrt{-\dot{x}^\mu\dot{x}^\nu g_{\mu\nu}}}, \qquad (38.34)$$

denoting as usual $\dot{x}^\mu = \frac{dx^\mu}{d\tau} = u^\mu$, using the on-shell relation $-\dot{x}^\mu\dot{x}^\nu g_{\mu\nu} = 1$, and finding the equations of motion by putting to zero the coefficient of the arbitrary δx^μ, we find the equation

$$-\frac{d}{d\tau}p_\mu + \frac{mc^2}{2}u^\rho u^\sigma\partial_\mu g_{\rho\sigma} = 0. \qquad (38.35)$$

Using the fact that $p_\mu = g_{\mu\nu}p^\nu$, dividing by mc^2 to replace p_μ by u_μ, we find

$$g_{\mu\nu}\frac{d}{d\tau}u^\nu = -\left(\frac{d}{d\tau}g_{\mu\nu}\right)u^\nu + \frac{1}{2}u^\rho u^\sigma\partial_\mu g_{\rho\sigma}. \qquad (38.36)$$

Using the chain rule to find $\frac{d}{d\tau}g_{\mu\nu} = \partial_\rho g_{\mu\nu}\dot{x}^\rho = \partial_\rho g_{\mu\nu}u^\rho$, then using the symmetry of $u^\rho u^\sigma$ under exchange of $(\rho\sigma)$, we find

$$g_{\mu\nu}\frac{d}{d\tau}u^\nu = -\frac{1}{2}(\partial_\rho g_{\mu\sigma} + \partial_\sigma g_{\mu\rho} - \partial_\mu g_{\rho\sigma})u^\rho u^\sigma \Rightarrow$$

$$\frac{d}{d\tau}u^\nu + \Gamma^\nu{}_{\rho\sigma}u^\rho u^\sigma = 0, \qquad (38.37)$$

which we define in an obvious way to be

$$\frac{Du^\nu}{d\tau} = 0. \qquad (38.38)$$

This is the geodesic equation for motion of a free particle in the gravitational field of curved spacetime.

- General relativity is the statement that physics is invariant under general coordinate transformations, $x'^\mu = x'^\mu(x)$.
- When trying to embed curved spaces into flat spaces of higher dimensionality, we need to consider both at least $d(d-1)/2$ dimensions and variable signature of the embedding space, which makes it not very useful.
- We need instead to consider spaces as intrinsically curved and defined by a general metric $g_{\mu\nu}(x)$, with line element $ds^2 = g_{\mu\nu}(x^\rho)dx^\mu dx^\nu$.
- We can think of the metric $g_{\mu\nu}(x)$ as a field, subject to a gauge invariance given by the general coordinate transformation.
- The infinitesimal general coordinate transformation is $\delta g_{\mu\nu} = \xi^\lambda\partial_\lambda g_{\mu\nu} + (\partial_\mu\xi^\lambda)g_{\lambda\nu} + (\partial_\nu\xi^\lambda)g_{\mu\lambda}$, which is a local version of translations – i.e., we have a gauge theory of translations.
- General relativity starts with the assumptions that gravity is geometry and geometry is sourced by matter.
- One then derives two physical principles, one principle being that physics is invariant under general coordinate transformations, and the equivalence principle that acceleration is the same as gravity, $m_i = m_g$, or that we can't distinguish being in a free-falling elevator from being weightless.
- GR tensors are obvious generalizations of SR tensors.
- Around any point on the manifold, in a neighborhood or on a tangent space, we have local Lorentz invariance.
- Covariant derivatives of tensors are defined using the Christoffel symbol $\Gamma^\mu{}_{\nu\rho}$, a "gauge field of gravity" that is a solution of the equation $D_\mu g_{\nu\rho}=0$ and is symmetric in $\nu\rho$.
- Free particles move on geodesics defined by the action $S = -mc^2\int ds$, with equation of motion $Du^\mu/d\tau = 0$.

Further Reading

See chapter 10 in [3] and chapters 8 and 9 in [40].

Exercises

(1) Consider the metric $g_{\mu\nu}$ for a sphere and a free particle moving on it. *Using the geodesic equation*, find the (free) geodesic motion.
(2) Repeat the exercise for the motion on the two-dimensional Lobachevski space, defined by

$$x^2 + y^2 - z^2 = -R^2. \tag{38.39}$$

(3)   Prove that the solution of $D_\mu g_{\nu\rho} = 0$ is

$$\Gamma^\mu{}_{\nu\rho} = \frac{1}{2} g^{\mu\sigma} \left(\partial_\rho g_{\nu\sigma} + \partial_\nu g_{\rho\sigma} - \partial_\nu g_{\rho\sigma} \right). \tag{38.40}$$

(4)   Considering the form of the infinitesimal transformation of coordinates,

$$\delta g_{\mu\nu}(x) = \xi^\lambda \partial_\lambda g_{\mu\nu}(x) + (\partial_\mu \xi^\rho) g_{\rho\nu} + (\partial_\nu \xi^\rho) g_{\mu\rho}, \tag{38.41}$$

show that indeed one can always find a system of coordinates such as to put the Christoffel symbol, defined now as

$$\Gamma^\mu{}_{\nu\rho} = \frac{1}{2} g^{\mu\sigma} \left(\partial_\rho g_{\nu\sigma} + \partial_\nu g_{\rho\sigma} - \partial_\sigma g_{\nu\rho} \right), \tag{38.42}$$

to zero.

39   The Einstein Action and the Einstein Equation

In the last chapter, we described the basics of general relativity as a theory of general coordinate invariance on general curved spacetimes. From two physical assumptions (gravity is geometry, and matter sources gravity), we obtained two physical principles: the principle that physics is invariant under general coordinate transformations and the equivalence principle. We also said we will also obtain an equation for the dynamics. In this chapter, we will study this equation, Einstein's equation, and the action that it comes from, Einstein's action.

We described the kinematics of curved spacetime and started to see how it could be related to general relativity. We have seen that the metric $g_{\mu\nu}$, which defines the line element $ds^2 = g_{\mu\nu}dx^\mu dx^\nu$, is not a good measure of the curvature of space. In fact, it is subject to general coordinate transformations, acting like a gauge theory of translations (a local form of translations). Moreover, we have local $SO(3, 1)$ Lorentz transformations in the tangent space, that also form a gauge theory. In fact, gravity in three dimensions is a gauge theory of the $ISO(2, 1)$ group (Poincaré – i.e., Lorentz plus translations), but in four dimensions it is not quite. Nevertheless, we have constructed a Christoffel symbol $\Gamma^\mu{}_{\nu\rho}$ like a kind of gauge field for $SO(3, 1)$ (for $SO(p, q)$ groups, the gauge field is in the adjoint; thus, it is of the form A_μ^{ab}, where the antisymmetric (ab) indices form the adjoint representation if a, b are fundamental indices), defining covariant derivatives that satisfy $D_\mu g_{\nu\rho} = 0$. But the solution $\Gamma^\mu{}_{\nu\rho}$ is symmetric in ν, ρ and identifies gauge and vector indices.

39.1   Riemann Tensor and Curvature

We must now construct a gauge-invariant action with a gauge covariant equation of motion, which means that as a first step, we must construct a gauge covariant field strength for the "gauge field" $\Gamma^\mu{}_{\nu\rho}$. For a group $SO(n)$, or $SO(p, q)$, the structure constants in $(A = (ab))$

$$F_{\mu\nu}^A = \partial_\mu A_\nu^A - \partial_\nu A_\mu^A + f^A{}_{BC}A_\mu^B A_\nu^C \tag{39.1}$$

are such that the field strength can be written as

$$F_{\mu\nu}^{ab} = \partial_\mu A_\nu^{ab} - \partial_\nu A_\mu^{ab} + A_\mu^{ac} A_\nu^{cb} - A_\nu^{ac} A_\mu^{cb}. \tag{39.2}$$

We leave the proof of this statement as an exercise.

In our case, then, we define the field strength of $(\Gamma^\mu{}_\nu)_\rho$ as

$$(R^\mu{}_\nu)_{\rho\sigma} = \partial_\rho(\Gamma^\mu{}_\nu)_\sigma - \partial_\sigma(\Gamma^\mu{}_\nu)_\rho + (\Gamma^\mu{}_\lambda)_\rho(\Gamma^\lambda{}_\nu)_\sigma - (\Gamma^\mu{}_\lambda)_\sigma(\Gamma^\lambda{}_\nu)_\rho. \tag{39.3}$$

This object is called the *Riemann tensor*, after the great mathematician Riemann who defined the modern concept of (Riemannian) manifolds as we understand them today.

Properties of the Riemann Tensor

From the explicit formula for the Christoffel symbol in terms of the metric, we saw that they are symmetric, $\Gamma^{\mu}{}_{\nu\rho} = \Gamma^{\mu}{}_{\rho\nu}$.

Similarly, from the explicit formula for the Riemann tensor as a field strength of $\Gamma^{\mu}{}_{\nu\rho}$, understood as a gauge field in the adjoint of $SO(1,3)$ – i.e., as an $(F^a{}_b)_{\mu\nu}$ – antisymmetric in both a, b and μ, ν, we have

$$(a)\ R_{\mu\nu\rho\sigma} = -R_{\nu\mu\rho\sigma}$$

$$(b)\ R_{\mu\nu\rho\sigma} = -R_{\mu\nu\sigma\rho}, \tag{39.4}$$

where, as expected, we have lowered the index with the metric, $R_{\mu\nu\rho\sigma} \equiv g_{\mu\lambda}R^{\lambda}{}_{\nu\rho\sigma}$.

On the other hand, we also have two symmetry properties that are not manifest,

$$(c)\ R_{\mu\nu\rho\sigma} = R_{\rho\sigma\mu\nu}$$

$$(d)\ R_{\mu[\nu\rho\sigma]} = 0. \tag{39.5}$$

We know that for a gauge theory, $[D_\mu, D_\nu] = F^a{}_{\mu\nu}T_a$. In the case of general relativity, this formula translates into an action on a covariant vector T_ρ of

$$[D_\mu, D_\nu]T_\rho = -(R^\sigma{}_\rho)_{\mu\nu}T_\sigma = (R_{\rho\sigma})_{\mu\nu}T^\sigma. \tag{39.6}$$

One can check this explicitly, using the formulas for the covariant derivative and the Riemann tensor.

Also from the gauge theory analogy, since for a YM gauge field we have the Bianchi identity

$$(D_{[\mu}F_{\nu\rho]})^a = 0, \tag{39.7}$$

we can translate that into general relativity in the obvious way, obtaining

$$D_{[\lambda}(R^\mu{}_\nu)_{\rho\sigma]} = 0 \Rightarrow D_\lambda R^\mu{}_{\nu\rho\sigma} + D_\sigma R^\mu{}_{\nu\lambda\rho} + D_\rho R^\mu{}_{\nu\sigma\lambda} = 0. \tag{39.8}$$

From the Riemann tensor, because of the identification of gauge with vector indices, we can construct contractions that are nontrivial. The first is the *Ricci tensor*, defined as

$$R_{\mu\nu} \equiv R^\lambda{}_{\mu\lambda\nu} = R^\lambda{}_{\mu\sigma\nu}\delta^\sigma_\lambda. \tag{39.9}$$

We can check that the Ricci tensor is symmetric, from its explicit formula.

We can further define the *Ricci scalar*,

$$\mathcal{R} \equiv R_{\mu\nu}g^{\mu\nu}. \tag{39.10}$$

This object is a scalar – i.e., it is invariant under general coordinate transformations, $R'(x') = R(x)$ – which means that it is a good *invariant* measure of the curvature of space.

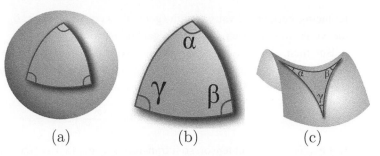

$$(a) \qquad\qquad (b) \qquad\qquad (c)$$

Figure 39.1 (a) A special triangle on a sphere, made from two meridian lines and a segment of the equator, has two angles of $90°$ ($\pi/2$), so the sum of the three angles is larger than $180°$ (π). (b) A similar triangle is drawn for a general curved space of positive curvature, emphasizing that the sum of the angles of the triangle exceeds $180°$ (π). (c) In a space of negative curvature, the sum of the angles of the triangle is below $180°$ (π).

Positive and Negative Curvature

1. In the case of *positive curvature*, $\mathcal{R} > 0$, the geometry of the manifold is such that the sum of the angles of a triangle is greater than π, $\alpha + \beta + \gamma > \pi$, as in Figure 39.1b. For instance, in the case of the two-sphere, by choosing a point at the North Pole and two points on the equator, we form a triangle from two meridian lines and an equator, and we have two $\pi/2$ angles, $\alpha = \beta = \pi/2$, so $\alpha + \beta + \gamma > \pi$ (see Figure 39.1a).

   Also, in this case, we break the Euclid postulate of Euclidean geometry that says that "two parallel lines, defined as lines perpendicular to the same line, never intersect, and remain at the same distance." In fact, as seen in the example above with the two-sphere, the meridian lines are parallel, yet they intersect at the North Pole. In general, parallel lines converge.

   The two-sphere S^2 is the two-dimensional space with Euclidean signature of constant positive curvature $\mathcal{R} > 0$. It is also a space of maximal symmetry, $SO(3)$. The curvature is (by dimensional analysis) $\mathcal{R} \propto 1/R^2$, where R is the radius of the sphere.

2. In the case of *negative curvature*, $\mathcal{R} < 0$, the geometry is such that the sum of the angles of a triangle is less than π, $\alpha + \beta + \gamma < \pi$, as in Figure 39.1c.

   In this case, we break the Euclid postulate, but in the other way: parallel lines still never intersect, but they *diverge*, so the distance between them increases.

   The basic example of a space with $\mathcal{R} < 0$ is two-dimensional Lobachevsky space, which is the two-dimensional space with Euclidean signature of constant negative curvature. It is also a space of maximal symmetry, $SO(2,1)$.

39.2 Turning Special Relativity into General Relativity and Einstein–Hilbert Action

In order to turn a specially relativistic theory into a generally relativistic theory, we need to first turn special relativity tensors into general relativity tensors. In particular, that means replacing normal derivatives with (generally) covariant ones, $\partial_\mu \to D_\mu$. Then we replace the Minkowski metric $\eta_{\mu\nu}$ with the general metric $g_{\mu\nu}$.

But then, in order to write actions, we need to redefine the integration measure as well, since $d^d x$ is not invariant under general coordinate transformations. Indeed, since $dx^\mu = (\partial x^\mu/\partial x'^\nu)dx'^\nu$, taking the determinant, we find

$$d^d x = \det\left(\frac{\partial x^\mu}{\partial x'^\nu}\right)d^d x'. \tag{39.11}$$

But $g_{\mu\nu}$ is a covariant tensor, so it transforms as $g'_{\mu\nu}(x') = (\partial x^\rho/\partial x'^\mu)(\partial x^\sigma/\partial x'^\nu)g_{\rho\sigma}$, and taking the determinant, we have ($g = \det g_{\mu\nu}$)

$$\det g'_{\mu\nu} \equiv g' = \left[\det\left(\frac{\partial x^\mu}{\partial x'^\nu}\right)\right]^2 g. \tag{39.12}$$

That means that an invariant integration measure is $d^d x\sqrt{-g}$, since

$$\sqrt{-g'}d^d x' = \sqrt{-g}d^d x. \tag{39.13}$$

We have put a minus sign inside the square root, since Minkowski signature is invariantly defined as $g \equiv \det g_{\mu\nu} < 0$.

Einstein–Hilbert Action

To write down an action, we first think of the action for the particle, which as we have seen is $-mc\int ds$, which is the integral of the invariant measure on the wordline of the particle. So a simple guess would be just the measure on the spacetime worldvolume,

$$S \sim \int \sqrt{-g}d^d x. \tag{39.14}$$

It just happens that it doesn't reproduce Newtonian gravity, as we will shortly see. Logically, we start considering the next simplest possibility, which is also the next one in terms of dimension, the Ricci scalar $\mathcal{R}$. Since Γ has one derivative, $[\Gamma] = 1$ (has mass dimension 1), and $\mathcal{R} \sim \partial\Gamma + \Gamma\Gamma$, so has mass dimension 2, $[\mathcal{R}] = 2$. Then we try

$$S = \frac{1}{16\pi G_N}\int d^d x\sqrt{-g}\mathcal{R}. \tag{39.15}$$

As we will see, this actually works and gives the correct Newtonian gravity in the weak gravity limit. The action was found by Einstein and also, independently, by Hilbert, so it is known as the *Einstein–Hilbert action*. The coefficient in front is chosen such that when we add matter, we reproduce the Newtonian gravity result. In the theorist's conventions, with $\hbar = c = 1$, when there is a single dimension (mass =1/length = energy = ...), the coefficient can be rewritten as

$$\frac{1}{16\pi G_N} = \frac{M_{\mathrm{Pl}}^{d-2}}{2}, \tag{39.16}$$

since it must have mass dimension $d - 2$ (the measure $d^d x\sqrt{-g}$ has dimension $-d$ and $\mathcal{R}$ has dimension 2, such that the action is dimensionless). Here M_{Pl} is called the *Planck mass*. In $d = 4$, we obtain the coefficient $M_{\mathrm{Pl}}^2/2$.

If we would use the mostly minus metric convention (for signature $(+---)$, we would have a minus sign in front of the action. This is so, since $\Gamma^\mu_{\ \nu\rho} \sim g^{-1}\partial g$ is invariant under scale transformations – i.e., rescalings of the metric $g_{\mu\nu} \to \lambda g_{\mu\nu}$ – and then so is $R^\mu_{\ \nu\rho\sigma} \sim \partial\Gamma + \Gamma\Gamma$, and its contraction, the Ricci tensor $R_{\mu\nu}$. In particular, for multiplication by minus of the metric, taking us from the mostly plus to the mostly minus one, $R_{\mu\nu}$ is invariant. But then $\mathcal{R} = R_{\mu\nu}g^{\mu\nu}$ changes by a minus sign.

We should note now that, as we will see, the Einstein–Hilbert action is found to work very well, which is why the resulting Einstein equation is so famous. But strictly speaking, there is nothing sacred about the Einstein–Hilbert action; it just happens to reproduce experiments well. On the contrary, general relativity is a powerful concept, at which we arrived logically, but it doesn't specify a single action. In fact, we could imagine adding terms of increasing dimension to it – for instance (at dimension 4), the invariants $\mathcal{R}^2, R^2_{\mu\nu}, R^2_{\mu\nu\rho\sigma}$. These terms actually do appear in specific combinations in a quantum theory of gravity, namely the only theory consistent at the perturbative level, string theory.

The reason that the action works so well is the fact that M_{Pl} is so large, about $10^{19} GeV$. Then, just by dimensional analysis, the $\mathcal{R}^2$ corrections come with powers of M_{Pl}, so that the dimensionless expansion parameter is $\mathcal{R}/M_{Pl}$. For a maximum curvature that we can observe, with a radius of curvature about, say, 10 km (for a black hole), we have still just $10\ km \sim [10^{-10}eV]^{-1}$, so the dimensionless ratio is still just 10^{-29}, absurdly small. But from a purely theoretical perspective, the Einstein–Hilbert action is not a necessary result of general relativity, an $\mathcal{R}^2$ action would work as well, until tested against experiment.

39.3 Einstein's Equations

We now derive the field equations of the gravitational action, Einstein's equations.

First we remember an argument we already made when discussing the Belinfante tensor. Since for an arbitrary matrix M, we have $\delta(M \cdot M^{-1}) = \delta(\mathbb{1}) = 0$, we obtain

$$\delta M \cdot M^{-1} + M \cdot \delta M^{-1} = 0. \tag{39.17}$$

Since, moreover, we have

$$\det M = e^{\text{Tr} \ln M}, \tag{39.18}$$

varying it, we get

$$\delta \det M = \delta e^{\text{Tr} \ln M} = e^{\text{Tr} \ln M} \text{Tr}(\delta M \cdot M^{-1}) = -\det M \, \text{Tr}(\delta M^{-1} \cdot M). \tag{39.19}$$

Applying this to the metric $g_{\mu\nu}$, we find

$$\frac{\delta g}{g} = -g_{\mu\nu}\delta g^{\mu\nu} \Rightarrow \frac{\delta\sqrt{-g}}{\sqrt{-g}} = -\frac{1}{2}g_{\mu\nu}\delta g^{\mu\nu}. \tag{39.20}$$

Now we can answer the question from before: why does the action $\sim \int d^dx\sqrt{-g}$ not work as an action for gravity? Because its variation is $\int d^dx\sqrt{-g}\delta g^{\mu\nu}(-g_{\mu\nu}/2)$, so its equation of motion would be the too-trivial equation $g_{\mu\nu} = 0$.

Now, we can consider also the variation of the Ricci scalar,

$$\delta\mathcal{R} = \delta(R_{\mu\nu}g^{\mu\nu}) = R_{\mu\nu}\delta g^{\mu\nu} + g^{\mu\nu}\delta R_{\mu\nu}. \qquad (39.21)$$

But the variation of $R_{\mu\nu}$ results in a total derivative term (a boundary term), as we now prove.

To do so, we use a trick that is used in many occasions: In order to calculate some invariant or covariant object, we go to a frame where the metric is locally flat, calculate, and put the result into a form that can be made covariant, and thus re-extended to be correct in any frame.

In the locally flat frame, the Christoffel symbol, which contains only the metric and its first derivative, is zero; indeed, we saw that $g_{\mu\nu} = \eta_{\mu\nu} + \mathcal{O}(\delta x^2)$. But then its first derivative, $\partial\Gamma$, is nonzero. Therefore, in the locally flat frame, the variation of the Ricci tensor is

$$\delta R_{\mu\nu} = \delta(\partial_\rho\Gamma^\rho{}_{\mu\nu}) - \delta(\partial_\nu\Gamma^\rho{}_{\mu\rho}). \qquad (39.22)$$

More generally, $\delta\Gamma$ is a nonzero tensor (the derivative is just the statement that we divide it by δx). In fact, it is easy to see how to turn the variation of the Christoffel tensor into a tensor, namely

$$\delta\Gamma^\mu{}_{\nu\rho} = \frac{1}{2}g^{\mu\lambda}(D_\rho\delta g_{\lambda\nu} + D_\nu\delta g_{\lambda\rho} - D_\lambda\delta g_{\nu\rho}). \qquad (39.23)$$

We can covariantize $\delta R_{\mu\nu}$ for free in the locally flat frame (where Γ is zero), obtaining

$$\delta R_{\mu\nu} = D_\rho\delta\Gamma^\rho{}_{\mu\nu} - D_\nu\delta\Gamma^\rho{}_{\mu\rho}. \qquad (39.24)$$

But since this is a covariant equation, it should be valid in any frame.

Since we have $D_\rho g_{\mu\nu} = 0$, we obtain

$$g^{\mu\nu}\delta R_{\mu\nu} = D_\rho(g^{\mu\nu}\delta\Gamma^\rho{}_{\mu\nu}) - D_\nu(g^{\mu\nu}\delta\Gamma^\rho{}_{\mu\rho})$$
$$= D_\mu(g^{\nu\rho}\delta\Gamma^\mu{}_{\nu\rho} - g^{\mu\nu}\delta\Gamma^\rho{}_{\nu\rho}) \equiv D_\mu U^\mu. \qquad (39.25)$$

But a total covariant derivative integrates to a boundary term, since

$$D_\mu U^\mu = \partial_\mu U^\mu + \Gamma^\mu{}_{\sigma\mu}U^\sigma \qquad (39.26)$$

and the Christoffel symbol is

$$\Gamma^\mu{}_{\sigma\mu} = \frac{1}{2}g^{\mu\lambda}(\partial_\mu g_{\lambda\sigma} + \partial_\sigma g_{\lambda\mu} - \partial_\lambda g_{\sigma\mu}) = \frac{1}{2}g^{\mu\lambda}\partial_\sigma g_{\lambda\mu} = \frac{\partial_\sigma\sqrt{-g}}{\sqrt{-g}}, \qquad (39.27)$$

which means that the Stokes's theorem becomes

$$\int_M d^dx\sqrt{-g}D_\mu U^\mu = \int_M d^dx\partial_\mu(\sqrt{-g}U^\mu) = \oint_{B=\partial M}\sqrt{-g}U^\mu d\Sigma_\mu. \qquad (39.28)$$

That means that for the purposes of the bulk equations of motion, we can drop it. It will only contribute to the boundary condition, which comes from the boundary term of the variation, but in that case, the gravitational action contains in any case another boundary term, called the Gibbons–Hawking boundary term. We will not describe it here, since we would need to introduce other concepts as well.

We still need to vary the $\sqrt{-g}$ measure of the action using (39.20). The (bulk) variation of the gravity action is then finally

$$\delta S_{\text{gravity}} = \frac{1}{16\pi G_N} \int d^d x \sqrt{-g}\, \delta g^{\mu\nu} \left[R_{\mu\nu} - \frac{1}{2} g_{\mu\nu} R \right]. \tag{39.29}$$

The condition for this to be equal to zero for any $\delta g^{\mu\nu}$ gives the bulk equation of motion for pure gravity,

$$R_{\mu\nu} - \frac{1}{2} g_{\mu\nu} R = 0. \tag{39.30}$$

This is the Einstein equation in vacuum.

We now can add matter in curved spacetime. We have already seen in Chapter 14 that the energy-momentum tensor can be defined in a curved spacetime, as the Belinfante tensor,

$$T_{\mu\nu} = -\frac{2}{\sqrt{-g}} \frac{\delta S_{\text{matter}}}{\delta g^{\mu\nu}}. \tag{39.31}$$

There we had defined the tensor around flat background, by putting at the end $g_{\mu\nu} = \eta_{\mu\nu}$, but here we don't need to do that.

Inverting the relation, we write the variation of the matter action as

$$\delta S_{\text{matter}} = -\frac{1}{2} \int d^d x \sqrt{-g}\, \delta g^{\mu\nu} T_{\mu\nu}. \tag{39.32}$$

Then the equations of motion of the combined gravity + matter sytem come from putting to zero the variation of the combined action,

$$\delta(S_{\text{gravity}} + S_{\text{matter}}) = \frac{1}{16\pi G_N} \int d^d x \sqrt{-g}\, \delta g^{\mu\nu} \left[R_{\mu\nu} - \frac{1}{2} g_{\mu\nu} R - 8\pi G_N T_{\mu\nu} \right] = 0, \tag{39.33}$$

which gives

$$R_{\mu\nu} - \frac{1}{2} g_{\mu\nu} R = 8\pi G_N T_{\mu\nu}. \tag{39.34}$$

This is *Einstein's equation with matter*. Here the tensor

$$R_{\mu\nu} - \frac{1}{2} g_{\mu\nu} R \equiv G_{\mu\nu} \tag{39.35}$$

is called the Einstein tensor.

39.4 Examples of Energy-Momentum Tensor

Example 1. Scalar Field

For the first example, we consider the massless free real scalar field with Minkowski space action

$$S_\phi^M = -\frac{1}{2} \int d^d x\, \partial_\mu \phi \partial_\nu \phi\, \eta^{\mu\nu}. \tag{39.36}$$

To put the action in curved space, we replace $\eta^{\mu\nu}$ with $g^{\mu\nu}$. We should also replace normal derivatives with covariant ones, just that *for a scalar* $D_\mu\phi = \partial_\mu\phi$ (there are no indices on which the Christoffel symbol can act). We finally need to use the invariant measure for integration for the action

$$S_\phi = -\frac{1}{2}\int d^d x\sqrt{-g}\partial_\mu\phi\partial_\nu\phi g^{\mu\nu}. \tag{39.37}$$

Then, using the Belinfante formula, we obtain

$$T^\phi_{\mu\nu} = \partial_\mu\phi\partial_\nu\phi - \frac{1}{2}g_{\mu\nu}(\partial_\rho\phi)^2. \tag{39.38}$$

Example 2. Electromagnetism

The next example is Maxwell's electromagnetism. Again, to write the Minkowski space action

$$S_A^M = -\frac{1}{4}\int d^d x F_{\mu\nu}F_{\rho\sigma}\eta^{\mu\rho}\eta^{\nu\sigma} \tag{39.39}$$

in a curved spacetime, we note that, since the Christoffel symbol $\Gamma^\sigma{}_{\mu\nu}$ is symmetric in μ, ν, we have

$$D_\mu A_\nu = \partial_\mu A_\nu - \Gamma^\sigma{}_{\mu\nu}A_\sigma \Rightarrow D_\mu A_\nu - D_\nu A_\mu = \partial_\mu A_\nu - \partial_\nu A_\mu. \tag{39.40}$$

That means that it suffices to replace $\eta^{\mu\nu}$ with $g^{\mu\nu}$ and use the correct invariant integration measure, namely

$$S_A = -\frac{1}{4}\int d^d x\sqrt{-g}F_{\mu\nu}F_{\rho\sigma}g^{\mu\rho}g^{\nu\sigma}, \tag{39.41}$$

but the field strength has the same formula as in flat space,

$$F_{\mu\nu}^A = \partial_\mu A_\nu^A - \partial_\nu A_\mu^A + g[A_\mu, A_\nu]^A. \tag{39.42}$$

Then the energy-momentum tensor, from the Belinfante formula, is

$$T_{\mu\nu} = -\frac{2}{\sqrt{-g}}\frac{\delta S}{\delta g^{\mu\nu}} = F_{\mu\rho}F_\nu{}^\rho - \frac{1}{4}g_{\mu\nu}F_{\rho\sigma}^2. \tag{39.43}$$

This restricts, for $g_{\mu\nu}$ becoming $\eta_{\mu\nu}$, to the same formula we obtained in Chapter 14.

Example 3. Perfect Fluid

We remember perfect fluids, which were defined in Chapters 7 and 19. A perfect fluid at rest (with no velocity) is characterized only by the energy density ρ and the pressure p. In the kinetic theory of gases, the pressure p is related to the average momentum flow through a surface.

We saw in Chapter 14 the interpretation of the components of the energy-momentum tensor. T_{00} is the energy density ρ. T_{ij}, the stress tensor, defines the spatial stresses of the system, or the flow of momentum component i through the surface j. In a rotationally invariant (perfect) fluid, there is only one quantity defining this, the pressure, so

$$T_{ij} = p\delta_{ij}. \tag{39.44}$$

All in all, the energy-momentum tensor for a perfect fluid at rest is

$$T_{\mu\nu} = \mathrm{diag}(\rho, p, p, p). \tag{39.45}$$

Later, we will be particularly interested in dust, or pressureless matter, with $p = 0$.

39.5 Interpretation of the Einstein Equation

We first notice that we can take the trace of the Einstein equation and find

$$\left(1 - \frac{d}{2}\right) R = 8\pi G_N T, \tag{39.46}$$

where $T \equiv T^{\mu}{}_{\mu}$. Replacing this relation back into the Einstein equation allows us to rewrite it as

$$R_{\mu\nu} = 8\pi G_N \left(T_{\mu\nu} - \frac{1}{d-2} T g_{\mu\nu} \right). \tag{39.47}$$

In both formulations, the original, (39.34), and the one in (39.47), the Einstein equation relates geometry (the Einstein tensor, as a function of the metric $g_{\mu\nu}$) on the left-hand side to matter (the energy-momentum tensor) on the right-hand side.

It would seem like this is the same as for any other field theory where we write an equation of motion for the field, where the field appears on the left-hand side of the equation and a fixed source (some "external" matter) appears on the right-hand side. For electromagnetism, the source is a matter current j_μ coupling to A_μ, and it is completely given as a function of spacetime, $j_\mu = j_\mu(x)$. The equation of motion then gives $A_\mu(x)$ as a function of the source $j_\mu(x)$.

But things are more complicated for gravity. We see that in all of the above examples, $T_{\mu\nu}$ depends on the metric $g_{\mu\nu}$ itself, which is now the variable supposed to be solved, appearing on the left-hand side of Einstein's equation.

That means that unlike any other classical field theory, the Einstein equation is solved by finding $g_{\mu\nu}$ and $T_{\mu\nu}$ at the same time; since we cannot fix $T_{\mu\nu}(x)$ from the beginning, we can only give an *ansatz* for $T_{\mu\nu}$ that includes the variable $g_{\mu\nu}$ to be solved.

That means that the Einstein equation is very difficult to solve. It is all the more difficult to solve since $R_{\mu\nu}$ is highly nonlinear in the metric (it involves several times the metric and its matrix inverse). On the contrary, in electromagnetism the equation of motion is linear in the variable A_μ, for instance. Even in Yang–Mills theory, the equation of motion has a nonlinear left-hand side, but the source (the right-hand side) can be completely specified.

Important Concepts to Remember

- The Riemann tensor $R^\mu{}_{\nu\rho\sigma}$ is like the field strength of the Christoffel symbol, the "gauge field of gravity," for an $SO(3,1)$ group.
- From it, we can define the symmetric Ricci tensor $R_{\mu\nu} = R^\sigma{}_{\mu\sigma\nu}$ and the Ricci scalar $R = R_{\mu\nu}g^{\mu\nu}$.
- Positive curvature $R > 0$ means that parallels intersect, and the space of constant positive curvature is the sphere.
- Negative curvature $R < 0$ means that parallels diverge, and the space of constant negative curvature is Lobachevsky space.
- To turn a special relativistic theory into a general relativistic one, we turn SR tensors to GR tensors, including ∂_μ into D_μ, and we use the invariant measure $\int d^d x \sqrt{-g}$.
- The Einstein–Hilbert action for gravity is $\frac{1}{16\pi G_N} \int d^x \sqrt{-g} R$, to which we add the matter action.
- The Einstein equation is $G_{\mu\nu} \equiv R_{\mu\nu} - \frac{1}{2}g_{\mu\nu}R = 8\pi G_N T_{\mu\nu}$.
- In curved space, the energy-momentum tensor is the Belinfante tensor, $T_{\mu\nu} = -\frac{2}{\sqrt{-g}}\frac{\delta S_{\text{matter}}}{\delta g^{\mu\nu}}$.
- For a perfect fluid, $T_{\mu\nu} = \text{diag}(\rho, p, p, p)$, and for dust $p = 0$.
- The solution for the Einstein equation gives $g_{\mu\nu}$ and $T_{\mu\nu}$ simultaneously, and we can only write an ansatz for the latter.

Further Reading

See chapter 11 in [3] and chapter 10 in [40].

Exercises

(1)   Prove that for an $SO(n)$ or $SO(p,q)$ group, f_{ABC} is such that $(A = (ab)$ is in the adjoint of the group, $a = 1, \ldots, n$ is in the fundamental of $SO(n)$),

$$F^{ab}_{\mu\nu} = \partial_\mu A^{ab}_\nu - \partial_\nu A^{ab}_\mu + A^{ac}_\mu A^{cb}_\nu - A^{ac}_\nu A^{cb}_\mu. \tag{39.48}$$

(2)   Check by direct substitution the symmetry properties

$$(a) \quad R_{\mu\nu\rho\sigma} = R_{\rho\sigma\mu\nu}$$
$$(b) \quad R_{\mu[\nu\rho\sigma]} = 0. \tag{39.49}$$

(3)   Check that for a "conformally flat" metric in d dimensions,

$$ds^2 = e^{2A(x)} dx^\mu dx^\nu \eta_{\mu\nu}, \tag{39.50}$$

the "Weyl tensor,"

$$C_{\mu\nu}{}^{\rho\sigma} = R_{\mu\nu}{}^{\rho\sigma} - 4S^{[\rho}_{[\mu}\delta^{\sigma]}_{\nu]}, \quad \text{where}$$

$$S_{\mu\nu} \equiv \frac{1}{2}\left(R_{\mu\nu} - \frac{R}{6}g_{\mu\nu}\right),$$ (39.51)

($S_{\mu\nu}$ is called the Schouten tensor) vanishes.

(4)   Consider adding to the Einstein–Hilbert Lagrangean a term (coming from quantum corrections)

$$\Delta\mathcal{L} = \frac{L^2}{16\pi G_N}R^2,$$ (39.52)

where R is the Ricci scalar. Calculate the resulting modification to Einstein's equation.

(5)   Consider a matter action in 2+1 dimensions for a $U(1)$ gauge field, composed of a Maxwell term and a Chern–Simons term. Calculate first the form it takes in curved space with metric $g_{\mu\nu}$, and then calculate the resulting (Belinfante) energy-momentum tensor.

40

Perturbative Gravity: Fierz-Pauli Action, de Donder Gauge and Other Gauges, Gravitational Waves

In the previous chapter, we defined the gravitational dynamics, in the form of the Einstein action and Einstein equation. We have also emphasized that it is a very nonlinear system – in general, hard to solve. That is why, in this chapter, we start the analysis of solutions to Einstein's equations with the simplest case we have, linearized gravity. We will find the action governing it corresponding to a spin 2 particle and found by Fierz and Pauli. Then we will study gauges – in particular, the de Donder gauge, which is the analog for gravity of the Lorenz gauge, giving just the KG equation. Finally, we will study gravitational waves – perturbatively examples and one nonperturbative example, the cylindrical wave of Einstein and Rosen.

40.1 Perturbative Gravity and Fierz-Pauli Action

In this chapter, we will study the four-dimensional case,

$$S_{\text{gravity}} = \frac{1}{16\pi G_N} \int d^4x \sqrt{-g} R. \tag{40.1}$$

Here, G_N is the Newton constant, appearing in the Newton potential,

$$U_N = -\frac{MG_N}{r}. \tag{40.2}$$

We also define $16\pi G_N \equiv \kappa_N^2$, and this definition is valid in any dimension. Then κ_N is the Newton coupling, defining the gravitational coupling between fields.

The Newtonian gravity limit is a limit of weak gravity, so weak fields, $g_{\mu\nu} - \eta_{\mu\nu} \ll 1$. It is also a nonrelativistic limit in the special sense, so for small velocities, $v \ll c$. But in this chapter, we will be interested in the case of weak gravity only, yet still a relativistic field theory. The nonrelativistic case will be treated at the beginning of the next chapter in more detail.

Fierz-Pauli Quadratic Action for Small Fields

Therefore, we consider the small field expansion as an expansion in κ_N – i.e.,

$$g_{\mu\nu} = \eta_{\mu\nu} + 2\kappa_N h_{\mu\nu}. \tag{40.3}$$

The factor of κ_N, with dimension -1, $[\kappa_N] = -1$ (since $[G_N] = -2$, so as to make the action dimensionless: $[d^4x] = -4$, and $[R] = 2$), is such as to make $h_{\mu\nu}$ of dimension 1, like a scalar field. With the normalization factor of 2, we will get just the kinetic term of a massless scalar.

We now expand the Einstein action to quadratic order. First, we expand $\sqrt{-g}$. The determinant of a 4×4 matrix is defined as

$$\det g_{\mu\nu} = \frac{1}{4!} \epsilon^{\mu\nu\rho\sigma} \epsilon^{\mu'\nu'\rho'\sigma'} g_{\mu\mu'} g_{\nu\nu'} g_{\rho\rho'} g_{\sigma\sigma'}. \tag{40.4}$$

Substituting the weak field expansion, we get

$$g \equiv \det g_{\mu\nu} = \frac{1}{4!} \epsilon^{\mu\nu\rho\sigma} \epsilon^{\mu'\nu'\rho'\sigma'} (\eta_{\mu\mu'} + 2\kappa_N h_{\mu\mu'}) \dots (\eta_{\sigma\sigma'} + 2\kappa_N h_{\sigma\sigma'}). \tag{40.5}$$

To calculate it, we need to use the general formula

$$\epsilon^{\mu_1 \dots \mu_k \mu_{k+1} \dots \mu_d} \epsilon_{\nu_1 \dots \nu_k \mu_{k+1} \dots \mu_d} = -k! \, (d-k)! \, \delta^{\mu_1 \dots \mu_k}_{\nu_1 \dots \nu_k}. \tag{40.6}$$

Here $\epsilon^{01\dots d-1} = +1$, and the indices are lowered with $\eta_{\mu\nu}$, and since one of the indices down must be 0, we get the minus sign from the right-hand side. In the sum over $\mu_{k+1} \dots \mu_d$ there are $(d-k)!$ permutations, and then $\mu_1 \dots \mu_d$ and $\nu_1 \dots \nu_k$ are equal, but not necessarily in the same order, so we have $k!$ terms, for a total factor of $-k! \, (d-k)!$. In particular, we have

$$\epsilon^{\mu\nu\rho\sigma} \epsilon_{\mu\nu\rho\sigma} = 4!, \qquad \epsilon^{\mu\nu\rho\sigma} \epsilon_{\mu'\nu\rho\sigma} = 3! \, \delta^{\mu}_{\mu'}. \tag{40.7}$$

We will only keep the zeroth- and first-order terms in $h_{\mu\nu}$, which are

$$-g = 1 + 2\kappa_N h^{\mu}_{\mu} + \mathcal{O}(h^2). \tag{40.8}$$

Here we have used that there are four terms with $h_{\mu\nu}$, and each comes with a $3! \, \eta^{\mu\nu}$ factor. Then finally,

$$\sqrt{-g} \simeq 1 + \kappa_N h, \tag{40.9}$$

where $h \equiv h^{\mu}_{\mu}$.

We next calculate the expansion of the Ricci scalar $R = g^{\mu\nu} R_{\mu\nu}$. From the formula for the Riemann tensor,

$$\begin{aligned} R_{\nu\rho} = R^{\mu}_{\nu\mu\rho} &= \partial_{\mu}\Gamma^{\mu}_{\nu\rho} - \partial_{\rho}\Gamma^{\mu}_{\nu\mu} + \Gamma^{\mu}_{\lambda\mu}\Gamma^{\lambda}_{\nu\rho} - \Gamma^{\mu}_{\lambda\rho}\Gamma^{\lambda}_{\nu\mu} \\ &= \partial_{\mu}\Gamma^{\mu}_{\nu\rho} + \Gamma^{\mu}_{\lambda\mu}\Gamma^{\lambda}_{\nu\rho} - (\mu \leftrightarrow \rho, \text{down}). \end{aligned} \tag{40.10}$$

The inverse metric is, in the κ_N expansion,

$$g^{\mu\nu} = \eta^{\mu\nu} - 2\kappa_N h^{\mu\nu} + \mathcal{O}(h^2). \tag{40.11}$$

Then the Christoffel symbol is, to order h^2,

$$\Gamma^{\mu}_{\nu\rho} = \kappa_N(\eta^{\mu\sigma} - 2\kappa_N h^{\mu\sigma})(\partial_{\nu}h_{\sigma\rho} + \partial_{\rho}h_{\sigma\nu} - \partial_{\sigma}h_{\nu\rho}). \tag{40.12}$$

We substitute this into the formula for the Ricci scalar, obtaining

$$\begin{aligned} R_{\nu\rho} = \; &\kappa_N(\eta^{\mu\sigma} - 2\kappa_N h^{\mu\sigma})(\partial_{\mu}\partial_{\nu}h_{\sigma\rho} + \partial_{\mu}\partial_{\rho}h_{\sigma\nu} - \partial_{\mu}\partial_{\sigma}h_{\nu\rho}) \\ &- 2\kappa_N^2(\partial_{\mu}h^{\mu\sigma})(\partial_{\nu}h_{\sigma\rho} + \partial_{\rho}h_{\sigma\nu} - \partial_{\sigma}h_{\nu\rho}) - (\mu \leftrightarrow \rho, \text{down}) \\ &+ \kappa_N^2 \eta^{\mu\sigma}(\partial_{\lambda}h_{\sigma\mu} + \partial_{\mu}h_{\sigma\lambda} - \partial_{\sigma}h_{\mu\lambda})\eta^{\lambda\tau}(\partial_{\nu}h_{\tau\rho} + \partial_{\rho}h_{\tau\nu} - \partial_{\tau}h_{\rho\nu}) - (\mu \leftrightarrow \rho, \text{down}). \end{aligned} \tag{40.13}$$

In (40.13), there are linear and quadratic terms in $h_{\mu\nu}$.

We need to organize the calculation, so we note that

$$R = g^{\nu\rho}R_{\nu\rho} = \eta^{\nu\rho}R_{\nu\rho}(\mathcal{O}(h)) - 2\kappa_N h^{\mu\nu}R_{\mu\nu}(\mathcal{O}(h)) + \eta^{\mu\nu}R_{\mu\nu}(\mathcal{O}(h^2)) + \mathcal{O}(h^3). \quad (40.14)$$

To continue, we first define the quantities (where raising and lowering of indices is done with the Minkowski metric)

$$h^{\mu}{}_{\mu} \equiv h, \quad \partial_{\mu}h^{\mu\rho} \equiv h^{\rho}. \quad (40.15)$$

The terms in $R_{\nu\rho}$ linear in h are the ones proportional to $\mu\sigma$, which become

$$R_{\nu\rho}(\mathcal{O}(h)) = \kappa_N(\partial^{\mu}\partial_{\nu}h_{\mu\rho} + \partial^{\mu}\partial_{\rho}h_{\mu\nu} - \partial^{\mu}\partial_{\mu}h_{\nu\rho}) - \kappa_N(\partial_{\rho}\partial_{\nu}h + \partial^{\mu}\partial_{\rho}h_{\mu\nu} - \partial_{\rho}\partial^{\mu}h_{\mu\nu})$$
$$= \kappa_N(\partial_{\nu}h_{\rho} + \partial_{\rho}h_{\nu} - \partial^{\mu}\partial_{\mu}h_{\nu\rho} - \partial_{\nu}\partial_{\rho}h). \quad (40.16)$$

Then we have

$$\eta^{\nu\rho}R_{\nu\rho}(\mathcal{O}(h)) = 2\kappa_N\partial^{\mu}(h_{\mu} - \partial_{\mu}h). \quad (40.17)$$

Then this, multiplied by 1 (the leading term in $\sqrt{-g}$), is the whole action linear in h, and is a total derivative, giving just a boundary term, which we drop. That means that, indeed, the action is quadratic in $h_{\mu\nu}$, as we wanted.

Moreover, then, the first quadratic term is

$$2\kappa_N h^{\nu\rho}R_{\nu\rho}(\mathcal{O}(h)) = 2\kappa_N^2 h^{\nu\rho}(\partial_{\nu}h_{\rho} + \partial_{\rho}h_{\nu} - \partial^{\mu}\partial_{\mu}h_{\nu\rho} - \partial_{\nu}\partial_{\rho}h), \quad (40.18)$$

and by partially integrating (under the integral sign), we obtain

$$2\kappa_N \int d^4x\, h^{\nu\rho}R_{\nu\rho}(\mathcal{O}(h)) = 2\kappa_N^2 \int d^4x\, [-2h_{\rho}^2 + (\partial_{\mu}h_{\nu\rho})^2 + h_{\rho}\partial^{\rho}h]. \quad (40.19)$$

Finally, we can calculate that

$$\int d^4x\, \eta^{\nu\rho}R_{\nu\rho}(\mathcal{O}(h^2)) = \int d^4x\, \kappa_N^2[(\partial_{\mu}h_{\nu\rho})^2 + 2h_{\mu}\partial^{\mu}h - (\partial_{\mu}h)^2 - 2(h_{\mu})^2], \quad (40.20)$$

which we leave as an exercise to prove.

Adding up the contributions, we obtain

$$2\kappa_N^2 S_{\text{gravity}} \simeq \int d^4x(1 + \kappa_N h)[\eta^{\nu\rho}R_{\nu\rho}(\mathcal{O}(h)) - 2\kappa_N h^{\nu\rho}R_{\nu\rho}(\mathcal{O}(h)) + \eta^{\nu\rho}R_{\nu\rho}(\mathcal{O}(h^2))]$$
$$+ \mathcal{O}(h^3)$$
$$= \int d^4x[\kappa_N h\eta^{\nu\rho}R_{\nu\rho}(\mathcal{O}(h)) - 2\kappa_N h^{\nu\rho}R_{\nu\rho}(\mathcal{O}(h)) + \eta^{\nu\rho}R_{\nu\rho}(\mathcal{O}(h^2))]$$
$$+ \mathcal{O}(h^3), \quad (40.21)$$

and substituting the various terms and partially integrating the total derivative in the first term to obtain $-2\kappa_N^2(\partial^{\mu}h)(h_{\mu} - \partial_{\mu}h)$, the action is

$$2\kappa_N^2 S_{\text{gravity}} = 2\kappa_N^2 \int d^4x \left[-h^{\mu}\partial_{\mu}h + (\partial_{\mu}h)^2 + 2h_{\mu}^2 - (\partial_{\mu}h_{\nu\rho})^2 \right.$$
$$\left. -h^{\mu}\partial_{\mu}h + \frac{1}{2}(\partial_{\mu}h_{\nu\rho})^2 + h^{\mu}\partial_{\mu}h - \frac{1}{2}(\partial_{\mu}h)^2 - h_{\mu}^2 \right]$$
$$= 2\kappa_N^2 \int d^4x \left[-\frac{1}{2}(\partial_{\mu}h_{\nu\rho})^2 + h_{\mu}^2 + \frac{1}{2}(\partial_{\mu}h)^2 - h^{\mu}\partial_{\mu}h \right]. \quad (40.22)$$

The resulting quadratic action,

$$S_{FP} = \int d^4x \left[-\frac{1}{2}(\partial_\mu h_{\nu\rho})^2 + h_\mu^2 - h^\mu \partial_\mu h + \frac{1}{2}(\partial_\mu h)^2 \right], \qquad (40.23)$$

is called the *Fierz-Pauli action*. Fierz and Pauli wrote this action as the action for a massless spin 2 field, subject to a gauge invariance (the equivalent of a spin 1 gauge field). In fact, one can prove that the only nonlinear extension of it, which respects gauge invariance and unitarity, is the Einstein–Hilbert action from which we derived it. Fierz and Pauli also showed how we can add a mass term to it, similarly to what we did for a gauge field, to find a Proca (massive vector) field. The mass term is

$$-\frac{m^2}{2} \int d^4x \, (h_{\mu\nu}h^{\mu\nu} - h^2). \qquad (40.24)$$

However, in this case, even for an infinitesimal mass, due to the fact that we change the number of degrees of freedom (since we lose gauge invariance when adding it), one finds that the deflection of light by the Sun (a classic test of general relativity, which we will study in Chapter 42) changes by a factor of 3/4. There have been many contradictory discussions in the literature about whether it is possible to construct a sensible theory of massive gravity by extending the mass term in (40.24), but we will not get into it.

40.2 Gauge Invariance, de Donder Gauge and Other Gauges

Let us verify that we still have gauge invariance for $h_{\mu\nu}$ in the Fierz-Pauli action. We could, of course, check on the final FP action. But it is easier to consider the gauge invariance of the Einstein–Hilbert action and take the same small κ_N limit. We have

$$\delta h_{\mu\nu} = \frac{\delta g_{\mu\nu}}{2\kappa_N} = \frac{\xi^\lambda}{2\kappa_N}\partial_\lambda g_{\mu\nu} + \frac{\partial_\mu \xi^\rho}{2\kappa_N}g_{\rho\nu} + \frac{\partial_\nu \xi^\rho}{2\kappa_N}g_{\mu\rho}$$

$$= \partial_\mu \left(\frac{\xi_\nu}{2\kappa_N}\right) + \partial_\nu \left(\frac{\xi_\mu}{2\kappa_N}\right) + \xi^\lambda \partial_\lambda h_{\mu\nu} + (\partial_\mu \xi^\rho)h_{\nu\rho} + (\partial_\nu \xi^\rho)h_{\mu\rho}. \qquad (40.25)$$

In the limit when κ_N is small, we obtain the gauge invariance

$$\delta h_{\mu\nu} = \partial_\mu \left(\frac{\xi_\nu}{2\kappa_N}\right) + \partial_\nu \left(\frac{\xi_\mu}{2\kappa_N}\right) \equiv \partial_\mu \xi'_\nu + \partial_\nu \xi'_\mu. \qquad (40.26)$$

This is indeed a gauge invariance for a symmetric tensor, just like the gauge symmetry for an antisymmetric tensor (two-form) was shown to be

$$\delta A_{\mu\nu} = 2\partial_{[\mu}\lambda_{\nu]} = \partial_\mu \lambda_\nu - \partial_\nu \lambda_\mu. \qquad (40.27)$$

This gauge invariance needs to be gauge fixed.

We can choose a gauge that is an analog of the Lorenz gauge in electromagnetism ($\partial^\mu A_\mu = 0$). This is called the *de Donder gauge*,

$$\partial^\mu \bar{h}_{\mu\nu} = 0, \quad \bar{h}_{\mu\nu} = h_{\mu\nu} - \eta_{\mu\nu}\frac{h}{2}. \qquad (40.28)$$

Using our notation, the de Donder gauge implies $h_\mu = \partial_\mu h/2$. Then we find that

$$h_\mu^2 - h^\mu \partial_\mu h + \frac{1}{2}(\partial_\mu h)^2 = \frac{1}{4}(\partial_\mu h)^2$$
$$(\partial_\rho \bar{h}_{\mu\nu})^2 = (\partial_\rho h_{\mu\nu})^2 + (\partial_\rho h)^2 - (\partial_\rho h)^2 = (\partial_\rho h_{\mu\nu})^2, \quad (40.29)$$

so the Fierz-Pauli action in the de Donder gauge becomes

$$S_{\text{FP, de D}} = \int d^4x \left[-\frac{1}{2}(\partial_\mu \bar{h}_{\nu\rho})^2 + \frac{1}{4}(\partial_\mu h)^2 \right]. \quad (40.30)$$

This separates the modes in the $\bar{h}_{\mu\nu}$ modes, with the usual scalar kinetic term, leading to the KG equation,

$$\Box \bar{h}_{\mu\nu} = 0, \quad (40.31)$$

and the trace mode h. This is just like in the Lorenz gauge for electromagnetism, where the equation was also KG, $\Box A_\mu = 0$.

The equation of motion is then just the wave equation, so we have *gravitational waves* as solutions to it. Before studying them further, consider other gauges.

We can use the four general coordinate transformations $x'^\mu(x^\rho)$ to fix four components of the metric, specifically

$$g_{00} = -1, \quad g_{0i} = 0, \quad (40.32)$$

obtaining the metric

$$ds^2 = -dt^2 + g_{ij}(x^\rho)dx^i dx^j. \quad (40.33)$$

This is called the *synchronous gauge*.

Another possibility for a gauge is in the case of weak fields $\kappa_N h_{\mu\nu} \ll 1$, *as well as for nonrelativistic motion*, $v^i \ll c$. In this case, we can put the metric in *Newtonian form*,

$$ds^2 \simeq -(1 + 2U_{\text{Newton}})dt^2 + (1 - 2U_{\text{Newton}})d\vec{x}^2, \quad (40.34)$$

where $d\vec{x}^2 = dr^2 + r^2 d\Omega_2^2$ is the flat space spatial metric, and thus recovering Newtonian gravity (as we will see in more detail in the next chapter). This is called the *Newtonian gauge*.

40.3 Gravitational Waves

In the de Donder gauge, the leading order Ricci tensor (40.16) becomes

$$R_{\mu\nu}(\mathcal{O}(h)) \simeq \frac{-2\kappa_N}{2}\Box h_{\mu\nu}. \quad (40.35)$$

Then the Einstein equation in vacuum, which is just $R_{\mu\nu} = 0$ (taking the trace of $G_{\mu\nu} = 0$ we obtain that $R = 0$ as well), becomes

$$\Box h_{\mu\nu} = 0. \quad (40.36)$$

Taking the trace for it, we obtain $\Box h = 0$ as well, so it is equivalent to $\Box \bar{h}_{\mu\nu} = 0$, which we have found before from the Fierz-Pauli action.

But in de Donder gauge, there is still a residual gauge symmetry – i.e., a coordinate ambiguity – just like in the Lorenz gauge for electromagnetism there was one, and we could also fix $A_0 = 0$. The gauge invariance $\delta h_{\mu\nu} = \partial_\mu \xi_\nu + \partial_\nu \xi_\mu$ implies (taking the trace) $\delta h = 2\partial^\mu \xi_\mu$, which then gives

$$\delta \bar{h}_{\mu\nu} = \partial_\mu \xi_\nu + \partial_\nu \xi_\mu - \eta_{\mu\nu}(\partial \cdot \xi). \tag{40.37}$$

Then we can find a gauge parameter that leaves the de Donder gauge invariant if we find

$$0 = \partial^\mu \delta \bar{h}_{\mu\nu} = \Box \xi_\nu + \partial_\nu(\partial \cdot \xi) - \partial_\nu(\partial \cdot \xi) = \Box \xi_\nu. \tag{40.38}$$

That means that we need a gauge parameter that satisfies the same equation ($\Box \xi_\mu = 0$) as $\bar{h}_{\mu\nu}$ ($\Box \bar{h}_{\mu\nu} = 0$), which allows us to fix four components of $\bar{h}_{\mu\nu}$. We can show that we can, in fact, fix $\bar{h}_{01}, \bar{h}_{02}, \bar{h}_{03}$ and $\bar{h}_{22} + \bar{h}_{33}$ (the proof is left as an exercise). Moreover, we have plane wave solutions, independent of y, z and moving in the x direction, which can, therefore, be written as $h_{\mu\nu}(t - x/c)$. For them, the gauge condition $\partial^\mu \bar{h}_{\mu\nu} = 0$ becomes

$$\dot{\bar{h}}_{1\mu} + \dot{\bar{h}}_{0\mu} = 0, \tag{40.39}$$

which means we can fix also h_{11}, h_{12}, h_{13} to zero, leaving as the only nonzero components $\bar{h}_{23}$ and $\bar{h}_{22} - \bar{h}_{33}$. These are two polarizations, which are transverse to the direction of propagation: just like $\partial^\mu A_\mu = 0$ means $p^\mu \epsilon_\mu(p) = 0$ – i.e., transverse polarization – $\partial^\mu \bar{h}_{\mu\nu} = 0$ means $p^\mu \epsilon_{\mu\nu} = 0$ – i.e., the same. Thus, there are two helicities, as expected.

We can now find the gravitational radiation field of a mass distribution, just like for electromagnetism we found the electromagnetic radiation field of a charge distribution. We first write the equation of motion with a source energy-momentum tensor $T_{\mu\nu}$, which *at this linearized level* is considered to be a fixed function $T_{\mu\nu}(x)$, since we use the Minkowski metric (at the full nonlinear level, as we said in the last chapter, we can only make an ansatz for $T_{\mu\nu}$, as a function of the unknown metric $g_{\mu\nu}$).

The Einstein equation becomes, for a traceless tensor $T \equiv T^\mu_{\ \mu} = 0$,

$$R_{\mu\nu} \simeq -\frac{1}{2}\Box(2\kappa_N \bar{h}_{\mu\nu}) = \frac{8\pi G_N}{c^4} T_{\mu\nu}. \tag{40.40}$$

But this is the same kind of equation as in the scalar and electromagnetic cases, namely KG equation with a source. The solution is then the same as for those (see Chapter 15), $q/(4\pi R)$ for a source q (solution of the spatial Poisson equation $\Delta_3 \phi = q\delta^3(x)$), just at the retarded time $t - R/c$ – i.e.,

$$2\kappa_N \bar{h}_{\mu\nu} = -\frac{4G_N}{c^4} \int \frac{dV}{R}(T_{\mu\nu})_{t-R/c}. \tag{40.41}$$

Far away from the source, we can take R out of the integral as R_0, the average distance to the source, and write

$$2\kappa_N \bar{h}_{\mu\nu} = -\frac{4G_N}{c^4 R_0} \int dV(T_{\mu\nu})_{t-R_0/c}. \tag{40.42}$$

But we can express $\int dV\, T_{ij}$ in terms of just the energy (mass) of the system, using the conservation of the energy-momentum tensor, considered *to leading order in the metric perturbation*. Separating spatial and temporal indices, $\partial_\nu T_\mu{}^\nu = 0$ becomes

$$\partial_i T_j{}^i + \partial_0 T_j{}^0 = 0$$
$$\partial_i T_0{}^i + \partial_0 T_0{}^0 = 0. \tag{40.43}$$

But to leading order, we raise and lower the indices on $T_{\mu\nu}$ with the Minkowski metric $\eta_{\mu\nu}$, so we obtain

$$\partial_i T_{ji} - \partial_0 T_{j0} = 0$$
$$\partial_i T_{0i} - \partial_0 T_{00} = 0. \tag{40.44}$$

Multiplying the first relation in (40.44) by x^k and integrating over a volume V, large enough so that it encompasses all the matter, we obtain

$$\partial_0 \int dV\, T_{j0}x^k = \int dV\, \partial_i(x^k T_{ji}) - \delta_i^k \int dV\, T_{ji}. \tag{40.45}$$

By our assumption of V encompassing all matter, after using Gauss's law to relate $\int dV \partial_i(\dots)$ by $\int_{\partial V}(\dots)$, the first term vanishes. Symmetrizing the indices, we then get

$$\int dV\, T_{jk} = -\frac{1}{2}\partial_0 \int dV\, (T_{j0}x^k + T_{k0}x^j). \tag{40.46}$$

Next, multiplying the second relation in (40.44) by $x^j x^k$ and integrating, we similarly obtain

$$\partial_0 \int dV\, T_{00}x^j x^k = \int dV\, \partial_i(T_{0i}x^j x^k) - \int dV\, (T_{0j}x^k + T_{0k}x^j), \tag{40.47}$$

and by the same assumption, the first term vanishes, and finally

$$\partial_0 \int dV\, T_{00}x^j x^k = -\int dV\, (T_{0j}x^k + T_{0k}x^j). \tag{40.48}$$

Combining (40.46) and (40.48), we find

$$\int dV T_{jk} = \frac{1}{2}\partial_0^2 \int dV\, T_{00}x^j x^k. \tag{40.49}$$

We can then finally write

$$\int \frac{dV}{R} T_{ij}\big|_{t-R/c} = \frac{1}{2R_0}\frac{\partial^2}{\partial x_0^2} \int dV\, T_{00}(t - R/c)x^i x^j, \tag{40.50}$$

where $x_0 = ct$. Here $T_{00} = \mu c^2$ is the energy density of the system. Moreover, if the wave is propagating in the direction 1, for the physical transverse traceless modes, $\bar{h}_{23}$ and $\bar{h}_{22} - \bar{h}_{33}$, we see that we can make the integral $\int dV\, T_{00}x^i x^j$ traceless, since the extra term will drop out in $\bar{h}_{23}$ and $\bar{h}_{22} - \bar{h}_{33}$. Then we find an expression that depends on the *quadrupole moment in the transverse traceless gauge*, Q_{ij}, defined by

$$Q^{ij} = \int dV \mu \left(x^i x^j - \frac{1}{3}\delta^{ij}r^2\right). \tag{40.51}$$

That means that *for the physical modes* (the two transverse and traceless modes), another way of saying it would be *if we impose a physical, transverse traceless gauge for* $\bar{h}_{ij}$, we finally have

$$2\kappa_N \bar{h}_{ij} = -\frac{2G_N}{c^2 R_0}\ddot{Q}^{ij}(t - R/c). \tag{40.52}$$

This formula was first obtained by Einstein in 1916.

So, while electromagnetic radiation is generated by (at least) dipoles, gravitational radiation is generated by (at least) quadrupoles in the multipole expansion. Otherwise, the formula for the radiation is similar to the one we found in Section 15.7 in Chapter 15, showing that the radiation field far away from the source is equal to the second time derivative of the multipole moment.

40.4 Exact Cylindrical Gravitational Wave (Einstein–Rosen)

We have finished with the analysis of the perturbative gravity. But to transition into nonperturbative gravity, we consider the question asked by Einstein and Rosen in a famous paper: can there be exact gravitational wave solutions, given that general relativity is so nonlinear? The answer was given in the affirmative for solutions with cylindrical symmetry, showing that gravitational waves are a real feature of general relativity. It is worth noting the history of the paper, however: Eintein first submitted the paper to the *Physical Review*, but the anonymous reviewer (now believed to be Robertson) pointed out that the conclusion in the first version of the paper was incorrect: it was claimed exactly the opposite, that there were no true gravitational waves, based on the same system of equations. Einstein refused to accept the reviewer's suggestions and withdrew the paper. However, he then reworked the paper by himself (Rosen had just returned to Russia), according to the reviewer's suggestions, and sent it to a less prestigious journal, where it was finally published. We here present a sketch of the proof of the solution. For more details, see the original paper.

We consider solutions of Einstein's equations in the vacuum, $R_{\mu\nu} = 0$, with cylindrical symmetry, so invariant under translations in z and under rotations in θ, where (r, θ) are polar coordinates and z is a coordinate transverse to the plane. This means that first, $g_{\mu\nu} = g_{\mu\nu}(r, t)$ only, but second, we can put the mixed metric components involving θ, z to zero,

$$g_{zi} = g_{\theta i} = 0. \tag{40.53}$$

That means that the only nonzero components are the diagonal $g_{rr}, g_{tt}, g_{zz}, g_{\theta\theta}$, and g_{0r}. But then we can use the two coordinate transformations remaining, $r' = r'(t, r)$ and $t' = t(t, r)$, to fix two metric components. We can choose

$$g_{0r} = 0, \quad g_{rr} = -g_{00}. \tag{40.54}$$

Then we denote

$$-g_{00} = g_{rr} = A = e^{\alpha}, \quad g_{\theta\theta} = C = e^{\gamma-\beta}, \quad g_{zz} = B = e^{\gamma+\beta}. \tag{40.55}$$

The metric finally becomes

$$ds^2 = e^{\alpha}(-dt^2 + dr^2) + e^{\gamma}(e^{\beta}dz^2 + e^{-\beta}d\theta^2). \qquad (40.56)$$

Further denoting $e^{\gamma} \equiv \sigma$, we find that the Einstein equations reduce to several equations. One of them is

$$(-\partial_t^2 + \partial_r^2)\sigma(r,t) = 0, \qquad (40.57)$$

i.e., the two-dimensional wave equation for $\sigma = e^{\gamma}$. Since $-\partial_t^2 + \partial_r^2 = (-\partial_t + \partial_r)(+\partial_t + \partial_r)$, the general solution of this equation is

$$\sigma(r,t) = f(r+t) + g(r-t). \qquad (40.58)$$

Another equation obtained is

$$(-\partial_t^2 + \partial_r^2)\alpha + \frac{1}{2}[(\partial_r\beta)^2 - (\partial_t\beta)^2 - (\partial_r\gamma)^2 + (\partial_t\gamma)^2] = 0, \qquad (40.59)$$

and there are several others. As a sanity check, we note that flat Minkowski space, with $\alpha = 0$ and $\beta = -\gamma = -\log r$ (so $ds^2 = -dt^2 + dr^2 + dz^2 + r^2 d\theta^2$), implying in particular that $\sigma = e^{\gamma} = r$, is a solution of equations (40.57) and (40.59).

We can consider more general solutions in a similar class, with $\sigma = Ar$ – i.e., for

$$f(x) = g(x) = \frac{A}{2}x. \qquad (40.60)$$

Then the Einstein equations reduce to only three equations:

$$-\partial_t^2\beta + \frac{1}{r}\partial_r\beta + \partial_r^2\beta = 0$$
$$\partial_r\alpha = \frac{r}{2}(\partial_r\beta^2 + \partial_t\beta^2) - \frac{1}{2r}$$
$$\partial_t\alpha = r\partial_r\beta\partial_t\beta. \qquad (40.61)$$

The first equation is just the wave equation, $\Box\beta = 0$, for a field with cylindrical symmetry (in this case, $\Box = -\partial_t^2 + \partial_z^2 + \partial_r^2 + \frac{1}{r}\partial_r + \frac{1}{r^2}\partial_\theta^2$ acts on $\beta(r,t)$ only). That means that β is the general solution of the four-dimensional wave equation with cylindrical symmetry. The second and third equations just define $\alpha(r,t)$ up to a constant, given a solution $\beta(r,t)$ of the first equation.

A sinusoidal gravitational wave is found at large distances ($r \gg 1$), when $1/r\partial_r \to 0$, so the equation for β is simply $-\partial_t^2\beta + \partial_r^2\beta = 0$, with the solution

$$\beta = \beta_0 + \sin\omega(t-r). \qquad (40.62)$$

This is a cylindrical gravitational wave that can be extended in the interior via the equations denoted in this section.

Important Concepts to Remember

- The Newtonian limit for gravity is $g_{\mu\nu} - \eta_{\mu\nu} = 2\kappa_N h_{\mu\nu} \ll 1$ and $v^i \ll c$, whereas just the first condition defines the weak field relativistic limit.
- In the weak field relativistic limit, we find the Fierz-Pauli action, quadratic in the fields, as a function of $h_{\mu\nu}$, with a canonical kinetic term, and $h^{\mu} = \partial_{\nu}h^{\mu\nu}$ and $h = h^{\mu}{}_{\mu}$.

- The gauge invariance in the weak field limit (for the Fierz-Pauli action) is $\delta h_{\mu\nu} = \partial_\mu \xi_\nu + \partial_\nu \xi_\mu$.
- The gauge invariance can be fixed by the de Donder gauge, $\partial^\mu \bar{h}_{\mu\nu} = 0$, where $\bar{h}_{\mu\nu} = h_{\mu\nu} - \eta_{\mu\nu}h/2$, leading to the action $\int [-\frac{1}{2}(\partial_\mu \bar{h}_{\nu\rho})^2 + \frac{1}{4}(\partial_\mu h)^2]$, and the equation of motion $\Box \bar{h}_{\mu\nu} = 0$.
- In the synchronous gauge, $ds^2 = -dt^2 + g_{ij}dx^i dx^j$, and in the Newtonian gauge, for the Newtonian limit, $ds^2 = -(1 + 2U_N)dt^2 + (1 - 2U_N)d\vec{x}^2$.
- Gravitational radiation, the solutions of $\Box \bar{h}_{\mu\nu} = 0$, are two transverse polarizations, $\bar{h}_{23}$ and $\bar{h}_{22} - \bar{h}_{33}$, for motion in x.
- Gravitational radiation far away from a mass distribution (source) is defined by the second time derivative of the retarded quadrupole moment $\ddot{Q}^{ij}(t - R/c)$.
- There are exact cylindrical gravitational waves (Einstein–Rosen), $ds^2 = e^\alpha(-dt^2 + dr^2) + e^\gamma(e^\beta dz^2 + e^{-\beta}d\theta^2)$, defined by $e^\gamma = Ar$ and a solution of the wave equation $\Box \beta(r,t) = 0$ in cylindrical coordinates.

Further Reading

See chapters 12 and 13 in [3], chapter 10 in [40], and the original paper on cylindrical waves [41].

Exercises

(1) Show that by a choice of coordinates, for the perturbative gravitational wave, we can make $\bar{h}_{01}, \bar{h}_{02}, \bar{h}_{03}$ and $\bar{h}_{22} + \bar{h}_{33}$ vanish.

(2) Consider a stellar object of ellipsoidal shape like the Earth, which contracts and expands on the N–S axis (the North and South Poles get closer, then farther, apart, etc.) so that there are changes only along the N–S axis. Does it emit gravitational radiation?

(3) Check that we have

$$\int d^4x \, \eta^{\nu\rho} R_{\nu\rho}(\mathcal{O}(h^2)) = \int d^4x \, \kappa_N^2 [(\partial_\mu h_{\nu\rho})^2 + 2h_\mu \partial^\mu h - (\partial_\mu h)^2 - 2(h_\mu)^2]. \quad (40.63)$$

Hint: Check first that the contribution of the terms on the first two lines in (40.13) to (40.63) vanishes.

(4) Calculate the quadratic term in the metric perturbation $h_{\mu\nu}$ around flat space in the quantity

$$aR^2 + bR_{\mu\nu}R^{\mu\nu}, \quad (40.64)$$

where a, b are constants, and R and $R_{\mu\nu}$ are the Ricci scalar and Ricci tensor, respectively. Is there a gauge that makes this term zero, or trivial?

(5)   Consider four equal masses m situated two on the Cartesian x axis on opposite sides of the origin, and two on the Cartesian axis y on opposite sides of the origin. The masses on the x axis have a *distance* to the origin given by

$$d_x = R[1 - \cos(\omega t)], \qquad\qquad (40.65)$$

and the ones on the y axis a distance to the origin given by

$$d_y = R[1 - \sin(\omega t)]. \qquad\qquad (40.66)$$

Calculate the gravitational wave perturbation in a transverse traceless gauge, $\bar{h}_{ij}$, at a distance $r \gg R$ in the z direction from the origin.

41

Nonperturbative Gravity: The Vacuum Schwarzschild Solution

In this chapter, we finally start the analysis of fully nonperturbative solutions to Einstein's gravity, with the most famous of them, the Schwarzschild solution. It was found by Schwarzschild in 1919, while he was fighting in the trenches of World War I, and it is the most general solution of the vacuum Einstein equations with spherical symmetry.

41.1 Newtonian Limit

We first want to understand its Newtonian limit. The Lagrangean for a nonrelativistic particle in a gravitational potential is

$$L = -mc^2 + m\frac{v^2}{2} - mU_N, \tag{41.1}$$

where we have added the potential rest energy of the particle. But in relativity (both the special and general kind), the action on the worldline of a particle, $x^\mu = x^\mu(t)$, is just

$$S = \int L\,dt = -mc^2 \int d\tau, \tag{41.2}$$

which means that in the Newtonian and nonrelativistic limit (weak fields and $v \ll c$),

$$d\tau = \left(1 - \frac{v^2}{2c^2} + \frac{U_N}{c^2}\right) dt. \tag{41.3}$$

Since we use $dx = v\,dt$ in the above, we have

$$ds^2 \simeq -c^2 dt^2 \left(1 + \frac{2U_N}{c^2}\right) + d\vec{x}^2, \tag{41.4}$$

and $d\vec{x}^2 = dr^2 + r^2 d\Omega_2^2$.

The solution we will be interested in is for the Einstein equations in vacuum, $R_{\mu\nu} = 0$ (since $T_{\mu\nu} = 0$), more precisely *outside a matter source* that is spherically symmetric. The matter must be nonrelativistic and non-interacting, so it looks like a fluid with $P = 0$ (dust) – i.e.,

$$T_{\mu\nu} = \text{diag}(\rho_m, 0, 0, 0). \tag{41.5}$$

Then $T \simeq -\rho$ (for weak fields, we contract with $\eta^{\mu\nu}$), so $T_{00} - \frac{1}{2}g_{00}T = \rho_m/2$. Then the Einstein equation

$$R_{\mu\nu} = 8\pi G_N \left(T_{\mu\nu} - \frac{1}{2}g_{\mu\nu}T\right) \tag{41.6}$$

becomes, for the 00 component,

$$R_{00} = -\frac{1}{2}\Box(2\kappa_N h_{00}) = -\frac{1}{2}\vec{\nabla}^2(2\kappa_N h_{00}) = 8\pi G_N \frac{\rho_m}{2}\,, \tag{41.7}$$

which means that we can put $2\kappa_N h_{00} = -2U_N$, since from it we obtain the usual local Gauss's law for gravity,

$$\vec{\nabla}^2 U_N = 4\pi G_N \rho_m. \tag{41.8}$$

Moreover, the ij components of the Einstein equation become

$$R_{ij} = -\frac{1}{2}\Box(2\kappa_N h_{ij}) = 8\pi G_N \frac{\rho_m}{2}\delta_{ij}, \tag{41.9}$$

which implies that we can put also

$$2\kappa_N h_{ij} = -2U_N\delta_{ij}. \tag{41.10}$$

Together with $2\kappa_N h_{00} = -2U_N$ (and the fact that no $\Box h_{0i} = 0$, so we can put h_{0i} to 0), we have

$$ds^2 \simeq -(1+2U_N)dt^2 + (1-2U_N)d\vec{x}^2, \tag{41.11}$$

which proves the statement we made in the last chapter about the metric above being the Newtonian limit of general relativity, in some gauge.

We consider a spherically symmetric solution, so $U_N = U_N(r)$ only. Moreover, we can do a coordinate transformation on r, to first-order

$$(1-2U_N(r))r^2 = r'^2, \tag{41.12}$$

and then to the leading order we can keep $dr^2 \simeq dr'^2$ (the change is higher order), to put the metric in the form

$$ds^2 \simeq -(1+2U_N(r))dt^2 + (1-2U_N(r))dr^2 + r^2 d\Omega_2^2. \tag{41.13}$$

This will be the small field limit of the static Schwarzschild solution.

41.2 Ansatz and Equations of Motion

Ansatz

We look for a spherically symmetric solution, so we can fix $g_{r\theta} = g_{r\phi} = g_{t\theta} = g_{t\phi} = 0$, and $g_{\mu\nu} = g_{\mu\nu}(r,t)$ (independent on θ and ϕ). Then we write

$$ds^2 = f(r,t)dt^2 + 2g(r,t)dr\,dt + h(r,t)dr^2 + p(r,t)d\Omega_2^2, \tag{41.14}$$

where $d\Omega_2^2 = d\theta^2 + \sin^2\theta d\phi^2$ is the metric on the spatial two-sphere. Moreover, now we still have two possible coordinate changes to perform,

$$r' = r'(r,t), \quad t' = t'(r,t), \tag{41.15}$$

and we can use them to put $g'(r',t') = 0$ and $p'(r',t') = r'^2$. Dropping the primes, and denoting $f = -e^\nu$, $h = e^\lambda$, we have finally the ansatz

$$ds^2 = -e^{\nu(r,t)}dt^2 + e^{\lambda(r,t)}dr^2 + r^2 d\Omega_2^2. \tag{41.16}$$

Christoffel Symbols

We now calculate the Christoffel symbols, using the formula

$$\Gamma^\mu_{\nu\rho} = \frac{1}{2}g^{\mu\sigma}(\partial_\nu g_{\sigma\rho} + \partial_\rho g_{\sigma\nu} - \partial_\sigma g_{\nu\rho}). \tag{41.17}$$

We note that we have a diagonal metric, so they become easier. First, since we need $\sigma = \mu$ only, it means that if $\mu \neq \nu \neq \rho$, $\Gamma^\mu_{\nu\rho} = 0$ (there are 12 such symbols). We also have (no summation over μ)

$$\Gamma^\mu_{\nu\mu} = \frac{1}{2}g^{\mu\mu}\partial_\nu g_{\mu\mu}, \tag{41.18}$$

Then, since ϕ is an isometry (the metric is independent of it), $\Gamma^\mu_{\phi\mu} = 0$ (there are four such symbols). We also have

$$\Gamma^\mu_{\nu\nu} = -\frac{1}{2}g^{\mu\mu}\partial_\mu g_{\nu\nu} \tag{41.19}$$

if $\mu \neq \nu$.

There are four possibilities for μ, and $4 \cdot 5/2 = 10$ for $\nu\rho$ (symmetric indices), for a total of 40 Christoffel symbols. They are

$$\Gamma^t_{tt}, \Gamma^t_{rr}, \Gamma^t_{rt}, \Gamma^t_{r\theta}, \Gamma^t_{r\phi}, \Gamma^t_{\theta\phi}, \Gamma^t_{\theta\theta}, \Gamma^t_{\phi\phi}, \Gamma^t_{t\theta}, \Gamma^t_{t\phi}$$
$$\Gamma^r_{rr}, \Gamma^r_{tt}, \Gamma^r_{tr}, \Gamma^r_{t\theta}, \Gamma^r_{t\phi}, \Gamma^r_{\theta\phi}, \Gamma^r_{\theta\theta}, \Gamma^r_{\phi\phi}, \Gamma^r_{r\theta}, \Gamma^r_{r\phi}$$
$$\Gamma^\theta_{\theta\theta}, \Gamma^\theta_{\theta\phi}, \Gamma^\theta_{\phi\phi}, \Gamma^\theta_{r\phi}, \Gamma^\theta_{t\phi}, \Gamma^\theta_{rt}, \Gamma^\theta_{rr}, \Gamma^\theta_{tt}, \Gamma^\theta_{\theta r}, \Gamma^\theta_{\theta r}$$
$$\Gamma^\phi_{\phi\phi}, \Gamma^\phi_{\theta\theta}, \Gamma^\phi_{\theta\phi}, \Gamma^\phi_{rt}, \Gamma^\phi_{r\theta}, \Gamma^\phi_{t\theta}, \Gamma^\phi_{rr}, \Gamma^\phi_{tt}, \Gamma^\phi_{\phi r}, \Gamma^\phi_{\phi t}. \tag{41.20}$$

Besides the $12 + 4 = 16$ vanishing symbols above, we also have

$$\Gamma^t_{\theta\theta} = -\frac{1}{2}g^{tt}\partial_t g_{\theta\theta} = 0 \quad \Gamma^t_{\phi\phi} = -\frac{1}{2}g^{tt}\partial_t g_{\phi\phi} = 0$$
$$\Gamma^t_{t\theta} = +\frac{1}{2}g^{tt}\partial_\theta g_{tt} = 0 \quad \Gamma^r_{r\theta} = +\frac{1}{2}g^{rr}\partial_\theta g_{rr} = 0$$
$$\Gamma^\theta_{\theta\theta} = +\frac{1}{2}g^{\theta\theta}\partial_\theta g_{\theta\theta} = 0 \quad \Gamma^\theta_{rr} = -\frac{1}{2}g^{\theta\theta}\partial_\theta g_{rr} = 0$$
$$\Gamma^\theta_{tt} = -\frac{1}{2}g^{\theta\theta}\partial_\theta g_{tt} = 0 \quad \Gamma^\theta_{\theta t} = +\frac{1}{2}g^{\theta\theta}\partial_t g_{\theta\theta} = 0$$
$$\Gamma^\phi_{\theta\theta} = -\frac{1}{2}g^{\phi\phi}\partial_\phi g_{\theta\theta} = 0 \quad \Gamma^\phi_{rr} = -\frac{1}{2}g^{\phi\phi}\partial_\phi g_{rr} = 0$$
$$\Gamma^\phi_{tt} = -\frac{1}{2}g^{\phi\phi}\partial_\phi g_{tt} = 0 \quad \Gamma^\phi_{\phi t} = \frac{1}{2}g^{\phi\phi}\partial_t g_{\phi\phi} = 0. \tag{41.21}$$

The only nonzero symbols are

$$\Gamma^t_{tt} = \frac{1}{2}g^{tt}\partial_t g_{tt} = \frac{\dot\nu}{2}$$
$$\Gamma^t_{rr} = -\frac{1}{2}g^{tt}\partial_t g_{rr} = \frac{\dot\lambda}{2}e^{\lambda-\nu}$$

$$\Gamma^t{}_{rt} = \frac{1}{2}g^{tt}\partial_r g_{tt} = \frac{\nu'}{2}$$

$$\Gamma^r{}_{rr} = \frac{1}{2}g^{rr}\partial_r g_{rr} = \frac{\lambda'}{2}$$

$$\Gamma^r{}_{tt} = -\frac{1}{2}g^{rr}\partial_r g_{tt} = \frac{1}{2}e^{\nu-\lambda}\nu'$$

$$\Gamma^r{}_{tr} = \frac{1}{2}g^{rr}\partial_t g_{rr} = \frac{\dot{\lambda}}{2}$$

$$\Gamma^r{}_{\theta\theta} = -\frac{1}{2}g^{rr}\partial_r g_{\theta\theta} = -re^{-\lambda}$$

$$\Gamma^r{}_{\phi\phi} = -\frac{1}{2}g^{rr}\partial_r g_{\phi\phi} = -re^{-\lambda}\sin^2\theta$$

$$\Gamma^\theta{}_{\phi\phi} = -\frac{1}{2}g^{\theta\theta}\partial_\theta g_{\phi\phi} = -\sin\theta\cos\theta$$

$$\Gamma^\theta{}_{\theta r} = +\frac{1}{2}g^{\theta\theta}\partial_r g_{\theta\theta} = \frac{1}{r}$$

$$\Gamma^\phi{}_{\theta\phi} = +\frac{1}{2}g^{\phi\phi}\partial_\theta g_{\phi\phi} = +\frac{\cos\theta}{\sin\theta}$$

$$\Gamma^\phi{}_{\phi r} = +\frac{1}{2}g^{\phi\phi}\partial_r g_{\phi\phi} = \frac{1}{r}. \tag{41.22}$$

Ricci Tensor Components

We use the formula from the last chapter,

$$R_{\nu\rho} = \partial_\mu\Gamma^\mu{}_{\nu\rho} - \partial_\rho\Gamma^\mu{}_{\nu\mu} + \Gamma^\mu{}_{\lambda\mu}\Gamma^\lambda{}_{\nu\rho} - \Gamma^\mu{}_{\lambda\rho}\Gamma^\lambda{}_{\nu\mu}. \tag{41.23}$$

For the tt component, we obtain

$$\begin{aligned}
R_{tt} &= \partial_\mu\Gamma^\mu{}_{tt} - \partial_t\Gamma^\mu{}_{t\mu} + \Gamma^\mu{}_{\lambda\mu}\Gamma^\lambda{}_{tt} - \Gamma^\mu{}_{\lambda t}\Gamma^\lambda{}_{t\mu} \\
&= \partial_t\Gamma^t{}_{tt} + \partial_r\Gamma^r{}_{tt} - \partial_t\Gamma^r{}_{tr} - \partial_t\Gamma^t{}_{tt} + \Gamma^t{}_{tt}(\Gamma^t{}_{tt} + \Gamma^r{}_{tr}) \\
&\quad + \Gamma^r{}_{tt}(\Gamma^r{}_{rr} + \Gamma^t{}_{rt} + \Gamma^\theta{}_{r\theta} + \Gamma^\phi{}_{r\phi}) - (\Gamma^t{}_{tt})^2 - (\Gamma^r{}_{tr})^2 - 2\Gamma^t{}_{tr}\Gamma^r{}_{tt} \\
&= \frac{1}{2}\partial_r(\nu'e^{\nu-\lambda}) - \frac{1}{2}\partial_t\dot{\lambda} + \frac{\dot{\nu}\dot{\lambda}}{4} - \frac{\dot{\lambda}^2}{4} - \frac{\nu'^2}{2}e^{\nu-\lambda} + \frac{\nu'e^{\nu-\lambda}}{2}\left(\frac{\lambda'}{2} + \frac{\nu'}{2} + \frac{2}{r}\right) \\
&= e^{\nu-\lambda}\left[\frac{\nu''}{2} - \frac{\nu'\lambda'}{4} + \frac{\nu'^2}{4} + \frac{\nu'}{r}\right] - \frac{\ddot{\lambda}}{2} + \frac{\dot{\nu}\dot{\lambda}}{4} - \frac{\dot{\lambda}^2}{4}. \tag{41.24}
\end{aligned}$$

For the rr component, we obtain

$$\begin{aligned}
R_{rr} &= \partial_\mu\Gamma^\mu{}_{rr} - \partial_r\Gamma^\mu{}_{r\mu} + \Gamma^\mu{}_{\lambda\mu}\Gamma^\lambda{}_{rr} - \Gamma^\mu{}_{\lambda r}\Gamma^\lambda{}_{r\mu} \\
&= \partial_r\Gamma^r{}_{rr} + \partial_t\Gamma^t{}_{rr} - \partial_r\Gamma^r{}_{rr} - \partial_r\Gamma^t{}_{rt} - \partial_r\Gamma^\theta{}_{r\theta} - \partial_r\Gamma^\phi{}_{r\phi} + \Gamma^r{}_{rr}(\Gamma^r{}_{rr} + \Gamma^t{}_{rt} + \Gamma^\phi{}_{r\phi} + \Gamma^\theta{}_{r\theta}) \\
&\quad + \Gamma^t{}_{rr}(\Gamma^t{}_{tt} + \Gamma^r{}_{tr}) - (\Gamma^r{}_{rr})^2 - (\Gamma^t{}_{rt})^2 - (\Gamma^\theta{}_{r\theta})^2 - (\Gamma^\phi{}_{r\phi})^2 - 2\Gamma^t{}_{rr}\Gamma^r{}_{rt} \\
&= \frac{1}{2}\partial_t(\dot{\lambda}e^{\lambda-\nu}) - \frac{1}{2}\partial_r\nu' - 2\partial_r\frac{1}{r} + \frac{\lambda'}{2}\left(\frac{\nu'}{2} + \frac{2}{r}\right) + \frac{\dot{\lambda}}{2}e^{\lambda-\nu}\left(\frac{\dot{\nu}}{2} - \frac{\dot{\lambda}}{2}\right) - \frac{\nu'^2}{4} - \frac{2}{r^2} \\
&= -e^{\lambda-\nu}\left[-\frac{\ddot{\lambda}}{2} - \frac{\dot{\lambda}^2}{4} + \frac{\dot{\lambda}\dot{\nu}}{4}\right] - \frac{\nu''}{2} - \frac{\nu'^2}{4} + \frac{\nu'\lambda'}{4} + \frac{\lambda'}{r}. \tag{41.25}
\end{aligned}$$

For the $\theta\theta$ component, we obtain

$$\begin{aligned}
R_{\theta\theta} &= \partial_\mu\Gamma^\mu{}_{\theta\theta} - \partial_\theta\Gamma^\mu{}_{\theta\mu} + \Gamma^\mu{}_{\lambda\mu}\Gamma^\lambda{}_{\theta\theta} - \Gamma^\mu{}_{\lambda\theta}\Gamma^\lambda{}_{\theta\mu} \\
&= \partial_r\Gamma^r{}_{\theta\theta} - \partial_\theta\Gamma^\phi{}_{\theta\phi} + \Gamma^r{}_{\theta\theta}(\Gamma^r{}_{rr} + \Gamma^t{}_{rt} + \Gamma^\theta{}_{r\theta} + \Gamma^\phi{}_{r\phi}) - (\Gamma^\phi{}_{\theta\phi})^2 - 2\Gamma^\theta{}_{r\theta}\Gamma^r{}_{\theta\theta}
\end{aligned}$$

$$= \partial_r(-re^{-\lambda}) - \partial_\theta\left(\frac{\cos\theta}{\sin\theta}\right) + (-re^{-\lambda})\left(\frac{\lambda'}{2} + \frac{\nu'}{2} + \frac{2}{r}\right) - \left(-\frac{\cos\theta}{\sin\theta}\right)^2 - 2(-re^{-\lambda})\frac{1}{r}$$

$$= e^{-\lambda}\left[-1 + \frac{r\lambda'}{2} - \frac{r\nu'}{2}\right] + 1. \tag{41.26}$$

For the $\phi\phi$ component, we obtain

$$R_{\phi\phi} = \partial_\mu\Gamma^\mu{}_{\phi\phi} - \partial_\phi\Gamma^\phi{}_{\phi\mu} + \Gamma^\mu{}_{\lambda\mu}\Gamma^\lambda{}_{\phi\phi} - \Gamma^\mu{}_{\lambda\phi}\Gamma^\lambda{}_{\phi\mu}$$
$$= \partial_r\Gamma^r{}_{\phi\phi} + \partial_\theta\Gamma^\theta{}_{\phi\phi} + \Gamma^r{}_{\phi\phi}(\Gamma^r{}_{rr} + \Gamma^t{}_{rt} + \Gamma^\theta{}_{r\theta} + \Gamma^\phi{}_{r\phi})$$
$$+\Gamma^\theta{}_{\phi\phi}\Gamma^\phi{}_{\theta\phi} - 2\Gamma^\phi{}_{r\phi}\Gamma^r{}_{\phi\phi} - 2\Gamma^\phi{}_{\theta\phi}\Gamma^\theta{}_{\phi\phi}$$
$$= \partial_r(-re^{-\lambda}\sin^2\theta) - \partial_\theta(\sin\theta\cos\theta) + (-re^{-\lambda}\sin^2\theta)\left(\frac{\lambda'}{2} + \frac{\nu'}{2} + \frac{2}{r}\right)$$
$$+(-\sin\theta\cos\theta)\left(+\frac{\cos\theta}{\sin\theta}\right) - 2\frac{1}{r}(-re^{-\lambda}\sin^2\theta) - 2\left(\frac{\cos\theta}{\sin\theta}\right)(-\sin\theta\cos\theta)$$
$$= e^{-\lambda}\sin^2\theta\left[-1 + \frac{r\lambda'}{2} - \frac{r\nu'}{2}\right] + \sin^2\theta. \tag{41.27}$$

That means that the Ricci scalar is

$$R = g^{rr}R_{rr} + g^{tt}R_{tt} + g^{\theta\theta}R_{\theta\theta} + g^{\phi\phi}R_{\phi\phi}$$
$$= e^{-\lambda}R_{rr} - e^{-\nu}R_{tt} + \frac{R_{\theta\theta}}{r^2} + \frac{R_{\phi\phi}}{r^2\sin^2\theta}$$
$$= e^{-\nu}\left[\ddot\lambda + \frac{\dot\lambda^2}{2} - \frac{\dot\lambda\dot\nu}{2}\right] + e^{-\lambda}\left[-\nu'' - \frac{\nu'^2}{2} + \frac{\nu'\lambda'}{2} + 2\frac{\lambda' - \nu'}{r} - \frac{2}{r^2}\right] + \frac{2}{r^2}. \tag{41.28}$$

We also obtain for the component tr of the Ricci tensor,

$$R_{tr} = \partial_\mu\Gamma^\mu{}_{tr} - \partial_t\Gamma^\mu{}_{r\mu} + \Gamma^\mu{}_{\lambda\mu}\Gamma^\lambda{}_{tr} - \Gamma^\mu{}_{t\lambda}\Gamma^\lambda{}_{r\mu}$$
$$= \partial_t\Gamma^t{}_{tr} + \partial_r\Gamma^r{}_{tr} - \partial_t(\Gamma^r{}_{rr} + \Gamma^t{}_{rt} + \Gamma^\phi{}_{r\phi} + \Gamma^\theta{}_{r\theta}) + \Gamma^t{}_{tr}(\Gamma^t{}_{tt} + \Gamma^r{}_{tr})$$
$$+\Gamma^r{}_{tr}(\Gamma^r{}_{rr} + \Gamma^t{}_{rt} + \Gamma^\phi{}_{r\phi} + \Gamma^\theta{}_{r\theta}) - \Gamma^t{}_{rt}\Gamma^t{}_{tt} - \Gamma^r{}_{rr}\Gamma^r{}_{tr} - \Gamma^t{}_{rr}\Gamma^r{}_{tt} - \Gamma^r{}_{rt}\Gamma^t{}_{tr}$$
$$= \frac{1}{2}\partial_r\dot\lambda - \partial_t\left(\frac{\lambda'}{2} + \frac{2}{r}\right) + \frac{\nu'\dot\nu}{4} - \frac{\nu'\dot\nu}{4} - \frac{\lambda'\dot\lambda}{4} + \frac{\dot\lambda}{2}\left(\frac{\lambda'}{2} + \frac{2}{r}\right) + \frac{\dot\lambda e^{\lambda-\nu}}{2} - \frac{\nu'e^{\nu-\lambda}}{2} - \frac{\dot\lambda\nu'}{4}. \tag{41.29}$$

leading to

$$R_{tr} = \frac{\dot\lambda}{r}. \tag{41.30}$$

41.3 Final Solution of Equations of Motion

Putting everything together, we obtain

$$G_{tt} = R_{tt} - \frac{1}{2}g_{tt}R = e^{\nu-\lambda}\left(-\frac{1}{r^2} + \frac{\lambda'}{r}\right) + \frac{e^\nu}{r^2} = 0$$

$$G_{rr} = R_{rr} - \frac{1}{2}g_{rr}R = \frac{1}{r^2} + \frac{\nu'}{r} - \frac{e^\lambda}{r^2} = 0. \tag{41.31}$$

Then

$$G_{tt} + G_{rr}e^{\nu-\lambda} = \frac{\nu' + \lambda'}{r}e^{\nu-\lambda} = 0, \qquad (41.32)$$

and $R_{tr} = \dot{\lambda} = 0$, implying together $\nu = -\lambda$, and that it is only a function of r.

Note that the remaining Einstein's equations are all solved by $\lambda = -\nu$.

Since at $r \to \infty$, where the fields are weak, we have

$$e^{\nu} = 1 + 2U_N = 1 - \frac{2MG_N}{r}, \qquad (41.33)$$

and we can check that this solves $G_{tt} = 0$, we have the final solution (Schwarzschild)

$$ds^2 = -dt^2 \left(1 - \frac{2MG_N}{r}\right) + \frac{dr^2}{1 - \frac{2MG_N}{r}} + r^2 d\Omega_2^2. \qquad (41.34)$$

The special radius $r = r_S = 2MG_N$ is called the *Schwarzschild radius*, and the locus $r = r_S$ the *event horizon*. Even though the metric looks singular at r_S, this is only because of a bad choice of coordinates. We can see this because the invariant measure of curvature, the Ricci scalar R, is calculated to be finite at the event horizon. In fact, there are coordinates, called Kruskal coordinates, where there are no singularities for the event horizon. Nevertheless, the presence of the horizon brings many unusual physical properties, like Hawking radiation and apparent loss of information, which will not be discussed here.

The solution we have found is the most general solution of the vacuum Einstein's equations with spherical symmetry. One can rigorously prove this, a statement known as *Birkhoff's theorem*. Here we have proven it more on physical grounds (when writing the ansatz). But this means that the gravitational field of any spherically symmetric mass distribution, like the gravitational field of the Earth, but *only outside the Earth* (where there is no matter) is described by it. If instead there is no matter all the way down to $r = r_S$, so the metric is valid until there, we say we have a *black hole*, see Figure 41.1.

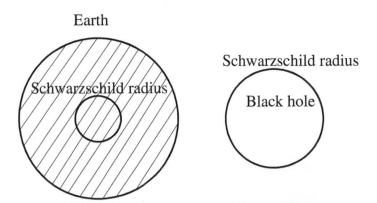

Figure 41.1   For the Earth, the Schwarzschild radius is inside the Earth, where the Schwarzschild metric is not valid anymore. For a black hole, the Schwarzschild radius is the limit of the black hole, the event horizon.

Further Reading

See chapter 12 in [3] and chapter 11 in [40].

Exercises

(1) Is

$$U_N = -\frac{M_1 G_N}{|\vec{x} - \vec{x}_1|} - \frac{M_2 G_N}{|\vec{x} - \vec{x}_2|} \tag{41.35}$$

a gravitational solution in the Newtonian limit? How about for the full Schwarzschild case? Why?

(2) What is wrong with the Schwarzschild solution with the opposite sign for

$$U_N = 1 - \frac{2MG_N}{r} \to 1 + \frac{2MG_N}{r}\ ? \tag{41.36}$$

(3) Check that the Schwarzschild solution $\dot{\lambda} = 0, \lambda = -\nu, e^\lambda = 1/(1 - 2MG_N/r)$ satisfies

$$R_{\theta\theta} - \frac{1}{2}g_{\theta\theta}R = 0 \quad \text{and} \quad R_{\phi\phi} - \frac{1}{2}g_{\phi\phi}R = 0. \tag{41.37}$$

(4) Consider the Euclidean signature version of the Schwarzschild solution, with $-dt^2 \to +dt^2$ in the metric, and the metric near the event horizon, $r = 2MG_N + \delta r$, $\delta r \ll 2MG_N$. What is the periodicity of the time coordinate t such that there is no singularity at $\delta r = 0$?

42 Deflection of Light by the Sun and Comparison with General Relativity

In this chapter, we will consider the first important test of general relativity: reproducing the deflection of light, coming from a distant star, by the Sun. This was the test to decide between general relativity and special relativity plus Newtonian gravity (that gives half the correct answer) that Einstein proposed, and was soon tested experimentally in 1919 by a team led by Arthur Eddington, who made observations of a total eclipse of the Sun in Brazil and in Africa. They could observe distant stars, whose light came close to the Sun, and compare the apparent positions of the starts on the sky with ones in the Sun's absence from the place. They found that general relativity is correct, though modern analyses of the data tend to say that the errors were too large to distinguish between them. Either way, since then, many more accurate observations have been made that agree with the general relativity result.

42.1 Motion of Light As Motion in a Medium with Small, Position-Dependent Index of Refraction

Light moves on null trajectories, with $ds^2 = 0$. In a curved spacetime, it will move on a *null geodesic*. For a massive particle (timelike geodesic), we have found the equation for the geodesic – i.e., the path of shortest distance – coming from the action $S = -mc^2 \int d\tau$, as

$$\frac{Du^\mu}{d\tau} = 0 \quad \leftrightarrow \quad \frac{du^\mu}{d\tau} + \Gamma^\mu{}_{\nu\rho} u^\nu u^\sigma = 0. \tag{42.1}$$

We can rewrite it by using $p^\mu = mu^\mu$ for nonzero mass.

For light, with $m = 0$, $d\tau = \sqrt{-ds^2} = 0$, so we cannot use it as a parameter. We must instead use another parameter λ on the worldline, called *affine parameter*. We define

$$p^0 \equiv \frac{dx^0}{d\lambda}, \tag{42.2}$$

and then, since $p^2 \equiv g_{\mu\nu} p^\mu p^\nu = 0$, we have also

$$p^i = \frac{dx^i}{d\lambda}. \tag{42.3}$$

The equation for a null geodesic is then the same one, but with τ replaced by λ, as we will show in more detail shortly, namely

$$\frac{Du^{\mu}}{d\lambda} = 0 : \quad \frac{d^2 x^{\mu}}{d\lambda^2} + \Gamma^{\mu}{}_{\nu\rho} \frac{dx^{\nu}}{d\lambda} \frac{dx^{\rho}}{d\lambda} = 0, \tag{42.4}$$

to which we need to add $p^2 = 0$.

In the case of $m \neq 0$,

$$p^{\mu} = \frac{\partial L}{\partial \dot{x}^{\mu}} = m g_{\mu\nu} u^{\nu} \tag{42.5}$$

is the canonically conjugate momentum to $x^{\mu}(\tau)$.

But for $m = 0$, the action vanishes for two reasons: the coefficient m vanishes, and $d\tau = 0$, so we need to write another action. The procedure we want to use is to go from the above *second-order action* to a *first-order action*. First order can mean first order in derivatives (turning an action second order in derivatives into one first order in them), or first order in fields (turning an action which is at least quadratic in fields into one that has also linear terms). It can be useful in a case like this to turn an action that is highly nonlinear (square root of a sum of terms) into one that is simple and quadratic. The only way we can do that is by introducing an auxiliary field, whose equation of motion is algebraic (has no derivatives), and when replacing it back into the action we get the original form.

In our case, we find a first-order action by introducing an auxiliary field $e(\tau)$ on the worldline parametrized by τ,

$$S = \frac{1}{2} \int d\tau \left[e(\tau)^{-1} \dot{X}^{\mu} \dot{X}^{\nu} g_{\mu\nu} - e(\tau) m^2 \right]. \tag{42.6}$$

We can check that solving the equation for $e(\tau)$,

$$2 \frac{\delta S}{\delta e(\tau)} = -e(\tau)^{-2} \dot{X}^2 - m^2 = 0 \Rightarrow e(\tau) = \frac{1}{m} \sqrt{-\dot{X}^{\mu} \dot{X}^{\nu} g_{\mu\nu}}, \tag{42.7}$$

and replacing in the action, we get back the previous one,

$$S = \frac{1}{2} \int d\tau \left[m \frac{\dot{X}^{\mu} \dot{X}^{\nu} g_{\mu\nu}}{\sqrt{-\dot{X}^{\mu} \dot{X}^{\nu} g_{\mu\nu}}} - m \sqrt{-\dot{X}^{\mu} \dot{X}^{\nu} g_{\mu\nu}} \right] = -m \int d\tau \sqrt{-\dot{X}^{\mu} \dot{X}^{\nu} g_{\mu\nu}}. \tag{42.8}$$

The difference is that now, for (42.6), we can take the $m \rightarrow 0$ limit, in the process also replacing τ with the affine parameter λ.

But, moreover, the action (42.6) has a symmetry, *reparametrization invariance*, which is the one-dimensional version of general coordinate invariance. The symmetry is for an arbitrary change $\tau' = \tau'(\tau)$, under which

$$e'(\tau') d\tau' = e(\tau) d\tau. \tag{42.9}$$

Then, indeed, the second term in the action is automatically invariant, and the first is also, since it contains $e(\tau) d\tau$ in the denominator (and the numerator is independent of both $d\tau$ and $e(\tau)$). But such an invariance (under a "gauge symmetry") means that we can fix a gauge for it, and we can choose to fix $e(\tau) = 1$. Finally, then, the gauge fixed first-order action for the massless particle is

$$S = \frac{1}{2} \int d\lambda \dot{X}^{\mu} \dot{X}^{\nu} g_{\mu\nu}(X^{\rho}(\lambda)). \tag{42.10}$$

Just like in the massive particle case, by varying the action just discussed, we find the same equation of motion, using the same steps as in Chapter 38,

$$\frac{\delta S}{\delta X^\mu(\lambda)} = -\frac{d}{d\lambda}(\dot{X}^\nu g_{\mu\nu}) + \frac{1}{2}(\partial_\mu g_{\nu\rho})\dot{X}^\nu \dot{X}^\rho = 0, \tag{42.11}$$

where the first term comes from the variation of X under the dot and the second from the variation in $g_{\mu\nu}(X)$, and implying in the same way

$$g_{\mu\nu}\frac{d}{d\lambda}u^\nu = -\left(\frac{d}{d\lambda}g_{\mu\nu}\right)u^\nu + \frac{1}{2}\partial_\mu g_{\nu\rho}u^\nu u^\rho = -g_{\mu\nu}\Gamma^\nu{}_{\rho\sigma}u^\nu u^\sigma \Rightarrow \frac{Du^\nu}{d\lambda} = 0. \tag{42.12}$$

We can now identify λ with t (the normal time parameter), which means, however, that now we must consider p^0 as an independent constant, different than $dx^0/d\lambda = 1$, so

$$\dot{X}^0 = 1, \quad \frac{p^i}{p^0} = \frac{dX^i}{dX^0} = \frac{dX^i}{dt}. \tag{42.13}$$

Then, in terms of p^μ, we can write the equation of motion as

$$\frac{dp_\mu}{dt} = \frac{1}{2}(\partial_\mu g_{\nu\rho})p^\nu \dot{x}^\rho. \tag{42.14}$$

The metric we are interested in is of the Newtonian limit type, i.e.,

$$ds^2 = -(1 + 2U_N)dt^2 + (1 - 2U_N)\delta_{ij}dx^i dx^j, \tag{42.15}$$

so

$$g_{00} = -(1 + 2U_N), \quad g_{ij} = (1 - 2U_N)\delta_{ij}. \tag{42.16}$$

Then the $\mu = 0$ component of the equation of motion gives, *for a static Newtonian potential*, $\partial_0 g_{\mu\nu} = 0$,

$$-\frac{d(g_{00}p^0)}{dt} = \frac{d}{dt}[(1 + 2U_N)p^0] = -\frac{1}{2}(\partial_0 g_{\nu\rho})p^\nu \dot{x}^\rho = 0. \tag{42.17}$$

The $\mu = i$ component, on the other hand, gives

$$\frac{d(g_{ij}p^j)}{dt} = \frac{d}{dt}[(1 - 2U_N)p^i] = \frac{1}{2}\partial_i[-(1 + 2U_N)]p^0 \dot{X}^0 + \frac{1}{2}\partial_i(1 - 2U_N)p^i \dot{X}^i. \tag{42.18}$$

Using (42.13), we obtain

$$\frac{d}{dt}[(1 - 2U_N)p^i] = -(\partial_i U_N)p^0\left[1 + \frac{\vec{p}^2}{(p^0)^2}\right]. \tag{42.19}$$

The condition that the geodesic is null, $p^2 = 0$, becomes

$$0 = p^\mu p^\nu g_{\mu\nu} = -(1 + 2U_N)(p^0)^2 + (1 - 2U_N)\vec{p}^2 \Rightarrow$$
$$\left(\frac{dX^i}{dt}\right)^2 = \frac{\vec{p}^2}{(p^0)^2} = \frac{1 + 2U_N}{1 - 2U_N} \simeq 1 + 4U_N, \tag{42.20}$$

but the second term is subleading, and can be neglected, if we multiply (42.20) by $\partial_i U_N$. Then we obtain finally for the i component of the equation of motion, in the Newtonian approximation (leading order)

$$\frac{1}{p^0}\frac{d}{dt}[(1-2U_N)p^i] = -2\partial_i U_N. \tag{42.21}$$

But together with the $\mu = 0$ component in (42.17), we can then write

$$\frac{d}{dt}\left[(1-4U_N)\frac{p^i}{p^0}\right] = -2\partial_i U_N. \tag{42.22}$$

But since $p^i/p^0 = dX^i/dt$, we finally find the equation of motion for the null geodesic in the Newtonian approximation,

$$\frac{d}{dt}\left[(1-4U_N)\frac{dX^i}{dt}\right] = -2\partial_i U_N. \tag{42.23}$$

This, however, is nothing but the equation of motion for light in a medium with variable index of refraction

$$n(\vec{x}) = \frac{c}{v(\vec{x})} = 1 - 2U_N(\vec{x}). \tag{42.24}$$

Indeed, Fermat's principle of least time in optics is equivalent to the action

$$S = \int n(\vec{x})ds = \int dt\sqrt{n^2(\vec{x})\frac{dx^i}{dt}\frac{dx^i}{dt}}, \tag{42.25}$$

which leads to the equation of motion (since $dx^i/dt = v = 1/n$ if $c = 1$)

$$\frac{d}{dt}\left[n^2(\vec{x})\frac{dx^i}{dt}\right] = \partial_i n(\vec{x}). \tag{42.26}$$

Thus, we have obtained a motion that can be described entirely using geometric optics. Then Snell's law is $n\sin\theta =$ constant, which for small angles is $n\theta =$ constant. Since $\theta = v_\perp/v_\parallel$, we get

$$-\frac{\delta\theta}{\theta} = \frac{\delta n}{n} \Rightarrow -\frac{\delta\theta}{\delta t} \simeq \frac{v_\perp}{v_\parallel}\frac{\delta n}{\delta t} \simeq \frac{v_\perp}{v_\parallel}\frac{\delta_\parallel n}{\delta t} \simeq \frac{\delta_\perp n}{\delta t}. \tag{42.27}$$

With $n(\vec{x}) = 1 - 2U_N(\vec{x})$, we get

$$\frac{\delta\theta}{\delta t} = \frac{\delta v_\perp/\delta t}{v_\parallel} \simeq -2\partial_\perp U_N, \tag{42.28}$$

and since $v_\parallel \simeq c = 1$ and $\delta t = \delta y/c = \delta y$, where y is the parallel direction to the original propagation direction and x is the one perpendicular, we obtain

$$-\frac{\delta\theta}{\delta y} = -2\partial_x U_N \Rightarrow -\theta = -2\int dy\partial_x U_N. \tag{42.29}$$

For a point source like the Sun, which is what we are interested in, with

$$U_N(r) = -\frac{MG_N}{r} = -\frac{MG_N}{\sqrt{x^2+y^2}}, \tag{42.30}$$

the total deflection angle, for light coming from $y = -\infty$ and reaching $y = +\infty$, we obtain

$$-\theta = -2MG_N x \int_{-\infty}^{+\infty} \frac{dy}{(x^2+y^2)^{3/2}} = \frac{4MG_N}{x}. \tag{42.31}$$

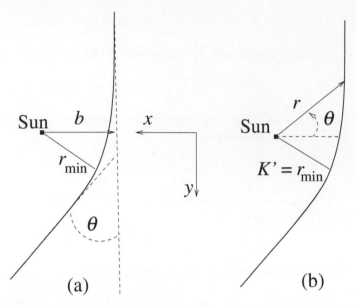

Figure 42.1 Deflection of light by the Sun. (a) Treatment in geometric optics, in (x, y) coordinates. (b) Treatment using Hamilton–Jacobi equation, in (r, θ) coordinates.

Here x was assumed to be approximately constant, the minimum distance from the Sun that the light passes at, approximately the same as the impact parameter (the distance from the Sun to the original direction of light, at $-\infty$), see Figure 42.1a.

42.2 Formal Derivation Using the Hamilton–Jacobi Equation

We can calculate more formally and precisely the equation of motion, having to make the Newtonian approximation only at the very end of the calculation, using the Hamilton–Jacobi equation reviewed in Chapter 1.

The Hamilton–Jacobi equation replaces the momentum p_μ by $\partial S/\partial x^\mu$, where S is Hamilton's principal function, in the Hamiltonian equation $H(\{x\}, \{p\}) = -\partial S/\partial t$, so

$$H\left(x_i, \frac{\partial S}{\partial x^i}\right) = -\frac{\partial S}{\partial t}. \tag{42.32}$$

However, in a generally relativistic theory, the Hamiltonian is the energy, part of p^μ, so the equation defining the Hamiltonian is really

$$p_\mu p_\nu g^{\mu\nu} = -m^2. \tag{42.33}$$

For light, $m = 0$, but the method can be used more generally for massive particles.
Therefore, we have the Hamiltoni–Jacobi equation

$$g^{\mu\nu} \frac{\partial S}{\partial x^\mu} \frac{\partial S}{\partial x^\nu} = 0. \tag{42.34}$$

In order for this to be a real Hamilton–Jacobi equation, we still need to substitute in

$$-\frac{\partial S}{\partial t} = H = E = \omega_0 \quad (\hbar = 1). \tag{42.35}$$

The solution of the equation will be to write a $S = S(x^\mu)$.

Finally, in the Hamilton–Jacobi method, to solve for the trajectory, we need to set $\partial S/\partial p_\mu$ (where $p_\mu = \partial S/\partial x^\mu$, as we said, and are themselves constants α_μ) equal to constants β^μ.

We now apply this method for the Schwarzschild metric,

$$ds^2 = -\left(1 - \frac{r_S}{r}\right)dt^2 + \frac{dr^2}{1 - \frac{r_S}{r}} + r^2(d\theta^2 + \sin^2\theta d\phi^2), \tag{42.36}$$

which, therefore has

$$g^{00} = -\left(1 - \frac{r_S}{r}\right)^{-1}, \quad g^{rr} = 1 - \frac{r_S}{r}, \quad g^{\theta\theta} = \frac{1}{r^2}. \tag{42.37}$$

Note that we are interested in deflection by the center (the Sun), so we choose (r, θ) to be the planar coordinates in the plane formed by the trajectory and the center. The angle ϕ would then take us out of this plane, that is why we don't consider it.

The Hamilton–Jacobi equation is then

$$-\left(1 - \frac{r_S}{r}\right)^{-1}\left(\frac{\partial S}{\partial t}\right)^2 + \left(1 - \frac{r_S}{r}\right)\left(\frac{\partial S}{\partial r}\right)^2 + \frac{1}{r^2}\left(\frac{\partial S}{\partial \theta}\right)^2 = 0. \tag{42.38}$$

Since we want to set

$$\frac{\partial S}{\partial t} = -E = -p^0, \quad \frac{\partial S}{\partial \theta} \equiv K = p^\theta, \tag{42.39}$$

we solve the H–J equation by separation of variables, writing

$$S = -Et + K\theta + S_r(r). \tag{42.40}$$

Moreover, we put $E = \omega_0$, and write $K \equiv K'\omega_0$. Then the H–J equation becomes an equation for $S_r(r)$,

$$-\left(1 - \frac{r_S}{r}\right)^{-1}\omega_0^2 + \left(1 - \frac{r_S}{r}\right)\left(\frac{\partial S_r}{\partial r}\right)^2 + \frac{K'^2\omega_0^2}{r^2} = 0, \tag{42.41}$$

easily integrated to

$$S_r(r) = \int dr \sqrt{\frac{\omega_0^2}{\left(1 - \frac{r_S}{r}\right)^2} - \frac{K^2/r^2}{1 - \frac{r_S}{r}}}. \tag{42.42}$$

In order to find the trajectory, as we said, according to the geneal Hamilton–Jacobi theory, we need to put $\partial S/\partial \omega_0$ to a constant in order to find $t = t(r)$, and $\partial S/\partial K$ to another constant, in order to find $\theta = \theta(r)$.

We find

$$\frac{\partial S}{\partial \omega_0} = -t + \omega_0 \int \frac{dr/\left(1 - \frac{r_S}{r}\right)^2}{\sqrt{\frac{\omega_0^2}{\left(1 - \frac{r_S}{r}\right)^2} - \frac{K^2/r^2}{1 - \frac{r_S}{r}}}} = -t_0, \tag{42.43}$$

and we can put the constant t_0 to zero for simplicity (otherwise, we just shift t by it) to finally find

$$t = \omega_0 \int \frac{dr/\left(1 - \frac{r_S}{r}\right)^2}{\sqrt{\frac{\omega_0^2}{\left(1-\frac{r_S}{r}\right)^2} - \frac{K^2/r^2}{1-\frac{r_S}{r}}}} = t(r), \tag{42.44}$$

which inverts to $r = r(t)$ (the trajectory as a function of time).

For deflection, however, we are interested in the trajectory $r(\theta)$, coming from putting $\partial S/\partial K$ to a constant, which we call θ_0 and can also put to zero for simplicity. We then obtain

$$\begin{aligned}
\theta &= K \int \frac{dr/r^2}{\left(1 - \frac{r_S}{r}\right)\sqrt{\frac{\omega_0^2}{\left(1-\frac{r_S}{r}\right)^2} - \frac{K^2/r^2}{1-\frac{r_S}{r}}}} \\
&= K' \int \frac{dr}{r^2\sqrt{1 - \frac{K'^2}{r^2}\left(1 - \frac{r_S}{r}\right)}} = \theta(r),
\end{aligned} \tag{42.45}$$

which inverts to $r(\theta)$. However, we can only solve equation (42.45) numerically for large r_S/r.

For (always) small r_S/r, however, which translates into small deflection angle, we can make an approximation and solve the integral. More precisely, we go back to the original form of $S_r(r)$ in (42.42) and make the change of coordinates

$$r(r - r_S) = r'^2, \tag{42.46}$$

which gives

$$S_r(r) = \omega_0 \int dr \sqrt{\left(1 - \frac{r_S}{r}\right)^{-2} - \frac{K'^2}{r'^2}} \simeq \omega_0 \int dr' \sqrt{1 + \frac{2r_S}{r'} - \frac{K'^2}{r^2}}, \tag{42.47}$$

where we have Taylor expanded in r_S/r inside the square root, assuming that K'^2/r'^2 can still be large. Now Taylor expanding in r_S/r the square root itself, we find

$$S_r(r) \simeq S_r^{(0)} + \omega_0 r_S \int^r \frac{dr'}{r'\sqrt{1 - \frac{K'^2}{r'^2}}} = S_r^{(0)} + \omega_0 r_S \cosh^{-1}(r/K'). \tag{42.48}$$

More precisely, we integrate between $r = R \to \infty$ and the inflection point, which we can easily see from the integral that is at $r = K'$. Indeed, the trajectory starts at $r = \infty$, goes to the minimum at $r = K'$, then back to infinity (see Figure 42.1b). Then the total variation of $\Delta S_r(r)$ is twice the variation from $r = R \to \infty$ to $r = K' -$ i.e.,

$$\Delta S_r(r) = \Delta S_r^{(0)} + 2\omega_0 r_S \cosh^{-1}(R/K'). \tag{42.49}$$

The equation for the trajectory is found as before, by putting $\partial S/\partial K'$ to a constant θ_0, leading to

$$\theta - \theta_0 \equiv \Delta\theta = -\frac{\partial \Delta S_r}{\partial K} = \frac{2r_S R}{K'\sqrt{R^2 - K'^2}} \to \frac{2r_S}{K'}. \tag{42.50}$$

The deflection angle is, therefore,

$$\Delta\theta = \frac{2r_S}{K'} = \frac{4MG_N}{K'},$$

(42.51)

where $K' = r_{\min}$, the minimum distance from the center along the geodesic, approximately equal to x from the Newtonian limit calculation.

42.3 Comparison with Special Relativity

In special relativity plus Newtonian gravity, we would use the sum of the two actions, so

$$S = -mc^2 \int d\lambda \sqrt{-\dot{X}^\mu \dot{X}^\nu \eta_{\mu\nu}} - m \int d\lambda \, U_N.$$

(42.52)

We obtain the equation of motion

$$\frac{d^2 X^i}{dt^2} = -m\partial_i U_N(\vec{x}) = \frac{1}{2}\partial_i n(\vec{x}),$$

(42.53)

which through the same geometric optics derivation gives the only half the deviation (due to the one half on the right-hand side),

$$\Delta\theta = \frac{r_S}{K'} = \frac{2MG_N}{K'}.$$

(42.54)

Thus indeed, there is a difference of a factor of 2 between the special relativistic and the general relativistic results, as advertised.

Important Concepts to Remember

- Light in a curved spacetime moves on null geodesics with equation of motion $Du^\mu/d\lambda = 0$, where λ is the affine parameter on the null worldline.
- We can write a first-order action for the massless particle that doesn't go to zero, $S = \frac{1}{2}\int d\tau \left[e^{-1}(\tau)\frac{dX^\mu}{d\tau}\frac{dX^\nu}{d\tau}g_{\mu\nu} - e(\tau)m^2\right]$.
- The action has reparametrization invariance and becomes for $m = 0$ and $\tau \rightarrow \lambda$, in the $e(\tau) = 1$ gauge, $S = \frac{1}{2}\int d\lambda \dot{X}^\mu \dot{X}^\nu g_{\mu\nu}$.
- The deflection of light by a point source (Sun) situated at a distance x from the incoming light ray is, for small angles, $\theta = 4MG_N/x$.
- The Hamilton–Jacobi equation for the massless particle in curved spacetime is $\frac{\partial S}{\partial x^\mu}\frac{\partial S}{\partial x^\nu}g^{\mu\nu} = 0$, with $-\frac{\partial S}{\partial t} = H = E$. From it, we can find the trajectory by putting $p_\mu = \partial S/\partial x^\mu$ and $\partial S/\partial p^\mu$ to constants.
- In special relativity plus Newtonian gravity, we get half the deflection, $\Delta\theta = 2MG_N/x$.

Further Reading

See chapter 12 in [3].

Exercises

(1) Consider two stars, S_1 and S_2, situated at the same angles on the sky, but at different distances from us. Light from a very distant object is deflected by the farthest object, S_1, passes between S_1 and S_2, and is deflected again such as to come to us, giving an image identical to the original one. Find the necessary condition for this scenario.

(2) Light from a distant object in the sky, behind a star situated at a distance d from us, is deflected such that we observe the light as a circle of angular diameter δ on the sky. What is the mass M of the star?

(3) Find perturbatively $r = r(t)$ for deflection of light by the Sun, from the Hamilton–Jacobi solution, similar to how we found $\theta(r)$ and $\Delta\theta$.

(4) Write explicitly the Hamilton–Jacobi equation for a *massive* particle being deflected in the Schwarzschild geometry, in a similar manner to the massless case (you don't need to solve it).

43

Fully Linear Gravity: Parallel Plane (pp) Waves and Gravitational Shock Wave Solutions

Until now, we have seen that general relativity is highly nonlinear, and, therefore, gravitational solutions are complicated and cannot be added up to give a new solution. But there is one important counterexample called parallel plane (pp) waves, for which we will see that two waves traveling in the same direction *can* be added up to give a new solution, so we do have a restricted linearity for this set of solutions. Note that we are not referring to the linearized gravity, for which, of course, we have linearity by construction: two weak gravitational waves add up to another weak gravitational wave. We are referring to the full Einstein's gravity theory.

We will further see that we can put such a pp wave in a background spacetime in some cases. Perhaps the most important application of pp wave is the Penrose limit, which is a limit on a curved background, where we look near a null geodesic, and we find that the metric is always a pp wave in flat spacetime. Another application is the gravitational shock wave, which is the limit spacetime obtained for a particle – or, more generally, a distribution of particles – moving at the speed of light. Finally, there is one example of a nontrivial nonlinear superposition of two gravitational solutions, the Khan and Penrose solutions, describing the superposition of two gravitational shock waves *moving in orthogonal directions* and *at all times*, even after the collision.

43.1 PP Waves

As the name suggests, parallel plane (pp) waves are solutions for which a plane moves (parallel to its initial position) in a perpendicular direction at the speed of light. They are usually considered to move in flat background, so the metric for the spacetime is

$$ds^2 = 2dx^+ dx^- + H(x^+, x^i)(dx^+)^2 + \sum_i dx^i dx^i. \tag{43.1}$$

This is called the *Brinkmann form* of the pp wave.

Here, the *lightcone coordinates* x^+, x^- are defined in terms of Minkowskian coordinates as

$$x^\pm = \frac{x \pm t}{\sqrt{2}}. \tag{43.2}$$

As we see, the metric depends on a single function $H(x^+, x^i)$ generally depending both on the lightcone direction x^+ (called "lightcone time") and the transverse directions x^i, and it moves on the $x^+ = $ constant direction (for instance, we will see later an example where $H(x^+, x^i) = \delta(x^+)h(x^i)$, in which case we only have $x^+ = 0$, or $x = -t$).

The metric $g_{\mu\nu}$ in the (x^+, x^-) directions is then in matrix form

$$G = \begin{pmatrix} H & 1 \\ 1 & 0 \end{pmatrix} \Rightarrow G^{-1} = \begin{pmatrix} 0 & 1 \\ 1 & -H \end{pmatrix}, \tag{43.3}$$

which means that $g^{--} = -H$, $g^{+-} = 1$, $g^{++} = 0$.

Christoffel Symbols and Ricci Tensors

We consider case by case for the upper index of the Christoffel symbol. We have

$$\Gamma^i{}_{\mu\nu} = \frac{1}{2}\delta^{ij}(\partial_\mu g_{\nu j} + \partial_\nu g_{\mu j} - \partial_j g_{\mu\nu}), \tag{43.4}$$

which gives

$$\Gamma^i{}_{++} = -\frac{1}{2}\partial_i H \tag{43.5}$$

and the rest are zero. Next we check that

$$\Gamma^+{}_{\mu\nu} = \frac{1}{2}g^{+-}(\partial_\mu g_{\nu-} + \partial_\nu g_{\mu-} - \partial_- g_{\mu\nu}) \tag{43.6}$$

has all components zero. Finally

$$\Gamma^-{}_{\mu\nu} = \frac{1}{2}g^{-+}(\partial_\mu g_{\nu+} + \partial_\nu g_{\mu+} - \partial_+ g_{\mu\nu}) \tag{43.7}$$

has as only nonzero components

$$\Gamma^-{}_{++} = \frac{1}{2}\partial_+ g_{++} = \frac{1}{2}\partial_+ H, \quad \Gamma^-{}_{+i} = \frac{1}{2}\partial_i g_{++} = \frac{1}{2}\partial_i H. \tag{43.8}$$

Since the only nonzero components are $\Gamma^-{}_{++}, \Gamma^-{}_{+i}$ and $\Gamma^i{}_{++}$, it is easy to see that we also have $\Gamma^\mu{}_{\lambda\mu} = 0$. That means that our formula for the Ricci tensor simplifies

$$R_{\nu\rho} = \partial_\mu \Gamma^\mu{}_{\nu\rho} - \partial_\rho \Gamma^\mu{}_{\nu\mu} + \Gamma^\mu{}_{\lambda\mu}\Gamma^\lambda{}_{\nu\rho} - \Gamma^\mu{}_{\lambda\rho}\Gamma^\lambda{}_{\nu\mu} = \partial_\mu \Gamma^\mu{}_{\nu\rho} - \Gamma^\mu{}_{\lambda\rho}\Gamma^\lambda{}_{\nu\mu}. \tag{43.9}$$

We first check that

$$\begin{aligned} R_{\nu-} &= \partial_\mu \Gamma^\mu{}_{\nu-} - \Gamma^\mu{}_{\lambda-}\Gamma^\lambda{}_{\nu\mu} = 0 \\ R_{\nu i} &= \partial_\mu \Gamma^\mu{}_{\nu i} - \Gamma^\mu{}_{\lambda i}\Gamma^\lambda{}_{\nu\mu} = 0, \end{aligned} \tag{43.10}$$

and finally we write

$$R_{\nu+} = \partial_\mu \Gamma^\mu{}_{\nu+} - \Gamma^\mu{}_{\lambda+}\Gamma^\lambda{}_{\nu\mu}, \tag{43.11}$$

splitting into

$$\begin{aligned} R_{i+} &= \partial_\mu \Gamma^\mu{}_{i+} - \Gamma^\mu{}_{\lambda+}\Gamma^\lambda{}_{i\mu} = 0 \\ R_{-+} &= \partial_\mu \Gamma^\mu{}_{-+} - \Gamma^\mu{}_{\lambda+}\Gamma^\lambda{}_{-\mu} = 0 \\ R_{++} &= \partial_\mu \Gamma^\mu{}_{++} - \Gamma^\mu{}_{\lambda+}\Gamma^\lambda{}_{+\mu} \\ &= \partial_i \Gamma^i{}_{++} = -\frac{1}{2}\partial_i^2 H. \end{aligned} \tag{43.12}$$

In conclusion, we see that the only nonzero component of the Ricci tensor is

$$R_{++} = -\frac{1}{2}\partial_i^2 H. \tag{43.13}$$

That immediately gives for two possible curvature invariants

$$R = g^{++}R_{++} = 0$$
$$R_{\mu\nu}R^{\mu\nu} = R_{++}R^{++} = R_{++}g^{+\mu}g^{+\nu}R_{\mu\nu} = 0. \tag{43.14}$$

We can, in fact, show that also the third obvious curvature invariant, $R_{\mu\nu\rho\sigma}R^{\mu\nu\rho\sigma}$, vanishes. We leave this as an exercise.

This then gives an important theorem of Horowitz and Steif: the pp wave is a solution not only of Einstein gravity, but also its quantum generalizations that involve extra curvature terms in the action made up of the three curvature invariants. Indeed, then the equations of motion will have the same $R_{\mu\nu}$ and R terms, but multiplied by functions of the curvature invariants, which all vanish.

The most important observation of this section is, however, that the Ricci tensor, giving Einstein's equation, is *the same as its linearized value* – namely, linear in the only relevant metric function, H, and acted upon by the flat space Laplacean ∂_i^2. That was previously true only at the linearized level, when we considered gravitational waves. But now that means that *as long as we keep the general pp wave ansatz* – i.e., for pp waves moving in the same direction – we have a linear problem: we can add two solutions of the *vacuum* Einstein's equations to make another solution. Indeed, now the Einstein's equations on the gravitational pp wave reduce to the single equation

$$R_{++} = -\frac{1}{2}\partial_i^2 H = 8\pi G_N T_{++}\,, \tag{43.15}$$

and all other components of $T_{\mu\nu}$ must vanish (since the other components of $R_{\mu\nu}$ vanish).

Thus, if H_1 solves $R_{++} \propto \partial_i^2 H_1 = 0$, and H_2 solves the same equation, then so does $H = H_1 + H_2$. Moreover, if H_1 solves the Einstein's equation with energy-momentum tensor $T_{++}^{(1)}$ and H_2 solves the Einstein's equation with $T_{++}^{(2)}$, then $H = H_1 + H_2$ solves the Einstein's equation with energy-momentum tensor $T_{++} = T_{++}^{(1)} + T_{++}^{(2)}$.

We observe that H can be harmonic in x^i if either T_{++} is independent of x^i, or with a delta function support.

43.2 The Penrose Limit

Penrose proved an important theorem that has many applications, in particular for the AdS/CFT correspondence. He showed that if we consider a generic gravitational background and a null geodesic in it, focusing in near the null geodesic, we obtain a space that is a gravitational pp wave in flat background.

To show this, Penrose considered a curved space with lightcone coordinates U, V; transverse coordinates Y^i; and a null geodesic in it defined by $V = Y^i = 0$ and $U = \tau$ being the affine parameter along the null geodesic. Then, by a change of the coordinates Y^i, we can put $g_{Ui} = 0$, by a change of the U coordinate, we can put $g_{UU} = 0$, and by a change of the V coordinate we can normalize $g_{UV} = 1$, to finally write the metric as

$$ds^2 = dV\left(dU + \alpha dV + \sum_i \beta_i dY^i\right) + \sum_{ij} C_{ij} dY^i dY^j. \tag{43.16}$$

Here, $\alpha = g_{VV}$, $\beta_i = g_{Vi}/2$ and $C_{ij} = g_{ij}$ are functions of all the coordinates.

To focus in near the null geodesic, we must get close to the $V = 0$, $Y^i = 0$ line, which can be done by rescaling the coordinates by dividing by a large quantity R and taking that to infinity. However, so that we don't get a singular limit, we need to have the same overall power of R in the metric for all metric components, which means that, since U is left invariant, V must be rescaled by $1/R^2$ instead of $1/R$. Then the rescaling of coordinates is

$$U = u, \quad V = \frac{v}{R^2}, \quad Y^i = \frac{y^i}{R}. \tag{43.17}$$

Moreover, in order to obtain a finite metric in the $R \to \infty$ limit, we must rescale the original metric by a factor of R^2, $ds^2 \to R^2 ds^2$; in other words, R was the original overall scale of the metric. Then, defining

$$C_{ij}(U = u, V = 0, Y^i = 0) \equiv g_{ij}(u), \tag{43.18}$$

we find the metric

$$ds^2 = 2dudv + g_{ij}(u)dy^i dy^j. \tag{43.19}$$

Now this doesn't look very much like the Brinkmann form of the pp wave. In fact, it is the *Rosen form* of the pp wave, and u, v, y^i are *Rosen coordinates*.

From Rosen to Brinkmann Coordinates

To change coordinates to the Brinkmann ones, we first write the metric $g_{ij}(u)$ in terms of e_i^a ("vielbeins"; see Chapter 46) as

$$g_{ij}(u) = e_i^a(u)e_j^b(u)\delta_{ab}, \tag{43.20}$$

define the inverse matrx e_a^i (such that $e_i^a e_a^j = \delta_i^j$ and $e_i^a e_b^i = \delta_b^a$), and then use the Lorentz type gauge invariance acting on the index a to put $e_i^a(u)$ into a form satisfying

$$\frac{d}{du}(e_{ai}(u))e_b^i(u) = \frac{d}{du}(e_{bi}(u))e_a^i(u). \tag{43.21}$$

With this relation, we can prove we have

$$g_{ij}dy^i dy^j = (dx^a + \dot{e}_i^a e_c^i x^c du)(dx^b + \dot{e}_j^b e_d^j x^d du)\delta_{ab}, \tag{43.22}$$

where we define the Brinkmann coordinates (x^+, x^-, x^a) from the Rosen coordinate (u, v, y^i) by

$$u = x^+$$
$$v = x^- - \frac{1}{2}\dot{e}_{ai}e_b^i x^a x^b$$
$$y^i = e_a^i x^a, \tag{43.23}$$

where the dot refers to d/du, and we, moreover, have

$$A_{ab} = \ddot{e}_{ai} e_b^i. \tag{43.24}$$

Then the Brinkmann form is

$$ds^2 = 2dx^+ dx^- + H(x^+, x^a)(dx^+)^2 + \sum_a dx^a dx^a, \tag{43.25}$$

and where

$$H(x^+, x^a) = A_{ab}(x^+) x^a x^b. \tag{43.26}$$

Note that for this Brinkmann form, we have

$$R_{++} = -\frac{1}{2} \partial_a^2 H = -\sum_a A_{aa} = -\ddot{e}_{ai} e_a^i. \tag{43.27}$$

Penrose Limit as Boost

We can interpret the Penrose limit as a combination of a boost and scaling, as follows. Consider a boost of parameter β in a direction x,

$$t' = \cosh \beta t + \sinh \beta x, \quad x' = \sinh \beta t + \cosh \beta x, \tag{43.28}$$

for which

$$\begin{aligned} x' - t' = e^{-\beta}(x - t) &\Rightarrow x'^- = e^{-\beta} x^- \\ x' + t' = e^{\beta}(x + t) &\Rightarrow x'^+ = e^{\beta} x^+. \end{aligned} \tag{43.29}$$

Next, choose $e^\beta = R$, where R is the overall scale of the metric; and take $R \to \infty$; and, finally, scale all the coordinates, t, x, y^i by $1/R$. Then we obtain the same Penrose scaling of the coordinates. That means that we can interpret the Penrose limit as boosting along a direction in the background, focusing in on that, and taking the scale of the background to infinity together with the boost.

Example: Penrose Limit of $AdS_3 \times S^3$

As an example of a Penrose limit, consider the product space $AdS_3 \times S^3$, where AdS_3 refers to a three-dimensional Anti-de Sitter (AdS) space. AdS is a space of constant negative curvature, obtained by an analytical continuation from the three-sphere. A three-sphere of unit radius has metric

$$ds^2 = \cos^2 \theta d\psi^2 + d\theta^2 + \sin^2 \theta d\phi^2, \tag{43.30}$$

which we leave to show as an exercise. Then the $AdS_3 \times S^3$ metric of overall scale R is

$$ds^2 = R^2(-\cosh^2 d\tau^2 + d\rho^2 + \sinh^2 d\alpha^2) + R^2(\cos^2 \theta d\psi^2 + d\theta^2 + \sin^2 \theta d\phi^2). \tag{43.31}$$

Consider a null geodesic defined to move along an equator of S^3, defined by $\theta = 0$ and parametrized by ψ, sitting at the center of AdS_3, $\rho = 0$, and having time coordinate τ.

We could follow exactly Penrose's prescription, but it is easier to work directly with a form that gives Brinkmann coordinates.

Indeed, we can consider a small θ and ρ in order to be near the null geodesic and approximate

$$ds^2 \simeq R^2[-(1+\rho^2)d\tau^2 + d\rho^2 + \rho^2 d\alpha^2] + R^2[(1-\theta^2)d\psi^2 + d\theta^2 + \sin^2\theta d\phi^2]. \quad (43.32)$$

Next, define lightcone coordinates for motion in ψ by (from the general theory)

$$\tilde{x}^{\pm} = \frac{\psi \pm \tau}{\sqrt{2}}. \quad (43.33)$$

Now the metric is not in the form from the Penrose theorem, since we are in a different gauge, defined in order to obtain the Brinkmann form directly. But the rescaling of coordinates needed for the Penrose limit must be the same, namely

$$\tilde{x}^+ = x^+, \quad \tilde{x}^- = \frac{x^-}{R^2}, \quad \rho = \frac{r}{R}, \quad \theta = \frac{y}{R}. \quad (43.34)$$

We now obtain that

$$R^2(d\psi^2 - d\tau^2) = 2R^2 d\tilde{x}^+ d\tilde{x}^- = 2dx^+ dx^-$$
$$R^2 d\tau^2 \simeq \frac{(dx^+)^2}{2}$$
$$R^2 d\psi^2 \simeq \frac{(dx^+)^2}{2}, \quad (43.35)$$

which means that in the $R \to \infty$ limit, the $AdS_3 \times S^3$ metric becomes

$$ds^2 = 2dx^+ dx^- - \frac{1}{2}(r^2 + y^2)(dx^+)^2 + d\vec{y}^2 + d\vec{r}^2, \quad (43.36)$$

where $d\vec{y}^2 = dy^2 + y^2 d\phi^2$ and $d\vec{r}^2 = dr^2 + r^2 d\alpha^2$.

Note that, in this case, we have

$$H = -\frac{1}{2}(r^2 + y^2) = \sum_{a,b} A_{ab} x^a x^b, \quad (43.37)$$

so $A_{ab} = -\frac{1}{2}\delta_{ab}$. In this case, we have

$$R_{++} = -\frac{1}{2}\partial_a^2 H = 2. \quad (43.38)$$

Thus, in this case, the energy-momentum tensor is constant, $8\pi G_N T_{++} = 2$, and H is harmonic.

43.3 Gravitational Shock Waves in Flat Space

An ideal gravitational shock wave is a wave (propagating at the speed of light) that is only located at $x^+ = 0$. In other words, the metric has a $\delta(x^+)$ factor in it. But that, in turn, is possible if the energy-momentum tensor has a $\delta(x^+)$ in it – in other words, if the source of the metric is a distribution of particles moving at the speed of light in the same direction and at the same time. In this case, we have already noted that H will also be harmonic. Indeed, we see from the Einstein's equation (43.15) that if

$$T_{++} = t_{++}(x^i)\delta(x^+),$$ (43.39)

then we have also

$$H(x^+, x^i) = h(x^i)\delta(x^+),$$ (43.40)

and the Einstein's equation becomes

$$\partial_i^2 h(x^i) = -16\pi G_N t_{++}(x^i),$$ (43.41)

and the other components of $t_{\mu\nu}$ vanish.

We want to understand how we can obtain such an energy-momentum tensor. It is logical to assume that a description in terms of a fluid, composed of some fluid particles, is adequate, so consider an ideal pressureless fluid (like is the case of free particles moving together)

$$T_{\mu\nu} = \rho u_\mu u_\nu.$$ (43.42)

Moreover, consider that a null velocity field, $u^\mu = (1, 1, 0, 0)$, so that $u_\mu u^\mu = 0$. Then we get the nonzero components

$$T_{00} = T_{11} = T_{01} = T_{10} = \rho.$$ (43.43)

We want to transform to null coordinates $x^\pm = (x \pm t)/\sqrt{2}$, so

$$x = \frac{x^+ + x^-}{\sqrt{2}}, \quad t = \frac{x^+ - x^-}{\sqrt{2}},$$ (43.44)

and

$$T_{++} = T_{\mu\nu}\partial_+ x^\mu \partial_+ x^\nu,$$ (43.45)

and similar formulas for all the other $T_{\mu\nu}$ components. We easily obtain

$$\begin{aligned} T_{++} &= \frac{T_{11}}{2} + \frac{T_{00}}{2} + T_{01} = 2\rho \\ T_{+-} &= \frac{T_{11}}{2} - \frac{T_{00}}{2} + \frac{T_{01}}{2} - \frac{T_{10}}{2} = 0 \\ T_{--} &= \frac{T_{11}}{2} + \frac{T_{00}}{2} - T_{01} = 0. \end{aligned}$$ (43.46)

That means that, indeed, a null velocity field results in a nonzero T_{++}, and all other components zero.

Point-Like Source

Consider then a single particle, situated at transverse coordinates $x^i = 0$ and moving on the $x^+ = 0$ line, thus with density

$$\rho = \frac{p}{2}\delta^{d-2}(x^i)\delta(x^+).$$ (43.47)

This then gives

$$T_{++} = p\delta^{d-2}(x^i)\delta(x^+) \Rightarrow t_{++} = p\delta^{d-2}(x^i), \qquad (43.48)$$

which means we need to solve the condition

$$\partial_i^2 h(x^i) = -16\pi G_N p\delta^{d-2}(x^i). \qquad (43.49)$$

For $d = 4$, the physical case, the solution is the harmonic function (the Green's function in two transverse dimensions)

$$h(x^i) = -4G_N p \ln \rho^2, \qquad (43.50)$$

where $\rho^2 = x^i x^i$ is the transverse radius squared, and in $d > 4$, we have (the Green's function in $d - 2$ transverse dimensions)

$$h(x^i) = \frac{16\pi G_N}{\Omega_{d-3}(d-4)} \frac{p}{\rho^{d-4}}, \qquad (43.51)$$

where Ω_{d-3} is the volume of the unit sphere in $d - 3$ dimensions.

General Shock Waves

Until now, we have considered a delta function shock wave, which is really what the term "shock wave" refers to. But note that from the point of view of the Einstein's equation, we didn't need to have a $\delta(x^+)$ profile at a single position ($x = -t$) in the propagation direction. We can as well use an arbitrary "wave function" $\Phi(x^+)$ in the x^+ direction – i.e., consider that the shock wave and the corresponding source are "spread out" in the propagation direction.

Indeed, we see that the Einstein's equation (43.15) is such that having a source with a wave function $\Phi(x^+)$,

$$T_{++} = t_{++}(x^i)\Phi(x^+) \qquad (43.52)$$

implies also a gravitational shock wave with the same wave function,

$$H(x^+, x^i) = h(x^i)\Phi(x^+), \qquad (43.53)$$

and then the Einstein's equation reduces to the same one as for the delta function shock wave,

$$\partial_i^2 h(x^i) = -16\pi G_N t_{++}(x^i). \qquad (43.54)$$

We can thus find the same solutions for $h(x^i)$ as in the delta function case. That is why from now on we consider the two cases together, and we make no distinction between the case with $\Phi(x^+)$ and the case with $\delta(x^+)$.

The Shock Wave Generated by a Planar Source and by a Graviton

Besides the point-like source, another important case is the case of a constant energy-momentum tensor in the plane transverse to the direction of propagation – i.e., $t_{++} =$ constant (independent of x^i). Then the Einstein equation reduces to

$$\partial_i^2 h(x^i) = -16\pi G_N t_{++}, \qquad (43.55)$$

whose solution is any

$$h(x^i) = A_{ij}x^i x^j \tag{43.56}$$

that satisfies

$$\operatorname{Tr} A = \sum_i A_{ii} = -8\pi G_N t_{++}. \tag{43.57}$$

The simplest solution is then for

$$h(x^i) = -8\pi G_N t_{++} \frac{1}{d-2} \sum_{i=1}^{d-2} x^i x^i. \tag{43.58}$$

But an important case is the solution for Einstein's equation in vacuum, with $t_{++} = 0$. This can be interpreted as a *graviton*, as a sourceless, pointlike gravitational shock wave. In this case, from the above equation, we must have

$$\sum_i A_{ii} = 0 \tag{43.59}$$

In the case $d = 4$, with transverse coordinates y and z, the simplest solution is

$$h(y,z) = A(y^2 - z^2), \tag{43.60}$$

and we can absorb the constant A in the definition of $\Phi(x^+)$, to write the solution

$$ds^2 = 2dx^+ dx^- + \Phi(x^+)(y^2 - z^2)(dx^+)^2 + dy^2 + dz^2. \tag{43.61}$$

In the case of these gravitons, with wave function $\Phi(x^+)$, the linearity of the pp wave solution means that, if we consider two gravitons with wave functions $\Phi_1(x^+)$ and $\Phi_2(x^+)$, then the graviton with linear combination wave function,

$$\Phi = \alpha\Phi_1 + \beta\Phi_2 \tag{43.62}$$

is also a solution of the same vacuum Einstein's equations.

43.4 Gravitational Shock Waves in Other Backgrounds

We can also put a gravitational shock wave in a background spacetime and, thus, study the behavior of gravitons in the background.

An example is for background metric of the "warped" type,

$$ds^2 = f(\vec{y})(-dx^+ dx^- + H(x^+, \vec{x}, \vec{y})(dx^+)^2 + g_{ij}(\vec{x})dx^i dx^j) + G_{\mu\nu}(\vec{y})dy^\mu dy^\nu, \tag{43.63}$$

and consider a delta function shock wave,

$$H(x^+, \vec{x}, \vec{y}) = h(\vec{x}, \vec{y})\delta(x^+). \tag{43.64}$$

We can check (left as an exercise) that in this case we have the nonzero Christoffel symbols

$$\Gamma^-{}_{+c} = -\partial_c H\,, \quad \Gamma^i{}_{++} = -\frac{1}{2}g^{ij}\partial_j H\,, \quad \Gamma^\mu{}_{++} = -\frac{1}{2}G^{\mu\nu}\partial_\nu(fH)\,, \tag{43.65}$$

where $c = (i, \mu)$.

In this case, again R_{++} is the only nonzero component of the Ricci tensor. *When the background (no H) Einstein's equations are satisfied*, we obtain

$$\begin{aligned}
R_{++} &= -\frac{1}{2}\frac{1}{\sqrt{g}}\partial_i\left(\sqrt{g}g^{ij}\partial_j H\right) - \frac{1}{2}\frac{f}{\sqrt{f^{2+n}G}}\partial_\mu\left(\sqrt{f^{2+n}G}\,G^{\mu\nu}\partial_\nu H\right) \\
&\quad + \frac{1}{2}(fH)\left[\frac{(\partial_\mu f)^2}{f^2} - \frac{\Delta_G f}{f}\right] \\
&= -\frac{f}{2}\Delta_{(x,y)}H + \text{linear in } g_{++} + \text{background,}
\end{aligned} \tag{43.66}$$

where $g_{++} = f(\vec{y})H$ is the "perturbation" in the metric and n is the number of $\vec{x}$ coordinates. As such, the last term, proportional with g_{++}, comes from the modification of the energy-momentum tensor due to g_{++}: there is a term proportional to the metric g_{ab}, which gets an additional contribution from g_{++}.

Finally then, the Einstein equation becomes

$$-\frac{f}{2}\Delta_{(x,y)}H(x,y) = 8\pi G_N t_{++}\,, \tag{43.67}$$

and we can consider the same kind of point-like energy-momentum tensor as in flat space,

$$t_{++} = p\delta(x^+)\delta^n(x^i)\delta^m(y^\mu)\,, \tag{43.68}$$

leading to the shock wave satisfying the equation

$$f(\vec{y})\Delta_{(x,y)}h(\vec{x}, \vec{y}) = -16\pi G_N p\delta^n(x^i)\delta^m(y^\mu)\,. \tag{43.69}$$

Again, as in flat space, we can actually consider a shock wave with profile $\Phi(x^+)$ instead of $\delta(x^+)$, so with

$$H(x^+, x^i, y^\mu) = h(x^i, y^\mu)\Phi(x^+)\,, \tag{43.70}$$

and nothing changes for the equation satisfied by $h(\vec{x}, \vec{y})$.

43.5 The Khan–Penrose Interacting Solution

We have seen that two graviton solutions, or vacuum gravitational shock waves with wave functions $\Phi_1(x^+)$ and $\Phi_2(x^+)$, but moving in the same direction (for $x^+ = $ constant), add up to give a new solution.

On the other hand, if the two waves move in different directions, they do not make in general another solution.

Consider two shock waves moving in the same coordinate x, but in opposite directions along it – i.e., one with $\Phi_1(x^+)$, and one with $\Phi_2(x^-)$. Then the waves will necessarily collide. If $\Phi_1(x^+) = \delta(x^+)$, $\Phi_2(x^-) = \delta(x^-)$, then the collision will occur at $x^+ = x^- = 0$. In this case, *before the collision* – i.e., for $x^+ < 0, x^- < 0$ – we can freely add the two solutions, as they do not interact, and write

$$ds^2 = 2dx^+ dx^- + (y^2 - z^2)[\Phi_1(x^+)(dx^+)^2 + \Phi_2(x^-)(dx^-)^2] + dy^2 + dz^2. \quad (43.71)$$

But after the collision, we don't know *a priori* the solution, since gravity is nonlinear again.

But Khan and Penrose wrote in 1971 an interacting metric for the whole space – that is, one of the very few existing interacting two-body metrics (which is not fixed by symmetry). We will not prove it, just write the solution for completeness.

The metric is easier to write in Rosen coordinates. Applying the general formalism to go from Brinkmann to Rosen coordinates, we obtain for a single delta function shock wave

$$ds^2 = -2dx^+ dx^- + f(x^+)dy^2 + g(x^+)dz^2, \quad (43.72)$$

with

$$f(x^+) = 1 + x^+ \theta(x^+), \quad g(x^+) = 1 - x^+ \theta(x^+). \quad (43.73)$$

We leave the proof as an exercise.

Then the Khan–Penrose solution is

$$ds^2 = -\frac{2t^2}{rw(pq+rw)^2}dx^+ dx^- + t^2 \left(\frac{r+q}{r-q}\right)\left(\frac{w+p}{w-p}\right)dy^2 + t^2 \left(\frac{r-q}{r+q}\right)\left(\frac{w-p}{w+p}\right)dz^2, \quad (43.74)$$

where

$$\begin{aligned} p &= x^+ \theta(x^+), \quad q = x^- \theta(x^-), \\ r &= \sqrt{1-p^2}, \quad w = \sqrt{1-q^2}, \\ t &= \sqrt{1-p^2-q^2}. \end{aligned} \quad (43.75)$$

One finds that the metric has a singularity on the line $(x^+)^2 + (x^-)^2 = 1$, with $x^+ > 0, x^- > 0$ and continued for $x^+ = 1$, $x^- < 0$ arbitrary, and for $x^- = 1$, $x^+ < 0$ arbitrary (see Figure 43.1). This matches the expectation that in the high-energy collision of two massless particles, here gravitons, a black hole (a gravitational object with a singularity) is formed, but the details of this process are not well understood, despite having an explicit metric.

Important Concepts to Remember

- Parallel plane (pp) waves are gravitational waves that have planes moving parallel to the initial position. In flat space, they are described as adding a term $H(x^+, x^i)(dx^+)^2$ to the metric.
- A pp wave linearizes gravity, since the only nonzero Ricci tensor component is $R_{++} = -\frac{1}{2}\partial_i^2 H$, matching the linearized expectation.
- Adding two pp waves moving in the same direction, we get another pp wave.
- The Penrose limit, focusing in near in a null geodesic in a curved spacetime, gives a pp wave in flat space, by a theorem by Penrose.
- The pp wave can be represented by the Brinkmann form, in terms of $H(x^+, x^i)$, or the Rosen form, in terms of $g_{ij}(u)$.

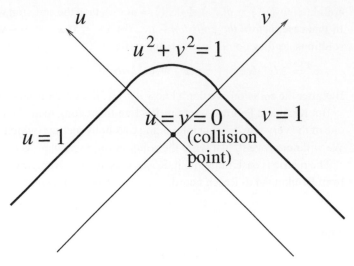

Figure 43.1 Khan–Penrose solution, with $u = x^+$ and $v = x^-$. There is a singularity line (thick line) in the future of the collision.

- A gravitational shock wave is sourced by a T_{++} distribution and a profile in the lightcone time direction x^+ of $\Phi(x^+)$ that can be $\delta(x^+)$ for a true shock wave.
- A pointlike source has $\delta(x^+)$ and $\delta^{d-2}(x^i)$, but the source can have any wave function $\Phi(x^+)$ and any shape in the transverse directions x^i. The equation of motion is $\partial_i^2 h(x^i) = -16\pi G_N t_{++}(x^i)$.
- For a constant t_{++}, we have a solution $h = A_{ij} x^i x^j$, with $\operatorname{Tr} A = -8\pi G_N t_{++}$, and for $t_{++} = 0$, representing a graviton, we have $\operatorname{Tr} A = 0$.
- We can put the shock waves in a gravitational background and generically obtain an equation of motion with the Laplacean Δ in the background instead of ∂_i^2.
- The Khan–Penrose solution describes the collision of two gravitational shock waves of the graviton type moving in the same x coordinate, but in opposite directions, before and after the collision.

Further Reading

See chapter 28 in [16]. The Khan–Penrose solution was found in [42].

Exercises

(1)   Prove that for the pp waves we have also

$$R_{\mu\nu\rho\sigma} R^{\mu\nu\rho\sigma} = 0. \tag{43.76}$$

(2)   Show that a three-sphere of radius 1, defined inside an Euclidean space by

$$x_1^2 + x_2^2 + x_3^2 + x_4^2 = 1, \qquad (43.77)$$

has a metric parametrized by three angles θ, ϕ, ψ as

$$ds^2 = \cos^2 \theta d\psi^2 + d\theta^2 + \sin^2 \theta d\phi^2. \qquad (43.78)$$

(3)   Calculate the Penrose limit for a four-dimensional space of Minkowskian signature, $\mathbb{R}_\tau \times S^3$, where the null geodesic again is on the equator at $\theta = 0$ and parametrized by ψ.

(4)   Show that the delta function gravitational shock wave in Rosen coordinates is given by (43.72) and (43.73).

44

Dimensional Reduction: The Domain Wall, Cosmic String, and BTZ Black Hole Solutions

In this chapter, we will describe two important solutions of relevance to cosmology, which have spatial extension in one or two directions. These are the cosmic string and domain wall, respectively. We will also describe an important solution in 2+1 dimensions, the BTZ black hole, which is simple enough to be easily obtained and has many applications.

44.1 The Domain Wall: Ansatz

We want to find a solution that is invariant under translations in two Euclidean coordinates, x and y, so the metric should be independent of them, $g_{\mu\nu} = g_{\mu\nu}(z, t)$. The solution should be generated by an infinitely thin distribution of matter, as a function of the third Euclidean coordinate z. Together with the fact that we want the metric at infinity to go over to the Minkowski metric, this means that we must have reflection invariance under the coordinate z, so for $z \to -z$. By coordinate transformations of the x and y coordinates, we can put $g_{xt} = g_{yt} = 0$, and by reflection invariance, we must also have $g_{zt} = 0$ (a $dz\,dt$ term is not invariant), while in the plane (x, y), we must have rotation invariance as well, leading to the ansatz

$$ds^2 = -D(z, t)dt^2 + A(z, t)(dx^2 + dy^2) + C(z, t)dz^2. \qquad (44.1)$$

At this point, we still have coordinate transformations $t' = t'(z, t)$, $z' = z'(z, t)$ available, so we can use them to put $C(z, t) = 1$ (another useful choice, used by Villenkin, is $C = D$). The final ansatz is then

$$ds^2 = -D(z, t)dt^2 + A(z, t)(dx^2 + dy^2) + dz^2. \qquad (44.2)$$

The energy density generating the sheet must be a infinitely thin, so

$$\rho = \sigma \delta(z). \qquad (44.3)$$

44.2 The Domain Wall: Einstein's Equation

We now calculate the equations of motion for this ansatz.

Christoffel Symbols

First, we calculate the Christoffel symbols using the usual formula $\Gamma^\mu{}_{\nu\rho} = \frac{1}{2}g^{\mu\sigma}(\partial_\nu g_{\sigma\rho} + \partial_\rho g_{\sigma\nu} - \partial_\sigma g_{\nu\rho})$. The metric has

$$g_{tt} = -D, \quad g_{ii} = A, \ i = x,y, \quad g_{zz} = 1. \tag{44.4}$$

We organize the Christoffel symbols by the upper index. We start with

$$\Gamma^z{}_{\mu\nu} = \frac{1}{2}g^{zz}(\partial_\mu g_{\nu z} + \partial_\nu g_{\mu z} - \partial_z g_{\mu\nu}) = \frac{1}{2}(\partial_\mu g_{\nu z} + \partial_\nu g_{\mu z} - \partial_z g_{\mu\nu}). \tag{44.5}$$

We see that the only nonzero components are

$$\Gamma^z{}_{tt} = -\frac{1}{2}\partial_z g_{tt} = \frac{D'}{2}$$
$$\Gamma^z{}_{ii} = -\frac{1}{2}\partial_z g_{ii} = -\frac{A'}{2}. \tag{44.6}$$

Next,

$$\Gamma^t{}_{\mu\nu} = \frac{1}{2}g^{tt}(\partial_\mu g_{t\nu} + \partial_\nu g_{t\mu} - \partial_t g_{\mu\nu}). \tag{44.7}$$

The only nonzero components are now

$$\Gamma^t{}_{tt} = \frac{1}{2}g^{tt}\partial_t g_{tt} = \frac{\dot{D}}{2D}$$
$$\Gamma^t{}_{ii} = -\frac{1}{2}g^{tt}\partial_t g_{ii} = \frac{\dot{A}}{2D}$$
$$\Gamma^t{}_{zt} = \frac{1}{2}g^{tt}\partial_z g_{tt} = \frac{D'}{2D}. \tag{44.8}$$

Next,

$$\Gamma^i{}_{\mu\nu} = \frac{1}{2}g^{ii}(\partial_\mu g_{i\nu} + \partial_\nu g_{i\mu} - \partial_i g_{\mu\nu}). \tag{44.9}$$

The only nonzero components are (no sum over i)

$$\Gamma^i{}_{zi} = \frac{1}{2}g^{ii}\partial_z g_{ii} = \frac{A'}{2A}$$
$$\Gamma^i{}_{ti} = \frac{1}{2}g^{ii}\partial_t g_{ii} = \frac{\dot{A}}{2A}. \tag{44.10}$$

Ricci Tensors

The tt component is (we have sum over i)

$$\begin{aligned}
R_{tt} &= \partial_\mu\Gamma^\mu{}_{tt} - \partial_t\Gamma^\mu{}_{t\mu} + \Gamma^\lambda{}_{tt}\Gamma^\mu{}_{\lambda\mu} - \Gamma^\mu{}_{t\lambda}\Gamma^\lambda{}_{t\mu} \\
&= \partial_z\Gamma^z{}_{tt} + \partial_t\Gamma^t{}_{tt} - \partial_t(\Gamma^t{}_{tt} + \Gamma^i{}_{ti}) + \Gamma^t{}_{tt}(\Gamma^t{}_{tt} + \Gamma^i{}_{ti}) \\
&\quad + \Gamma^z{}_{tt}(\Gamma^t{}_{zt} + \Gamma^i{}_{zi}) - (\Gamma^t{}_{tt})^2 - (\Gamma^i{}_{ti})^2 - 2\Gamma^z{}_{tt}\Gamma^t{}_{tz} \\
&= \frac{D''}{2} - \frac{\ddot{A}}{A} + \frac{\dot{A}^2}{2A^2} + \frac{\dot{A}\dot{D}}{2AD} - \frac{D'^2}{4D} + \frac{A'D'}{2A}. \tag{44.11}
\end{aligned}$$

The zz component is

$$
\begin{aligned}
R_{zz} &= \partial_\mu \Gamma^\mu{}_{zz} - \partial_z \Gamma^\mu{}_{z\mu} + \Gamma^\lambda{}_{zz}\Gamma^\mu{}_{\lambda\mu} - \Gamma^\mu{}_{z\lambda}\Gamma^\lambda{}_{z\mu} \\
&= -\partial_z(\Gamma^t{}_{zt} + \Gamma^i{}_{zi}) - (\Gamma^t{}_{zt})^2 - \sum_i (\Gamma^i{}_{zi})^2 \\
&= -\frac{D''}{2D} + \frac{D'^2}{4D^2} - \frac{A''}{A} + \frac{A'^2}{2A^2}.
\end{aligned}
\tag{44.12}
$$

The 11 and 22 components are

$$
\begin{aligned}
R_{11} &= \partial_\mu \Gamma^\mu{}_{11} - \partial_1 \Gamma^\mu{}_{1\mu} + \Gamma^\lambda{}_{11}\Gamma^\mu{}_{\lambda\mu} - \Gamma^\mu{}_{1\lambda}\Gamma^\lambda{}_{1\mu} \\
&= \partial_z \Gamma^z{}_{11} + \partial_t \Gamma^t{}_{11} + \Gamma^z{}_{11}(\Gamma^t{}_{zt} + \Gamma^i{}_{zi}) + \Gamma^t{}_{11}(\Gamma^t{}_{tt} + \Gamma^i{}_{ti}) - 2\Gamma^z{}_{11}\Gamma^1{}_{1z} - 2\Gamma^t{}_{11}\Gamma^1{}_{1t} \\
&= -\frac{A''}{2} + \frac{\ddot{A}}{2D} - \frac{\dot{A}\dot{D}}{4D^2} - \frac{A'D'}{4D} = R_{22}.
\end{aligned}
\tag{44.13}
$$

Finally then, the Ricci scalar is

$$
\begin{aligned}
R &= R_{tt}g^{tt} + R_{zz}g^{zz} + R_{ii}g^{ii} = -\frac{R_{tt}}{D} + R_{zz} + \frac{2R_{11}}{A} \\
&= -\frac{D''}{D} + \frac{D'^2}{2D^2} - 2\frac{A''}{A} + \frac{A'^2}{2A^2} - \frac{A'D'}{AD} - \frac{\dot{A}^2}{2A^2 D} + 2\frac{\ddot{A}}{AD} - \frac{\dot{A}\dot{D}}{AD^2}.
\end{aligned}
\tag{44.14}
$$

Einstein Tensors

The Einstein tensors are then found to be

$$
\begin{aligned}
G_{tt} &= R_{tt} - \frac{1}{2}g_{tt}R = R_{tt} + \frac{DR}{2} \\
&= -\frac{A''}{A}D + \frac{A'^2}{4A^2}D + \frac{\dot{A}^2}{4A^2} \\
G_{zz} &= R_{zz} - \frac{1}{2}g_{zz}R = R_{zz} - \frac{1}{2}R \\
&= \frac{A'^2}{4A^2} + \frac{A'D'}{2AD} + \frac{\dot{A}^2}{4A^2 D} - \frac{\ddot{A}}{AD} + \frac{\dot{A}\dot{D}}{2AD^2} \\
G_{ii} &= R_{11} - \frac{1}{2}g_{11}R = R_{11} - \frac{A}{2}R \\
&= -\frac{A}{2}\left[-\frac{D''}{D} + \frac{D'^2}{2D^2} - \frac{A''}{A} + \frac{A'^2}{2A^2} - \frac{A'D'}{2AD} - \frac{\dot{A}^2}{2A^2 D} + \frac{\ddot{A}}{AD} - \frac{\dot{A}\dot{D}}{2AD^2} \right].
\end{aligned}
\tag{44.15}
$$

44.3 The Domain Wall Solutions

Perturbative Nonrelativistic Solution

We consider first the solution with a nonrelativistic source – i.e., with pressure $P \simeq 0$, so $T_\mu{}^\nu = \mathrm{diag}(-\rho, 0, 0, 0)$. Moreover, we find the perturbative solution away from the sheet at $z = 0$. On the sheet, we impose the boundary conditions

$$
D(0) = A(0) = 1, \quad D'(0) = A'(0) = 0,
\tag{44.16}
$$

the first being a normalization, and the second from reflection symmetry.

We can choose a *static solution* in this case (nonrelativistic means we can be in the rest frame, where the solution should be static), so $\dot{A} = \dot{D} = 0$. If we are at small z, we can also choose $A'', D'' \gg A', D'$, or their powers.

That means that near $z = 0$, the equations of motion become

$$
\begin{aligned}
G_{tt} &\simeq -A'' = 8\pi G_N \sigma \delta(z) \\
G_{zz} &\simeq 0 = 0 \\
G_{ii} &\simeq \frac{D'' + A''}{2} = 0.
\end{aligned}
$$
(44.17)

Then the equations reduce to $A'' = -D''$ and

$$
D'' = 8\pi G_N \sigma \delta(z).
$$
(44.18)

Together with the boundary condition $D(0) = 1$, we have the solution

$$
D(z) \simeq 1 + 4\pi G_N \sigma |z|,
$$
(44.19)

since

$$
\partial_z |z| - \theta(z) - \theta(-z) \Rightarrow \partial_z^2 |z| = 2\delta(z),
$$
(44.20)

where $\theta(z)$ is the Heaviside function (1 for $z \geq 0$ and 0 for $z < 0$). Moreover, since $A(0) = 1$ also, we have

$$
A(z) \simeq 1 - 4\pi G_N \sigma |z|,
$$
(44.21)

giving the full domain wall solution close to $z = 0$ as

$$
ds^2 \simeq -[1 + 4\pi G_N \sigma |z|]dt^2 + [1 - 4\pi G_N |z|](dx^2 + dy^2) + dz^2.
$$
(44.22)

Exact Relativistic Solution

We can check that the full Einstein's equations derived earlier have an exact solution, in the case considered by Vilenkin, of a relativistic source. More precisely, a source that has a constant potential, $V(\phi) = \Lambda$. This gives an energy-momentum tensor $T_{\mu\nu} = \Lambda g_{\mu\nu}$, so $T^\mu{}_\nu = \Lambda \delta^\mu_\nu$, implying $P = -\rho$. But the pressure is only valid in the directions parallel to the sheet, so we have $T^\mu{}_\nu = \text{diag}(-\rho, -\rho, -\rho, 0)$. The equations of motion are then

$$
G_{tt} = 8\pi G_N \sigma D(z)\delta(z), \quad G_{ii} = -8\pi G_N \sigma A(z)\delta(z), \quad G_{zz} = 0.
$$
(44.23)

One can check that the exact solution is

$$
ds^2 = -(1 - 2\pi G_N \sigma |z|)^2 dt^2 + dz^2 + (1 - 2\pi G_N \sigma |z|)^2 e^{4\pi G_N \sigma t}(dx^2 + dy^2),
$$
(44.24)

though we will not do it here.

44.4 Cosmic String: Ansatz

We next consider the cosmic string solution. This is a solution with spatial extension along one direction, z, along which it must be translationally invariant, and rotationally invariant in the transverse plane parametrized by r and θ, so $g_{\mu\nu} = g_{\mu\nu}(r,t)$ only. Moreover, we consider a static solution, so $g_{\mu\nu} = g_{\mu\nu}(r)$ only. Invariance under z and θ implies also $g_{\mu z} = g_{\mu\theta} = 0$, so

$$ds^2 = -A(r)dt^2 + D(r)dr^2 + B(r)d\theta^2 + C(r)dz^2 + E(r)drdt. \qquad (44.25)$$

But we can use the transformations of three of the coordinates (r,t,θ,z) to fix a gauge where $D = C = 1$ and $E = 0$.

The energy-momentum tensor is $T^\mu{}_\nu = \text{diag}(-\rho, P, 0, 0)$, where the pressure, only along the parallel z coordinate, is taken to come also from a relativistic source, with $V(\phi) = \Lambda =$ constant, so $P = -\rho$. Moreover, the source is infinitely thin, so

$$T^0{}_0 = T^z{}_z = -\mu\delta(x)\delta(y) = -\mu\delta^2(r). \qquad (44.26)$$

Then $T = T^\mu{}_\mu = 0$.

Then ρ and $P_z = -\rho$ will balance each other out in the Newton potential, giving $U_N = 0$, and it will therefore be consistent to use the last remaining coordinate transformation to fix $A(r) = 1$ also. The final ansatz is therefore

$$ds^2 = -dt^2 + dr^2 + dz^2 + B(r)d\theta^2. \qquad (44.27)$$

The metric components are thus

$$g_{rr} = g_{zz} = 1, g_{tt} = -1, g_{\theta\theta} = B(r). \qquad (44.28)$$

44.5 Cosmic String: Einstein's Equations

Christoffel Symbols

With only one nontrivial component, $g_{\theta\theta} = B(r)$, most Christoffel symbols are 0. Since $\partial_t g_{\mu\nu} = 0$ and $g_{t\mu} = 0$ for $\mu \neq t$, whereas $\partial_\mu g_{tt} = 0$, we find $\Gamma^t{}_{\mu\nu} = 0$. Since $\partial_z g_{\mu\nu} = 0$ and $g_{z\mu} = 0$ for $\mu \neq z$, whereas $\partial_\mu g_{zz} = 0$, we find $\Gamma^z{}_{\mu\nu} = 0$. Since

$$\Gamma^r{}_{\mu\nu} = \frac{1}{2}g^{rr}(\partial_\mu g_{r\nu} + \partial_\nu g_{r\mu} - \partial_r g_{\mu\nu}), \qquad (44.29)$$

it follows that the only nonzero symbol is

$$\Gamma^r{}_{\theta\theta} = -\frac{1}{2}g^{rr}\partial_r g_{\theta\theta} = -\frac{B'}{2}. \qquad (44.30)$$

Since

$$\Gamma^\theta{}_{\mu\nu} = \frac{1}{2}g^{\theta\theta}(\partial_\mu g_{\theta\nu} + \partial_\nu g_{\theta\mu} - \partial_\theta g_{\mu\nu}), \qquad (44.31)$$

it follows that the only nonzero symbol is

$$\Gamma^\theta{}_{r\theta} = \frac{1}{2}g^{\theta\theta}\partial_r g_{\theta\theta} = \frac{B'}{2B}. \qquad (44.32)$$

Ricci Tensors

The Ricci tensors are

$$
\begin{aligned}
R_{tt} &= \partial_\mu \Gamma^\mu{}_{tt} - \partial_t \Gamma^\mu{}_{t\mu} + \Gamma^\lambda{}_{tt}\Gamma^\mu{}_{\lambda\mu} - \Gamma^\mu{}_{t\lambda}\Gamma^\lambda{}_{t\mu} = 0 \\
R_{zz} &= \partial_\mu \Gamma^\mu{}_{zz} - \partial_z \Gamma^\mu{}_{z\mu} + \Gamma^\lambda{}_{zz}\Gamma^\mu{}_{\lambda\mu} - \Gamma^\mu{}_{z\lambda}\Gamma^\lambda{}_{z\mu} = 0 \\
R_{rr} &= \partial_\mu \Gamma^\mu{}_{rr} - \partial_r \Gamma^\mu{}_{r\mu} + \Gamma^\lambda{}_{rr}\Gamma^\mu{}_{\lambda\mu} - \Gamma^\mu{}_{r\lambda}\Gamma^\lambda{}_{r\mu} \\
&= -\partial_r \Gamma^\theta{}_{r\theta} - (\Gamma^\theta{}_{r\theta})^2 = -\frac{B''}{2B} + \frac{B'^2}{4B^2} \\
R_{\theta\theta} &= \partial_\mu \Gamma^\mu{}_{\theta\theta} - \partial_\theta \Gamma^\mu{}_{\theta\mu} + \Gamma^\lambda{}_{\theta\theta}\Gamma^\mu{}_{\lambda\mu} - \Gamma^\mu{}_{\theta\lambda}\Gamma^\lambda{}_{\theta\mu} \\
&= \partial_r \Gamma^r{}_{\theta\theta} + \Gamma^r{}_{\theta\theta}\Gamma^\theta{}_{r\theta} - 2\Gamma^r{}_{\theta\theta}\Gamma^\theta{}_{r\theta} \\
&= -\frac{B''}{2} + \frac{B'^2}{4B} = B R_{rr}.
\end{aligned}
\tag{44.33}
$$

The Ricci scalar is

$$
R = g^{tt}R_{tt} + g^{zz}R_{zz} + g^{rr}R_{rr} + g^{\theta\theta}R_{\theta\theta} = R_{rr} + \frac{R_{\theta\theta}}{B} = -\frac{B''}{B} + \frac{B'^2}{2B^2} = 2R_{rr}. \tag{44.34}
$$

If we write $B \equiv D^2$, we obtain

$$
R = -\frac{2D''}{D}. \tag{44.35}
$$

Then the Einstein tensors are

$$
G_{tt} = R_{tt} + \frac{1}{2}R = -\frac{D''}{D}
$$

$$
G_{zz} = R_{zz} - \frac{1}{2}R = +\frac{D''}{D}
$$

$$
G_{rr} = R_{rr} - \frac{1}{2}R = 0
$$

$$
G_{\theta\theta} = R_{\theta\theta} - \frac{1}{2}g_{\theta\theta}R = 0. \tag{44.36}
$$

The Einstein's equations for the ansatz reduce then to a single equation,

$$
-\frac{D''}{D} = 8\pi G_N \rho = 8\pi G_N \mu \delta^2(\vec{r}). \tag{44.37}
$$

44.6 Cosmic String Solution via Dimensional Reduction

The solution of the equation (44.37) for $r \neq 0$ is obviously

$$
D(r) = kr, \tag{44.38}
$$

where k is a constant, which we need to find.

In order to find it, we use an observation: The ansatz we have is effectively dimensionally reducing gravity on the z direction, down to 2+1 dimensions. Indeed, $g_{\mu\nu}$ is independent of z, and $g_{zz} = 1$, which means that R_{ij}, $i,j \neq z$, is independent of z, and $R_{zz} = 0$. That means that we can consider the 2+1 dimensional gravity instead, and the equations of motion will be the same. Note, however, that if we do so, now the energy-momentum tensor is not traceless anymore, but rather

$$T^\mu{}_\nu = \text{diag}(-\rho, 0, 0). \tag{44.39}$$

Since moreover $\rho = \sigma \delta^2(\vec{r})$, this is the pointlike energy-momentum tensor that generates a black hole in 2+1 dimensions.

As we said, the calculation of the Einstein equations is the same, with

$$G_{tt} = R_{tt} - \frac{1}{2} g_{tt} R = \frac{R}{2} = -\frac{D''}{D} = 8\pi G_N \mu \delta^2(\vec{r})$$
$$G_{rr} = G_{\theta\theta} = 0. \tag{44.40}$$

However, gravity in 2+1 dimensions is special. To see this, we count the number of degrees of freedom of gravity in d spacetime dimensions. The metric $g_{\mu\nu}$ is a symmetric tensor, so it has $\frac{d(d+1)}{2}$. But on it, we can act with d general coordinate transformations, with infinitesimal parameters ξ^μ. These act as gauge transformations and allow us to fix d components of the metric, leaving $d(d-1)/2$ independent ones. The equations of motion (Einstein equations) remove more components. To see how many, we use a trick. We have seen that in de Donder gauge, $\partial^\mu \bar{h}_{\mu\nu} = 0$, the linearized equations of motion (we don't need to consider the full nonlinear ones, the nonlinearities will not influence the number of degrees of freedom) are simply Klein-Gordon, $\Box h_{ij} = 0$, which don't reduce the number of degrees of freedom. But there are d gauge conditions (for the index ν) in the de Donder gauge, so we reduce the degrees of freedom by a further d. Note that we could have worked without fixing a gauge, but the analysis would have been even more complicated. Thus, all in all we have

$$\frac{d(d-3)}{2} = \frac{(d-1)(d-2)}{2} - 1 \tag{44.41}$$

independent components. The second form of (44.41) means that we can consider the metric components as symmetric, traceless, and transverse (i.e., with $d - 2$ values for the index, instead of d). Either way, in $d = 3$, we obtain zero components! (and we can check that indeed, in $d = 4$ we obtain two components, which we have already identified previously).

That means that gravity in three dimensions has no propagating degrees of freedom, and is trivial. We can fix a gauge for any solution of the Einstein equation in which the metric is *locally* whatever we want it to be – for example, flat space. But that doesn't say that the metric is flat *globally*.

Indeed, now considering the solution we have found in 3+1 dimensions, dimensionally reduced, namely

$$ds^2 = -dt^2 + dr^2 + k^2 r^2 d\theta^2, \tag{44.42}$$

where as usual $\theta \in [0, 2\pi]$, by the coordinate transformation $\theta' = k\theta$, becomes the flat space metric in polar coordinates,

$$ds^2 = -dt^2 + dr^2 + r^2 d\theta'^2, \tag{44.43}$$

just that with $\theta' \in [0, 2\pi k]$. But then, only if $k = 1$ do we have a metric that is flat globally. If not, we have a *cone*, obtained by cutting a wedge of "deficit angle" $2\pi(1 - k)$ out of flat

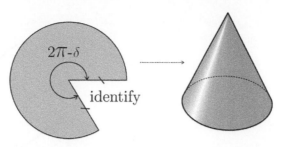

Figure 44.1 A flat cone is obtained by cutting out an angle $2\pi\delta$ from flat space, so that the polar angle takes values $\theta \in [0, 2\pi - \delta]$, and then identifying the cut. The the cone is flat everywhere, except at the tip, $r = 0$.

space (centered at $r = 0$), and identifying the edges of the cut, as in Figure 44.1. The cone is then flat everywhere, *except at the tip* $r = 0$. That is exactly what we have found earlier.

It remains to calculate k.

We integrate the equation of motion (44.37) over a volume $\int_V dV = \int d\theta \int r\,dr$ and obtain

$$\int d\theta \int_0^\infty r\,dr \frac{\frac{d^2}{dr^2}D/k}{r} = -8\pi G_N\mu. \tag{44.44}$$

The left-hand side becomes, under the *asssumption* that $(D/k)'(r=0)=0$ and $(D/k)'(r=\infty) = 1$,

$$\int d\theta[D'(r=\infty) - D'(r=0)] = \int d\theta = 2\pi k. \tag{44.45}$$

However, that cannot be quite correct, since for $k = 1$ we must obtain 0, which means that the equation must really be

$$2\pi(k-1) = -8\pi G_N\mu \Rightarrow k = 1 - 4G_N\mu. \tag{44.46}$$

Then indeed the solution we find is

$$ds^2 = -dt^2 + dz^2 + dr^2 + (1 - 4G_N\mu)^2 r^2 d\theta^2, \tag{44.47}$$

which was first found by Gott, and has a deficit angle of $\Delta\theta = -4G_N\mu$. Moreover, the singularity in the (r, θ) plane is a (useful) abstraction, and Gott found a smooth solution (nonconical) that becomes the above cone in the limit of delta function source.

Finally, the deflection of light by the cosmic string is very simple: the motion in flat space is a straight line, and the cone is just flat space without a wedge. When the straight line hits one edge of the wedge, it is then "transferred" (by the identification of the edges) to the other edge, so that it make the same angle with the edge- (see Figure 44.2). But this amounts to, from the point of view of the flat space from there the light ray came, a deflection angle equal to the deficit angle,

$$\Delta\theta = -8\pi G_N\mu. \tag{44.48}$$

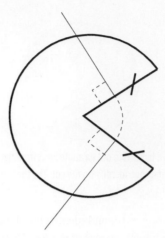

Deflection of light by the cosmic string: The line is straight along the cone (with the identification), but when representing it as a plane with a wedge cut out, it looks like we have a deflection equal to the deficit angle.

44.7 Cosmic String: Alternative Weak Field Derivation

We can also find the solution in the perturbative (weak field) case, by using the linearized Einstein equations,

$$\Box \tilde{h}_{\mu\nu} = -16\pi G_N \left(T_{\mu\nu} - \frac{1}{2}\eta_{\mu\nu}T \right), \tag{44.49}$$

where $\tilde{h}_{\mu\nu} = 2\kappa_N h_{\mu\nu} = g_{\mu\nu} - \eta_{\mu\nu}$.

The perturbative energy-momentum tensor is

$$T_{\mu\nu} \simeq \text{diag}(\rho, -\rho, 0, 0), \tag{44.50}$$

which gives $T = -2\rho$, so we have

$$\Box \tilde{h}_{00} = -16\pi G_N \left(\rho + \frac{T}{2} \right) = 0, \tag{44.51}$$

with the solution $\tilde{h}_{00} = 0$.

Moreover, now

$$\Box \tilde{h}_{ij} = -16\pi G_N \left(-\frac{T}{2} \right) \delta_{ij} = -16\pi G_N \rho \delta_{ij} = -16\pi G_N \sigma \delta^2(\vec{r})\delta_{ij}. \tag{44.52}$$

We want a static solution, so $\Box \tilde{h}_{ij} = \Delta \tilde{h}_{ij}$, and then we obtain just the Poisson equation in two dimensions. But its solution, the Green's function, in two dimensions is $\ln r/(2\pi)$. To check that, we integrate over a disk D

$$\int_D dV \Delta \frac{\ln r}{2\pi} = \oint_{S^1} d\vec{S} \cdot \vec{\nabla} \frac{\ln r}{2\pi} = 2\pi r \int \partial_r \frac{\ln r}{2\pi} = 1, \tag{44.53}$$

where we have used Stokes's theorem. Then, finally, we have

$$\tilde{h}_{ij} = -\frac{16\pi G_N \mu}{2\pi} \ln \frac{r}{r_0} \delta_{ij}, \tag{44.54}$$

and the cosmic string metric is

$$ds^2 \simeq -dt^2 + dz^2 + \left(1 - 8G_N\mu \ln \frac{r}{r_0}\right)(dr^2 + r^2 d\theta^2). \tag{44.55}$$

This doesn't look like the same solution as the exact one, but that is simply due to a change of coordinates.

Indeed, defining $A = 8G_N\mu \ln r/r_0$, consider the coordinate change

$$(1 - A)r^2 = (1 - 8G_N\mu)r'^2 \Rightarrow$$
$$dr' = \frac{dr\left[\sqrt{1-A} - \frac{8G_N\mu}{2\sqrt{1-A}}\right]}{\sqrt{1 - 8G_N\mu}} \simeq dr\sqrt{1-A}, \tag{44.56}$$

where we have used the fact that $G_N\mu \ll 1$. Finally, we obtain

$$ds^2 \simeq -dt^2 + dz^2 + dr'^2 + (1 - 8G_N\mu)r'^2 d\theta^2, \tag{44.57}$$

i.e., the metric is the same as the one found before.

44.8 The BTZ Black Hole Solution in 2+1 Dimensions and Anti–de Sitter Space

We already saw that in 2+1 dimensions gravity has no degrees of freedom, and locally we can put the metric to a flat one. Then a black hole solution to the vacuum Einstein's equations was equal to a cone, which is a slice of flat space with a global identification, so that there is a singularity at $r = 0$.

But it took a work in 1992 by Bañados, Teitelboim (now Bunster), and Zanelli (BTZ) to realize that if we add a *cosmological constant*, we can have nontrivial black hole solution *with a horizon* and characterized by a mass, just like in 3+1 dimensions. The mass will be still related to a deficit angle, which is present just like in the vacuum case.

A cosmological constant can be easily described as just a constant term Λ in the energy-momentum tensor,

$$8\pi G_N T_{\mu\nu} = \Lambda g_{\mu\nu}. \tag{44.58}$$

From the definition of the energy-momentum tensor, $T_{\mu\nu} = -\frac{2}{\sqrt{-g}}\frac{\delta S_m}{\delta g^{\mu\nu}}$, we see that we can obtain the cosmological constant by adding to the Einstein–Hilbert action a term (remember that $\delta\sqrt{-g} = -\frac{1}{2}\sqrt{-g}g_{\mu\nu}\delta g^{\mu\nu}$)

$$\frac{1}{16\pi G_N}\int d^D x\sqrt{-g}(2\Lambda). \tag{44.59}$$

The solution we are looking for will have the analogous ansatz to the Schwarzschild solution in 3+1 dimensions, so we choose

$$ds^2 = -f(r)dt^2 + \frac{dr}{f(r)} + r^2 d\phi^2.$$ (44.60)

That means that the nontrivial components of the metric are

$$g_{tt} = -f(r), \quad g_{rr} = \frac{1}{f(r)}, \quad g_{\phi\phi} = r^2.$$ (44.61)

Christoffel Symbols

Among the

$$\Gamma^r{}_{\mu\nu} = \frac{1}{2}g^{rr}(\partial_\nu g_{\mu r} + \partial_\mu g_{\nu r} - \partial_r g_{\mu\nu}),$$ (44.62)

we see the only nonzero components are

$$\Gamma^r{}_{rr} = \frac{1}{2}g^{rr}\partial_r g_{rr} = -\frac{1}{2}\frac{f'}{f}$$

$$\Gamma^r{}_{tt} = -\frac{1}{2}g^{rr}\partial_r g_{tt} = \frac{1}{2}ff'$$

$$\Gamma^r{}_{\phi\phi} = -\frac{1}{2}g^{rr}\partial_r g_{\phi\phi} = -rf.$$ (44.63)

Among the

$$\Gamma^t{}_{\mu\nu} = \frac{1}{2}g^{tt}(\partial_\nu g_{\mu t} + \partial_\mu g_{\nu t}),$$ (44.64)

we see that the only nonzero component is

$$\Gamma^t{}_{rt} = \frac{1}{2}g^{tt}\partial_r g_{tt} = \frac{1}{2}\frac{f'}{f}.$$ (44.65)

Among the

$$\Gamma^\phi{}_{\mu\nu} = \frac{1}{2}g^{\phi\phi}(\partial_\nu g_{\mu\phi} + \partial_\mu g_{\nu\phi}),$$ (44.66)

we see that the only nonzero component is

$$\Gamma^\phi{}_{r\phi} = \frac{1}{2}g^{\phi\phi}\partial_r g_{\phi\phi} = \frac{1}{r}.$$ (44.67)

We can now also calculate that among the sums $\Gamma^\mu{}_{\lambda\mu}$ the only nonzero one is with $\lambda = r$, namely

$$\Gamma^\mu{}_{r\mu} = \Gamma^r{}_{rr} + \Gamma^t{}_{rt} + \Gamma^\phi{}_{r\phi} = \frac{1}{r}.$$ (44.68)

Ricci Tensors

For the Ricci tensors, we find the nonzero values

$$R_{rr} = \partial_\mu \Gamma^\mu{}_{rr} - \partial_r \Gamma^\mu{}_{r\mu} + \Gamma^\lambda{}_{rr}\Gamma^\mu{}_{\lambda\mu} - \Gamma^\mu{}_{r\lambda}\Gamma^\lambda{}_{r\mu}$$

$$= \partial_r \Gamma^r{}_{rr} - \partial_r \frac{1}{r} + \Gamma^r{}_{rr}\frac{1}{r} - (\Gamma^r{}_{rr})^2 - (\Gamma^t{}_{rt})^2 - (\Gamma^\phi{}_{r\phi})^2$$

$$= -\frac{1}{2f}\left(f'' + \frac{f'}{r}\right)$$

$$R_{tt} = \partial_\mu \Gamma^\mu{}_{tt} + \Gamma^\lambda{}_{tt}\Gamma^\mu{}_{\lambda\mu} - \Gamma^\mu{}_{t\lambda}\Gamma^\lambda{}_{t\mu}$$
$$= \partial_r \Gamma^r{}_{tt} + \Gamma^r{}_{tt}\Gamma^\mu{}_{r\mu} - 2\Gamma^t{}_{tr}\Gamma^r{}_{tt}$$
$$= \frac{f}{2}\left(f'' + \frac{f'}{r}\right)$$

$$R_{\phi\phi} = \partial_\mu \Gamma^\mu{}_{\phi\phi} + \Gamma^\lambda{}_{\phi\phi}\Gamma^\mu{}_{\lambda\mu} - \Gamma^\mu{}_{\phi\lambda}\Gamma^\lambda{}_{\phi\mu}$$
$$= \partial_r \Gamma^r{}_{\phi\phi} + \Gamma^r{}_{\phi\phi}\Gamma^\mu{}_{r\mu} - 2\Gamma^\phi{}_{\phi r}\Gamma^r{}_{\phi\phi}$$
$$= -rf'. \tag{44.69}$$

We can then calculate also the Ricci scalar as

$$R = g^{rr}R_{rr} + g^{tt}R_{tt} + g^{\phi\phi}R_{\phi\phi} = fR_{rr} - \frac{R_{tt}}{f} + \frac{R_{\phi\phi}}{r^2} = -f'' - 2\frac{f'}{r}. \tag{44.70}$$

The BTZ Black Hole Solution

The Einstein equation is now

$$R_{\mu\nu} - \frac{1}{2}g_{\mu\nu}R = \Lambda g_{\mu\nu} \Rightarrow R_{\mu\nu} = \left(\Lambda + \frac{R}{2}\right)g_{\mu\nu}. \tag{44.71}$$

Substituting the Ricci tensor, Ricci scalar, and metric, we get from R_{rr} and R_{tt} the same equation,

$$-\frac{1}{2}\left(f'' + \frac{f'}{r}\right) = \Lambda - \frac{1}{2}\left(f'' + 2\frac{f'}{r}\right), \tag{44.72}$$

which thus reduces to

$$\frac{f'}{2} = \Lambda r \tag{44.73}$$

and implies $\frac{f''}{2} = \Lambda$. The equation for $R_{\phi\phi}$ is now

$$-\frac{f'}{r} = \Lambda - \frac{1}{2}\left(f'' + 2\frac{f'}{r}\right), \tag{44.74}$$

which reduces to the same $f'' = 2\Lambda$ equation.

Since Λ has mass dimension 2 (as does R), we can define a scale l by the condition

$$\Lambda = \frac{1}{l^2}. \tag{44.75}$$

Then the Einstein equations on the ansatz reduce to the simple equation

$$f' = \frac{2}{l^2}r, \tag{44.76}$$

with general solution

$$f(r) = \frac{r^2}{l^2} + C. \tag{44.77}$$

The constant C will be related to the mass of the black hole, and will be, in fact, proportional to it, as in the Schwarzschild solution in 3+1 dimensions. Moreover, for the solution to have a *horizon*, we need $C < 0$. Since in 2+1 dimensions, G_N has mass dimension -1, so that $\frac{R}{16\pi G_N}$ has mass dimension 3, C has to be proportional to $-G_N M$. It turns out that the numerical prefactor is 8, so, in fact,

$$f(r) = \frac{r^2}{l^2} - 8G_N M, \qquad (44.78)$$

for the final *BTZ solution*

$$ds^2 = -\left(\frac{r^2}{l^2} - 8G_N M\right) + \frac{dr^2}{\frac{r^2}{l^2} - 8G_N M} + r^2 d\phi^2. \qquad (44.79)$$

We see that the solution has an event horizon (apparent singularity) at $r = l\sqrt{8G_N M}$.

Anti–de Sitter Space

If $M > 0$, or $C < 0$, we obtain a black hole solution. For the previous $f(r)$, for any M, we obtain

$$R = -f'' - 2\frac{f'}{r} = -\frac{6}{l^2}. \qquad (44.80)$$

That means that locally, the space has constant negative curvature, appearing in the case of a (negative) cosmological constant. Such a space is called (three-dimensional) *Anti–de Sitter space*, or AdS space, in this case AdS_3.

Note that the space was already considered as an example for the Penrose limit leading to pp wave.

But, in fact, due to the global issues, it turns out that we obtain the standard AdS space only for $8G_N M = -1$, or

$$f(r) = \frac{r^2}{l^2} + 1. \qquad (44.81)$$

Moreover, one can show that the AdS space *in any dimension* (so AdS_D) is written as

$$ds_D^2 = -\left(\frac{r^2}{l^2} + 1\right) dt^2 + \frac{dr^2}{\frac{r^2}{l^2} + 1} + r^2 d\Omega_{D-2}^2. \qquad (44.82)$$

One can also define AdS_3 space (and by generalization, any D-dimensional AdS space) as an embedding, by an analytical continuation from the sphere, namely by using the constraint

$$-X_{-1}^2 - X_0^2 + X_1^2 + X_2^2 = -l^2 \qquad (44.83)$$

inside the space with two timelike directions,

$$ds^2 = -dX_{-1}^2 - dX_0^2 + dX_1^2 + dX_2^2, \qquad (44.84)$$

just like a three-sphere is defined by a constraint

$$Y_1^2 + Y_2^2 + Y_3^2 + Y_4^2 = R^2 \qquad (44.85)$$

in an Euclidean space

$$ds^2 = dY_1^2 + dY_2^2 + dY_3^2 + dY_4^2. \qquad (44.86)$$

This definition also emphasizes the symmetries (making the group $SO(2,2)$ for rotations in the flat space with two time and two space coordinates) of AdS space, and it makes it clearer where the scale l of the constant negative curvature comes from.

We leave as a (rather nontrivial) exercise to find the relation between the embedding coordinates X_{-1}, X_0, X_1, X_2 and the coordinates (t, r, ϕ) for AdS_3.

Important Concepts to Remember

- The domain wall ansatz is $ds^2 = -D(z,t)dt^2 + A(z,t)(dx^2 + dy^2) + dz^2$.
- For a nonrelativistic infinitely thin domain wall source $T^\mu{}_\nu = \mathrm{diag}(-\rho, 0, 0, 0)$, with $\rho = \sigma\delta(z)$, the solution near $z = 0$, $D(z) \simeq 1 + 4\pi G_N\sigma|z|$, $A(z) \simeq 1 - 4\pi G_N\sigma|z|$.
- The exact solution (Vilenkin) for the relativistic domain wall with source $T^\mu{}_\nu = \mathrm{diag}(-\rho, -\rho, -\rho, 0)$ is $D = (1 - 2\pi G_N\sigma|z|)^2$, $A = (1 - 2\pi G_N|z|)^2 e^{4\pi G_N\sigma t}$.
- The cosmic string ansatz for a relativistic string, with $T^\mu{}_\nu = \mathrm{diag}(-\rho, 0, 0, -\rho)$, is $ds^2 = -dt^2 + dz^2 + dr^2 + B(r)d\theta^2$.
- One can dimensionally reduce to 2+1 dimensions, which gives the black hole in 2+1 dimensions.
- The number of degrees of freedom in d dimensions is $d(d-3)/2$, which means that in three dimensions there are no propagating degrees of freedom, and we can put the metric to the flat one, though not globally.
- The black hole solution in 2+1 dimensions, lifting to the cosmic string, is the cone, flat space with $\theta \in [0, 2\pi k]$, or equivalently with $B(r) = k^2 r^2$ and $\theta \in [0, 2\pi]$.
- Deflection of light by the cosmic string is with the defect angle $\Delta\theta = -8\pi G_N\mu$.
- At weak field, the cosmic string can be put to the form $ds^2 = -dt^2 + dz^2 + (1 - 8G_N\mu \ln r/r_0)(dr^2 + r^2 d\theta^2)$.
- The BTZ black hole has a horizon and exists in the presence of a negative cosmological constant Λ.
- The BTZ black hole of mass M corresponds to $f(r) = \frac{r^2}{l^2} - 8G_N M$ in the Schwarzschild-like ansatz and has a horizon at $r = l\sqrt{8G_N M}$.
- A cosmological constant is a constant term in the action, $\int 2\Lambda/(16\pi G_N)$, leading to a constant term $\Lambda g_{\mu\nu}$ in the energy-momentum tensor $8\pi G_N T_{\mu\nu}$.
- The solution with constant negative curvature, arising in the presence of a negative cosmological constant, is called Anti-de Sitter space (AdS space), and has $f(r) = \frac{r^2}{l^2} + 1$ in any dimension.
- AdS space can be obtained by a sphere-like embedding in a flat space with two time-like coordinates, with $R^2 = -l^2$.

Further Reading

See chapter 11 in [40]. The perturbative domain wall and cosmic string solutions were found by Vilenkin in [43]. The exact domain wall solution was found in [44], and the exact cosmic string solution in [45]. The BTZ black hole was found in [46], and it is reviewed in [47].

Exercises

(1)   Check that Villenkin's solution solves the Einstein equations we derived.

(2)   Use the deficit angle picture to calculate the geodesic, in Euclidean coordinates at infinity, passing near the cosmic string.

(3)   Find the relation between the embedding coordinates X_{-1}, X_0, X_1, X_2 of the constraint (44.83) and the (t, r, ϕ) coordinates in the derivation of AdS space.

(4)   Find the coefficient c such that the combination

$$R_{\mu\nu}R^{\mu\nu} + cR^2 \tag{44.87}$$

vanishes on the BTZ solution.

45

Time-Dependent Gravity: The Friedmann-Lemaître-Robertson-Walker (FLRW) Cosmological Solution

In this chapter we will describe the cosmological solution of Friedmann, Lemaître, Robertson, and Walker, describing an expanding Universe. At very large distances (much larger than galaxies, and even than clusters of galaxies, the largest structures in the Universe), the Universe looks homogenous (invariant under spatial translations) and isotropic (invariant under spatial rotations). At smaller scales, we have structures (planets, stars, galaxies, etc.). Then it was natural to ask what is the solution of Einstein's equations with these properties, sourced by a matter distribution in the form of a fluid with energy density ρ and perhaps pressure p?

The answer was given by Friedmann, Lemaître, Robertson, and Walker, though it is worth noting that, although his work was very early, Lemaître's contribution got forgotten for almost a hundred years, and only now his name is rightfully associated with the cosmological solution. It is also worth noting that Einstein initially thought that an expanding Universe was not consistent (and with observations), and he introduced the "cosmological constant" term in the Einstein's equation (with the opposite sign to the one experimentally observed today) with the sole purpose of finding a static solution of the Einstein's equation, what is now known as "Einstein's static Universe," a highly special and unnatural case. When later is was proven experimentally by Hubble that the Universe does indeed expand, he retracted the idea of the cosmological constant, calling it a mistake; we now know that we *do* have a cosmological constant, just that it leads to an even *accelerated* expansion.

So space expands uniformly through the Universe, leading to the "Hubble law" that nearby the expansion (the velocity of the receding away of a point) is proportional to the distance from the point. For a closed space like a sphere, we can picture this as having a spherical balloon and filling it with air so that it expands. Then every point will move away from any other at a speed proportional to the distance between them, as in Figure 45.1.

45.1 Metric Ansatz

First, a gauge choice for general coordinate transformations, allowing us to put $g_{0i} = 0$, implies the metric

$$ds^2 = -f(x^\mu)dt^2 + dl^2. \tag{45.1}$$

Then, homogenity in space means $f = f(t)$ only, and then we can redefine $f(t)dt^2 \equiv dt'^2$, and we will drop the prime for simplicity, thus

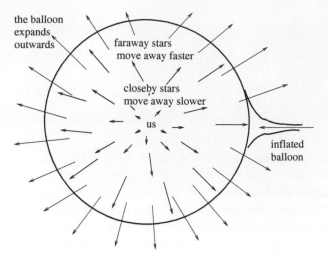

the balloon expands outwards

faraway stars move away faster

closeby stars move away slower

us

inflated balloon

Figure 45.1 The expansion of the Universe is like the expansion of an inflated balloon. Because of it, points move faster apart as the distance between them increases.

$$ds^2 = -dt^2 + dl^2. \tag{45.2}$$

Then homogeneity and isotropy for the spatial metric dl^2 implies

$$dl^2 = a^2(t) f(r^2 = x^2 + y^2 + z^2)(dx^2 + dy^2 + dz^2). \tag{45.3}$$

Indeed, (45.3) is manifestly isotropic, though not homogenous. At first sight, it looks like it would not be homogenous. But, in fact, we can impose for the space to have constant curvature. This is an invariant measure of homogeneity (the local invariant is the same at any point). Using *comoving coordinates*, i.e., coordinates that move together with the space (in the expanding balloon picture, imagine drawing a grid on the balloon and then filling it with air), the metric of the constant curvature space is

$$dl^2 = a^2(t) \frac{dx^2 + dx^2 + dz^2}{\left[1 + \frac{k}{4}(x^2 + y^2 + z^2)\right]^2}. \tag{45.4}$$

We can change to spherical coordinates by

$$x = \tilde{r} \sin\theta \cos\phi$$
$$y = \tilde{r} \sin\theta \sin\phi$$
$$z = \tilde{r} \cos\theta, \tag{45.5}$$

after which the metric becomes

$$dl^2 = a^2(t) \frac{d\tilde{r}^2 + \tilde{r}^2(d\theta^2 + \sin^2\theta d\phi^2)}{\left[1 + \frac{k}{4}\tilde{r}^2\right]^2}. \tag{45.6}$$

Here and before, $k = 0, \pm1$, with $k = 0$ corresponding to a flat space, $k = +1$ to a closed space like a sphere, and $k = -1$ to an open hyperbolic space. In fact, we can check that the Ricci scalar of the space is

$$R = \frac{k}{a^2} ,$$

(45.7)

(which we leave as an exercise), consistent with the above.

After a further coordinate transformation,

$$\frac{\tilde{r}}{1 + \frac{\tilde{r}^2}{4}} = \sin r, \quad k = +1$$

$$\frac{\tilde{r}}{1 - \frac{\tilde{r}^2}{4}} = \sinh r, \quad k = -1$$

$$\tilde{r} = r, \quad k = 0,$$

(45.8)

the metric becomes

$$dl^2|_{k=0} = a^2(t)[dr^2 + r^2(d\theta^2 + \sin^2 \theta d\phi^2)]$$
$$dl^2|_{k=+1} = a^2(t)[dr^2 + \sin^2 r(d\theta^2 + \sin^2 \theta d\phi^2)]$$
$$dl^2|_{k=-1} = a^2(t)[dr^2 + \sinh^2 r(d\theta^2 + \sin^2 \theta d\phi^2)].$$

(45.9)

The geometry of space is different now. Circles, parametrized by θ (i.e., for $dr = d\psi = 0$), have a different relation of the circumference (perimeter) over the radius than in Euclidean space. We obtain $dl = a(t)f(r)d\theta$, where $f(r)$ is $\sin r, \sinh r$, or r, depending on k. Integrating over θ, we obtain

$$l = 2\pi a(t)\sin r < 2\pi a(t), \quad k = +1$$
$$l = 2\pi a(t)\sinh r > 2\pi a(t), \quad k = -1$$
$$l = 2\pi a(t)r, \quad k = 0.$$

(45.10)

For a closed space ($k = +1$) like a sphere, it is easy to see that $l < 2\pi r$ (putting $a(t) = 1$). Indeed, consider an equator on the sphere. With respect to the radius R of the sphere, its length is $2\pi R$. But *on the sphere*, the "radius" of the equator is the meridian line from the North Pole to the equator, of length $r = \pi R/2$, so $l = 4r < 2\pi r$. For the hyperbolic space, it is harder to understand intuitively the fact that $l > 2\pi r$, but the same logic applies.

Also, two-spheres behave differently, having a different relation for their area to the radius than in Euclidean space. Now

$$dl^2 = a^2 f^2(r)(d\theta^2 + \sin^2 \theta d\phi^2) \Rightarrow d\Omega_2 = a^2(t)f^2(r)d\theta \sin \theta d\phi ,$$

(45.11)

which integrates to

$$A(r) = 4\pi a^2 \sin^2 r, \quad k = +1$$
$$A(r) = 4\pi a^2 \sinh^2 r, \quad k = -1$$
$$A(r) = 4\pi a^2 r^2, \quad k = 0.$$

(45.12)

Finally, we use a coordinate transformation to *FLRW coordinates*,

$$r' = \sin r, \quad k = +1$$
$$= \sinh r, \quad k = -1$$
$$= r, \quad k = 0,$$

(45.13)

after which the spatial metric becomes

$$dl^2 = a^2(t) \left[\frac{dr'^2}{1 - kr'^2} + r'^2 (d\theta^2 + \sin^2 \theta d\phi^2) \right]. \tag{45.14}$$

As before, the spacetime metric is $ds^2 = -dt^2 + dl^2$. In the following, we drop the prime on r'.

45.2 Christoffel Symbols and Ricci Tensors

Christoffel Symbols

We calculate the Christoffel symbols for this metric and obtain first

$$\Gamma^t_{\mu\nu} = \frac{1}{2} g^{tt} (\partial_\mu g_{t\nu} + \partial_\nu g_{t\mu} - \partial_t g_{\mu\nu}) = -\frac{1}{2} (\partial_\mu g_{t\nu} + \partial_\nu g_{t\mu} - \partial_t g_{\mu\nu}). \tag{45.15}$$

We can easily see from (45.15) that $\Gamma^t_{t\mu} = 0 = \Gamma^t_{ij}$, where $t \neq i \neq j$. The nonzero terms are

$$\Gamma^t_{rr} = \frac{1}{2} \partial_t g_{rr} = \frac{a\dot{a}}{1 - kr^2}$$

$$\Gamma^t_{\theta\theta} = \frac{1}{2} \partial_t g_{\theta\theta} = a\dot{a}r^2$$

$$\Gamma^t_{\phi\phi} = \frac{1}{2} \partial_t g_{\phi\phi} = a\dot{a}r^2 \sin^2 \theta. \tag{45.16}$$

Next,

$$\Gamma^r_{\mu\nu} = \frac{1}{2} g^{rr} (\partial_\mu g_{r\nu} + \partial_\nu g_{r\mu} - \partial_r g_{\mu\nu}) = \frac{(1 - kr^2)}{2a^2} (\partial_\mu g_{r\nu} + \partial_\nu g_{r\mu} - \partial_r g_{\mu\nu}). \tag{45.17}$$

We easily see that $\Gamma^r_{tt} = \Gamma^r_{t\theta} = \Gamma^r_{t\phi} = \Gamma^r_{r\theta} = \Gamma^r_{r\phi} = \Gamma^r_{\theta\phi} = 0$. Then,

$$\Gamma^r_{rr} = \frac{1}{2} \partial_r \ln g_{rr} = +\frac{kr}{1 - kr^2}$$

$$\Gamma^r_{tr} = \frac{1}{2} \partial_t \ln g_{rr} = \frac{\dot{a}}{a}$$

$$\Gamma^r_{\theta\theta} = -\frac{1}{2} g^{rr} \partial_r g_{\theta\theta} = -r(1 - kr^2)$$

$$\Gamma^r_{\phi\phi} = -\frac{1}{2} g^{rr} \partial_r g_{\phi\phi} = -r(1 - kr^2) \sin^2 \theta. \tag{45.18}$$

Then,

$$\Gamma^\theta_{\mu\nu} = \frac{1}{2} g^{\theta\theta} (\partial_\mu g_{\theta\nu} + \partial_\nu g_{\theta\mu} - \partial_\theta g_{\mu\nu}) = \frac{1}{2a^2 r^2} (\partial_\mu g_{\theta\nu} + \partial_\nu g_{\theta\mu} - \partial_\theta g_{\mu\nu}). \tag{45.19}$$

We easily see that $\Gamma^\theta_{tt} = \Gamma^\theta_{rr} = \Gamma^\theta_{rt} = \Gamma^\theta_{t\phi} = \Gamma^\theta_{r\phi} = \Gamma^\theta_{\theta\theta} = \Gamma^\theta_{\theta\phi} = 0$, and

$$\Gamma^\theta_{t\theta} = \frac{1}{2} \partial_t \ln g_{\theta\theta} = \frac{\dot{a}}{a}$$

$$\Gamma^\theta_{r\theta} = \frac{1}{2} \partial_r \ln g_{\theta\theta} = \frac{1}{r}$$

$$\Gamma^\theta_{\phi\phi} = -\frac{1}{2} g^{\theta\theta} \partial_\theta g_{\phi\phi} = -\sin \theta \cos \theta. \tag{45.20}$$

Finally,

$$\Gamma^\phi{}_{\mu\nu} = \frac{1}{2}g^{\phi\phi}(\partial_\mu g_{\phi\nu} + \partial_\nu g_{\phi\mu} - \partial_\phi g_{\mu\nu}). \tag{45.21}$$

Since ϕ is an isometry, $\partial_\phi g_{\mu\nu} = 0$, and we easily see that the only nonzero components are $\Gamma^\phi{}_{\phi\mu}$, but for $\mu \neq \phi$, so

$$\begin{aligned}
\Gamma^\phi{}_{t\phi} &= \frac{1}{2}\partial_t \ln g_{\phi\phi} = \frac{\dot{a}}{a}\\
\Gamma^\phi{}_{r\phi} &= \frac{1}{2}\partial_r \ln g_{\phi\phi} = \frac{1}{r}\\
\Gamma^\phi{}_{\theta\phi} &= \frac{1}{2}\partial_\theta \ln g_{\phi\phi} = \frac{\cos\theta}{\sin\theta}.
\end{aligned} \tag{45.22}$$

Ricci Tensors

The tt component is

$$\begin{aligned}
R_{tt} &= \partial_\mu \Gamma^\mu{}_{tt} - \partial_t \Gamma^\mu{}_{t\mu} + \Gamma^\mu{}_{\lambda\mu}\Gamma^\lambda{}_{tt} - \Gamma^\mu{}_{t\lambda}\Gamma^\lambda{}_{t\mu}\\
&= -\partial_t(\Gamma^r{}_{tr} + \Gamma^\theta{}_{t\theta} + \Gamma^\phi{}_{t\phi}) - (\Gamma^r{}_{tr})^2 - (\Gamma^\theta{}_{t\theta})^2 - (\Gamma^\phi{}_{t\phi})^2\\
&= -3\left[\frac{d}{dt}\left(\frac{\dot{a}}{a}\right) + \frac{\dot{a}^2}{a^2}\right] = -3\frac{\ddot{a}}{a}.
\end{aligned} \tag{45.23}$$

The rr component is

$$\begin{aligned}
R_{rr} &= \partial_\mu \Gamma^\mu{}_{rr} - \partial_r \Gamma^\mu{}_{r\mu} + \Gamma^\mu{}_{\lambda\mu}\Gamma^\lambda{}_{rr} - \Gamma^\mu{}_{r\lambda}\Gamma^\lambda{}_{r\mu}\\
&= \partial_t \Gamma^t{}_{rr} + \partial_r \Gamma^r{}_{rr} - \partial_r(\Gamma^r{}_{rr} + \Gamma^\theta{}_{r\theta} + \Gamma^\phi{}_{r\phi}) + \Gamma^t{}_{rr}(\Gamma^r{}_{tr} + \Gamma^\theta{}_{t\theta} + \Gamma^\phi{}_{t\phi})\\
&\quad + \Gamma^r{}_{rr}(\Gamma^r{}_{rr} + \Gamma^\theta{}_{r\theta} + \Gamma^\phi{}_{r\phi}) - (\Gamma^r{}_{rr})^2 - (\Gamma^\theta{}_{r\theta})^2 - (\Gamma^\phi{}_{r\phi})^2 - 2\Gamma^t{}_{rr}\Gamma^r{}_{tr}\\
&= \frac{2(k + \dot{a}^2) + a\ddot{a}}{1 - kr^2}.
\end{aligned} \tag{45.24}$$

The $\theta\theta$ component is

$$\begin{aligned}
R_{\theta\theta} &= \partial_\mu \Gamma^\mu{}_{\theta\theta} - \partial_\theta \Gamma^\mu{}_{\theta\mu} + \Gamma^\mu{}_{\lambda\mu}\Gamma^\lambda{}_{\theta\theta} - \Gamma^\mu{}_{\theta\lambda}\Gamma^\lambda{}_{\theta\mu}\\
&= \partial_t \Gamma^t{}_{\theta\theta} + \partial_r \Gamma^r{}_{\theta\theta} - \partial_\theta \Gamma^\phi{}_{\theta\phi} + \Gamma^t{}_{\theta\theta}(\Gamma^r{}_{tr} + \Gamma^\theta{}_{t\theta} + \Gamma^\phi{}_{t\phi})\\
&\quad + \Gamma^r{}_{\theta\theta}(\Gamma^r{}_{rr} + \Gamma^\theta{}_{r\theta} + \Gamma^\phi{}_{r\phi}) - (\Gamma^\phi{}_{\theta\phi})^2 - 2\Gamma^t{}_{\theta\theta}\Gamma^\theta{}_{t\theta} - 2\Gamma^r{}_{\theta\theta}\Gamma^\theta{}_{r\theta}\\
&= r^2(2(k + \dot{a}^2) + a\ddot{a}).
\end{aligned} \tag{45.25}$$

The $\phi\phi$ component is

$$\begin{aligned}
R_{\phi\phi} &= \partial_\mu \Gamma^\mu{}_{\phi\phi} - \partial_\phi \Gamma^\mu{}_{\phi\mu} + \Gamma^\mu{}_{\lambda\mu}\Gamma^\lambda{}_{\phi\phi} - \Gamma^\mu{}_{\phi\lambda}\Gamma^\lambda{}_{\phi\mu}\\
&= \partial_t \Gamma^t{}_{\phi\phi} + \partial_r \Gamma^r{}_{\phi\phi} + \partial_\theta \Gamma^\theta{}_{\phi\phi} + \Gamma^r{}_{\phi\phi}(\Gamma^r{}_{rr} + \Gamma^\theta{}_{r\theta} + \Gamma^\phi{}_{r\phi})\\
&\quad \Gamma^t{}_{\phi\phi}(\Gamma^r{}_{tr} + \Gamma^\theta{}_{t\theta} + \Gamma^\phi{}_{t\phi}) - 2\Gamma^\phi{}_{\phi\phi}\Gamma^\phi{}_{\theta\phi} - 2\Gamma^\phi{}_{r\phi}\Gamma^r{}_{\phi\phi} - 2\Gamma^\phi{}_{t\phi}\Gamma^t{}_{\phi\phi}\\
&= [(2(k + \dot{a}^2) + a\ddot{a}]r^2 \sin^2\theta = R_{\theta\theta}\sin^2\theta.
\end{aligned} \tag{45.26}$$

Raising the first index, we obtain

$$\begin{aligned}
R^t{}_t &= g^{tt}R_{tt} = 3\frac{\ddot{a}}{a}\\
R^r{}_r &= g^{rr}R_{rr} = \frac{2(k + \dot{a}^2)}{a^2} + \frac{\ddot{a}}{a}
\end{aligned}$$

$$R^\theta{}_\theta = g^{\theta\theta}R_{\theta\theta} = \frac{2(k+\dot{a}^2)}{a^2} + \frac{\ddot{a}}{a}$$
$$R^\phi{}_\phi = g^{\phi\phi}R_{\phi\phi} = \frac{2(k+\dot{a}^2)}{a^2} + \frac{\ddot{a}}{a}. \tag{45.27}$$

45.3 Solution to the Einstein's Equations

The Einstein's equations are written as

$$R^\mu{}_\nu = 8\pi G_N\left(T^\mu{}_\nu - \frac{1}{2}\delta^\mu_\nu T\right). \tag{45.28}$$

But the source of the cosmological solution is, as we said, a perfect fluid with energy-momentum tensor

$$T^\mu{}_\nu = \text{diag}(-\rho, p, p, p), \tag{45.29}$$

so then $T = 3p - \rho$. As we see, the Einstein's equations reduce to just two equations (since the $\theta\theta$ and $\phi\phi$ equations are the same as the rr one),

$$3\frac{\ddot{a}}{a} = -4\pi G_N(\rho + 3p)$$
$$\frac{2(k+\dot{a}^2)}{a^2} + \frac{\ddot{a}}{a} = 4\pi G_N(\rho - p). \tag{45.30}$$

Dividing the first equation by 3 and subtracting it from the second, we get, together with the first equation,

$$\frac{\ddot{a}}{a} = -\frac{4\pi G_N}{3}(\rho + 3p)$$
$$\frac{\dot{a}^2}{a^2} = -\frac{k}{a^2} + \frac{8\pi G_N}{3}\rho. \tag{45.31}$$

The second equation is called the *Friedmann equation*, and the first is the *second Friedmann equation*, or the *acceleration equation*.

The Friedman equation is an equation for

$$H \equiv \frac{\dot{a}}{a}, \tag{45.32}$$

called the *Hubble constant* (though it is not a constant in time, $H = H(t)$), giving the Hubble law $\dot{a} = Ha$. The acceleration equation gives the acceleration $\ddot{a}$, but it is not an independent equation. It is derived from the Friedmann equation, using the equation for the conservation of energy.

In flat space, the conservation equation for the energy-momentum tensor is $\partial_\mu T^\mu{}_\nu = 0$, which for $\nu = 0$ (energy component) gives for a perfect fluid just $\dot{\rho} = 0$. But in curved space, the conservation law is

$$D_\mu T^\mu{}_\nu = \partial_\mu T^\mu{}_\nu + \Gamma^\mu{}_{\mu\rho}T^\rho{}_\nu - \Gamma^\rho{}_{\mu\nu}T^\mu{}_\rho = 0. \tag{45.33}$$

This can be seen in several ways. First, we can take D_μ of the Einstein's equations and, since the left-hand side is zero by the Bianchi identity (after two contractions and using the symmetry properties)

$$D_\lambda R^\mu{}_{\nu\rho\sigma} + D_\sigma R^\mu{}_{\nu\lambda\rho} + D_\rho R^\mu{}_{\nu\sigma\lambda} = 0 \Rightarrow D_\mu R^\mu{}_\nu - \frac{1}{2}D_\nu R = 0, \tag{45.34}$$

we obtain that the right-hand side is also zero – i.e., the conservation law. Another way to see it is to realize we need a tensor equation, and $\partial_\mu T^\mu{}_\nu = 0$ is not; we need to covariantize it. Yet another way is to realize that, as we said when defining the energy-momentum tensor, it is the current associated with the symmetry of translations, so by considering *local translations* – i.e., general coordinate transformation invariance – we obtain a gauge theory of translations, in which the current is covariantly conserved, $D_\mu j^{\mu a} = 0$, obtaining again the same relation.

The $\nu = 0$ conservation law for the diagonal energy-momentum tensor is now

$$\partial_0 T^0{}_0 + T^0{}_0 \sum_\mu \Gamma^\mu{}_{0\mu} - \sum_\mu \Gamma^\mu{}_{0\mu} T^\mu{}_\mu = 0, \tag{45.35}$$

which becomes just

$$-\dot\rho - 3\frac{\dot a}{a}(\rho + p) = 0. \tag{45.36}$$

Taking the time derivative of the Friedmann equation, and using the equation obtained above, we obtain

$$2\frac{\dot a \ddot a}{a^2} - 2\frac{(\dot a^2 + k)\dot a}{a^3} = \frac{8\pi G_N}{3}\dot\rho = -8\pi G_N(\rho + p)\frac{\dot a}{a}, \tag{45.37}$$

which, after dividing by $2\dot a/a$, becomes just the acceleration equation, with the Friedmann equation subtracted from it. That means that the acceleration equation is indeed dependent of the Friedmann equation, once the conservation of the energy-momentum tensor is taken into account.

Important Concepts to Remember

- The Universe is homogenous and isotropic on large scales.
- The homogenous and isotropic expanding metric is $ds^2 = -dt^2 + a^2(t)[dr^2 + r^2(d\theta^2 + \sin^2\theta d\phi^2)]/(1 + kr^2/r)^2$, which has a constant scalar $R = k/a^2$, so $k = +1$ for closed Universe, $k = 0$ for flat one and $k = -1$ for an open hyperbolic Universe.
- In FLRW coordinates, the FLRW metric is $ds^2 = -dt^2 + a^2(t)[dr^2/(1-kr^2) + r^2(d\theta^2 + \sin^2\theta d\phi^2)]$.
- The Einstein's equations for the FLRW metric reduce to the Friedmann equation $H^2 = -k/a^2 + (8\pi G_N/3)\rho$ and the acceleration equation that follows from it and the conservation equation $\dot\rho + r\dot a/a(\rho + P) = 0$.

Further Reading

See chapter 14 in [3] and chapter 12 in [40].

Exercises

(1)   Calculate the Ricci scalar for the spatial FLRW metric and show that it is

$$R = \frac{k}{a^2(t)}. \tag{45.38}$$

(2)   Find $a = a(t)$ for:

- matter: $p = 0$.
- radiation: $p = \rho/3$.
- cosmological constant: $p = -\rho$.

(3)   For the case in exercise 2, find the conformal time η for $k = 0$ (flat three-dimensional space), defined by $ds^2 = a^2(\eta)[d\eta^2 + d\vec{x}^2]$, and the resulting $a(\eta)$.

(4)   If we have a combination of matter, radiation, and cosmological constant, adding up to a total that gives a flat Universe, with $k = 0$, giving the densities today as $\rho_{m,0}$, $\rho_{r,0}$ and $\rho_{\Lambda,0}$, find the time evolution of the densities of these three fluid components.

46 Vielbein-Spin Connection Formulation of General Relativity and Gravitational Instantons

In this chapter, we consider solutions of the vacuum Einstein's equations in Euclidean space – i.e., gravitational instantons. These will be found similarly to the Yang–Mills instantons, using a self-duality condition for the Riemann tensor. To derive them, we start with a discussion of an alternative formulation of general relativity, using *vielbeins and spin connections*, which define something more similar to the gauge fields needed for instatons.

46.1 Vielbein-Spin Connection Formulation of General Relativity

In the standard formulation of general relativity described in this book until now, the basic variable is the metric field $g_{\mu\nu}(x^\rho)$. This then defines the Christoffel symbol $\Gamma^\mu_{\nu\rho}$, which acts somewhat like a "gauge field of gravity." Then the Riemann tensor was defined in a way that made this analogy more obvious by being like the "field strength" of the Christoffel symbol (except that "space" and "gauge" indices are the same).

But there is a way to formulate general relativity in which it is more obvious the connection to gauge theory. It is a kind of "first-order formulation" for gravity, in the sense presented before, when discussing dualities: we introduce an extra, auxiliary field that makes the theory linear in derivatives instead of having the two derivatives of the Einstein formulation. The new form is in terms of the "vielbein" $e^a_\mu(x)$ and the "spin connection" $\omega^{ab}_\mu(x)$.

The word "vielbein" comes from German, where "viel" means "many" and "bein" means leg. It is a generalization of the case in four dimensions, where it is known as "vierbein." where "vier" means "four." One also uses "einbein," "zweibein", "dreibein," etc., for other dimensions, from "ein, zwei, drei," etc., meaning "one, two, three," etc.

On a small enough scale, any curved space looks locally flat, which means it will have Lorentz invariance. Thus we have a "local Lorentz invariance," with parameter $\lambda^a_b(x^\rho)$, where $a, b = 0, 1, 2, 3$ are indices in a fundamental representation of the $SO(1, 3)$ Lorentz group, and this invariance is made manifest by its action on the vielbein e^a_μ, as

$$e^a_\mu(x) \to e'^a_\mu(x) = \lambda^a_b(x^\rho)e^b_\mu(x^\rho). \tag{46.1}$$

The vielbein is defined as a sort of "square root of the metric," by

$$g_{\mu\nu}(x) = e^a_\mu(x)e^b_\nu(x)\eta_{ab}. \tag{46.2}$$

Therefore, a is a "flat" index, acted upon by a local Lorentz transformation, infinitesimally

$$\delta_{L.L.} e^a_\mu(x) = \delta\lambda^a{}_b(x) e^b_\mu(x), \tag{46.3}$$

and μ is a "curved" index, acted upon by a general coordinate transformation, infinitesimally

$$\delta_\xi e^a_\mu(x) = (\xi^\rho \partial_\rho) e^a_\mu + (\partial_\mu \xi^\rho) e^a_\rho. \tag{46.4}$$

In this way, we see that the general coordinate transformation on the metric is the standard one,

$$\delta_\xi g_{\mu\nu}(x) = (\xi^\rho \partial_\rho) g_{\mu\nu} + (\partial_\mu \xi^\rho) g_{\rho\nu} + (\partial_\nu \xi^\rho) g_{\mu\rho}. \tag{46.5}$$

The description in terms of the metric $g_{\mu\nu}(x)$ is equivalent to the one in terms of the vielbein $e^a_\mu(x)$. It would seem like that is not the case, because e^a_μ has a priori d^2 components, which is 16 in four dimensions, compared to $d(d+1)/2$, 10 in four dimensions, for the symmetric matrix $g_{\mu\nu}$. But while both are still subject to general coordinate transformations, e^a_μ has an extra local Lorentz invariance with parameters $\Lambda^a{}_b(x)$, which can be used to fix $d(d-1)/2$ components, six in four dimensions. So we have the same number of degrees of freedom.

But while the vielbein is a sort of gauge field, since it has a vector index μ and a group index a in $SO(3,1)$, the index is in the fundamental representation instead of the adjoint.

There is one more object that is more similar to the gauge field but whose introduction means considering a first-order formulation. This is the *spin connection* $\omega^{ab}_\mu(x)$, where (ab) is antisymmetric, so it defines an index in the adjoint of the $SO(3,1)$ group. The spin connection is, as the name suggests, a "connection" (mathematical term for gauge field) for the action on spinor fields. That is, the (general relativity) covariant derivative acting on spinors is obtained by multiplying ω^{ab}_μ with the generator of the Lorentz group in the spinor representation, which is $\Gamma_{ab} \equiv \frac{1}{2}[\Gamma_a, \Gamma_b]$, where Γ^a are the gamma matrices. Thus, the covariant derivative on a spinor ψ is – similar to the usual formula for a gauge field – $D_\mu \phi = \partial_\mu \phi + A^a_\mu T_a \phi$,

$$D_\mu \psi \equiv \partial_\mu \psi + \frac{1}{4} \omega^{ab}_\mu \Gamma_{ab} \psi. \tag{46.6}$$

From this definition, which means that $D_\mu \psi$ is covariant under general coordinate transformations (transforms as $\partial_\mu \phi$), we can infer that ω^{ab}_μ acts as a gauge field for the local Lorentz group, with fundamental index a.

But we already had enough degrees of freedom to describe gravity just from e^a_μ, which means that ω^{ab}_μ is extra, and it must be some sort of auxiliary field for a first-order formulation. That is to say, the equation of motion for ω must give a fixed form in terms of e^a_μ, $\omega = \omega(e)$. In fact, this is true *in the absence of dynamical fermions* (which means fermions with a kinetic term); otherwise we need to modify $\omega(e)$ to an $\omega(e, \psi)$, but we will not consider that.

The function $\omega = \omega(e)$ is defined as the solution of the "vielbein postulate," or "no torsion constraint,"

$$T^a_{\mu\nu} \equiv 2D_{[\mu} e^a_{\nu]} = 2\partial_{[\mu} e^a_{\nu]} + 2\omega^{ab}_{[\mu} e^b_{\nu]} = 0. \tag{46.7}$$

Another version of the vielbein postulate is the condition without antisymmetrization,

$$D_\mu e_\nu^a \equiv \partial_\mu e_\nu^a + \omega_\mu^{ab} e_\nu^b - \Gamma^\rho_{\ \mu\nu} e_\rho^a = 0. \qquad (46.8)$$

The antisymmetric part is the condition in (46.8), since $\Gamma^\rho_{\ \mu\nu}$ is symmetric in $\mu\nu$. The symmetric part fixes also the Christoffel symbol in terms of the vielbein.

The solution of the vielbein postulate is

$$\omega^{ab}(e) = \frac{1}{2} e^{a\nu}(\partial_\mu e_\nu^b - \partial_\nu e_\mu^b) - \frac{1}{2} e^{b\nu}(\partial_\mu e_\nu^a - \partial_\nu e_\mu^a) - \frac{1}{2} e^{a\rho} e^{b\nu}(\partial_\rho e_{c\sigma} - \partial_\sigma e_{c\rho}) e_\mu^c. \qquad (46.9)$$

The proof of this is left as an exercise.

T^a is called "torsion" and is a sort of field strength for the vielbein; the vielbein postulate says that this field strength vanishes. But more easily understood is the field strength of the spin connection,

$$R_{\mu\nu}^{ab}(\omega) = \partial_\mu \omega_\nu^{ab} - \partial_\nu \omega_\mu^{ab} + \omega_\mu^{ac} \omega_\nu^{cb} - \omega_\nu^{ac} \omega_\mu^{cb}. \qquad (46.10)$$

Indeed, as we can see, this is the same formula as the field strength for a gauge field, in the adjoint of an $SO(p,q)$ group, like the Lorentz group. That means that, unlike the Riemann tensor defined in terms of the Christoffel symbol $\Gamma^\mu_{\ \nu\rho}(g)$, now space and gauge indices are separate, and we have a well-defined field strength.

But, as we could guess from the formulas, the two field strengths, $R^{ab}(\omega(e))$ and $R^\mu_{\ \nu\rho\sigma}(\Gamma(g(e)))$ are related in the obvious way:

$$R_{\rho\sigma}^{ab}(\omega(e)) = e_\mu^a e^{-1,\nu b} R^\mu_{\ \nu\rho\sigma}(\Gamma(g(e))). \qquad (46.11)$$

Here $e^{-1,\nu b}$ is the (matrix) inverse of the vielbein, $e^{-1,\nu b} e_\mu^a = \eta^{ab}$.

The formula above means that the field strength of ω, $R^{ab}(\omega)$, is just the Riemann tensor with two indices "flattened" with the vielbein. Finally, that means that the Ricci scalar can be obtained from the field strength by acting with the (inverse) vielbein,

$$R = R_{\mu\nu}^{ab} e_a^{-1\ \mu} e_b^{-1\ \nu}. \qquad (46.12)$$

The defining relation of the vielbein, $g_{\mu\nu} = e_\mu^a \eta_{ab} e_\nu^b$, as a matrix relation is $g = e\eta e$, which means that

$$-\det g = (\det e)^2, \qquad (46.13)$$

so that, finally, we can rewrite the Einstein–Hilbert action in terms of the vielbein as

$$S_{\text{EH}} = \frac{1}{16\pi G_N} \int d^4x (\det e) R_{\mu\nu}^{ab}(\omega(e)) e_a^{-1\ \mu} e_b^{-1\ \nu}. \qquad (46.14)$$

This action is still a second-order action, since it is just an action in terms of the vielbein, instead of the equivalent metric. The vielbein satisfies the vielbein postulate, which gives $\omega = \omega(e)$. To have a first-order formulation, we need to introduce an independent auxiliary field, and by solving its equation of motion, we go back to the second-order formulation.

But we have prepared the ground to make it easy to guess which field is this auxiliary field: it is the spin connection, ω_μ^{ab}. If we treat it as an independent field, instead of the solution to the vielbein postulate, we obtain *the first-order formulation for gravity*, also known as the *Palatini formulation*, in terms of $(e_\mu^a, \omega_\mu^{ab})$. Then it turns out that the equation of motion for ω is exactly the vielbein postulate, with solution $\omega = \omega(e)$.

To prove that, we first note that, from the definition of the determinant,

$$\det e_\mu^a = \frac{1}{4!} \epsilon^{\mu\nu\rho\sigma} \epsilon_{abcd} e_\mu^a e_\nu^b e_\rho^c e_\sigma^d, \tag{46.15}$$

we can also obtain the relation

$$(\det e) e_a^{-1\,[\mu} e_b^{-1\,\nu]} = \frac{1}{4} \epsilon^{\mu\nu\rho\sigma} \epsilon_{abcd} e_\rho^c e_\sigma^d. \tag{46.16}$$

In other dimensions, a simple generalization of this relation leads to a similar relation.

Using it, we can rewrite the Einstein–Hilbert action as

$$\begin{aligned}
S_{\text{EH}} &= \frac{1}{16\pi G_N} \frac{1}{4} \int d^4 x \epsilon^{\mu\nu\rho\sigma} \epsilon_{abcd} R_{\mu\nu}^{ab}(\omega) e_\rho^c e_\sigma^d \\
&= \frac{1}{16\pi G_N} \int \epsilon_{abcd} R^{ab} \wedge e^c \wedge e^d,
\end{aligned} \tag{46.17}$$

where the last equality is in form language.

Thus rewritten, it is obvious that the equation of motion for ω is

$$\epsilon_{abcd} \epsilon^{\mu\nu\rho\sigma} (D_\nu e_\rho^c) e_\sigma^d = 0, \tag{46.18}$$

which becomes just the "vielbein postulate,"

$$T_{[\mu\nu]}^a \equiv 2 D_{[\mu} e_{\nu]}^a = 0. \tag{46.19}$$

We note, however, that the action for gravity is not exactly the Yang–Mills action for the spin connection ω_μ^{ab} – i.e., with gauge group $SO(3, 1)$. In fact, it is not even gauge invariant off-shell – i.e., if ω and e are abitrary. Only if $\omega = \omega(e)$, and we identify flat (a) and curved (μ) indices, thus going back to the Einstein formulation, with local Lorentz transformation identified with general coordinate transformations, then that is true.

But, in this specific case, we can write the local Lorentz transformation on the spin connection as a local Lorentz ($SO(3, 1)$) gauge transformation with parameter $\Lambda^a_{\;b}$ – i.e., as

$$(\omega'^a_{\;b})_\mu = (\Lambda^{-1})^a_{\;c}(\omega^c_{\;d})_\mu \Lambda^d_{\;b} + (\Lambda^{-1})^a_{\;c} \partial_\mu \Lambda^c_{\;b}. \tag{46.20}$$

Indeed, we can check that in this case, the no-torsion constraint (vielbein postulate) $De^a = 0$, defining $\omega = \omega(e)$, is left invariant, which we leave as an exercise to prove. Then, by the fact that $R(\omega)$ is the Yang–Mills curvature of ω, we know that the curvature transforms covariantly,

$$(\mathcal{R}'^a_{\;b})_{\mu\nu} = (\Lambda^{-1})^a_{\;c}(\mathcal{R}^c_{\;d})_{\mu\nu} \Lambda^d_{\;b}. \tag{46.21}$$

This suggests, however, that many of the techniques one uses in Yang–Mills theory can be applied. We will see that one important such application is to find gravitational analogues of the instaton solutions of Yang–Mills theory.

46.2 Taub-NUT Solutions

An important solution of the flat space Einstein's equations, $R_{\mu\nu} = 0$, is the solution generically called Taub-NUT space, first discovered by Taub, and then rediscovered and extended in a different context by Newman, Tamburino, and Unti (note that the correct order of authors is alphabetic: N, T, U, but "NUT" sounds better, so it caught on; Gibbons and Hawking even defined later on the notion of "nut" and "bolt" type singularities). I will not present the derivation of these solutions since they are quite involved – just the final result.

The solution found by Taub is

$$ds^2 = -U(t)^{-1}dt^2 + (2l)^2 U(t)(d\psi + \cos\theta d\phi)^2 + (t^2 + l^2)(d\theta^2 + \sin^2\theta d\phi^2), \quad (46.22)$$

where

$$U(t) = -1 + \frac{2(mt + l^2)}{t^2 + l^2} = \frac{l^2 - t^2 + 2mt}{t^2 + l^2}, \quad (46.23)$$

and m and l are two constants. Here θ, ϕ, ψ are Euler angles on S^3, with ranges $0 \le \psi \le 4\pi$, $0 \le \theta \le \pi$ and $0 \le \phi \le 2\pi$.

Consider "left-invariant one-forms" σ_i on S^3 (thought of as $SU(2)$; see Chapter 3), defined by

$$g^{-1}dg = i\sum_a \sigma_a \frac{\tau_a}{2}, \quad (46.24)$$

where τ_a are the Pauli matrices. Since, as we saw in Chapter 3, $e^{i\frac{\vec{\alpha}\cdot\vec{\tau}}{2}} = \cos\left|\frac{\vec{\alpha}}{2}\right| + i\frac{\vec{\alpha}\cdot\vec{\sigma}}{|\vec{\alpha}|}\sin\left|\frac{\vec{\alpha}}{2}\right|$, we can write

$$e^{i\frac{\theta\tau_1}{2}} = \cos\frac{\theta}{2} + i\tau_1\sin\frac{\theta}{2} = \begin{pmatrix} \cos\frac{\theta}{2} & i\sin\frac{\theta}{2} \\ +i\sin\frac{\theta}{2} & \cos\frac{\theta}{2} \end{pmatrix} = \tilde{R}_1(\theta)$$

$$e^{i\frac{\phi}{2}\tau_3} = \begin{pmatrix} e^{i\frac{\phi}{2}} & 0 \\ 0 & e^{-i\frac{\phi}{2}} \end{pmatrix} = \tilde{R}_3(\phi), \quad (46.25)$$

using the notation of Chapter 3. Then, a generic matrix element of $S_3 \simeq SU(2)$ is

$$g = e^{i\frac{\phi}{2}\tau_3}e^{i\frac{\theta}{2}\tau_1}e^{i\frac{\psi}{2}\tau_3}, \quad (46.26)$$

and then from the definition

$$g^{-1}dg = i\frac{\sigma_1}{2}\tau_1 + i\frac{\sigma_2}{2}\tau_2 + i\frac{\sigma_3}{3}\tau_3, \quad (46.27)$$

we get the left-invariant one-forms in terms of the Euler angles,

$$\begin{aligned} \sigma_1 &= \sin\psi d\theta - \cos\psi\sin\theta d\phi \\ \sigma_2 &= \cos\psi d\theta + \sin\psi\sin\theta d\phi \\ \sigma_3 &= d\psi + \cos\theta d\phi. \end{aligned} \quad (46.28)$$

Then, if in the metric element (ϕ, θ, ψ) appears only in the combinations $\sigma_1, \sigma_2, \sigma_3$, the metric has the $SU(2)$ invariance of S^3.

But the metric (46.22) can in fact, be written in terms of them, since

$$\sigma_1^2 + \sigma_2^2 = d\theta^2 + \sin^2\theta d\phi^2 = d\Omega_2^2 , \tag{46.29}$$

which is the metric on the two-sphere, and

$$\sigma_1^2 + \sigma_2^2 + \sigma_3^2 = d\Omega_3^2 \tag{46.30}$$

is the metric on the three-sphere (foliated in terms of two-spheres), so (46.22) can be rewritten as

$$ds^2 = -U(t)^{-1}dt^2 + (2l)^2 U(t)\sigma_3^2 + (t^2 + l^2)(\sigma_1^2 + \sigma_2^2). \tag{46.31}$$

The metric then has topology and symmetries of $\mathbb{R}_t \times S^3$. As it is, the metric describes a cosmological model, since it depends on time t. It also has an apparent singularity at $U = 0$ – i.e., at $t = m \pm \sqrt{m^2 + l^2}$.

But Newman, Tamburino, and Unti (NUT) have shown how to extend this metric, finding good coordinates that allow its extension. Moreover, like in the Schwarzschild case, depending on what side of the apparent singularity we are so that U is either positive or negative, we change what is "time" and what is "radius." So exchanging the names for t and r, we obtain a *static solution*, which is a generalization of the Schwarzschild case that depends on another parameter, l.

If m is understood as an "electric" type mass, then l is thought of as a "magnetic" type mass in a sense that will be clarified shortly.

The NUT metric can be written in the form (now r and t notations are interchanged with respect to the Taub metric)

$$ds^2 = -f(r)\left[dt + 4l\sin^2\frac{\theta}{2}d\phi\right]^2 + \frac{dr^2}{f(r)} + (r^2 + l^2)(d\theta^2 + \sin^2\theta d\phi^2)$$
$$f(r) = 1 - \frac{2(mr + l^2)}{r^2 + l^2} = \frac{r^2 - l^2 - 2mr}{r^2 + l^2}. \tag{46.32}$$

Note that this form emphasizes that at $l = 0$ we obtain the Schwarzschild solution; thus, m is indeed normal ("electric") mass. Then l is a new parameter, which will be called "magnetic" mass since it is another parameter of the static solution.

NUT actually presented the metric in the opposite signature $(+ - --)$ – that is, they presented $-ds^2$ – as

$$ds^2 = -U(r)\left[dt + 4l\sin^2\frac{\theta}{2}d\phi^2\right]^2 + \frac{dr^2}{U(r)} - (r^2 + l^2)(d\theta^2 + \sin^2\theta d\phi^2)$$
$$U(r) = -1 + \frac{2(mr + l^2)}{r^2 + l^2} = \frac{l^2 - r^2 + 2mr}{r^2 + l^2} = -f(r), \tag{46.33}$$

which emphasizes instead the connection with the Taub metric.

Moreover, NUT have another generalization, with

$$U(r) = -\epsilon + \frac{2(mr + \epsilon l^2)}{r^2 + l^2} , \tag{46.34}$$

and $\epsilon = (+1, 0, -1)$, but only the $\epsilon = +1$ case is considered in detail. Moreover, though it is not completely obvious, we can also consider l to be imaginary instead of real.

But one can also consider analytical continuations of the Taub-NUT metric for a solution with Euclidean signature. But, by definition, a solution of the vacuum Einstein's equations in Euclidean signature is called a gravitational instanton. Like the instantons in gauge theory, the gravitational instantons should have applications for the quantum version of the gravitational theory – i.e., for *quantum gravity*, as defining transition amplitudes between states.

46.3 Hawking's Taub-NUT Gravitational Instanton

The simplest is Hawking's analytical continuation of the NUT metric. Analytical continuation in time (to Euclidean space) is done by putting $t = i\tau$, which turns the Minkowski metric $ds^2 = -dt^2 + d\vec{x}^2$ into the Euclidean one, $ds^2 = d\tau^2 + d\vec{x}^2$. If, moreover, we consider an imaginary "magnetic mass," of modulus equal to the normal ("electric") mass m,

$$l = \pm im, \tag{46.35}$$

then, by $r = R + m$, we have

$$f(r) = \frac{r^2 + m^2 - 2mr}{r^2 - m^2} = \frac{(r-m)^2}{r^2 - m^2} = \left(1 + \frac{2m}{R}\right)^{-1} \equiv V(R)^{-1}, \tag{46.36}$$

and the metric becomes

$$\begin{aligned}
ds^2 &= +f(r)\left[d\tau \pm 4m\sin^2\frac{\theta}{2}d\phi\right]^2 + \frac{dr^2}{f(r)} + (r^2 - m^2)d\Omega_2^2 \\
&= V(R)^{-1}\left[d\tau \pm 4m\sin^2\frac{\theta}{2}d\phi\right]^2 + V(R)(dR^2 + R^2 d\Omega_2^2),
\end{aligned} \tag{46.37}$$

or yet, if $d\vec{x} \cdot d\vec{x} = dR^2 + R^2 d\Omega_2^2$,

$$ds^2 = V(R)^{-1}\left[d\tau \pm 4m\sin^2\frac{\theta}{2}d\phi\right]^2 + V(R)d\vec{x} \cdot d\vec{x}. \tag{46.38}$$

Moreover, we can redefine τ and ϕ such as to put the metric in the form

$$ds^2 = V(R)^{-1}(d\tau + \vec{\omega} \cdot d\vec{x})^2 + V(R)d\vec{x} \cdot d\vec{x}, \tag{46.39}$$

provided (in order to solve Einstein's equations)

$$\vec{\nabla} \times \vec{\omega} = \vec{\nabla}V. \tag{46.40}$$

But this implies $\Delta V = \vec{\nabla} \cdot \vec{\nabla}V = 0$, which has the more general solution

$$V(R) = 1 + \sum_i \frac{2m_i}{|\vec{R} - \vec{R_i}|}. \tag{46.41}$$

In fact, we will see that, in this case, the Einstein equation reduces to a self-duality equation, which on the ansatz becomes just $\vec{\nabla} \times \vec{\omega} = \vec{\nabla}V$.

Before that, we note that there exists another gravitational instanton solution, found by Page. It is obtained from the Taub metric (46.22) by the "analytical continuation" $l = iN$ and $U \to -U$, and renaming t as r. Then we have

$$ds^2 = U(r)^{-1}dr^2 + 4N^2 U(r)(d\psi + \cos\theta d\phi)^2 + (r^2 - N^2)d\Omega_2^2$$
$$U(r) = \frac{r^2 - 2mr + N^2}{r^2 - N^2} = \frac{(r - r_+)(r - r_-)}{(r - N)(r + N)}$$
$$r_\pm = m \pm \sqrt{m^2 - N^2}. \tag{46.42}$$

In the "extremal" case $m = N$, when $r_\pm = m$, and, thus,

$$U(r) = \frac{(r - m)^2}{r^2 - m^2} = \frac{r - m}{r + m}, \tag{46.43}$$

the metric becomes

$$ds^2 = \frac{r + m}{r - m}dr^2 + 4m^2\frac{r - m}{r + m}\sigma_3^2 + (r^2 - m^2)(\sigma_1^2 + \sigma_2^2)$$
$$= \frac{r + m}{r - m}dr^2 + (r^2 - m^2)\left[\sigma_1^2 + \sigma_2^2 + \left(\frac{2m}{r + m}\right)^2\sigma_3^2\right]. \tag{46.44}$$

By the equivalence of the Taub and NUT constructions, this instanton with $m = N$ is the same as the metric found by Hawking.

46.4 The Eguchi–Hanson Metric and Yang–Mills-Like Ansatz

Eguchi and Hanson made the powerful observation that we can use the same formalism as in Yang–Mills theory to search for instanton solutions.

Yang–Mills Instanton

One can start by noting that in Yang–Mills theory, the self-duality condition (which solves the Euclidean space Yang–Mills equations, as we saw),

$$F_{\mu\nu} = \pm\frac{1}{2}\epsilon_{\mu\nu\rho\sigma}F_{\rho\sigma}, \tag{46.45}$$

can be solved by the ansatz

$$A_\nu = \rho(r)g^{-1}\partial_\mu g, \tag{46.46}$$

where g is an $SU(2)$-valued group element,

$$g = \frac{1}{r}(t - i\vec{x}\cdot\vec{\tau}), \quad r^2 = t^2 + \vec{x}^2, \tag{46.47}$$

if one satisfies the equation

$$\rho'(r) + \frac{2}{r}\rho(\rho - 1). \tag{46.48}$$

The proof is left as an exercise.

The solution of this equation is

$$\rho(r) = \frac{r^2}{r^2 + a^2},$$

(46.49)

giving the usual (BPST-'t Hooft) instanton.

Preliminaries in Gravity

We start by using form language on the vielbein-spin connection formulation of general relativity to emphasize the similarity with the Yang–Mills case.

We define first vielbein and spin connection one-forms in the obvious way,

$$e^a \equiv e^a_\mu dx^\mu , \quad \omega^{ab} \equiv \omega^{ab}_\mu dx^\mu,$$

(46.50)

as well as the curvature two-form and (no-)torsion two-form

$$R^{ab} = d\omega^{ab} + \omega^{ac} \wedge \omega^{cb}$$
$$T^a = De^a = de^a + \omega^{ab} \wedge e^b = 0,$$

(46.51)

Taking the exterior derivative of the no-torsion constraint, we get

$$0 = d\omega^{ab} \wedge e^b - \omega^{ab} \wedge de^b = d\omega^{ab} \wedge e^b - \omega^{ab} \wedge (-\omega^{bc} \wedge e^c) = R^{ab} \wedge e^b.$$

(46.52)

Defining the curvature tensor with all flat indices (which is the same as the Riemann tensor with all flattened indices, as we saw),

$$R^a{}_{bcd} \equiv (R^a{}_b)_{\mu\nu} e_c^{-1\,\mu} e_d^{-1\,\nu};$$

(46.53)

the identity is the antisymmetry identity for the (flattened) Riemann tensor,

$$R^a{}_{[bcd]} = 0.$$

(46.54)

But now we can define the curvature two-form dual in the flat (gauge – i.e., local Lorentz) indices ab,

$$\tilde{R}^{ab} \equiv \frac{1}{2} \epsilon^{ab}{}_{cd} R^{cd}.$$

(46.55)

Then the Einstein's equations in vacuum,

$$R_{\mu\nu} \equiv R^\rho{}_{\mu\rho\nu} = 0$$

(46.56)

become, with flat indices,

$$R^{ac}{}_{bc} = R^{ca}{}_{cb} = R^\mu{}_\nu e^a_\mu e_b^{-1\,\nu} = 0,$$

(46.57)

where we have used the antisymmetry of the Riemann tensor in both first and last pairs of indices.

But in the equation

$$0 = \tilde{R}^a_b \wedge e^b \Rightarrow 0 = \tilde{R}^a{}_{b[\rho\sigma} e^b_{\tau]} = \frac{1}{2} \epsilon^a{}_{bcd} R^{cd}{}_{[\rho\sigma} e^b_{\tau]}$$

(46.58)

we see that is, in fact, equivalent with the Einstein equation $R^{ac}{}_{bc} = 0$. Comparing with the Riemann tensor identity $R^a{}_b \wedge e^b = 0$, we see then that the Einstein's equation in vacuum is equivalent with the (anti)self-duality condition

$$R^a{}_b = \pm \tilde{R}^a_b. \tag{46.59}$$

Of course, like in the Yang–Mills case, this is only possible in Euclidean space for real fields; in Minkowski space, the condition is only satisfied for complex fields.

Still, unlike the Yang–Mills case, the Riemann tensor has two derivatives, so the self-duality condition is still second order in derivatives. Yet, also unlike Yang–Mills, we can impose a further self-duality on the spin connection without extra constraints.

Indeed, note that the curvature of the spin connection can be written explicitly as

$$
\begin{aligned}
R^2{}_3 &= d\omega^2{}_3 + \omega^2{}_0 \wedge \omega^0{}_3 + \omega^2{}_1 \wedge \omega^1{}_3 \\
R^0{}_1 &= d\omega^0{}_1 + \omega^0{}_2 \wedge \omega^2{}_1 + \omega^0{}_3 \wedge \omega^3{}_1 ,
\end{aligned}
\tag{46.60}
$$

plus two more pairs of formulas obtained from the cyclic permutations of $(1,2,3)$ in (46.60). But since the (anti)self-duality condition is $R^0{}_1 = \pm R^2{}_3$ and the cyclic permutations of $(1,2,3)$, we find that it is satisfied provided we have a self-duality of the spin connection as well,

$$\omega^0{}_i = \pm \frac{1}{2}\epsilon_{ijk}\omega^{jk}. \tag{46.61}$$

We can rewrite this condition in a similar way to the one on the curvature, by defining the dual spin connection (in gauge, or local Lorentz indices)

$$\tilde{\omega}^a_b \equiv \frac{1}{2}\epsilon_{abcd}\omega^c{}_d, \tag{46.62}$$

which means that the condition is now

$$\omega^a{}_b = \pm \tilde{\omega}^a_b. \tag{46.63}$$

We saw that the condition is sufficient, but it is also necessary, since we can always transform the spin connection by the (Euclidean) Lorentz transformation $\Lambda \in O(4)$, as in (46.20), after which the curvature transforms covariantly. Since the covariant transformation of $R^a{}_b$ preserves the self-duality of R constraint, but the gauge transformation of ω doesn't preserve the self-duality of the spin connection, we can always make a gauge transformation to put the spin connection in self-dual form as well.

Now the self-duality condition on the spin connection is a condition that is first order in derivatives (on the vielbein) because of the no-torsion constraint $(de^a + \omega^{ab} \wedge e^b = 0)$. Thus, again, we have reduced solving for the instanton to a set of differential equations first order in derivatives.

46.5 Instanton Ansätze and Solutions

One can consider the (Euclidean signature) ansatz, inspired by (46.44),

$$ds^2 = f^2(r)dr^2 + \frac{r^2}{4}(\sigma_1^2 + \sigma_2^2 + g^2(r)\sigma_3^2). \tag{46.64}$$

The vielbein one-forms corresponding to it are (the metric $g = e^a e^a$):

$$e^a = \left(f(r)dr, \frac{r}{2}\sigma_1, \frac{r}{2}\sigma_2, \frac{rg(r)}{2}\sigma_3 \right). \tag{46.65}$$

Then the spin connections that satisfy the no-torsion constraint $de^a + \omega^a{}_b \wedge e^b = 0$ are (as we can easily check)

$$\omega^1{}_0 = \frac{1}{rf(r)}e^1 , \quad \omega^2{}_0 = \frac{1}{rf(r)}e^2 , \quad \omega^3{}_0 = \left(\frac{1}{rf(r)} + \frac{g'}{f(r)g(r)} \right) e^3$$
$$\omega^2{}_3 = \frac{g(r)}{r}e^1 , \quad \omega^3{}_1 = \frac{g(r)}{r}e^2 , \quad \omega^1{}_2 = \frac{2 - g^2(r)}{rg(r)}e^3 . \tag{46.66}$$

Then, we impose antiself-duality on the previous spin connections, which leads to the equations

$$f(r)g(r) = 1 , \quad g(r) + rg'(r) = f(r)(2 - g^2(r)), \tag{46.67}$$

where the first equation comes from $\omega^1{}_0 = \omega^2{}_3$ and $\omega^2{}_0 = \omega^3{}_1$, and the second comes from $\omega^3{}_0 = \omega^1{}_2$.

The solution of these equations is

$$g^2(r) = f^{-2}(r) = 1 - \frac{a^4}{r^4}. \tag{46.68}$$

Thus, the Eguchi–Hanson instanton solution is

$$ds^2 = \left[1 - \frac{a^4}{r^4} \right]^{-1} dr^2 + \frac{r^2}{4}(\sigma_1^2 + \sigma_2^2) + \frac{r^2}{4}\left[1 - \frac{a^4}{r^4} \right] \sigma_3^2. \tag{46.69}$$

At $r \to \infty$, the metric becomes

$$ds^2 \to dr^2 + \frac{r^2}{4}(\sigma_1^2 + \sigma_2^2 + \sigma_3^2) = dr^2 + r^2 d\Omega_3^2, \tag{46.70}$$

which is *locally* flat Euclidean space. But globally it is not, since in order to avoid a singularity at $a \neq 0$, we need to have $\psi \sim \psi + 2\pi$, which is only half the periodicity required for the sphere.

That means that the Eguchi–Hanson instanton is *Asymptotically Locally Euclidean (ALE)*. We refer to such a space as an ALE space.

Gibbons and Hawking have classified possible *removable* singularities in metrics (the singularity is only apparent and can be removed by a convenient choice of coordinates). They have found two possibilities:

- A *nut*-type singularity is one that is actually like the origin of space in polar or spherical coordinates – i.e., one that looks like $\mathbb{R}^4$ close by. This is the singularity obtained for the Euclidean Taub-NUT solution in (46.44). Defining "proper Euclidean time" near the apparent singularity at $r = m$ by

$$d\tau^2 \equiv \frac{dr^2}{4}\left(\frac{r + m}{r - m} \right), \tag{46.71}$$

one can calculate that the Euclidean Taub-NUT metric at $r \simeq m$, or $r - m \equiv \epsilon \to 0$, is

$$ds^2 \simeq d\tau^2 + \tau^2(\sigma_1^2 + \sigma_2^2 + \sigma_3^2). \tag{46.72}$$

More generally, a nut singularity is one where, if the metric is written as

$$ds^2 = d\tau^2 + a^2(\tau)\sigma_1^2 + b^2(\tau)\sigma_2^2 + c^2(\tau)\sigma_3^2, \tag{46.73}$$

near the apparent singularity at $\tau = 0$, we have $a^2 \simeq b^2 \simeq c^2 \simeq \tau^2$.

- A *bolt*-type singularity is one like the the Eguchi–Hanson instanton, which becomes approximately $\mathbb{R}^2 \times S^2$, so near the singularity at $\tau = 0$, we have $a^2 \simeq b^2 =$ finite, and $c^2 \simeq n^2\tau^2$, with $n \in \mathbb{N}$ (integer). Indeed, the Eguchi–Hanson metric becomes near the singularity

$$ds^2 \simeq d\tau^2 + \tau^2\sigma_3^2 + a^2(\sigma_1^2 + \sigma_2^2). \tag{46.74}$$

46.6 Gibbons–Hawking Multi-Instanton Solution

One can make a transformation of coordinates on the Eguchi–Hanson metric that takes it to a form similar to the Hawking multi-Taub-NUT one:

$$ds^2 = V(R)^{-1}(d\psi + \vec{\omega} \cdot d\vec{x})^2 + V(R)d\vec{x} \cdot d\vec{x}, \tag{46.75}$$

where, as before, $R^2 = \vec{x} \cdot \vec{x}$, $\vec{\nabla}V = \pm\vec{\nabla} \times \vec{\omega}$, but now we have

$$V(R) = \sum_{i=1,2} \frac{1}{|\vec{x} - \vec{x}_i|}. \tag{46.76}$$

Note that we have only two points and that we don't have the 1 in $V(R)$.

In the case $a = 0$ and $\vec{x}_i = 0$, so $V(R) = 1/R$, the transformation of coordinates is as follows: Define

$$R \equiv \frac{r^2}{4} \Rightarrow r = 2\sqrt{R} \Rightarrow dr = \frac{dR}{\sqrt{R}}, \tag{46.77}$$

and then consider R, θ, ϕ as spherical coordinates for the $\mathbb{R}^3$ parametrized by $\vec{x}$. Since in spherical coordinates

$$\vec{\nabla}\times\vec{\omega} = \frac{1}{r\sin\theta}\left(\frac{\partial}{\partial\theta}(\omega_\phi\sin\theta) - \frac{\partial\omega_\theta}{\partial\phi}\right)\hat{r} + \frac{1}{r}\left(\frac{1}{\sin\theta}\frac{\partial\omega_r}{\partial\phi} - \frac{\partial}{\partial r}(r\omega_\phi)\right)\hat{\theta} + \frac{1}{r}\left(\frac{\partial}{\partial r}(r\omega_\theta) - \frac{\partial\omega_r}{\partial\theta}\right)\hat{\phi}, \tag{46.78}$$

equating this with $\vec{\nabla}V = -\frac{1}{r^2}\hat{r}$ gives a solution

$$\omega_\phi = \frac{\cos\theta}{r\sin\theta}, \omega_r = \omega_\theta = 0 \Rightarrow \vec{\omega}\cdot d\vec{x} = \cos\theta d\phi. \tag{46.79}$$

Then the Eguchi–Hanson metric (for $a = 0$) becomes

$$ds^2 = \frac{dR^2}{R} + R(\sigma_3^2 + \sigma_1^2 + \sigma_2^2), \tag{46.80}$$

which matches the Gibbons–Hawking form, which equals

$$ds^2 = R(d\psi + \cos\theta d\phi)^2 + \frac{1}{R}(dR^2 + R^2 d\Omega_3^2). \tag{46.81}$$

Gibbons and Hawking wrote the multicenter solution, which unifies the multi-instanton version of the Euclidean Taub-NUT and the Eguchi–Hanson form, with the same ansatz for the metric,

$$ds^2 = V(R)^{-1}(d\psi + \vec{\omega} \cdot d\vec{x})^2 + V(R)d\vec{x} \cdot d\vec{x}, \tag{46.82}$$

where, still, $R^2 = \vec{x} \cdot \vec{x}$ and $\vec{\nabla}V = \pm\vec{\nabla} \times \vec{\omega}$, but now the function V is more general,

$$V = \epsilon + \sum_{i=1}^{n} \frac{2m_i}{|\vec{x} - \vec{x}_i|}, \tag{46.83}$$

and $\epsilon = 0$ or 1. The solution is free of singularities only if $m_i = m$ (all masses equal). In fact, imposing the ansatz (46.83), we can find that the condition to have a self-dual spin connection (which implies a self-dual Riemann tensor and, in turn, the Einstein equation, as we saw) is $\vec{\nabla}V = \pm\vec{\nabla} \times \vec{\omega}$. In turn, this implies $\vec{\nabla} \cdot \vec{\nabla}V - \Delta V = 0$, which has the general solution written earlier.

Then $\epsilon = 1$ is the Hawking Euclidean Taub-NUT multi-instanton metric, and $\epsilon = 0$ is the Eguchi–Hanson multi-instanton. The $\epsilon = 0$ is Asymptotically Locally Euclidean (ALE), just like the simple Eguchi–Hanson metric, so it makes sense to be considered for a (quantum) theory in flat space, whereas $\epsilon = 1$ is not.

Important Concepts to Remember

- We can rewrite general relativity in terms of the vielbein $e^a_{\ \mu}$, with one "flat" (local Lorentz) and one "curved" (spacetime) index, which is the square root of the metric, $g_{\mu\nu} = e^a_{\ \mu} e^b_{\ \nu} \eta_{\mu\nu}$.
- For the covariant derivative on spinors $D_\mu \psi = \partial_\mu \psi + \omega^{ab}_\mu \Gamma_{ab}/4$, we have a "spin connection" $\omega^{ab}_\nu(e)$, satisfying the "vielbein postulate" or "no-torsion constraint" $T^a \equiv D_{[\mu} e^a_{\ \nu]} = \partial_{[\mu} e^a_{\ \nu]} + \omega^{ab}_{[\mu} e^b_{\ \nu]} = 0$.
- If the spin connection is considered independent instead of $\omega = \omega(e)$, we have a first-order formulation of general relativity or Palatini formalism.
- In the Palatini formalism, the field strength of ω, $R^{ab}(\omega) = d\omega^{ab} + \omega^{ac} \wedge \omega^{cb}$ becomes the Riemann tensor under substitution of $\omega = \omega(e)$, and the resulting Einstein–Hilbert action, $S = \frac{1}{16\pi G_N} \int \epsilon_{abcd} R^{ab} \wedge e^c \wedge e^d$, has $T^a = 0$ as the ω equation of motion.
- The Taub metric, a solution of the vacuum equations $R_{\mu\nu} = 0$, is a cosmological model with $U(t)$ and an apparent singularity at $U(t) = 0$, removed by going to NUT coordinates, where it can be turned into a static metric with $U = U(r)$ (the notion of time and space get interchanged accross the apparent singularity).
- The Euclidean version of Taub-NUT, or "Taub-NUT space," is a gravitational instanton, relevant for quantum gravity.
- The Hawking gravitational instanton is obtained by analytically continuing to Euclidean space by $t = i\tau$ and $l = \pm im$ (imaginary "magnetic" mass l, equal to the "electric" mass m) and can be written in terms of a V satisfying $\Delta V = 0$, with solution $V = 1 + 2m/R$ that can be generalized to multi-instantons.

- A gravitational instanton ansatz can be made by analogy with Yang–Mills, putting an (anti)self-dual Riemann tensor, $R^{ab} = \pm \frac{1}{2}\epsilon^{abcd}R_{cd}$, which is obtained by an (anti)self-dual spin connection, $\omega^{ab} = \frac{1}{2}\epsilon_{abcd}\omega^{cd}$.
- The Eguchi–Hanson gravitational instanton is obtained by the self-dual spin connection ansatz, and is an Asymptotically Locally Euclidean (ALE) space.
- Removable apparent singularities are of two types: "nut" like the Taub-NUT case, where locally the space becomes $\mathbb{R}^4$ in spherical coordinates near the singularity, and "bolt" like the Eguchi–Hanson solution, where locally the space becomes $\mathbb{R}^2 \times S^2$ in polar coordinates near the singularity.
- The Gibbons–Hawking multi-instanton has the most general V, with $V = \epsilon + \sum_i 2m_i/|\vec{x} - \vec{x}_i|$, and $\epsilon = 0$ for Euclidean Taub-NUT and $\epsilon = 0$ for Eguchi–Hanson.

Further Reading

The Taub-NUT metric was found by Taub in [48] and then generalized by Newman, Tamburino and Unti in [49]. An explanation of their connection was given in [50]. For an analysis of the topology of the Taub-NUT metric, see Hawking and Ellis's book [51]. Hawking noted that Taub-NUT space can be made into a gravitational instanton in [52], and Page wrote a variant for it [53]. The Eguchi–Hanson metric was described in [54], and turned into multi-instantons by Gibbons and Hawking in [55]. The gravitational instantons were reviewed in [56, 57].

Exercises

(1)   Prove that (46.9) is the solution of the vielbein postulate.

(2)   Prove that the gauge (local Lorentz) transformation of the spin connection leaves invariant the no-torsion constraint.

(3)   Prove that the self-duality condition for Yang–Mills, on the ansatz $A_\mu = \rho(r)g^{-1}\partial_\mu g$, gives the equation (46.48).

(4)   Prove that the equation of motion for the vielbein in the vielbein-spin connection formulation of the Einstein–Hilbert action (46.17) reduces, on the vielbein postulate $\omega = \omega(e)$, to the usual Einstein equation in vacuum.

References

[1] H. Goldstein, C. Poole, and J. Safko, *Classical mechanics*, 3rd edition, Addison-Wesley, 2002.

[2] H. Georgi, *Lie algebras in particle physics*, Perseus Books, 1999.

[3] L. D. Landau and E. M. Lifshitz, *Course of theoretical physics, vol 2: The classical theory of fields*, 4th edition, Elsevier Butterworth Heinemann, 2007.

[4] M. E. Peskin and D. V. Schroeder, *An introduction to quantum field theory*, Avalon Publishing, 1995.

[5] M. Burgess, *Classical covariant fields*, Cambridge University Press, 2002.

[6] L. D. Landau and E. M. Lifshitz, *Course of theoretical physics*, vol. 6: Fluid mechanics, 2nd edition, Elsevier Butterworth Heinemann, 2007.

[7] T. Bohr, M. H. Jensen, G. Paladin, and A. Vulpiani, *Dynamical systems approach to turbulence*, Cambridge University Press, 1998.

[8] O. Babelon, D. Bernard, and M. Talon, *Introduction to classical integrable systems*, Cambridge University Press, 2003.

[9] M. D. Schwartz, *Quantum field theory and the standard model*, Cambridge University Press, 2014.

[10] G. Sterman, *An introduction to quantum field theory*, Cambridge University Press, 1993.

[11] S. Weinberg, *The quantum theory of fields, vol. I: Foundations*, Cambridge University Press, 1995.

[12] D. W. F. Alves, C. Hoyos, H. Nastase, and J. Sonnenschein, "Knotted solutions, from electromagnetism to fluid dynamics," *Int. J. Mod. Phys.* A **32**, no. 33, 1750200 (2017) [arXiv:1707.08578 [hep-th]].

[13] R. Rajaraman, *Solitons and instantons: An introduction to solitons and instantons in quantum field theory*, North Holland, 1982.

[14] N. Manton and P. Sutcliffe, *Topological solitons*, Cambridge University Press, 2004.

[15] H. Nastase, "DBI skyrmion, high energy (large s) scattering and fireball production," [arXiv:hep-th/0512171].

[16] H. Nastase, *String theory methods for condensed matter physics*, Cambridge University Press, 2017.

[17] H. Nastase and J. Sonnenschein, "More on Heisenberg's model for high energy nucleon-nucleon scattering," *Phys. Rev. D* **92**, 105028 (2015) [arXiv:1504.01328 [hep-th]].

[18] L. Alvarez-Gaume and S. F. Hassan, "Introduction to S duality in N=2 supersymmetric gauge theories: A pedagogical review of the work of Seiberg and Witten," *Fortsch. Phys.* **45**, 159 (1997) [hep-th/9701069].

[19] E. W. Kolb and M. S. Turner, *The early universe*, Westview Press, 1990.

[20] S. R. Coleman, "Q balls," *Nucl. Phys. B* **262**, 263 (1985) Erratum: [*Nucl. Phys. B* **269**, 744 (1986)].

[21] T. D. Lee and Y. Pang, "Nontopological solitons," *Phys. Rept.* **221**, 251 (1992).

[22] C. H. Taubes, "Arbitrary N-vortex solutions to the first order Landau-Ginzburg equations," *Commun. Math. Phys.* **72**, 277 (1980).

[23] T. M. Samols, "Vortex scattering," *Commun. Math. Phys.* **145**, 149 (1992).

[24] N. S. Manton and J. M. Speight, "Asymptotic interactions of critically coupled vortices," *Commun. Math. Phys.* **236**, 535 (2003) [arXiv:hep-th/0205307].

[25] A. Mohammed, J. Murugan, and H. Nastase, "Looking for a Matrix model of ABJM," *Phys. Rev. D* **82**, 086004 (2010) [arXiv:1003.2599 [hep-th]].

[26] A. Mohammed, J. Murugan, and H. Nastase, "Abelian-Higgs and vortices from ABJM: Towards a string realization of AdS/CMT," *JHEP* **1211**, 073 (2012) [arXiv:1206.7058 [hep-th]].

[27] J. L. Gervais, A. Jevicki, and B. Sakita, "Collective coordinate method for quantization of extended systems," *Phys. Rept.* **23**, 281 (1976).

[28] N. H. Christ and T. D. Lee, "Quantum expansion of soliton solutions," *Phys. Rev. D* **12**, 1606 (1975).

[29] J. L. Gervais and A. Jevicki, "Point canonical transformations in path integral," *Nucl. Phys. B* **110**, 93 (1976).

[30] J. L. Gervais and A. Jevicki, "Quantum scattering of solitons," *Nucl. Phys. B* **110**, 113 (1976).

[31] A. Jevicki and B. Sakita, "The quantum collective field method and its application to the planar limit," *Nucl. Phys. B* **165**, 511 (1980).

[32] H. Nastase, M. A. Stephanov, P. van Nieuwenhuizen, and A. Rebhan, "Topological boundary conditions, the BPS bound, and elimination of ambiguities in the quantum mass of solitons," *Nucl. Phys. B* **542**, 471 (1999) [arXiv:hep-th/9802074].

[33] G. V. Dunne, "Aspects of Chern-Simons theory," [arXiv:hep-th/9902115].

[34] S. Rao, "An Anyon primer," [arXiv:hep-th/9209066].

[35] E. Witten, "Three lectures on topological phases of matter," *Riv. Nuovo Cim.* **39**, no. 7, 313 (2016) [arXiv:1510.07698 [cond-mat.mes-hall]].

[36] G. W. Moore and N. Read, "Nonabelions in the fractional quantum Hall effect," *Nucl. Phys. B* **360**, 362 (1991).

[37] P. K. Townsend, K. Pilch, and P. van Nieuwenhuizen, "Selfduality in odd dimensions," *Phys. Lett.* **136B**, 38 (1984) Addendum: [*Phys. Lett.* **137B**, 443 (1984)].

[38] J. Murugan and H. Nastase, "A nonabelian particle-vortex duality in gauge theories," *JHEP* **1608**, 141 (2016) [arXiv:1512.08926 [hep-th]].

[39] J. Murugan, H. Nastase, N. Rughoonauth, and J. P. Shock, "Particle-vortex and Maxwell duality in the $AdS_4 \times \mathbb{CP}^3$/ABJM correspondence," *JHEP* **1410**, 51 (2014) [arXiv:1404.5926 [hep-th]].

[40] J. Peebles, *Principles of physical cosmology*, Princeton University Press, 1993.

[41] A. Einstein and N. Rosen, "On gravitational waves," *J. Franklin Inst.*, **223**, 43 (1937).

[42] K. A. Khan and R. Penrose, "Scattering of two impulsive gravitational plane waves," *Nature* **229**, 185 (1971).

[43] A. Vilenkin, "Gravitational field of vacuum domain walls and strings," *Phys. Rev. D* **23**, 852 (1981).

[44] A. Vilenkin, "Gravitational Field of Vacuum Domain Walls," *Phys. Lett.* **133B**, 177 (1983).

[45] J. R. Gott, III, "Gravitational lensing effects of vacuum strings: Exact solutions," *Astrophys. J.* **288**, 422 (1985).

[46] M. Banados, C. Teitelboim, and J. Zanelli, "The black hole in three-dimensional space-time," *Phys. Rev. Lett.* **69**, 1849 (1992) [arXiv:hep-th/9204099].

[47] S. Carlip, "The (2+1)-dimensional black hole," *Class. Quant. Grav.* **12**, 2853 (1995) [gr-qc/9506079].

[48] A. H. Taub, "Empty space-times admitting a three parameter group of motions," *Annals Math.* **53**, 472 (1951).

[49] E. Newman, L. Tamburino, and T. Unti, "Empty space generalization of the Schwarzschild metric," *J. Math. Phys.* **4**, 915 (1963).

[50] C. W. Misner, "The flatter regions of Newman, Unti and Tamburino's generalized Schwarzschild space," *J. Math. Phys.* **4**, 924 (1963).

[51] S. W. Hawking and G. F. R. Ellis, *The large scale structure of space-time*, Cambrigde University Press, 1973.

[52] S. W. Hawking, "Gravitational instantons," *Phys. Lett. A* **60**, 81 (1977).

[53] D. N. Page, "Taub-NUT instanton with an horizon," *Phys. Lett.* **78B**, 249 (1978).

[54] T. Eguchi and A. J. Hanson, "Asymptotically flat selfdual solutions to Euclidean gravity," *Phys. Lett.* **74B**, 249 (1978).

[55] G. W. Gibbons and S. W. Hawking, "Gravitational multi-instantons," *Phys. Lett.* **78B**, 430 (1978).

[56] T. Eguchi and A. J. Hanson, "Selfdual solutions to Euclidean gravity," *Annals Phys.* **120**, 82 (1979).

[57] T. Eguchi and A. J. Hanson, "Gravitational instantons," *Gen. Rel. Grav.* **11**, 315 (1979).

[58] V. Rubakov, *Classical theory of gauge fields*, Princeton University Press, 2002.

Index